Malta and Water (AD 900 to 1900)

Irrigating a semi-arid landscape

Keith Buhagiar

BAR International Series 2829

2016

Published in 2016 by
BAR Publishing, Oxford

BAR International Series 2829

Malta and Water (AD 900 to 1900)

ISBN 978 1 4073 1629 1

COVER IMAGE *A perched aquifer gallery and its fronting subterranean reservoir at Wied Liemu in the territory of Rabat (Malta). The light source is emitted by a small remotely operated submersible engaged in the exploration and mapping of this gallery system during one of the past sessions of the Malta/Gozo Cistern Exploration Project. Source: the author.*

Printed in England

BAR
PUBLISHING

BAR titles are available from:

BAR Publishing
122 Banbury Rd, Oxford, OX2 7BP, UK
EMAIL info@barpublishing.com
PHONE +44 (0)1865 310431
FAX +44 (0)1865 316916
www.barpublishing.com

"In Malta, one may say without much exaggeration, that water is as precious as wine, and everywhere, the preservation of the purity of water is at least as essential as the conservation of the flavour of wine."

Osbert Chadwick, *Report on the Water-Supply of Malta*, (Malta, 1884), iii.

Acknowledgements

The obligations incurred in the writing of this book are many. The completion of this work would have been difficult without the academic supervision of Prof. Anthony Bonanno. His expertise, advice, guidance and patience proved indispensable. Furthermore, Prof. Bonanno was a constant source of encouragement, especially during disheartening moments and when the difficulties encountered appeared insurmountable. Equally valuable was the input of the late Prof. Godfrey Wettinger who singled out Late Medieval and Early Modern archival documentation which was directly relevant to my area of specialisation as well as making available his time for lengthy discussion on my research topic.

A very special word of thanks goes to Dr Richard Reece, formerly of the Institute of Archaeology, University College London. It is through his direct intervention that the Trust provided financial assistance allowing me to carry out specialised research work in the various Oxford University libraries, provided funding for a preceding orientation trip, as well as generously covering the cost for additional colour printing. Furthermore, Dr Reece kindly accepted to review an earlier version of this book and provided me with indispensable feedback.

Dr Richard Reece and Prof. Mario Buhagiar of the University of Malta were always an unwavering source of inspiration and readily provided advice and encouragement. Furthermore, my interest in medieval archaeology was sparked off by Prof. Mario Buhagiar's passion for this research area. It was him who stirred in me a fascination for rock-cut dwellings and architecture, the subject of my MA thesis. This eventually paved the way for my interest in hydrology and the water management systems discussed in this book.

I am also indebted to Rev. Dr Charles Mallia, Archbishop's Delegate for Maltese Church Schools and Dr Rose Anne Cuschieri, Director for Educational Services at the Secretariat for Catholic Education (Malta), for their unwavering support and encouragement, and for taking active interest in my field of specialisation.

The input of other experts in fields relating to Maltese archaeology, hydrology, and geology was equally indispensable. Dr Nicholas Vella, Head of Department of Classics and Archaeology at the University of Malta, always provided me with advice when necessary. The same applies to Dr Timmy Gambin, also of the Department of Classics and Archaeology at the University of Malta, who apart from roping me into the Malta Cistern Project, made available all the data gathered by this research project. The expert advice of Ing. Dirk De Ketelaere, Dr Gordon Knox and Ing. Marco Cremona, all of the Malta Water Association, on matters concerning the current water situation in Malta was providential. Dr Joseph Mangion, formerly Director of the Malta Resources Authority and Mr Manuel Sapiano, currently Chief Technical Officer (Energy and Projects) at the Sustainable Energy and Water Conservation Unit in the Office of the Prime Minister, granted me direct access to the Malta Resources Authority water archives. Mr Sapiano always readily addressed my queries and provided me with Geographic Information System data on Globigerina Limestone coastal distribution and the predictable water percolation trends within. A proper understanding of the impervious qualities of the Maltese Globigerina Limestone deposits was only possible because of the invaluable assistance given by Geotechnical Engineer Adrian Mifsud and Dr Saviour Xerri.

The assistance of Mr Mario Vassallo was of instrumental importance in the location of the majority of the Mġarr (Malta) springs. Mr Anthony Buhagiar, Mr Alexander Camilleri and Mrs Ann Gingell Littlejohn proofread earlier versions of this text. Furthermore, it was through the kind intervention of Mrs Gingell Littlejohn that contact was established with Mrs Victoria Tabone, one of the landlords of Gomerino Estate at Rabat (Malta). Apart from granting me access to her property, Mrs Tabone also made available the estate's *cabreo* manuscript. Fr Eugene P. Teuma took care of the organisation and logistics necessary to access the Għeriexem perched aquifer gallery at Rabat (Malta), whilst Mr Joe Micallef granted me access to the Wied Ħażrun perched aquifer gallery, amongst others, within his properties. A word of gratitude also goes to Mr Charles Buhagiar, Mr Philip Micallef and all farmers I came into contact with, without whose co-operation and permission to access their property this work would not have been possible.

My appreciation and gratitude also goes to Dr Anthony Luttrell, Prof. Paul Arthur, Prof. Stanley Fiorini, Prof. Joseph Pace Asciak, Prof. Chris Gerrard, Prof. Joseph Brincat, Dr Albert Ganado, Dr Joan Abela, Dr Anton Bugeja, Dr Alfred Baldacchino, Dr Charles Grech, Dr George Said-Zammit, Dr Ryan Farrugia, Dr Reuben Grima, Architect Michael Ellul, Mrs Marion Pace Asciak, Mr Edwin Lanfranco, Mr George Azzopardi, Mr Daniel Borg, Mr Timothy Tabone, Mr Alex Camilleri, Ms Annalise Falzon, Mr Mario Gauci, Mr Martin Galea de Giovanni, Mr Dean Galea and other scholars and researchers who provided me with advice and information on aspects relating to my field of specialisation. I am also deeply indebted to Ms Maroma Camilleri and her supporting staff at the National Library of Malta, as well as Mr Charles Farrugia and supporting staff at the National Archives of Malta. Apart from facilitating access to *cabrei* manuscripts and other archival documentation, both Ms Camilleri and Mr Farrugia granted permission to reproduce all images obtained from the National Library and the National Archives. I am also indebted to Mgr. Laurence Gatt for granting me access to the Bishop's Curia Archives at Floriana, Malta and to Canon John Azzopardi and Mr Mario Gauci of the Cathedral Museum Archives for facilitating my research there.

Field-research in South Italy was made possible by financial aid made available by the Marquis Scicluna Trust Fund. Research in the Calabria territory was facilitated by logistical support provided by Dott. Luciana De Rose, Dipartimento LISE - Laboratorio di Cartografia Storica (HiCaL), of the Università della Calabria. I am also deeply indebted to Ing. Bartolomeo Schembri and his wife Gemma Battaglia, Antonio Iabichino, as well as Ignazio Fiorilla, Disaster Manager for the *Protezione Civile del Comune di Scicli* and his team, for providing logistical support, enabling the smooth running of the various field-research sessions carried out at Scicli and in other areas of south-east Sicily. Dr Sandra Lucore and Dr Andrew Walthall were indispensable in helping me understand the dynamics and workings of the subterranean water systems at Morgantina, Sicily, as were the fellow excavation team members from Princeton University.

Finally, a very special word of gratitude is due to my wife, Dr Sarah Pulé of the University of Malta, my son Alan and our respective families. It is their patience and unwavering support for my work which enabled me to complete this project.

Acknowledgements for Figures

Cathedral Archives (Mdina)	5.48; 7.3
Cistern Exploration Project (Malta/Gozo)	3.10; 3.11; 7.12
Curia Archives (Malta)	5.119; 5.120
Department of Classics & Archaeology, University of Malta	3.7; 3.8
National Archives (Malta)	4.1; 4.2; 4.4; 4.6; 4.7; 4.8; 4.9; 5.47; 5.86; 5.87; 6.16; 6.25; 6.42; 6.43; 6.62; 6.69; 6.91; 6.96; 6.114
National Library (Malta)	5.114; 6.1; 6.2; 6.3; 6.4; 6.6; 6.7; 6.9; 6.13; 6.15; 6.21; 6.35; 6.40; 6.41; 6.44; 6.45; 6.49; 6.50; 6.51; 6.52; 6.53; 6.56; 6.64; 6.66; 6.67; 6.68; 6.99; 6.101; 6.103; 6.105; 6.106; 6.113; 6.115; 6.116; 6.120; 6.121; 6.122; 6.124; 6.125; 6.126
Adrian Mifsud	2.8; 2.9; 2.10; 2.11
Manuel Sapiano	5.102
Victoria Tabone	4.5

Contents

List of Figures

Chapter 3

Chapter 4

Chapter 5

Chapter 6

Chapter 7

Chapter 8

Abbreviations

ACM	Archivum Cathedralis Melitae
AIM	Archivum Inquisitionis Melitae
AOM	Archivum Ordinis Melitae
AWWA	American Water Works Association
CEM	Curia Episcopalis Melite
FAO	Food and Agriculture Organisation of the United Nations
GOV	Government
IGMI	Istituto Geographico Militare: ente cartografico dello Stato (Italia)
Lib.	Library
MAR	Museum Annual Reports (Malta)
MCM	Museum of the Cathedral, Mdina
MEPA	Malta Environment and Planning Authority
MRA	Malta Resources Authority
MS.	Manuscript
NAM	National Archives (Malta)
NLM	National Library (Malta)
Not.	Notary
NSO	National Statistics Office (Malta)
PW	Public Works
Treas.	Treasury
Univ.	Università

Abstract

This book synthesises archaeological and historical research in order to investigate Maltese water management technology in the Medieval, Early Modern and Modern periods, more specifically between AD 900 and AD 1900. Maltese terrestrial geological formations and stratification are a determining factor in conditioning the formation of subterranean aquifers, water-harvesting and storage, landscape development and utilisation. Central to this publication are reservoirs, cisterns, wells and perched aquifer galleries. These water features have for centuries provided farmers tilling arable land with a supplementary water source other than the limited and erratic seasonal rainfall.

The data and conclusions presented in this book are the result of extensive personal field and archival research and include an assessment of the available documentary sources of evidence including place-names and cartographic sources. An emphasis is placed on Upper Coralline Limestone perched aquifer subterranean galleries. Comparative research suggests that a number of these share common characteristics with *qanat* technology of the Islamic and the Roman worlds and in a Maltese context, were possibly part of a new agricultural and technological package introduced during the Muslim or the post-Muslim period between the eleventh and the fourteenth centuries AD. The influence other perched aquifers located in Globigerina Limestone formations had on past settlement patterns and other archaeologically-significant remains in the vicinity was also investigated.

There was continuous demand for water captured from the various Maltese perched aquifers, both for agricultural and domestic consumption, right until the latter part of the nineteenth century. As from the 1870s, however, an understanding of the mechanism behind the workings of the mean-sea-level aquifer led to a dramatic shift in the water management and distribution policies practised in the archipelago and resulted in an ever-decreasing reliance on perched aquifer systems. The twentieth-century contamination of most local perched aquifer sources with nitrates, a by-product of modern agricultural activity, made these no longer fit for human consumption. However, the Maltese perched aquifers are still a precious aid to agricultural activity taking place in an otherwise semi-arid context. So are the Upper Coralline Limestone perched aquifer galleries. Many of these still supply the agricultural estates in which they are placed with a perennial water supply.

Chapter One

Introduction and Methodology

1.0 Introduction

This book combines archaeological and historical data to examine water capture across the Maltese islands with a particular emphasis on Upper Coralline Limestone perched aquifer galleries. During the periods under review, water management was an essential practice in order to ensure water capture and its supply to places where and when necessary. In terms of supply, the primary focus of this book is on field irrigation, but it has to be clarified from the onset that water distribution and allocation for agricultural usage falls outside the scope of this study. Whilst modern agrarian technology has drastically depleted both perched and mean-sea-level aquifers, Malta provides a useful case study because older hydrology-related interventions within its agrarian landscape remain functional and mostly unchanged.

As early as 1995, site visits to Wied San Niklaw (ME7) in the territory of Mellieħa and the Ta' Baldu area (RA43) of Wied ir-Rum (RA56) in Rabat (Malta) territory led by M. Buhagiar had already introduced me to 2 perched aquifer gallery systems located within these water-carved valleys (Fig. 1).[1] At this early stage, however, the significance of these hydrological systems, their landscape context and the geological stratification of the deposits within which these were excavated had still not been properly understood. Instrumental in stimulating my curiosity and interest in past water retrieval and management systems was field-research conducted between 1999 and 2002 in the north and north-west sectors of Malta south of the Great Fault in part fulfilment of my Master of Arts degree in Archaeology (Buhagiar K. 2002). Onsite research held during this period not only permitted me to gain familiarity and understanding of the various cave-settlement typologies, their distribution and the landscape geomorphological characteristics in which they were placed, but also introduced me to an ever increasing number of fresh water springs, the landscape context of which always appeared to be closely related to the troglodytic dwellings under investigation.

It was evident from the outset that the archaeological significance of many of these perennial water systems originating from within a series of man-excavated galleries had never been properly examined. Their antiquity was suspected by some, but there was no evidence to showcase tangibly their cultural significance. As many of the perched aquifer galleries included in this volume have remained constantly in use and subjected to periodic maintenance since their inception, it is unlikely for these to contain stratified deposits in their interior. The archaeological investigation of the terraced land fronting the galleries might yield more promising results, but this has still to be undertaken. This study, therefore, attempts to place the discussed water systems into their proper context through a detailed analysis of their archaeological and historical potential. Field data were supplemented with archival, historical and cartographic research. The importance of this research method as well as the adoption of a retrogressive approach was indicated to me by A. Bonanno and by N.C. Vella, both from the Department of Classics and Archaeology at the University of Malta.[2] Apart from yielding stringent evidence for the presence of perched aquifer galleries within the Maltese landscape already in the Knights' period, the employment of this methodological approach allowed for a case to be made for the medieval significance of a number of the water systems discussed below. Indeed, working by elimination did expose relevant data which could otherwise have been easily ignored.

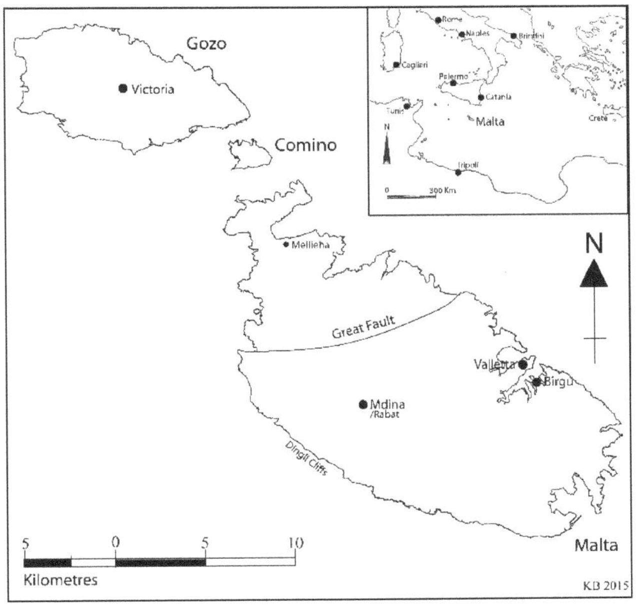

Figure 1.1 Location map for the Maltese archipelago.

[1] M. Buhagiar, founder and first Head of the History of Art Department at the University of Malta, has researched and published extensively on Maltese Late Medieval art history and archaeology.

[2] In 2005 N.C. Vella had already pointed out the importance of this methodological approach, especially when carrying out archaeology-related landscape research. See Hunt and Vella 2005, 61-62; Vella and Spiteri 2009, 19.

Field-research carried out in other countryside localities of the Maltese archipelago since 2002 was not only instrumental in revealing the existence of more identical perched aquifer retrieval systems, but also revealed different techniques of water extraction and harvesting systems which were of a potential archaeological interest. Apart from field irrigation, water retrieved from a number of Upper Coralline Limestone perched aquifer galleries was in the Early Modern and Modern periods tapped for domestic consumption. This aspect is taken into account particularly in chapter four, where the presented data show the resilient nature of these water extraction systems even during periods of extended drought.

At an early stage of this study, personal field, archival and geology-related research had already resulted in the detection of a series of small and highly localised aquifers mainly located in Middle and Lower Globigerina Limestone deposits. Published scholarly data on the porosity levels of these deposits are, however, limited. Another factor that still has to be appropriately addressed is the fact that the chemical composition of these Globigerina Limestone strata and their permeability levels appear to fluctuate even in a small area.[3] This field of specialisation has only recently been taken up by Geotechnical Engineer Adrian Mifsud for further scientific assessment and investigation as one of the principal research areas for his Ph.D. thesis. This work is still in progress.

Similar water galleries to those recorded and investigated within the Maltese context can also be found in the Calabria region of South Italy, in Sicily and North Africa. Field-research was, therefore, extended to Calabria and the southern heel of Italy, and Sicily. Whereas water gallery presence in the Salento peninsula still needs to be further investigated and ascertained, water galleries feature in the Calabrian landscape.[4] A number of perched aquifer galleries in Sicily were also personally investigated. These largely similar Italian and Sicilian water systems appear to be very sparsely documented and have so far only attracted very limited scholarly attention. In this respect, a thorough comparative analysis of the Calabrian, Sicilian and Maltese water systems is still a *desideratum*.

1.1 Aims and Objectives

The aims of this study are to: (i) Review Maltese water management in the Modern, Early Modern and Late Medieval periods with specific emphasis on water capture and storage; and (ii) Place Maltese water capture systems into their appropriate cultural context.

The objectives of this study are to: (i) Employ a retrogressive analytical approach to identify Late Medieval water capture systems; (ii) Relate Late Medieval water capture technology to Maltese geological stratigraphy with particular reference to Upper Coralline Limestone and Globigerina Limestone aquifers; (iii) Contextualise the mean-sea-level and perched aquifers in terms of their archaeological and historical significance; (iv) Categorise the nature of water sources mentioned in the historical documentary sources and attempt to correlate them to existing ones; and (v) Investigate the context of the Maltese *giardino* agricultural framework.

1.2 Malta and its Water in Context

This study concentrates on the central Mediterranean hydraulic context, placing specific emphasis on Maltese water capture systems, but for the sake of completion this section name-checks pioneers of water studies and expounds their principal assertions. The wider Mediterranean framework is taken into account by giving prominence to the concept of "hydraulic archaeology" and the archaeological research methods employed in the study of Spanish hydraulic systems.

The work of Karl A. Wittfogel, a German-American historian specialising in Oriental studies provides a natural point of departure for the investigation of hydraulic systems and the communities these serviced. Wittfogel's theory centres upon "hydraulic societies" (Wittfogel 1955, 43-57). It not only attempted to universally portray direct state intervention and monopoly on the water management systems of preindustrial societies in China, India, Peru and Mexico but also argued that this caused the rise, and determined the character, of ancient states. Central to this theory is the co-ordination of mass labour, the maintenance of water systems, water allocation and crop harvesting and storage (Wittfogel 1967A; Price 1994, 187-204; Mabry 2000, 285-86).[5]

Wittfogel's concept offered a widespread and testable model for the origins of state structure but became exposed to criticisms and refutations. Its testability provided an opportunity for alternative hydraulic theories to be proposed (Bell 1994, 104-5; Mabry 2000, 286-87). A principal opponent of the Wittfogelian paradigm was Robert McCormick Adams whose main argument asserted that large-scale irrigation projects were the consequence and not the cause of dynastic states (Adams

[3] Personal communication by Geotechnical Engineer A. Mifsud.

[4] The present writer investigated the initial tract of a water gallery at Zungri, Calabria. A number of water galleries were also observed to be present within the Calabria region by R. Farrugia, a specialised paediatrician by profession, during his frequent trekking expeditions to this South Italian region.

[5] A.M. Watson similarly maintains that in the early Islamic period, Muslim-led states often assumed direct responsibility for the funding and maintenance of irrigation projects. See Watson 1974, 16, 27.

R. 1966).[6] Wittfogel's hydraulic theory was also refuted by Timothy Earle (Earle 1978; Earle 1980, 1-28). Taking the Halelea district of Hawaii as a model, Earle similarly concluded that its irrigation systems were small-scale in nature and that there was "virtually no evidence that Hawaiian irrigation would have required centralized management for construction, maintenance, or water distribution" (Earle 1978, 193-94).[7] Taking stock of the above refutations to Wittfogel's theory, it does not mean that small-scale irrigation systems can never be managed by a centralised state structure. The centralised management of such a venture, on the other hand, implies that the cost involved would outweigh the benefits of such an arrangement (Mabry 2000, 292).

J.B. Mabry discards a significant portion of Wittfogel's theory but points out that Wittfogel was right when asserting that the "complexities of large-scale water management have been a significant pressure that helped shape the organizations and evolutionary directions of those societies depending on it" (Mabry 2000, 289). Wittfogel's assumption that large-scale irrigation systems can only be managed by centrally controlled hierarchies is also challenged by Mabry and induced discussion on the relationship between the size of the irrigation systems and the complexity of the management structures involved (Mabry 2000, 289). Eva and Robert Hunt, amongst others, found no conclusive relationship between the size of the water system and the structure of authority (Hunt 1978, 69-123; Hunt 1988, 335-55). Nonetheless, R. Hunt estimated that a hydraulic system larger than one hundred hectares in size had a much higher probability of being regulated by a "unified administrative authority structure" (Hunt 1988, 347). On the other hand, any correlation between the scale of the hydraulic system(s) and central authority is by no means straightforward (Wilson 2004, 117-18). Also worth considering is the concept of "scalar stress" which may be caused by the number of individuals involved in the irrigation process other than the size of the hydraulic area or the number of people this supports (Mabry 2000, 290).[8] It is also possible that water-related managerial difficulties increase exponentially with the size of the hydraulic system (Mabry 2000, 290). Whatever the case is, the longevity of Wittfogel's hydraulic hypothesis is proof that it cannot be outrightly confirmed or refused (Harrower 2009, 66-67).

In Spain, the concept of "hydraulic archaeology" and "hydraulic systems" developed through the investigation of water-related systems mainly located in the eastern regions of the Iberian Peninsula and the Balearic Islands as from the 1970s (Barceló 1995; Kirchner and Navarro 1993; Glick and Kirchner 2000; Kirchner 2009). The objective of hydraulic archaeology is to reconstruct the design of a hydraulic system and to detect alterations and additions that this might have been exposed to throughout its existence (Glick and Kirchner 2000, 275-76; Kirchner 2008, 474-75). A principal point of departure was T.F. Glick's study on the society and irrigation of medieval Valencia (Glick 1970).[9] Basing himself on medieval period textual sources, the analysis of place-name evidence and the interpretation of the collected data, Glick laid the foundation for further studies and investigation (Glick 1970; Glick 1979; Retamero 2008, 136; Kirchner 2009, 151-52).

Glick's assertion is that the irrigation communities of Valencia were only exposed to localised governing factors (Glick 1970, 31-51). Within this environment, technical transmission formed part of peasant knowledge, therefore providing no scope for the presence of large-scale centralised administration (Watson 1974, 25; Watson 1983; Barceló 1998, 5-13; Glick and Kirchner 2000, 267, 327-29). Taking the al-Andalus situation as a case in point, Andalusian peasants inhabiting areas which contained either surface or subsurface aquifers which were easily tapped, could engage themselves in the construction of new hydraulic systems (Glick and Kirchner 2000, 292-93; Kirchner 2009, 160-61).[10] The majority of Andalusian hydraulic systems are Muslim in foundation and it appears that the excavation of *qanat*-type galleries formed part and parcel of the hydraulic transformation of these landscapes (Kirchner and Navarro 1993, 159; Glick and Kirchner 2000, 268, 305-7; De Meulemeester 2005, 609-15; Bertrand and Sánchez Viciana 2009, 151-78; Kirchner 2009, 151-52, 154-55; Ortega Ortega 2010, 123-46).[11] There is evidence to suggest that many irrigated areas consciously avoided aquifer exhaustion and that demographic growth was kept in check through kin segmentation (Glick and Kirchner 2000, 292-93; Kirchner 2008, 472-73; Kirchner 2009, 160-61). Design is a central concept to peasant hydraulics and required provision for the conveyance of water from its source to distribution canals through the employment of gravity-fed systems as well as the planning of the necessary irrigable space and water storage tanks. Tapping a new water source necessitates the implementation of a new design and the creation of a new system where its water delivery grid is at times linked to a pre-extant hydrological system (Hunt 1988, 343-44; Glick and Kirchner 2000, 269-71; Kirchner 2008, 473).

Even though Glick questioned if the Spanish *huertas* or gardens were the result of a centralised authority, evidence shows that a number of these were, in their

[6] A review by Wittfogel on the Adams 1966 publication brings to the fore the divergent nature of the theories formulated by both scholars. See Wittfogel 1967B, 90-92.

[7] See also Leach 1979, 473-75; Bell 1994, 108-12; Mabry 2000, 286.

[8] On the concept of "scalar stress" see Friesen 1999, 21-37; Alberti 2014, 1-15.

[9] For an essay on water usage in Spain during the medieval period see Matrín-Bueno and Reklaityte 2008, 201-42.

[10] The majority of Andalusian hydraulic systems are Muslim in foundation and little is known of pre-Islamic water-related systems. See Kirchner and Navarro 1993, 159; Glick and Kirchner 2000, 268; De Meulemeester 2005, 609-15; Kirchner 2009, 151-52, 154-55.

[11] For an investigation of Roman period irrigation possibilities in Spain see Butzer *et al* 1985, 479-509; Lloris 2006, 147-97.

formative years, only modest land-holdings and were the result of direct peasant intervention within the landscape (Glick 2007; Glick and Kirchner 2000, 295-300). Glick and Barceló also worked upon the concept of "fragility" of Spanish hydraulic systems. This implies that any modification to the original hydrological layout can impair the efficiency of the system and adversely impact water allocation organisation (Barceló 1996, 29-31; Kirchner 2009, 160). Of particular interest is the notion that the available hydraulic space influences the location where the farming community managing that hydraulic unit resides. Such units of habitation fall outside the irrigated perimeter and do not occupy any cultivable space (Kirchner and Navarro 1993, 160-61; Glick and Kirchner 2000, 293).

M. Barceló focused on the archaeological aspect of irrigation (Retamero 2008, 137). As a pioneer in this field of research, he not only attempted to determine the origins of irrigated areas but also endeavoured to link these to neighbouring communities which were possibly making use and benefiting from the water-yield of such systems (Barceló 1986, 22). Barceló's focus, therefore, included not only "hydraulic units" but the whole hydraulic system as this probably contains the social code of the community living off it (Kirchner and Navarro 1993, 160; Kirchner 2008, 482-83). Another project embarked upon by Barceló was the creation of a corpus of irrigated Andalusian areas containing an estimate of the population size of rural communities that built these hydraulic systems (Kirchner and Navarro 1993, 168). The work of P. Cressier, on the other hand, primarily concentrated upon the examination of the archaeological strata of Muslim period settlements in the al-Andalus territory. Of particular significance is the way Cressier managed to reconstruct changes experienced by the urban framework through the employment of cadastral records and without carrying out any subsurface investigation of these areas (Cressier 1983, 89-124; Cressier 1988, 123-34; Cressier 1993, 38-43).

A study by C. Gerrard on irrigation in the Aragón region illustrates how many of Spain's medieval hydraulic structures were engaged in water capture, storage and distribution. There are various methods of water capture but these mainly centred on the interception of surface stream water by means of diversion dams or gravity dams, which spanned the width of the whole watercourse. Water originating from *qanat* drainage systems was also captured for later usage. Water storage was achieved through the excavation/construction of reservoirs, especially in areas where water flow was unpredictable. From this point onwards water was channelled to areas which necessitated irrigation by means of gravity-fed irrigation channels which closely followed the contours of the terrain (Gerrard 2011, 7-16).[12]

The methodology employed in the study of the hydraulic landscape of the al-Andalus territory and other areas of Spain consisted in a combination of fieldwork, historical documentation, place-name evidence, aerial photography and the utilisation of cartographic sources (Kirchner and Navarro 1993, 163; Glick and Kirchner 2000, 276; Kirchner 2008, 474-75; Kirchner 2009, 151-52). Old maps, when available, together with other archival sources of evidence were also utilised. This approach is not excavation-oriented and seeks to examine the landscape and its hydrological context through an analysis of visible surface remains (Gerrard 2011, 4). Data were first collected through the employment of archaeological field surveys. These primarily targeted the reconstruction of hydraulic systems and took note of the original articulation of the water source(s), primary and secondary distribution canals, field construction and other water-related features (Kirchner and Navarro 1993, 162-64; Glick and Kirchner 2000, 276-77; Kirchner 2009, 154-55). An effective procedure employed for the mapping of irrigated areas and settlement nuclei consisted in the utilisation of a series of enlarged aerial photographs covering the area of interest. These were overlain by transparent films on which any water-related data could be marked through the use of permanent markers (Kirchner and Navarro 1993, 163; Glick and Kirchner 2000, 277). The second step of research involved the investigation of feudal period documentation and other useful archival sources of evidence (Glick and Kirchner 2000, 277). The gathered information was then processed and compared to the fieldwork and surveying data – a process which allowed a more holistic evaluation of the hydraulic landscape under scrutiny. It is only at this stage that the drawing-up of a plan illustrating the original hydraulic design and any subsequent transformations was produced. In this stage of research, it was often found necessary to revisit the landscape area under investigation in order to verify specific aspects which might have gone unnoticed in the first survey (Kirchner and Navarro 1993, 164-65). It is the hydraulic survey and the plotting of the collected data which make it possible to establish chronology and phases of construction (Kirchner 2009, 159).

A number of hydraulic systems have been in continuous use since their inception, but despite episodes of reconstruction, these systems have in many instances reproduced their original appearance. The employed research methodology also took this variable into consideration and section and elevation drawings of all hydraulic units that make up the system were as a consequence incorporated into the field plans. This was done in an attempt to establish which hydraulic systems may be reproducing original models (Kirchner and Navarro 1993, 166). The enlargement of agricultural systems was determined by the original design framework which included water-yield and adequate slopes for the employment of gravity-fed water

[12] See also Kirchner 2008, 471-502.

distribution (Retamero 2008, 140-41; Kirchner 2009, 155-56).[13]

As a concluding note, the research methodology adopted in this book closely parallels research methods employed in the study of Spanish hydraulic systems.[14] In both, there is no engagement with archaeological excavation, and both concentrate on archival documentation and old cartographic sources, and the material traces of evidence which result from the transformation of the landscape in which these hydraulic systems are located. By investigating water capture, storage and distribution systems, the studies on Spanish hydraulic units mentioned in this section adopted a more exhaustive approach to the concept of hydraulic archaeology. The emphasis of this study, however, is water capture. Because prior to this study the dating of Maltese water capture systems was unknown, this work focuses almost exclusively on the association different types of water capture systems have to the different aquifer typologies from which these originate. Water storage is whenever possible also taken up, but together with water distribution and allocation, this will be investigated further in future research.

1.3 Water and the Concept of Risk

There is also a widening body of research on hazard, vulnerability, risk and resilience which is relevant to this project (Hunt 1978; Wilhite 1993; Iglesias A. 2003; Iglesias E. *et al* 2003; Cancelliere *et al* 2009; Celaya *et al* 2009; Casado Sáenz and Montoya 2009; Garrido and Gómez-Ramos 2009; Iglesias A. *et al* 2009; Wilhite 2009; Maliva and Missimer 2012; Gerrard and Petley 2013). Some of this work is specific to water and the threat of drought, relating, in particular to work undertaken in Spain and elsewhere (Cressier 1983; Barceló 1986; Cressier 1988; Bruins 1993; Barceló 1995; Calatrava and Garrido 2005; Below *et al* 2007; Retamero 2008; Bertrand and Sánchez Viciana 2009; Rossi *et al* 2009; Ortega Ortega 2010; Gerrard 2011). Concepts such as risk, hazard and vulnerability are commonly used in association with natural calamities such as drought or floods.[15] "Hazard" infers a situation of potential harm or damage. "Risk" is commonly defined as an existing threat to a system, be it human life, the environment or cultural heritage. The concept of "risk" is similar in perception to that of "hazard" but differs in that it not only implies a potential danger but asserts a real threat. "Vulnerability" is generally defined as the degree of susceptibility to damage caused by a hazardous phenomenon (Tsakiris 2009, 82-88; Cancelliere *et al* 2009, 93-108; Iglesias A. *et al* 2009, 153-54). The ability of a community or a hydraulic system to withstand the challenge posed by a natural hazard, possibly without interruption, is commonly referred to as "resilience" (ASME Innovative Technologies Institute and AWWA 2010, 4, 19).

Drought is a normal component of all climate regimes and does not discriminate between low and high rainfall areas. Because drought is a regional phenomenon, its characteristics will vary according to which climate regime it occurs in (Wilhite 2009, 4). Episodes of drought involve a natural reduction in the precipitation of a region or area for an extended period of time (Wilhite 1993, 4). The dynamics of drought are complex and of all natural hazards, are the least understood. Furthermore, start and end dates of episodes of drought are frequently difficult to determine.[16] Periods of drought are commonly long in duration and influence extensive areas (Below *et al* 2007, 329-30; Maliva and Missimer 2012, 27-30). The lack of a standardised quantitative definition of drought makes an accurate assessment of its impact difficult to establish. It is only during the past decades that various experimental methodological approaches aimed at measuring drought have been proposed (Below *et al* 2007, 330; Wilhite 2009, 5; López-Barrero and Iglesias A. 2009, 26-27; Tsakiris and Pangalou 2009, 69-70; Maliva and Missimer 2012, 32).[17]

Drought can be classified into meteorological, hydrological, agricultural and socio-economical.[18] Whereas in the period covered by this study, meteorological drought was on various occasions the culprit leading to a scarcity in water availability,[19] Maltese society is presently afflicted by socio-economical drought. This is due to a disproportionate demographic growth in comparison to the archipelago's natural hydrological resources. During the past decades, mean-sea-level aquifer extraction levels have exceeded the aquifer recharge rate resulting in aquifer depletion (Sapiano *et al* 2006; Sapiano 2008; Sapiano *et al* 2008).[20]

Drought can severely impact areas with high dependence on subsistence agriculture and cause famine and widespread mortality (Below *et al* 2007, 329-30).

---

[13] In a more recent study, the statistical analysis of 160 hydraulic spaces in the al-Andalus territory was undertaken, and permitted the formulation of a morphological typology for the irrigated spaces present within. See Kirchner 2008, 476-77.

[14] See section 1.6.

[15] A drawback in the employment of such terminology is that these have different meanings not only for the lay person but also for scientists and engineers attempting to quantify these concepts. See Tsakiris 2009, 81.

[16] A timeframe of 2 to 3 months is usually required to establish that drought has officially set in and usually develops in a gradual way. See Wilhite 1993, 3, 8; Bruins 1993, 133-34; Below *et al* 2007, 330; Wilhite 2009, 4-5; Tsakiris and Pangalou 2009, 69.

[17] On the usefulness of drought indices for the planning and managing of drought crises in semi-arid zones see Tsakiris and Pangalou 2009, 69-77, 79.

[18] Meteorological drought refers to a lack of precipitation; hydrological drought to a shortage of surface water; agricultural drought to a drastic reduction of soil moisture; and socioeconomical to human induced drought. See Wilhite 1993, 5-7; Garrido and Gómez-Ramos 2009, 135.

[19] For documented instances of drought in nineteenth-century Malta refer to chapter four, sections 4.1 and 4.2.

[20] A human-induced drought similarly affected Israel in the early 1990s. See Bruins 1993, 140-41.

When drought sets in, the agricultural sector is often the first to be affected and results in a reduction of crop yields (Iglesias A. 2003, 108; Iglesias, A. *et al* 2009, 154; Wilhite 2009, 4). The demands for water by communities in affected areas for domestic, agricultural and industrial consumption increase drought severity, but these can vary amongst different communities even in the same region, depending on societal vulnerability and resilience (Wilhite 1993, 4, 9-10; Wilhite 2009, 3-4). Generally, greater resilience is found in communities frequently exposed to natural hazards, probably because these would have evolved more sophisticated coping skills when dealing with hazardous events (Gerrard and Petley 2013, 1071-72). Drought becomes a disaster when the local environment, communities, their resultant economy and agricultural practices are negatively affected (Below *et al* 2007, 328-29; Wilhite 2009, 6-7; Iglesias A. *et al* 2009, 153-54). Structural problems which expose societal vulnerability may consist in a lack of adequate hydraulic infrastructure and inadequate management (Iglesias A. *et al* 2009, 154). Furthermore, in drought-prone regions, another drought event is likely to take place before that same area or region fully recovers from the previous one (Wilhite 2009, 6-7). The impact of drought on society may endure for years. Data gathered between 1900 and 2004 indicate that over eleven million deaths worldwide were drought related (Below *et al* 2007, 339).

Drought impact on agricultural activity depends on the type of agriculture being practised. In the case of rain-fed agriculture, drought impact is commonly severe. Irrigation agriculture, on the other hand, copes much better with episodes of drought, especially when water originates from subterranean aquifers (Watson 1974, 15-16; Rossi *et al* 2009, 211-12). This does not mean that groundwater is unaffected by episodes of prolonged drought, but healthy aquifers buffer and delay the influence of drought even when in continuous use for irrigation and other purposes. In this sense, groundwater utilisation which does not exceed natural recharge rates, provides reliability and resilience (Casado Sáenz and Montoya 2009, 221-40). In the Mediterranean region, water scarcity and drought are frequent and it is the utilisation of reliable ground water sources which greatly mitigate this (Casado Sáenz and Montoya 2009, 222).[21]

Restrictions on water supplies in time of drought also cause pressure on towns and cities and place a considerable amount of strain on river water, when available, and aquifers, thus creating stress and conflict between different individuals and groups. Such water-related disputes happen whenever the required water needs are not met by the hydrological resources of a region or an area. These disputes can worsen or be resolved according to drought duration. During periods of remission from drought when water is more abundant, conflict is commonly forgotten (Shatanawi *et al* 2008A, 159-72; Shatanawi *et al* 2008B, 205-12; Tamoh *et al* 2008, 131-42; Celaya *et al* 2009, 161-62, 164).

Mitigation strategies employed to negotiate the risk of drought and water shortage in Malta would include the construction of aqueducts in the seventeenth and nineteenth centuries to supply urban areas like Valletta, Birgu, Bormla and Senglea with a secure water supply. In rural areas water shortage risk would be partly mitigated through the construction of reservoirs and cisterns and the tapping of aquifers.

Archaeologists and geographers considering "risk" in other parts of the Mediterranean include C. M. Gerrard and D. N. Petley (Gerrard and Petley 2013). Overall, "risk" literature is interesting because it emphasises the extent to which human populations put themselves at risk. It also presents strategies resorted to in order to mitigate and adapt to hazards (Gerrard and Petley 2013, 1065-66). Droughts have severely impacted ancient cultures and were either entirely or partially responsible for societal collapse (Maliva and Missimer 2012, 31, 517-23; Cook 2013, 89-102). Farmers have traditionally developed some informal strategies to cope with weather risks through actions taken either before or after the risk event takes place. Some of these strategies include changing labour allocations, conservation tillage in order to protect soil moisture and the varying of cropping practices (Garrido and Gómez-Ramos 2009, 137; Gerrard and Petley 2013, 1068-70). In a number of Mediterranean basins, the annual water allotment to farmers varies on a seasonal level, thus mechanisms are resorted to by the farming communities to compensate for this (Calatrava and Garrido 2005, 119-42; Iglesias, E. *et al* 2003, 211-29; Garrido and Gomez-Ramos 2009, 139). In areas of southeast Sicily, when water resources are limited, priority is given to the irrigation of fruit orchards (Rossi *et al* 2009, 211-12).

1.4 Chronology

"Early Medieval", "High Medieval", and "Late Medieval" are terms which this book recurrently makes use of. As the timeframe covered by these lacks standardisation, an explanation of the chronological sequence for Malta employed in the framework of this study is necessary. "Early Medieval" covers the mid-sixth to the tenth centuries AD and encompasses the Byzantine and the early Muslim periods. "High Medieval" comprises the late tenth till the early thirteenth centuries AD and, in a local chronological framework, involved the late Muslim and Norman periods. "Late Medieval" encompasses the late thirteenth century until the arrival

[21] This also applies to the Maltese context. The resilient nature of Upper Coralline Limestone aquifers is showcased by the drought of 1834-1841. See chapter four, section 4.1.

of the Hospitaller Knights of the Order of St John in Malta in 1530.[22]

The post-1530 period is referred to as the "Early Modern" period, ending in 1798 with the capitulation of Malta by the Hospitaller Knights to the French forces. The term "Modern" spans the French, British and contemporary periods. A commonly employed pre-medieval term is "Late Antiquity", which approximately extends from the fourth to the seventh centuries AD.[23]

1.5 Landscape Theory and Research

A fully-fledged study on Maltese landscape development has still to be written (Vella and Spiteri 2009, 17). In its absence, inspiration can be obtained from a series of landscape and field-research studies which, even though dealing with foreign contexts, are, at least in part, relevant to the local situation.[24] The term "landscape" is difficult to define (Roberts 1987, 76; Layton and Ucko 1999, 1). It is a subjective notion and current term usage associates it with both the "environment", generally shaped by human action, as well as a "representation", which signifies meanings attributed to a given setting (Layton and Ucko 1999, 1). Similarly, the European Landscape Convention defines landscape as "an area, as perceived by people, whose character is the result of the action and interaction of natural and/or human factors" (Europe Treaties Series (ETS) No. 176, 2000).[25] In order to place landscape study in a more archaeologically-oriented context, it is probably best to speak in terms of a "social landscape". The term "social" places an emphasis on human-related activity, whereas "landscape" gives a geomorphological dimension to this concept (Gosden and Head 1994, 113). This quite aptly typifies contemporary archaeological landscape studies which commonly tend to approach this topic from two divergent viewpoints: (a) the physical and (b) the cultural approach.

Physical landscape study places an emphasis on the geomorphological framework which allows human societies to exist within (Roberts 1987, 79; Gosden and Head 1994, 114). It is thus the landscape context and its features that attracted and initiated human attention and intervention in an area, with the end result being that tracts of it were subsequently modified because of human activity (Roberts 1987, 89).[26] Applying physical landscape development to a local context, it appears that it was water availability, fertile soil and the shelter offered by valleys which probably encouraged field construction and the excavation of perched aquifer galleries in specifically selected areas of north and north-west Malta. This resulted in the formation of a *giardino*, or horticultural, framework, the landscape context of which still has to be satisfactorily investigated.[27]

Landscapes are also commonly identified as a cultural image. In this model, artistic representation together with individual perception are significant factors in the understanding of a given landscape (Gosden and Head 1994, 113; Carman 1998, 96; Layton and Ucko 1999, 1; Attema 2002, 19, 26; Johnson 2007, 3; Favia 2012, 97-108). Perceptible cultural traces in a given landscape can yield data on matters related to water-harvesting and management. Elements associated to societal, religious and economic considerations, amongst others, can also be retrieved through an appropriate interpretation of cultural landscape contexts.[28] It is the cultural approach to landscape studies that allows the implementation of retrogressive analytical methodology. That "the past is a foreign country" is a notion which might similarly apply to cultural landscape interpretation.[29] Earlier landscape occupants might have perceived their surroundings in a different manner from the present-day scholars attempting to reconnect with it (Johnson 2007, 4).

Malta's limited land surface area, a steady demographic growth and a semi-arid climate, coupled with limited agricultural resources, have during the past millennium been a major contributing factor in altering the island's landscape from a natural into a largely man-made one (Grech 1999, 89-104; Sultana and Gauci 1999, 109-112). Stretches of coastal cliffs and difficult to access inland ravines are amongst the only surviving natural *refugia* where the majority of the surviving endemic flora and fauna species are found.[30] The more accessible sites, such as garigue karst land and valleys, have during the course of centuries been radically altered and exploited (Sultana and Falzon 2002, 24-25). It is my view that any archaeology-oriented landscape study, especially if carried out in Maltese territory, should ideally take into account the site's physical setting and cultural context (Gosden and Head 1994, 115-16; Aston 1997, 91; Johnson 2007, 202). Personal experience has demonstrated that both landscape research methods are closely intertwined and, when working in the field, these are sometimes difficult to separate and compartmentalise.

[22] For a similar chronological classification of the Maltese medieval period see Dalli 2006, 11-19; Docter *et al* 2012, 113.
[23] This chronological classification was discussed with A. Bonanno and M. Buhagiar, both of the University of Malta.
[24] Central to this are studies by Vita-Finzi 1978; Macready and Thompson 1985; Aston 1997; Fowler 1998; Leveau *et al* 1999; Attema *et al* 2002; Rippon 2000; Rippon 2006; Faulkner *et al* 2010; Rippon 2012; Rippon 2013.
[25] See http://conventions.coe.int/Treaty/en/Treaties/Html/176.htm
[26] Apart from investigating the geomorphology of a site or area, the study of floral distribution can also yield valuable evidence in this respect. See Gosden and Head 1994, 114.
[27] P. Saliba's studies involving an evaluation of rubble walls and field-terracing within the Maltese landscape succeed in shedding more evidence in this respect. See Saliba 2008.
[28] For a discussion on possible cultural landscape applications within a Maltese context see Vella and Spiteri 2009, 16-17.
[29] For an in-depth discussion on this matter see Lowenthal 2003, 3-74.
[30] *Refugia* is a biological term commonly used to denote an area in which a population of organisms can survive through a period of unfavourable conditions.
See http://oxforddictionaries.com/definition/english/refugium. See also Sammut Carbone 2002, 45-46.

A combination of these practices would thus help achieve a better understanding of the landscape area under review.

Subsurface scientific excavation is central to most archaeology-related studies and an ideal situation should preferably involve a combination of this and other field practices. However, in instances where little is known about the existence of archaeologically-relevant features in a landscape, it is best for any subsurface investigation to be preceded by a proper assessment of the above-ground evidence through systematic field-research. Archaeological field surveys and studies relating to surface topography are often pivotal in exposing a site's landscape context and setting. These additionally permit the identification of formerly unknown archaeologically-significant remains. In such instances, archaeologically-relevant landscapes can be regarded as palimpsests in which are preserved traces of earlier patterns of occupation and usage (Roberts 1987, 91; Aston 1997, 39-40). It also has to be realised that a number of historically and archaeologically important landscapes are in many ways directly linked to the national or regional heritage of a place or area. I consider this to be the case in several sites located in the north and north-west sector of Malta, a number of which are situated at Wied ir-Rum (RA56) and the adjoining Wied Ħażrun (RA55) in the territory of Rabat (Malta). This will be discussed separately in other sections of the book.

An analysis of available historical documentation is important because it can help establish the *terminus post quem* or the *terminus ante quem* of hydrological features. This allows archaeologically significant items in a landscape for which there is historical documentary coverage to be placed within a more concrete timeframe (Roberts 1987, 91; Aston 1997, 39-40; Vella and Spiteri 2009, 16).[31] In this book, the relevance of such an approach is showcased by a number of subterranean water galleries at Wied ir-Rum and Wied Ħażrun and which the scrutiny of Knights' period *cabrei* and other documentation has helped place in a more secure context.[32]

As early as 1991, A.T. Luttrell had already alluded to the importance of assessing the available written and the non-written record, especially in matters relating to the study of medieval Malta, where historical source material is scarce or non-existent (Luttrell 1991, 33-34). He pointed out that the non-written evidence does not only involve archaeology as a practice, but can also be extended to the study of place-names, surviving architectural features, frescoes and coins amongst other elements having a possible cultural significance (Luttrell 1991, 34). Subterranean rock-excavated spaces, especially the perched aquifer galleries, wells and cisterns, cannot be excluded from this list. The subterranean nature of these water-related spaces provided a more favourable context for their survival than that faced by other above-ground features. Any attempt at analysing their significance and importance through the employment of any available historically-relevant source material and archaeological field-research hence becomes even more meaningful.[33] Recent landscape-oriented studies have already successfully embraced this approach, the employment of which in a Maltese context was first highlighted in 2003 by C. Hunt and N.C. Vella in their publication accounting for field-research carried out in a field at Mistra in Malta (Hunt and Vella 2005).[34]

Many of the above research possibilities have during the past decades been conjoined into an investigative approach known as "historic landscape characterisation" (Rippon 2012). This involves an interdisciplinary methodology which takes note of local history, historical geography, landscape archaeology, cartographic sources, place-name evidence, vernacular architecture, aerial photography and satellite imagery and seeks to integrate the wide-ranging archaeological and documentary material onto a digital platform using Geographic Information Systems (GIS) software (Oosthuizen 2006, 77-79; Rippon 2006; Rippon 2012; Rippon 2013). The advantages of this approach is the integrated way in which all data on a site or area can be tagged with increasing numbers of attributes and can lead to a holistic study of landscape archaeology and history (Rippon 2006, 17, 20-26; Rippon 2012, 54-55). Historic landscape characterisation has been successfully employed within a British context. Its application within a Mediterranean situation will probably necessitate adjustments and modifications based on the area within which research is being carried out (Turner and Crow 2010).

1.6 Methodology

This book is the result of non-excavational data gathered during the course of numerous field-research sessions carried out in north and north-west Malta as well as in Gozo and Comino. The limited knowledge and availability of information on the hydrological systems central to the discussion of this study necessitated the adoption of an approach which was suited to the local context. The research method employed also took into consideration the fact that any ceramic-oriented investigation would have proven to be problematic. Apart

[31] *Terminus post quem* refers to the earliest date that can be established with certainty for any particular land feature or object. *Terminus ante quem* refers to the latest date any land feature or object can be associated with.

[32] For an account on *cabrei*-type documentation see chapter six, section 6.

[33] On the mapping of wells and cisterns see Luttrell, 1991, 42-43.

[34] See also Camilleri 2003, on the study of apiaries; Saliba 2008, on the investigation of rubble walls; Vella E. 2010 on the study of the Maltese *Girna* in northern Malta; and Magro Conti and Saliba 2007 when assessing the significance of cart-ruts in ancient landscapes.

from the fact that extensive ploughing in shallow soil deposits probably led to the destruction of a number of High or Late Medieval archaeological sites throughout the Maltese territory, an official local ceramic typology for these periods has not yet been established (Luttrell 2002B, 3-6). Progress in this respect has been undoubtedly achieved during the last two decades but to date, only short preliminary studies on the Maltese Medieval ceramic typologies have been published (Molinari and Cutajar 1999, 9-16; Cutajar 2001, 79-85; Bruno and Cutajar 2002, 109-38; Cutajar 2004, 55-64; Bruno and Cutajar 2013, 15-29; Hahs 2010).

Taking the above factors into consideration, the best method to contextualise the water capture systems under investigation constituted the adoption of a retrogressive analytical approach, which involves the removal of known dated features within a landscape in order to gain better knowledge of its components at an earlier period (Rippon 2012, 3-5; Vermeulen 2004, 125-27; Oosthuizen 2006, 77-79; Cousins 2010, 27; Greene and Moore 2010, 83; Turner and Crow 2010, 216-20).[35] In theory, the main value of adopting a retrogressive approach to landscape analysis lies in that it helps to recover evidence of "human activity from periods that might otherwise be missed or wrongly assigned if one started one's analysis forward from some fixed point" (Cousins 2010, 25). There is no fixed starting point nor agenda for this method of investigation and only on completion of a landscape study will it be known how far back one is able to go. For this approach to succeed and be able to view a particular landscape from the viewpoint of its occupants at the time, all available/known historical data have to be utilised (Oosthuizen 2006, 77-79; Cousins 2010, 25-27).

The method of research employed in this book closely parallels that adopted in the study of Spanish hydrological systems,[36] but also took stock of the local contextual setting within which the studied water capture systems are placed. Data collection methodology centred on extensive field-research carried out in areas of the Maltese territory which were potentially relevant to my field of specialisation. Cases in point are Wied ir-Rum (RA56) and Wied Ħażrun (RA55), both in the territory of Rabat (Malta). This approach enabled me to document past human interaction with the landscape through the investigation of evidence detectible at surface level,[37] and was indeed successful in singling out a number of potential Late Medieval sites. A two-tiered fieldwork approach was adhered to in all instances, with equal importance being placed on the geological composition of the landscape context as well as its historical and archaeological potential.[38] The ability of detecting perched aquifer levels and water availability during field-reconnaissance sessions, together with a good working knowledge on the water retention potential of the Maltese geological deposits was an indispensable asset.

The second stage of research involved the scrutiny of British period archival evidence and cartographic data, eighteenth and seventeenth-century *cabreo* data and other material dating to the Knights period, as well as the employment, when available, of Late Medieval data, place-name evidence, vernacular architecture and other geological and geographic data. An analysis of the collected data constituted the third phase of research. This led to the identification of areas/territories for which there was adequate historical and archival coverage enabling the implementation of a retrogressive analytical approach.[39] The fourth and final phase of research centred on a detailed field investigation of the identified sites, during which sessions, any available historical and cartographic data were brought into check with the available landscape and its hydrological features. It was during this phase of research that the location of water capture systems, associated data and any related water storage features were plotted and analysed in greater detail.

This research method made it possible to "peel away" modern and Early Modern interventions several sites were subjected to in order to better perceive and attempt to contextualise the water capture strategies employed within these land-holdings during the period under investigation. In landscape areas for which there was adequate historical coverage, this approach furnished me with sufficient evidence to discern landscape changes over a given timeframe. As a matter of fact, it was the principal catalyst conditioning the choice of sample areas, the data for which are presented in chapters five to seven. Once brought into check, a comparative analysis of the archival and field evidence allowed the peeling away, in a metaphorical manner, of the top-most layers in order to tentatively uncover earlier period cultural contexts (Vella and Spiteri 2009, 16-17; Greene and Moore 2010, 83).

As the primary scope of this study was the contextualisation of water capture systems and their tentative dating, it is only these, together with other relevant features such as water storage reservoirs which were accurately mapped.[40] Field boundaries were only taken note of when comparing the demarcation of

[35] The importance of documentary sources in the study of the archaeology of landscapes was also outlined by Lightfoot and Miller 1996, 78-101; Aston 1997, 39-40; Johnson 2007, 148; Cousins 2010, 25-27.

[36] See section 1.2. A similar course of action was also partly resorted to in Rippon 2006, 17-18.

[37] The advantages of this methodological approach are discussed in Jones B. 1985, 2.

[38] The importance of a geological approach to field archaeology had already been singled out in 1978. See Vita-Finzi 1978, 10. A similar methodological approach was followed when investigating settlement and water-related issues of selected Greco-Roman cities in the western, central and eastern Mediterranean. See Crouch 2004, 27-241.

[39] This approach was also hinted at by Jones B. 1985, 3.

[40] For a similar investigative approach see Aston 1997, 91.

properties and estates included in the field-research with *cabrei* documentary sources of evidence and other data. Similarly, the investigation and mapping of water distribution systems, field terracing as well as the collection and investigation of the ceramic evidence fell outside the scope of this study. Nonetheless, the gathered data, an analysis of which is attempted in chapters five to seven, provide a platform for future detailed investigation of areas and agricultural estates which can be more holistically studied through the adoption of the historic landscape characterisation methodology discussed in section 1.5 above. Furthermore, landscape research, if carried out in a diligent and scientific manner, establishes a sound basis upon which later sub-surface investigation can be carried out (Mercer 1985, 9).

The excavation of vertical shafts connecting qanat galleries with the surface (see chapter three, section 3.1.6), commonly resulted in the concentric piling of the excavated debris around the shaft access points at surface level. These spoil rings are commonly referred to as qanat "doughnuts" or "molehills" (Cressey 1958, 30-31; Wilson and Mattingly 2003, 236), but as they do not occur locally, their study could not be undertaken.

The occurrence of calcite deposits was identified inside a number of Maltese Upper Coralline Limestone perched aquifer galleries. Even though calcite samples can be dated through the employment of a specialised method technically known as Carbon-14 dating, no funding was available for such an endeavour during the course of this study. Furthermore, Francisco Javier Santos Arevalo of the Centro Nacional de Aceleradores (CNA) of the Universidad de Sevilla, Seville, Spain, was briefed about the Maltese situation and personally communicated that even though it is technically possible to date such calcite deposits, the principal difficulty for samples originating from a Maltese context concerns the interpretation of the obtained Carbon-14 result(s). The tentative dating of a select number of calcite samples will be given priority in future research work as soon as financial support is made available.

As this book focuses on central Mediterranean water capture systems, field-research was also carried out in the Calabria and Salento regions of Southern Italy as well as in the central and south-east Sicilian provinces of Enna, Syracuse and Ragusa. Malta was an Island satellite in the lee of Sicily throughout most of the medieval period and therefore it is only natural that cultural and architectural influences as well as technology would filter locally because of Malta's close geographic association to South Italy and Sicily (Metcalfe 2002, 289-317; Dalli 2008, 245-58; Buhagiar K. 2012B, 95-118). Field-research in Calabria and Sicily involved the partial investigations of water galleries, thus allowing the comparison of these water capture systems with the ones in Malta. Particularly significant was field-research carried out at Morgantina in the territory of Enna (Sicily), in collaboration with archaeologists from Princeton University.[41] The water galleries at Morgantina will be discussed in more detail in chapter three, section 3.1.6.

1.7 Methodological Limitations

The retrogressive analytical approach adopted for the purpose of this study provided enough historical coverage to show that the majority of the water features mentioned below already formed part of the Maltese hydrological landscape during the Knights period. Historical coverage drastically decreased with earlier period documentation, but retrogressive analysis still made it possible to link the hydrological features of a number of sites mentioned in this publication with the Late Medieval period.

There is little to comment on aspects dealing with the historic, archival and cartographic research phase of this book. The identification, investigation and processing of this material was in many instances an uneventful and straight forward procedure. The only major limitation encountered involved the absence of plans and field boundary data when dealing with early seventeenth-century cadastral records.[42] In instances when sites were not again mentioned in eighteenth-century and later documentation, the location of these estates/properties was difficult to establish with certainty.

It was, however, the field-research aspect of this study, which presented a number of major limitations which had to be dealt with. A primary variable that should always be given its due consideration and weight prior to any field-reconnaissance sessions taking place in a local context is hunting and bird-trapping. The spring and autumn hunting and bird-trapping seasons are to be avoided at all costs, during which period hunters and bird-trappers are quite likely to display an added measure of hostility. The Maltese and Gozitan countryside is, therefore, largely inaccessible during the hunting seasons. A prevailing sentiment amongst the local hunting and bird-trapping community is that since Malta's accession to the European Union (EU) in 2004, matters relating to this topic are being unjustly scrutinised by the EU, to the detriment of these local and traditional "pastimes".

Farmers are generally much more tolerant and helpful, especially if informed in advance of any field-related activity taking place in their area and if the sowing and the harvesting seasons are avoided. Apart from having a higher field visibility, the summer months, when most fields are left fallow, were commonly indicated as the best time during which field activity could be carried out. Personal experience has also shown how farmers can be much more collaborative when approached by only one

[41] This collaboration was made possible through the kind intervention of S. Lucore of the American School at Athens.
[42] See chapter six, sections 6 and 6.1.

field-researcher. Surveying and other archaeology-related equipment was frequently observed to arouse suspicion as to the real or perceived intentions of my presence. Photographic equipment, GPS locational hardware and other surveying-related gear were, therefore, kept concealed and hardly made use of when farmers were present. The same applied to the onsite use of a staff - a number of photographs presented herein should have ideally been provided with a scale, but apart from being a bulky and cumbersome equipment item, especially when transporting in challenging terrain, farmers did not respond well to such an object.

It was realised early on that the best way of winning over the farmers' confidence was to introduce myself as a historian and naturalist studying water availability. No reference to archaeology was made and emphasis was always placed on reassuring the farmers in question that nothing adverse would result from granting access to their property. The fear of anything relating to archaeology is, therefore, another variable which cannot be overlooked.

Experience has shown that many farmers in the Rabat and Dingli area of Malta are genuinely fearful and overly concerned that any archaeologically-relevant remains in their property will lead to unwarranted external interference. A typical case in point is the site of Simblija (RA41), which is located in the point of intersection between Wied ir-Rum (RA56) and Wied Ħażrun (RA55) in the Rabat (Malta) territory. The area first became the focus of scholarly interest and study in 1995, when a three-arched structure which was built to abut the rock ledge was identified as a probable Late Medieval rural church. The fronting square or *misraħ*, together with a number of adjoining caves, were also identified as being of historical and archaeological significance (*Saliba et al 2002*). Conservation measures and upgrading were consequently undertaken at Simblija and the site was presented to the public in 2003.

There is ample documentation in hand to show that Simblija and the adjoining tracts of Wied ir-Rum and Wied Ħażrun are Government-owned property. Furthermore, the tenants at Simblija were very adequately compensated for any lost agricultural-related storage space that the change in use and function of the site entailed. As early as 2005, however, a number of tenants were already restricting visitor access to the area. At a slightly later stage, the approach to Simblija from the Wied Ħażrun side was closed off and signage indicating the site's whereabouts removed. Since 2008, the level of hostility has been significantly stepped-up and visitors attempting access to Simblija are very often threatened with court action.[43]

Archival, cartographic and field-research have led to the identification of equally important historically and archaeologically-relevant remains at the nearby site of Tal-Callus (RA39), located on the southern side of Wied ir-Rum valley. The identified remains are discussed at length in chapter four, section 6.1.7 and are solely being mentioned here in order to forewarn about the attitude which might hinder any potential collaboration between the agricultural and the cultural heritage stakeholders. Following the personal identification of the Tal-Callus site and the singling out of a number of potentially important features, efforts were undertaken to contact the land tenants in order to obtain access to the property's interior and pursue further field-investigation. My presence was always shunned and the tenants were very reluctant to speak to me. Well-informed sources have recently brought to my attention the fact that the Tal-Callus tenants are well aware of the historical and archaeological potential of the property they lease from the Government and fear that this might undergo a process similar to that experienced by the Simblija tenants.[44]

Whereas such field studies are typically carried out by a team of field archaeologists, most of the field-research, the results and observations of which are presented in this book, were the outcome of personal, individual fieldwork. Carrying out field-research alone has its disadvantages and perils. Furthermore, subterranean perched aquifer galleries should be ideally investigated by a team composed of both archaeologists and speleologists and not by a single individual, as was mainly the case in this instance. For the purpose of this publication, the notion of being a lone field-researcher nonetheless bore fruit. A single person roaming the countryside in an organised and systematic manner draws much less attention than a larger field-research team. In granting permission to access their properties, farmers are often discouraged by the presence of a large party. To the farmers' psyche, a large organised group as well as the presence of instrumentation brings about a higher degree of formality than a lone researcher. This, therefore, often made it easier to win over the farmers' confidence, and frequently helped to ease their concern with the result that they divulged more water-related information.

In rural areas where entry had previously been denied, field-research was discreetly carried out between the hours of one and three thirty in the afternoon during the months of July and August. Most fields are left vacant during this peak time of high solar radiation. Coupled with the excessively high temperatures that Malta experiences during the summer season, this leaves the

[43] A number of court arraignments have taken place over access rights to Simblija. In December 2010, a team of cyclists from the Weasels Cycling Club were arraigned in court over complaints by N. Ciantar after cycling through the stretch of country lane which from Simblija gives access to Wied Ħażrun. Refer to the court sitting chaired by Magistrate C.S. Herrera dating to 27 July 2011. The case against the cyclists was dropped on the premise that no clear signage was affixed by N. Ciantar.

[44] Personal communication by D.M. Galea, an archaeology graduate from the Department of Classics and Archaeology at the University of Malta.

countryside almost totally devoid of any agriculture-related activity in this two-and-a-half-hour time-window.

Unfortunately, a substantial portion of my field-research coincided with the commencement, in 2010, of the installation of water meters in most privately-owned boreholes which during the previous two years were registered with the Malta Resources Authority (MRA).[45] Whilst borehole metering did not impact the area of research, as the water retrieved from the perched aquifer galleries was not subjected to this metering process, any enquiries about water always tended to be regarded with an added measure of suspicion. Farmers and private individuals became on occasion extremely wary when admitting to a water presence within their property. The fact that not all private boreholes were registered with the MRA most likely caused some concern in relation to the true nature of my enquiries.

Dense reed growth and uncovered water reservoirs in a rural landscape frequently indicate the presence of perched aquifer galleries. Geomorphological experience gathered over a span of fifteen years of field-research also enabled me to easily identify perched aquifer levels within the local landscape. Whenever possible, the interior of galleries was also personally investigated. This was not achievable in all instances either due to matters related to land ownership or to a high water level and/or thick sedimentation deposits in the gallery system. In other instances, the principal access point to galleries was no longer reachable either due to cave-ins or to cliff-face detachment.

The interior of perched aquifer galleries was frequently investigated alone – a matter which hindered the accurate surveying and planning of these water systems. Even if the service of a team of experienced surveyors was to be employed, the confined nature of these water retrieval systems, which frequently follow a non-linear course, makes their surveying a difficult and time-consuming task. More accurate results can be achieved through the employment of a number of digital surveying techniques, including 3D laser scanning and digital photogrammetry.[46] The flooded sections of the perched aquifer galleries of Għar Ilma (GO13) in Gozo and that of Lunzjata in the territory of Rabat (Malta) have been successfully surveyed and planned through the employment of experimental hardware consisting of a remotely operated submersible equipped with sonar scanners (Clark *et al* 2008, 662-67; White *et al* 2010).[47] Very encouraging results were obtained through this method, which will, in the future, hopefully be employed in the surveying and investigation of other perched aquifer systems.

Similarly, the Malta Cistern Project permitted the investigation of a number of potentially important cisterns and wells located at Mdina, the Gozo Citadel and elsewhere.[48] Apart from the water conservation measures made possible by such an investigative approach (as otherwise, cisterns under investigation would have to be drained of their water content), subterranean spaces too narrow for human access can now be chartered for investigation. The importance of the study of Maltese wells and cisterns was already pointed out by A.T. Luttrell in 1975, where it was hoped that these could be classified according to different typological styles and perhaps dated (Luttrell 1975B, 12). Whilst it is probable that a substantial number of wells and subterranean water storage tanks investigated by the Malta Cistern Project team are centuries old, the data available to hand make it difficult to reach conclusions on cistern typological development. Many of the investigated cisterns are bottle-shaped in nature, but the mapping hardware utilised so far only makes it possible to produce sonar image scans of areas in cisterns containing at least a metre-depth of water. Empty cisterns and subterranean features located above the cistern's water level cannot possibly be mapped through the employment of this method. Cisterns are multi-period features which remained in use over a long period of time. At this stage of research, therefore, dating is a difficult task. Cisterns and wells located within securely documented contexts are obviously much more easily datable.[49]

1.8 Published Sources, Theses and Dissertations[50]

The point of departure for any study on Maltese geology and geography should certainly include the Bowen-Jones *et al* publication (Bowen-Jones *et al* 1962). Written in the early 1960s, this reference work is still largely relevant, especially when providing an introductory background to Maltese geomorphology, its spatial distribution and weather and climatic data. A more recently published text containing upgraded geological data for Malta and the central Mediterranean region within which it is placed is

[45] Refer to Legal Notice 241 of 2010 on Groundwater Abstraction (Metering) Regulations. See http://mra.org.mt/wp-content/uploads/2012/08/40.Groundwater-Abstraction-Metering-Regulations.pdf

[46] Such systems have already been successfully employed in the surveying and production of 3D models of rock-excavated subterranean environments elsewhere. See Powell 2009, 20-25; Alshawabkeh and Bal'awi 2010, 124-45.

[47] See also http://users.csc.calpoly.edu/~zwood/MaltaMapping/index.html

[48] The Malta Cistern Project could only materialise thanks to the valuable assistance and logistical support provided by T. Gambin of the Department of Classics and Archaeology at the University of Malta. I am truly grateful to T. Gambin for involving me in this project and for passing on data retrieved during the various seasons of well and cistern investigation.

[49] Cases in point are wells or cisterns located within the Tas-Silġ and the San Pawl Milqi sites. In the former instance, a number of water-related features have been partly investigated by the Malta Cistern Project team, but this will be discussed in further detail in chapter three, section 3.2.3.

[50] This section is not intended to provide a comprehensive list of the bibliographical sources consulted in the writing of this book. It is, rather, a brief account mentioning some important texts found useful in the research phase of this publication.

by M. Pedley *et al* (Pedley *et al* 2002). More specific data targeting the geomorphological characteristics of specific Maltese limestone strata are found in scholarly publications by J.A. Cassar, S. Scerri and P. Gatt amongst others (Cassar 2004; Cassar 2010; Scerri 2005; Gatt P.A. 2006). The essay by T. Zammit on the water supply of the Maltese Islands remains an ideal point of departure on the topic of past water management and includes a historiography of sources known to the author at the time of writing (Zammit T. 1924). The work of T.O. Morris provides a good detailed overview of water management strategies adopted by the British until the early 1950s (Morris 1953). This also contains some reflections on the low permeability levels of the Middle and Lower Globigerina Limestone deposits. During the 1950s minor perched aquifers had already been detected in Middle and Lower Globigerina Limestone strata by the local water authorities but not deemed fit for commercial exploitation.[51]

A number of historical sources were of direct relevance to my field of specialisation and were scrutinised for any hydrological-relevant data. High and Late Medieval sources of evidence are scarce (Luttrell 1975B, 1), but a scrutiny of notarial deeds, the earliest of which date to the 1460s, together with other evidence, did yield data which were directly relevant to the area of research. Another important source of evidence is Jean Quintin d'Autun's description of Malta published in Lyons in 1536 (Quintinus Haeduus 1536). It provided an insight into the Late Medieval local water situation. For the purpose of this publication, H.C.R. Vella's translation of this work was the source which I primarily consulted (Vella H.C.R. 1980). The original Latin text was also made use of.

G.F. Abela's *Della Descrittione di Malta* dating to 1647 (Abela 1647), and the revised edition of this work entitled *Malta Illustrata* published in two volumes by Count Ciantar in 1771 and 1780 respectively, were also widely consulted (Ciantar 1772).[52] Abela's interest in archaeology, place-name evidence and natural history, amongst other fields of research, rendered sections of his work directly relevant to my field of specialisation.[53] Abela's *Parte di Malta Inabitata* was one of these most applicable sections. Apart from giving an insight into agricultural intensification efforts, seemingly taking place in Malta during the first half of the seventeenth century in a probable attempt to increase local crop productivity levels, this section also betrays Abela's familiarity with both the rural terrain and the *cabrei* NLM. Treas. B. 289 and 290 manuscripts, a number of which are of a more or less coeval date. Occupying the post of Vice-Chancellor of the Order of the Knights of St. John, it is likely that Abela would have had direct access to this data. Water-related material mentioned in the Ciantar 1772 edition was also taken note of (Ciantar 1772). Taking into consideration that the Ciantar publication dates to the closing decades of the Knights' period in Malta, more weight was placed on the 1647 text by Abela.

From an academic, historical and archaeological perspective on medieval Malta, the 1975 publication edited by A.T. Luttrell entitled "Medieval Malta: Studies on Malta before the Knights", was found to be an important point of departure (Luttrell 1975A). Papers authored by A.T. Luttrell and H. Bresc, while not appearing to offer any direct relevance to my area of specialisation, were essential in facilitating the placing of my findings in their proper social and economic context (Luttrell 1975B, 1-70; Bresc 1975, 126-62). The same applies to A.T. Luttrell's essay entitled "The Making of Christian Malta", which discusses developments in the medieval field of specialisation since the publication of the "Medieval Malta" text in 1975 (Luttrell 1975B, 1-70; Luttrell 2002B, 1-17). Other useful texts are G. Wettinger's comprehensive volume on Maltese place-names (Wettinger 2000), which provides evidence on the first recorded instances for the *għajn* (spring), *bir* (well/cistern) and other water-related toponyms, as well as his essay on "Agriculture in Malta in the Late Middle Ages" (Wettinger 1982, 1-48), and his paper entitled "The Lost Villages and Hamlets of Malta" (Wettinger 1975, 181-216).

The Documentary Sources of Maltese History series was of pivotal importance when assessing the relevance of fourteenth and fifteenth-century *Cancelleria Regia* documentation in possession of the State Archives at Palermo, Sicily. This material was transcribed by S. Fiorini and published between 1999 and 2007 (Fiorini 1999B; Fiorini 2004; Fiorini 2007). Late Medieval documentation by notaries Paolo Bonello and Giacomo Zabbara was likewise transcribed by Fiorini and published between 1996 and 2005 (Fiorini 1996; Fiorini 1999A; Fiorini 2005). This extensive documentation targets a wide-ranging number of topics, including a number of water-related entries.

A number of unpublished specialist dissertations were also found to be of particular relevance to this study. The most important of these was B. Blouet's unpublished Ph.D. thesis (Blouet 1963). Discussing landscape changes in Malta during the Knights' period, it provides the reader with a comprehensive analysis of archival sources, a number of which specifically deal with countryside-related activity and development. Three other dissertations attempting to interpret data contained in the Knights' and British period *cabrei* manuscripts are in many ways complementary to this book. The first two are

[51] See chapter two, sections 2.4.2 and 2.6.2.

[52] Ciantar's work, whilst still a very valid contribution, has to be approached with caution as it contains a number of errors. Luttrell partly attributes this to the fact that Ciantar went blind before the issuing of the first volume and died before the publication of the second. See Luttrell 1975B, 2.

[53] For details concerning Abela's scholarly and antiquarian activities in Malta see Bonanno 1984, 27-37. Abela's work is not free of defects but personal field-research demonstrated that matters relating to rural Malta and water sources are generally reliable in nature. See Luttrell 1975B, 2-3.

MA dissertations by P.C. Saliba and D. Borg, both from the University of Malta (Saliba 2008; Borg 2011). Saliba's work attempted an evaluation of the rubble walls and field-terracing present in the Maltese landscape, through the partial utilisation of the above-mentioned sources of evidence. Borg's study documented landscape changes in the Mġarr (Malta) area in the Early Modern period. The third is a Ph.D. thesis submitted at the University of Aberdeen in 2001 entitled "A Forest History of the Maltese Islands to 1800", by C.F. Grech (Grech 2001). None of these works deals directly with water management strategies. Their use of *cabrei* data exposes the multi-disciplinary relevance and potential of these aforementioned archival sources. E. Vella's unpublished MA dissertation on a landscape archaeology approach to the corbelled huts or *giren* of northern Malta is another study which similarly attempts to place these structures in their appropriate landscape context using a retrogressive analytical approach (Vella E. 2010).

In another Ph.D. thesis by R. Grima on the landscape context of monumentality in Late Neolithic Malta, it was already realised that *għajn* and *bir* place-name spatial distribution is largely influenced by local geology. Whilst the majority of *għajn* place-names are located in areas containing Upper Coralline Limestone deposits, *bir* place-names commonly occur in the remaining areas of the Maltese archipelago (Grima 2004, 341-42; Grima 2005, 68-70).

1.9 Cartographic Sources

A number of cartographic sources proved to be of crucial importance in the execution of this study. Basic topographic, spatial and place-name data for the Maltese archipelago can be obtained through the use of the 1:25000-scale maps. The third revised edition dating to 1983 based on photographic data gathered by *Istituto Geografico Militare* of Florence (IGMI) was made use of. For field-research and property boundary identification, the use of more detailed maps having a scale of 1:2500 was indispensable. Produced by the Malta Environment and Planning Authority (MEPA), the latest version of these maps can be easily accessed and viewed on-line through the MEPA map server webpage.[54] This updated digital cartographic version supersedes earlier editions published in 1968 and 1974 and it is on this digitised map version of the archipelago that all ten-figure map references presented in this text are based. The spatial map referencing system taken up by MEPA makes use of the Universal Transverse Mercator (UTM) projection and the European 1950 datum.[55]

The cartographic representations found to be best suited for the purpose of this research were the late nineteenth and early twentieth-century 25-inch survey sheets for Malta produced by the School of Military Engineering at Chatham in the United Kingdom. A series of 152 survey sheets accurately documented the topography of the island of Malta. Gozo, Comino and the minor islands, however, were never included in this survey. Of particular importance to my field of research was the fact that these survey sheets, albeit only for Malta, also marked the location of wells or cisterns and tanks. One major drawback is that no distinction was applied between wells and cisterns. Similarly, the absence of contour lines makes it difficult to discern the height and shape of the land without any appropriate familiarity with the actual terrain.

Worthy of mention in this section is a cartographic database compiled in the early 1940s, now in the possession of the Malta Resources Authority.[56] This contains data of privately-owned water sources found in the archipelago, be they springs or wells, registered with the Government authorities in 1943. Technically referred to as the "Survey of Private Groundwater Abstraction Sources",[57] the collected data are an extremely useful tool in regard to discerning the quantity and location of privately-owned boreholes which were excavated locally from the latter half of the nineteenth century till the 1940s. The information contained within helped to establish if the numerous "wells" indicated on the late nineteenth and early twentieth-century 25-inch survey sheets of Malta are actually tapping an aquiferous source or whether they are water cisterns. Furthermore, a number of datasheet entries provide useful information: the circular or rectangular shape of the well-shafts, the approximate water-yield per well, and more importantly, the distinction between the older wells and those excavated in the opening decades of the twentieth century. These data will certainly feature in future water-related studies. Many of the Upper Coralline Limestone galleries mentioned in the datasheets postdate the turn of the twentieth century and steer clear of cliffs and ravines, thus having no connection with the perched aquifer galleries discussed below.

The 1943 database, therefore, provided little or no tangible data on the perched aquifer galleries. Datasheets for the Wied ir-Rum (RA56), Misraħ Suffara and Ħofret ir-Rizz (RA27), all in the Rabat (Malta) territory, were consulted as were those of other areas. A number of entries do relate to perched aquifer galleries discussed in this book, but apart from being referred to as "old springs", no mention of the excavated water tunnel tapping the perched aquifer source was made. A typical

[54] See http://www.mepa.org.mt/mepa-mapserver

[55] In the spatial referencing system adopted by MEPA, the leading "4" in the Easting and the leading "39" in the Northing have been removed. Taking map reference 27904 88504 as an example, in order to convert this to the full UTM spatial referencing system, it will have to read 427904E, 3988504N.

[56] Access to this database was granted by the then Director of the Malta Resources Authority, J. Mangion and M Sapiano, Chief Technical Officer (Water), Sustainable Energy and Water Conservation Unit (Ministry for Energy and Health).

[57] Personal communication by M. Sapiano.

case in point is spring no. 2476, marked in coloured ink on Malta Survey Sheet 85 published in 1922. This corresponds to the Ta' Baldu Spring located in the uppermost reaches of Wied ir-Rum in the territory of Rabat (Malta). The datasheet for this entry classifies it as an "old spring" and specifies that "water flows by gravity from natural opening in rocks." The "delivery and storage arrangements" section of this same datasheet again mentions that water is derived "from natural rock opening into a recipient, latter discharge [*sic*] into stone channels to site of irrigation".[58] In this particular instance, no direct reference to the perched aquifer water gallery at Ta' Baldu (RA43) was made, suggesting a lack of familiarity with the dynamics of such water systems during this period. On the other hand, it was noted that the spring flow varied according to season and that it provided a water-yield of ca. eighteen thousand gallons of water per year.

Another example is the Għajn Tejtes spring (RA20), which in the "Survey of Private Groundwater Abstraction Sources" corresponds to spring no. 2463.[59] This entry not only completely ignores the existence of a series of three perched aquifer galleries from which the Għajn Tejtes spring water originates, but only points to the uncovered reservoir in which the spring water is stored. Għajn Tejtes is simply referred to as an "old spring" where the "spring water flows by gravity from natural rocks". A perched aquifer gallery in the vicinity, located beneath the recently identified rock-excavated church of St James, portrays an identical situation (Buhagiar K. 2013, 98-107). The 1943 database classified this as spring no. 2468 and only states that "water flows by gravity from natural openings in the rock".

Other useful tenement maps are those included in the 1860s "Descriptive Plans of the Crown Property" *cabrei* compiled between 1861 and 1867. These were the result of a surveying exercise which included all Government-owned properties.[60] The survey was completed by 1860 (Saliba 2008, 134), and the resultant plans were assembled into four large hard-bound leather registers, more technically referred to as *cabrei*,[61] between 1861 and 1867.[62] Reference to water sources is made in these volumes and these were particularly useful in the interpretation of contemporaneous and earlier period documentation presented in chapters four to seven.

Another cartographic representation for the archipelago is the Geological Map of the Maltese Islands having a scale of 1:25,000 and for which there are two sheets (Debono and Xerri 1993, Geological Map of Malta, Gozo and Comino, Sheets 1 and 2). Malta is included in sheet 1, whilst Gozo and Comino feature in sheet 2. It is only through the detailed study and analysis of this map that the location and spatial distribution of the water-bearing strata were identified and understood.

The comparison of cartographic and geological data with satellite imagery of the Maltese archipelago was achieved through the employment of the public shareware geographic information software Google Earth. This software provided complete aerial coverage of the archipelago, the representative images for which are of a fair ground resolution quality. This research method was only possible as from 2010, when the ground resolution quality of satellite imagery data for the Maltese islands was significantly increased, making it possible to get a sufficiently clear perception of land features from a relatively low eye altitude of 350 metres above surface level. Water reservoirs and reed growths, indicative of a water presence within the landscape, could be easily detected with the use of the aforementioned software. This method also provided a better spatial perception of the land area covered in the older map sources prior to the carrying out of any field-related research. Google Earth software was also directly made use of whilst conducting field-research. In countryside areas with adequate network coverage, third generation communications technology (3G) allowed access to the Google Earth satellite imagery online database directly through the employment of a global positioning systems (GPS) enabled smart phone. The employment of this system not only allowed the author of this study to readily locate water-related features detected during the pre-field-research sessions, but to accurately plot any archaeologically-relevant feature identified during the field sessions.

1.10 Archival Sources

The research work necessary for the formulation and writing of this book was primarily carried out at the National Library in Valletta, the National Archives at Rabat (Malta), the Notarial Archives, the Mdina Cathedral Archives and the Curia Archives in Floriana. All yielded important data, but the bulk of the archival research was carried out in the first two of the above-mentioned venues. The Treasury A and B series at the National Library yielded significantly relevant data for the Knights' period. The Treasury B volumes, which mostly consist of property books (*cabrei*) dating to 1654-1810, were found to be of particular relevance. Numerous Treasury B manuscripts were investigated. In order to come to an assessment of the water situation in the century following the arrival of the Knights' in Malta, it was decided to primarily concentrate on *cabrei* sources

[58] Refer to the Survey of Private Groundwater Abstraction Sources.
[59] Refer to the Survey of Private Groundwater Abstraction Sources.
[60] This work was carried out under the direction of E.L. Galizia, the Chief Surveyor of the Department of Land Revenue and Public Works.
[61] See chapter five, section 6.
[62] The 1866 and 1867 Descriptive Plans of the Crown Property *cabrei* were issued under the direction of the Honourable Giovanni Vella, Collector of Land Revenue. All Crown Property *cabrei* volumes were formerly housed at the Chief Draughtsman's Office at Floriana, but since the last quarter of 2010 they came to form part of the National Archives of Malta collection.

covering the seventeenth and the first half of the eighteenth centuries.

The National Archives were made use of chiefly in matters relating to the British period. Apart from being the repository where the Descriptive Plans of the Crown Property volumes are now housed, nineteenth-century Dispatch manuscripts, amongst other sources of evidence, were found to contain valuable, previously untapped, data on the hydrological situation and developments taking place in the local context as from the mid-1830s.

1.11 Book Structure

As mentioned in section 1 of this chapter, this book adopts a retrogressive methodological approach in order to tentatively establish by elimination, possible modes of water management strategies resorted to during the Late Medieval period in Malta. My field and archival research, observations, analyses of the collected data and conclusions are presented in the following seven chapters:

Chapter two focuses on the geological makeup of the Maltese archipelago, its climate and an analysis of the water-bearing characteristics of several strata found within. Apart from the mention of the perched aquifer forming above the impermeable Blue Clay stratum, this chapter makes a case for the existence of other minor perched aquifers located in Globigerina Limestone deposits and their exploitation since antiquity.

Chapter three is an overview of different water management techniques employed in a Mediterranean context since antiquity. The various sections within this chapter tackle the diffusion of water management technology and discuss issues relating to rain-water runoff and harvesting, *qanat* and *cuniculi* systems, together with various modes of water-harvesting systems observed in a Maltese context, but which are unfortunately difficult to date.

Chapter four deals with the changing Maltese hydrological landscape from 1798 to the present. The aim of this chapter is to examine and investigate water-related changes taking place locally during this period. Until the first half of the eighteenth century, Malta's hydrological framework was still primarily dependent on a series of perched aquifer springs dotting the countryside in north-west Malta, which in the early seventeenth century were connected to the Wignacourt Aqueduct system. This model rapidly changed in the ensuing decades when the dynamics behind the workings of the mean-sea-level aquifer were grasped. Perched aquifer gallery excavation in Upper Coralline Limestone deposits taking place during this period is also investigated. Additionally, a section of this chapter presents the relevance of data emanating from the 1860s Descriptive Plans of the Crown Property *cabrei* in the study of water management technology during this period.

The following two chapters discuss the water scenario during the Knights' period. Chapter five deals with the early Knights' period cartographic sources, an analysis of hydrology-related aspects imparted by Fra Gian Francesco Abela and the examination of other seventeenth-century archival evidence which makes direct reference to the availability of perched aquifer galleries in Malta. Chapter six takes into consideration the *cabrei* manuscript sources for the seventeenth and the eighteenth century and the water data divulged therein.

The High and Late Medieval hydrological situation for the archipelago is discussed in chapter seven. The Knight Commissioners' report of 1524 and Jean Quintin's description of Malta published in 1536 proved to be an ideal point of departure, from which I worked my way through the earlier period documentation. A scrutiny of the water-related place-name evidence till 1560 was also attempted, as well as an assessment of the *Cancelleria Regia* and Late Medieval documentation of the period.

Chapter eight synthesises the principal findings of this study and attempts an interpretation of the presented data.

Appendix 1 comprises a gazetteer database containing data on the sites mentioned in the text. The listed sites are classified according to the district in which they are located and in order to facilitate usability, sites were assigned an identification number to allow cross-referencing with the text. Taking the site of *Ghadira di Torbet Gharfagna* in the Attard district as an example, in the gazetteer this was assigned the identification number of "AT1" and is referred to in the text as *Ghadira di Torbet Gharfagna* (AT1).

Readers of this book should also be alerted to the fact that a number of place-names reproduced herein were transcribed from the original documentary sources. Divergent place-name transliteration should not be considered as the consequence of typographical errors – it is the reflection of unstandardized norms. In view of this and in order to facilitate readability, a comparative table of place-names summarising the occurrence of the same toponym in different centuries is included in Appendix 2. The transliteration of archival sources in either Italian or Latin which are quoted in sections of the text presented below is also governed by the above characteristics. A philological appreciation of Italian, Maltese and Latin words/terms made use of in this text is attempted in Appendices 3 to 5.

As a concluding comment to this section, this study also employs the use of a significant number of figures and plates. These not only indicate site location, but also showcase the field and archival research undertaken in part-fulfilment of this publication.

Chapter Two

Geomorphological, Climatic and Hydrological Considerations

2.0 Geology

The Maltese archipelago, the largest islands of which, in descending order of size are Malta, Gozo and Comino, lies in the central Mediterranean Sea, 93.3 km south of Sicily. Central to the discussion of this book is the island of Malta, which occupies a total land surface area of 153 km², and has a maximum length and width of 27.4 km and 14.5 km respectively (Fig. 1.1). Geological deposits are almost exclusively sedimentary in formation, and started to form in a marine environment thirty to ca. six million years BP. The archipelago owes its origin to prolonged stress between the European and African continents, where plate tectonic activity completely reshaped the central Mediterranean basin into a series of horst and graben formations. Tectonic activity also uplifted several portions of the Sicilian-Tunisian platform on which Malta lies a couple of hundred metres above sea level at the highest points (Bowen-Jones *et al* 1962, 25; Zammit-Maempel 1977, 18; Pedley *et al* 2002, 1, 18-29, 63).

2.1 Climatic Considerations: Present-day and past

The climate of the Maltese archipelago is characteristically Mediterranean and is typified by hot dry summers and relatively warm wet winters (Bowen-Jones *et al* 1962, 48-49). The annual temperature range is approximately 15°C and an average annual precipitation of 550 mm makes rainfall insufficient and erratic and creates regular drought conditions (Skinner *et al* 1997, 188; Sapiano *et al* 2006, 21) (Fig. 2.1). Maltese rainfall values conform to those of a semi-arid climate which, on the other hand, experiences high seasonal and interannual variability. Some years are excessively wet, whilst others are exceptionally dry. Data gathered between 1947 and 2004 for the Luqa meteorological station show how annual precipitation can vary by as much as 300 mm more, or 250 mm less than the rainfall mean (Sapiano *et al* 2006, 21; Sapiano *2008,* 80-81).

Since rainfall readings for the Valletta weather station date back to 1841, these enable the computation of a reliable local climatic scenario for the past two centuries. Rainfall between September and November is fairly reliable in nature and is often characterised by heavy downpours, but progressively decreases during the spring months (Park 1977, 508-9). On average the wettest month is December. July is the driest month during which drought has been registered in most years (Bowen-Jones *et al* 1962, 48-49) (Fig. 2.1). There are also regional precipitation variations. Areas in north and north-west Malta receive an average annual rainfall of 635 mm, whilst the precipitation levels for the Marsaxlokk area in south-east Malta are of 508 mm or less (Park 1977 508-9). Downpours are often localised and as much as 16 per cent of the rainwater is lost as surface runoff and forty per cent through evaporation (Azzopardi 1995, 75). The warmest months are July and August, during which temperatures occasionally reach ca. 40°C.

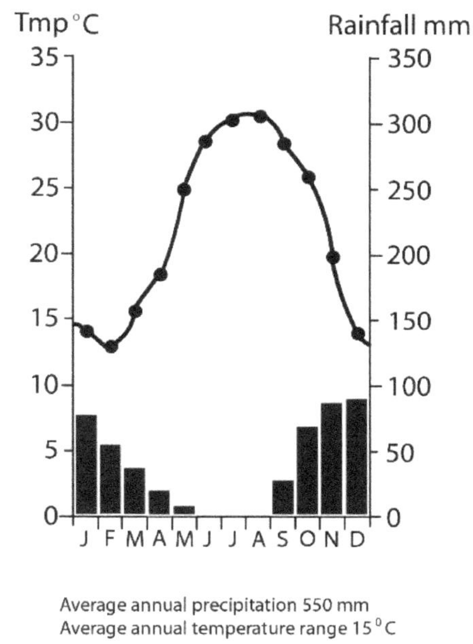

Fig. 2.1 Climate graph showing current rainfall and temperature trends for the Maltese Islands (after Waugh 2003, 215).

The main water-harvesting and management difficulties resulting from the observed rainfall patterns in the archipelago include: (a) high-intensity, but short-duration rainfall events; (b) years of poor rainfall, resulting in inadequate groundwater recharge; (c) considerable rainwater runoff in years of above-average precipitation; and (d) seasonal rainfall scarcity which coincides with higher than normal water requirements (Sapiano *et al* 2006, 22-23).

The occurrence of climate shifts in the Mediterranean region since antiquity is still subject to controversy and debate. Nonetheless, basing oneself on sedimentation analysis, archaeological and historical evidence, it appears that Mediterranean climatic conditions in Late Antiquity were not appreciably different from those of the present-day and were subjected to temporary and minor wetter, colder and drought interludes (Vita-Finzi 1969, 112-14; Grove and Rackham 2001, 145; Bolle 2003, 22-23; Carroll *et al* 2012, 37; Lionello *et al* 2012, xlv-xlvi; McCormick *et al* 2012, 169-220).[1]

Recent local sediment analysis and radio-carbon dating from Marsa suggest an increase in sediment erosion, possibly the effect of more rainfall, at around 1000 years BP. This was followed by a warmer period, which in north Tunisia resulted in rising temperatures and an increase in humidity levels at around 900 to 700 years BP (Fenech 2007, 112).

In another study, the winter mean climate reconstruction between 1500 and 1995 was attempted for the larger Mediterranean area. No drastic winter temperature and rainfall changes were detected even though fluctuations were registered. The coldest winter was in 1891, the warmest in 1772, the wettest in 1684 and the driest in 1609 (Luterbacher and Xoplaki 2003, 133-53; Luterbacher *et al* 2006, 27-97; Camuffo *et al* 2010, 169-99). Local annual rainfall statistics gathered between 1851 and 1950 did not register any rainfall decrease during the course of a century. An analysis of Maltese weather data has, however, detected a shift from the second half of the rainy season (January to March) to the first half (September to December) in terms of monthly precipitation (Hyde 1955, 93). Furthermore, in the 1950s it was observed that Malta was experiencing an extended dry season (Hyde 1955, 93).

2.2 Changes to Malta's Demographic Base over the Centuries

The dearth of available documentary evidence can only provide a very approximate estimate of Maltese population figures in the Late Medieval period. The most important study remains to date G. Wettinger's analysis of the militia list of 1419 which gives a list of able bodied men for compulsory military service (Wettinger 1969, 80-106). The militia list data coupled with other documentary sources of evidence intimates that the population figure for the island of Malta from 1419 to 1480 was in the region of 8,335 and 10,633 individuals (Wettinger 1969; Fiorini 1983; Fiorini 1986; Fiorini 1993). A temporary demographic decline may have set in around 1450. The population numbers increased to ca. 17,000 in 1530, of which 12,000 probably lived on Malta. The population of the archipelago in the 1240s is also estimated at 10,000 (Wettinger 1969, 83; Fiorini 1993, 113-14, 125-26; Blouet 2004, 56-57; Dalli 2006, 28-29). Taking an average population base of 10,000 individuals, it is likely that for most of the Late Medieval period, the Maltese archipelago had a population density of ca. 32 persons per km².

A number of population estimates were carried out in the archipelago during the Knights period. Malta had a population base of 32,290 in 1590; 43,798 in 1617; and 51,750 in 1632. For 1726 and 1784, Malta's population is estimated at 64,353 and 87,669 individuals respectively (Central Office of Statistics Malta (COS) 1957, iv-vi; Fiorini 1983, 325-43; Fiorini 1986, 41-100; Blouet 2004, 81-87).

The average population density for the nineteenth century was of 350 people per km². Censuses carried out during this period show that Malta had a population of 93,054 in 1807; 114,499 in 1842; and 123,496 in 1851 (COS 1957, iv-vi).[2] It was during the course of the twentieth century that the Maltese population registered a dramatic increase. In 1901 the archipelago's population stood at 184,742, and at 211,564 individuals in 1911 (National Statistics Office Malta (NSO) 2010, 3). Data gathered in the 2011 census show that during that year, the Maltese population stood at 417,432 individuals and had a population density of 1,325 persons per km² (NSO 2014, xi-xiv).

2.3 Crop Types

A general overview of documented crops cultivated in the Maltese archipelago during the Late Medieval period is presented in this section. A more in-depth analysis of crop types in the context of Late Medieval period agriculture is attempted in chapter eight, section 8.1. Farming formed the basis of Maltese economy in the period under review. Dry and irrigation farming were both practised but there are indications to show that dry farming predominated (Wettinger 1982, 3). Irrigation farming commonly took place in areas having *ġnien* and *għajn* place-names and frequently included vegetable, fruit tree, vine and cotton cultivation (Blouet 1963, 54-58; Wettinger 1982, 3-4, 23-27; Fiorini 1993, 115-16, 174-75; Dalli 2006, 158).[3] Until 1530, cotton formed the basis of the Maltese economy and was the principal cash crop exported to a number of major neighbouring ports (Wettinger 1982, 15-21).

Products synonymous with dry farming included the cultivation of wheat and barley. The mixture of these two crop types was locally known as *maħlut* (Wettinger 1982, 13; Fiorini 1993, 174). Cumin was another cash crop the economic importance of which was only secondary to cotton (Wettinger 1982, 21-22). This was grown in areas dedicated to dry farming, as also appears to have been the case with flax cultivation (Wettinger 1982, 21-22). The retting of flax was, however, frequently associated with

[1] For a detailed analysis of current Mediterranean climatic data see Lionello *et al* 2006, 1-16; Lionello *et al* 2012, xlvii-xlviii; Ulbrich *et al* 2012, 301-34.

[2] For a study on nineteenth-century Maltese demography see Agius 2013.

[3] The Latin equivalent of *ġnien* is *viridarium* (pl. *viridaria*). See Wettinger 1982, 27-29.

pools of stagnant water commonly found in marshland areas located in bays and inlets. Of particular significance is a public proclamation dating to 28 July 1519 (Wettinger 1982, 29-31). As this forbade the retting of flax in *giardino* (*ġnien*) areas such as Ġnien is-Sultan (RA13) and Għeriexem (RA26), one is led to conclude that prior to this date, flax retting was also widespread in *giardino* areas containing spring water sources.

In areas of Upper and Lower Coralline Limestone deposits, it appears that large stretches of garigue karstland, commonly located on promontories flanking the sides of valleys, were commonly utilised for rough grazing for the upkeep of herds of sheep and goat. These areas also provided farming folk with thistles, thyme and thorns, commonly used as fire fuel (Blouet 1963, 55; Wettinger 1982, 3-4, 31-35; Fiorini 1993, 159). It is probable that many areas of garigue karstland were developed into arable land during the course of the Early Modern and the Modern periods as a result of pressures stemming from demographic growth (Wettinger 1982, 31-32).

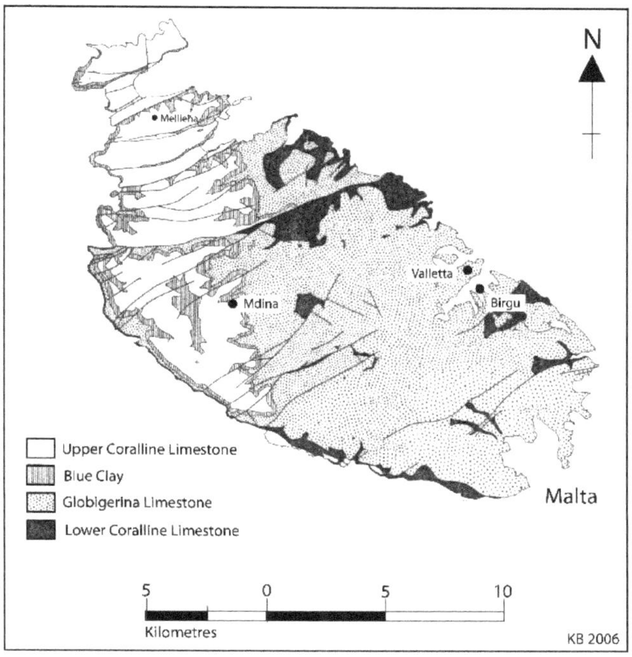

Fig 2.2 Geological map of Malta (after Bowen-Jones *et al* 1962, 24).

2.4 Maltese Geological Formations, Properties and Stratification

Sedimentary deposits form in underwater environments. Even though it is estimated that these constitute seven per cent of the world crust by volume, they cover seventy-five per cent of its surface area. Sedimentary deposits document life on earth and record the composition, climate and topography of former landmasses as well as the physical, chemical and biological conditions of oceans that no longer exist (Dolgoff 1996, 213-14, 226). Sediments also provide the means of documenting geological deposition in a chronological sequence. As is the case with archaeological stratification, the oldest layer of a given sequence is the lowest and is overlain in turn by successively younger layers (Dolgoff 1996, 213-14).

Maltese sedimentary deposits are biochemical in nature. Biochemical sediments predominate in warm underwater environments teeming with life, but have a scarcity of land-derived sediments (Dolgoff 1996, 216-17, 222-23). The Maltese limestones were primarily derived from shell matter and other marine organisms deposited on the continental margins to form carbonate shelves (Pedley et al 2002, 41-42). Calcium carbonate concentrations found within sedimentary deposits form because of increasing water temperature and minimal water pressure. This explains why high calcium carbonate concentrations form in shallow warm tropical and subtropical seas, but are rarely found in great quantity in areas of higher latitudes and deep waters (Dolgoff 1996, 221-23). Calcium carbonate also forms an important part in the cementation process of limestone deposits (Dolgoff 1996, 224-25). Due to the presence of rounded grains of calcium carbonate known as ooliths, Maltese limestones are frequently classified as being of an oolitic nature (Pedley *et al* 2002, 61-62).[4]

Maltese limestones display an unexpected variety of coloured sediments which range from a dark greyish-green colour, to buff, yellow, white and red. The bluish colour of several deposits is generally due to the presence of minerals such as oxides, iron or clay particles (Cachia 1988, 4). Four distinct rock layers constitute the basic geology (Figs. 2.2, 2.3), and when undisturbed by land faulting, the horizontal stratification from bottom to top reads as follows: (a) Lower Coralline Limestone; (b) Globigerina Limestone; (c) Blue Clay; and (d) Upper Coralline Limestone (Pedley *et al* 2002, 35-36). Based on micro-chemical similarities, this classification takes the Greensand layer as being the lowermost stratum of the Upper Coralline Limestone deposit and contradicts older publications which regarded this as being a completely distinct entity from Upper Coralline Limestone.[5]

The porosity levels of the Maltese rock deposits are discussed in section 2.6, hereunder. While examining the impermeable qualities of the Blue Clay deposit, particular emphasis is placed on the water-bearing properties of the various Globigerina Limestone deposits. Marly deposits located within the latter are responsible for the formation of localised aquifers, which when detected and tapped by

[4] The analysis of a number of Maltese Globigerina Limestone specimens revealed percentages of calcium carbonate which vary from 63.2 to as much as 94.7 per cent and as low as 30 per cent in specimens retrieved from the Upper Globigerina beds. See Cachia 1988, 10-12.

[5] For a classification of Maltese geological deposition in which Greensand formations are considered to be a separate rock layer see Bowen-Jones *et al* 1962, 25-29. See also section 2.4.4 below.

past societies, might have encouraged agriculture and settlement.

2.4.1 *Lower Coralline Limestone*

Lower Coralline Limestone is usually a dense, semi-crystalline bedded limestone and its colour ranges from white to red and buff (Bowen-Jones *et al* 1962, 26). This rock-type is hard and well-jointed, and readily allows the percolation of water (Bowen-Jones *et al* 1962, 43; Waugh 1995, 180). The Lower Coralline Limestone sequence is at least 140 m thick, but mostly lies below sea level. Exposed Lower Coralline Limestone deposits are a feature common to the coastal cliff-regions along the western coast of Malta. With the exception of sizeable outcrops in the Mosta-Naxxar area and Attard, these rarely occur in inland locations (Morris 1952, 54-55; Pedley *et al* 2002, 43). Four subdivisions of this rock stratum exist. Listed from an upper to a lower sequence these are the: (a) Mara Member; (b) Xlendi Member; (c) Attard Member; and (d) Magħlaq Member (Morris 1952, 54-55; Pedley *et al* 2002, 43). Lower Coralline Limestone is of a typically carboniferous configuration (Pedley *et al* 2002, 43-45).

2.4.2 *Globigerina Limestone*

Globigerina Limestone outcrops cover considerable portions of land in central and south-eastern Malta and western Gozo (Figs. 2.2, 2.3). The quality of Globigerina deposits varies significantly, even within the same quarry. Likewise, its thickness varies considerably, from ca. 205 m in south-eastern Malta, to the thinnest layer (22 m) recorded on Gozo (Zammit-Maempel 1977, 24-25). The economic value of this rock horizon has been recognised since antiquity and it retains its place as the main building material (Cassar 2010, 13-25). The Maltese Globigerina deposits bear a striking resemblance to Horner Schichten deposits in Austria and the Italian Pecten-Fauna of Schio which belong to the Burdigalian era of the Lower Miocene period (Hyde 1955, 35).

It was not until a few decades ago that Globigerina Limestone strata classification was standardised. In the mid-nineteenth century, T.A. Spratt referred to this geological deposit as "sandstone" and divided it into five substrata of differently coloured varieties (Spratt 1854, 5-8). Leith Adams named this rock deposit "Calcareous Sandstone" (Adams A.L. 1870, 124-30), whilst "Globigerina Limestone" was the name assigned by J. Murray (Murray 1890, 459-74). Osbert Chadwick simply identified this deposit as the "yellow rock" (Hyde 1955, 34).

A conflicting number of horizons in the Globigerina Limestone deposit were identified by past specialists in the subject. C.H. Colson and J.H. Cooke identified eight and nine substrata respectively (Cooke 1896, 502-11). C. Rizzo subdivided the deposit into twelve distinct layers which he classified into three different sub-formations: Upper, Middle and Lower Globigerina Limestone (Rizzo 1932, 12-14). F.A. Bather and C.T. Trechmann subdivided the Globigerina deposit into four distinct layers which from top to bottom read: (a) White rotten limestone; (b) Hard grained limestone; (c) Semi-crystalline limestone; and (d) Yellowish-soft limestone (cited in Hyde 1955, 34).

The three widely accepted geological subdivisions for the Globigerina deposit in use today subscribe to Rizzo's classification mentioned above (Debono and Xerri 1993, Geological Map of Malta, Sheet 1). Thin phosphatic nodule beds mark the division between the Middle and Upper Globigerina Limestone layers. Phosphatic nodule beds can be visually described as an agglomeration of rounded or oval, smooth, pebble-like concretions with a chocolate brown colour which have a glazed surface (Zammit-Maempel 1977, 24-25; Pedley *et al* 2002, 49). Being marly in nature, Globigerina deposits are less permeable than the Maltese Lower and Upper Coralline Limestone formations and were in the nineteenth and early twentieth centuries considered as being of lesser value for water absorption (Morris 1952, 53). The porosity level of the Maltese Globigerina strata can reach values of up to 4.7 per cent (Cassar 2010, 13-14). The thickest Globigerina Limestone beds are located in the Ħamrun region on the east coast of Malta (Morris 1952, 55-56).

The characteristic elements of Upper Globigerina are its fine-grained texture and yellow-orange colour. It contains abundant petropods and scattered phosphatic nodules. Pockets of light grey or blue argillaceous marl have also

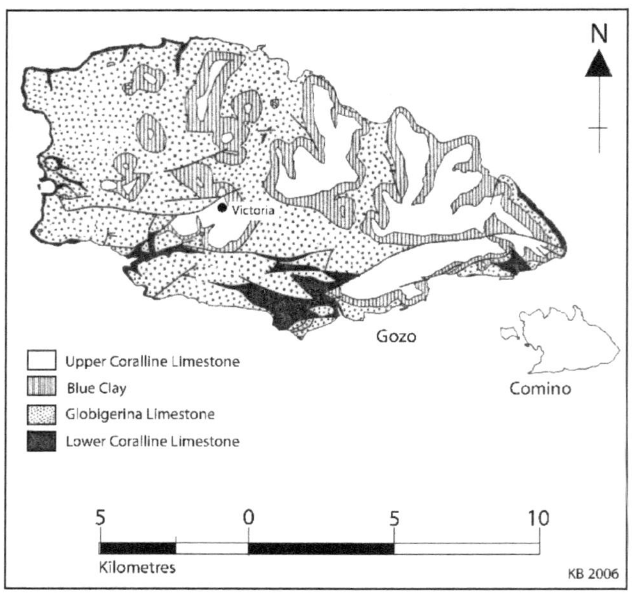

Fig. 2.3 Geological map of Gozo (after Bowen-Jones *et al* 1962, 24).

been identified.[6] Upper Globigerina is similar in appearance and consistency to Lower Globigerina, but is inferior in working quality when quarried (Zammit-Maempel 1977, 25). Some flaggy beds are heat-resistant and were used in the past for the building of small cooking stoves known locally as *kwiener*.[7] The Upper Globigerina deposit varies in thickness from 8 to 26 m (Morris 1952, 55-56). The transition towards the overlying clays is in instances very gradual, and no exact border-line can be established (Hyde 1955, 35). Its lithology is identical to that of the lower portion of the Blue Clay deposit (Morris 1952, 56).

Middle Globigerina has a thickness ranging between 15 to 38 m and consists of a phosphatic nodule bed with limestone pebbles and abundant corals, fish teeth and other fossils. Bluish-grey marls abound in this deposit and in some areas chalky limestone and chert nodules are also present (Morris 1952, 56; Zammit-Maempel 1977, 25; Pedley *et al* 2002, 49).

Ferruginous and phosphatic nodules and abundant casts of fossils and fish teeth are characteristic of the Lower Globigerina Limestone layer (Bowen-Jones *et al* 1962, 27). This has an average thickness of 40 m and is much sought after by the construction industry. Lower Globigerina contains little argillaceous deposits and where the Middle Globigerina marls have weathered away, it is well-cemented and strengthened (Morris 1952, 56). An easily recognised seam of yellow-brown hard, irregular phosphatic nodules generally marks the transition towards the underlying Lower Coralline Limestone. Where this is missing, a *Scutella Bed* forms a clear boundary line (Hyde 1955, 35). This stone horizon, which was formerly also referred to as "third quality stone", is identified as *Franka* by the local masons and can be easily shaped by saws, axes and chisels (Pedley *et al* 2002, 52; Cassar 2010, 14). The permeability of Lower Globigerina is generally low, but water percolation is aided by joints and fissures which have developed within (Morris 1952, 56).

2.4.3 Blue Clay

From a hydrological perspective, Blue Clay is one of the most important rock horizons available locally. It is due to its presence and impervious nature that an easily accessible water table, commonly referred to as the "perched aquifer" forms at the base of Upper Coralline Limestone formations (Morris 1952, 55). With a maximum thickness of 70 m, water stored above this rock horizon has probably been recognised as a vital and easily accessible resource since antiquity (Zammit-Maempel 1977, 26; Pedley *et al* 2002, 35, 55, 81). The perched water table was until the latter half of the nineteenth century the only reliable water source available in the archipelago.[8] Blue Clay is not evenly distributed, with its largest concentrations being present in the north and north-west Malta (Fig. 2.2). It is also worthwhile noting that in proportion to Malta's surface area, Gozo has more abundant clay deposits (Fig. 2.3). Blue Clay erosion produces 45° slopes, which often make the apparent thickness of such outcrops seem far greater than is actually the case (Zammit-Maempel 1977, 26) (Fig. 2.4).

Fig. 2.4 Exposed Blue Clay slopes at Ġnejna Bay in the territory of Mġarr, Malta. Source: the author.

2.4.4 Upper Coralline Limestone

The transition from the Upper Coralline deposits to the Blue Clay layer is generally well-defined and often forms a clear line of demarcation. Upper Coralline Limestone is the youngest rock formation, for which four subdivisions have been identified. All are carboniferous in nature and listed in sequence from top to bottom these are: (a) Ġebel Imbark Member; (b) Tal-Pitkal Member; (c) Mtarfa Member; and (d) Għajn Melel Member (Debono and Xerri 1993, Geological Map of Malta, Sheet 1). All Upper Coralline strata are permeable and numerous fissures within allow the percolation of rain water into the underlying deposits (Bowen-Jones *et al* 1962, 43).

The Ġebel Imbark Member, which has an average thickness varying between four and 25 m, is made up of hard, pale-grey carbonates with sparse fauna inclusions. The Tal-Pitkal Member is composed of pale-grey and brownish-grey coarse grained wackstones and packstones containing a diversity of fauna inclusions. Its thickness varies from 30 to 50 m (Debono and Xerri 1993, Geological Map of Malta, Sheet 1) (Fig. 2.5).

The Mtarfa Member deposit is on the other hand constituted of massive to thickly bedded carbonate mudstones and wackstones and is chalky in nature. The

[6] The pockets of light grey or blue argillaceous marl are rich in clay content and readily dissolve when exposed to water. It is possible for this geological layer to retain small quantities of localised, exposed subterranean water deposits. See Morris 1952, 55-56; Bowen-Jones *et al* 1962, 27; Debono and Xerri 1993, Geological Map of Malta, Sheet 1.

[7] Sing. *kenur*. See Zammit-Maempel 1977, 25. One widely used such geological deposit is located close to Xewkija in Gozo and was formerly used in the building of ovens.

[8] See chapter four, section 4.2.

Fig. 2.5 Exposed Tal-Pitkal Member (A), and Għajn Melel Member (B) deposits at Ta' Kuljat, Gozo. The fronting field contains Blue Clay deposits (C). A perched aquifer gallery (B) tunnels into the Għajn Melel Member in order to tap a perched aquifer located at that level. The extracted water is stored into the fronting, circular-shaped reservoir. Source: the author.

thickness of this deposit varies from 12 to 16 m and due to its friable disposition can easily be excavated (Debono and Xerri 1993, Geological Map of Malta, Sheet 1). Mtarfa Member deposits have a distinctively whitish colour, a characteristic which makes these rock outcrops easily identifiable, even from a distance. Past personal research had already detected the presence of numerous cliff-face settlements excavated in this deposit together with a surprising density of perched aquifer galleries which tap the aquifer in order to extract a perennial water flow (Buhagiar K. 2012A, 157-61) (Fig. 2.6). As a matter of fact, Mtarfa Member deposits commonly overly Blue Clay deposits directly and therefore give direct access to the perched aquifer.

The Għajn Melel Member forms the lowermost Upper Coralline Limestone subdivision (Fig. 2.5). Modern geological maps append the glauconite-rich Greensand formation to this member (Debono and Xerri 1993, Geological Map of Malta, Sheet 1). Greensand deposits can be easily identified by their brittle and coarse nature and are unevenly distributed throughout the archipelago.

Fig. 2.6 A cave settlement excavated into the brittle Mtarfa Member deposits at Il-Qlejgha tal-Bahrija in the territory of Rabat, Malta. Source: the author.

These deposits are orange-brown in colour and when an abundance of glauconite is present, they attain a greenish colour, a characteristic which has given rise to the name of this rock stratum. Its average thickness is of ca. eight metres, but in south-west Malta it is ca. 15 m thick (Bowen-Jones *et al* 1962, 29; Pedley *et al* 2002, 56-58). Greensand weathers by forming a thick scum on the surface, beneath which cavities may form. Large detached rock-boulders from this layer are often found on the underlying clay and Globigerina Limestone slopes (Bowen-Jones *et al* 1962, 31).

2.5 Landforms and Relief

The Maltese Islands were subjected to considerable tectonic movement over the past millennia. Micro-plate movement has shaped the relief of the land into a series of horst and graben zones (Bowen-Jones *et al* 1962, 31; Dart *et al* 1993, 1156-66; Pedley *et al* 2002, 39). The near-vertical walls of fault planes contain ample evidence of slickensiding, the smooth surfaces of which suggest that this land movement is of a relatively recent geological date (Bowen-Jones *et al* 1962, 32).

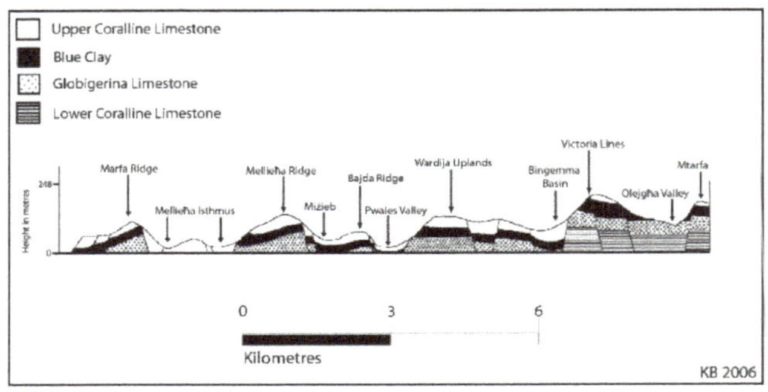

Fig. 2.7 North-South geological cross-section of northern Malta (after Bowen-Jones *et al* 1962, 41).

Two main types of land movement have shaped Maltese topography: (a) fault lines spanning in a north-west to a south-east direction; and (b) fault lines crossing from a north-east to a south-westerly direction (Schembri 1993, 27-39). The largest and most impressive of these is the Great Fault which follows a south-west to north-east axis and divides Malta into two distinct regions (Figs. 1.1, 2.2). Starting at Ras ir-Raħeb near Fomm ir-Riħ and passing through Binġemma, Mosta and Naxxar, this fault line finally meets the sea on the east coast of Malta close to Madliena (Pedley *et al* 2002, 39).

The area north of the Great Fault is scarred by a series of parallel ridges and valleys and contains both the steepest slopes and the largest expanses of near-level land in Malta (Fig. 2.7). The prevailing geological deposit in this region is Upper Coralline Limestone. In the area south of the Great Fault, Globigerina Limestone deposits prevail. The eastern section of this area consists of a series of valleys carved by streams descending to the central plains (Bowen-Jones *et al* 1962, 26-29). This down-cutting process commenced sometime during the past five million years, but more accurate data are so far unavailable (Pedley *et al* 2002, 87).

Land faulting is directly responsible for the fragmentation of substantially large subterranean water deposits into a series of aquifers (Morris 1952, 80-108). Faulting also encourages water percolation between deposits and is a factor which warrants consideration when assessing water availability and landscape transformation due to human agency (Pedley *et al* 2002, 36).[9]

Fig. 2.8 A freshly exposed marly bed located within Lower Globigerina Limestone deposits at Sliema. Source: A. Mifsud.

2.6 The Hydrological Retention Potential of Maltese Geological Formations

The permeability of Maltese geological formations is not determined by rock porosity alone, but also depends on size, shape, pore continuity, geometrical proportions and general arrangement of the grain particles (Bonello 1988, 62-63). Water permeability in limestone beds has been found to be more pronounced when taking place parallel to the deposit's bedding planes and not perpendicular to it. Permeability is furthermore enhanced by the presence of large-sized grains and lower grain sphericities. A lack of uniformity in grain size, a reduction in pore compaction and deposition of mineral matter (cementation), are all factors which encourage permeability (Morris 1952, 53; Bonello 1988, 71-72).

Bedding planes, joints and fissures typically present in Upper Coralline Limestone deposits allow the free downward percolation of rainwater until impermeable strata, such as the Blue Clay stratum, are encountered (Morris 1952, 53-55) (Fig. 2.5). There are instances where Globigerina horizons are composed of thick and relatively impervious marly beds through which water percolation is slow (Figs. 2.8, 2.9, 2.10, 2.11). Globigerina marly deposits not only limit water percolation within, but also encourage water loss as surface runoff. The ability of Maltese Limestones to retain water and encourage or discourage percolation varies considerably and rapidly, even within the same geological formation (Bonello 1988, 71-72).[10] It is the localised presence of such hydrological deposits that this section of the study investigates. Had localised aquifers in Upper Coralline and Globigerina Limestone deposits been identified during the archipelago's prehistoric and historic phases, these would certainly have been a determining factor in encouraging the settlement and landscape transformation of an area.

2.6.1 The perched and mean-sea-level aquifers

Malta's natural water sources are derived from the "perched" and the "mean-sea-level" aquifers (Sapiano 2008, 82). The Upper Coralline Limestone perched aquifers consist of water deposits which collect above impervious Blue Clay horizons and are mainly located in north and north-west Malta (Fig. 2.2). The majority of these aquifers are breached by either water-carved valleys or horst and graben landforms (Bowen-Jones *et al* 1962, 43-44) (Fig. 2.7). Land faulting and subsidence has thus led to the creation of a series of largely parallel aquiferous blocks which from a hydrological standpoint are independent from each other (Sapiano *et al* 2006, 28). Excluding small outliers such as Il-Qarraba and Għajn Tuffieħa, both in the territory of Mġarr (Malta), together with Fort Mosta in central Malta, and Il-Ħofra and the St Leonard area to the south-east of

Fig. 2.9 Water collecting in a freshly excavated Middle Globigerina Limestone deposit at Birkirkara, Malta. Source: A. Mifsud.

[9] See also section 2.6.1.

[10] The study of Globigerina rock samples obtained from quarries in Mqabba, Siġġiewi, Qrendi, Kirkop and Tal-Balal close to the San Ġwann and B'Kara settlements established that the Globigerina Limestone samples from the Siġġiewi quarries had the lowest porosity and permeability levels. See also Bowen-Jones *et al* 1962, 43; Morris 1952, 53-54.

Malta, Upper Coralline Limestone formations occupy ca. 36.32 km^2 of Malta's surface area. The Upper Coralline perched aquifers occupy an area which is equivalent to ca. 31.69 km^2 (Morris 1952, 55).

From a contemporary commercial viewpoint, Upper Coralline perched aquifers are constrained by their small size and minimal storage capacity and thus have a low abstraction potential (Sapiano 2008, 82). Notwithstanding, until the mid-nineteenth century, it was the Upper Coralline perched aquifers which provided the archipelago with its principal source of water supply, even during episodes of extended drought (Sapiano *et al* 2006, 37; Buhagiar, K. 2012A, 157-59).[11] One mode of Upper Coralline perched aquifer water extraction was through the utilisation of water galleries which mainly tunnel into Mtarfa Member rock outcrops in order to extract a perennial water source. These feature prominently in this book.

Fig. 2.10 Freshly quarried bluish-grey marl chippings located in a Middle Globigerina Limestone deposit at Birkirkara, Malta. Source: A. Mifsud.

The mean-sea-level aquifer is composed of a body of rainwater which percolates down to sea level and which due to its lesser density, settles above the heavier salt water. In the 1940s, the mean-sea-level aquifer was composed of a ca. four to five-metre-thick convex-shaped hydraulic head of fresh water, the thickest section of which was located towards the centre of Malta. Uncontrolled extraction from the mean-sea-level aquifer has in recent years caused a large-scale depletion of this subterranean resource (Bowen-Jones *et al* 1962, 43; Sapiano *et al* 2006, 27). The total volume of water stored in the mean-sea-level aquifer has in the last decade been tentatively established at 1,500 cubic hectometres (Sapiano *et al* 2006, 29).

Rainfall and water leakages from the Upper Coralline perched aquifers contribute towards the recharge of the archipelago's mean-sea-level aquifer system. Globigerina and Lower Coralline Limestone formations are the main rock deposits in which the mean-sea-level aquifer is located (Schembri and Baldacchino 1998, 41-43).[12] Due to the low permeability levels registered in Globigerina Limestone deposits, it is believed that the main recharge points of the mean-sea-level aquifer are through the highly permeable but limited Lower Coralline Limestone outcrops (Bowen-Jones *et al* 1962, 43). On the other hand, fissures and micro fractures in Globigerina Limestone formations also appear to play a vital role in the recharge of the mean-sea-level aquifer (Sapiano *et al* 2006, 28).

In areas which experienced significant land faulting, such as at Pwales Valley in north-west Malta, the mean-sea-level aquifer is in actual fact located within Upper Coralline Limestone deposits (Sapiano *et al* 2006, 28) (Fig. 2.7). The island of Comino is another case in point. As a result of extensive rift faulting in the Malta-Gozo channel, only Upper Coralline Limestone and Quaternary-period deposits are visible above surface level on Comino. Such conditions make the submerged Blue Clay horizon

Fig. 2.11 A pile machine boring through marly Middle Globigerina Limestone deposits at Mriehel, Malta. Water originating from a localised aquifer accumulating in this Middle Globigerina deposit collects within the excavated pile shaft. Source: A. Mifsud.

[11] For a discussion on the water extraction potential of the Upper Coralline Limestone perched aquifers during the early British period see sections 4.1 and 4.2.

[12] The best quality mean-sea-level aquifer deposits are usually located within Lower Coralline Limestone strata. A mean-sea-level aquifer within Globigerina Limestone strata often yields a second-class water supply. See Morris 1952, 11, 22.

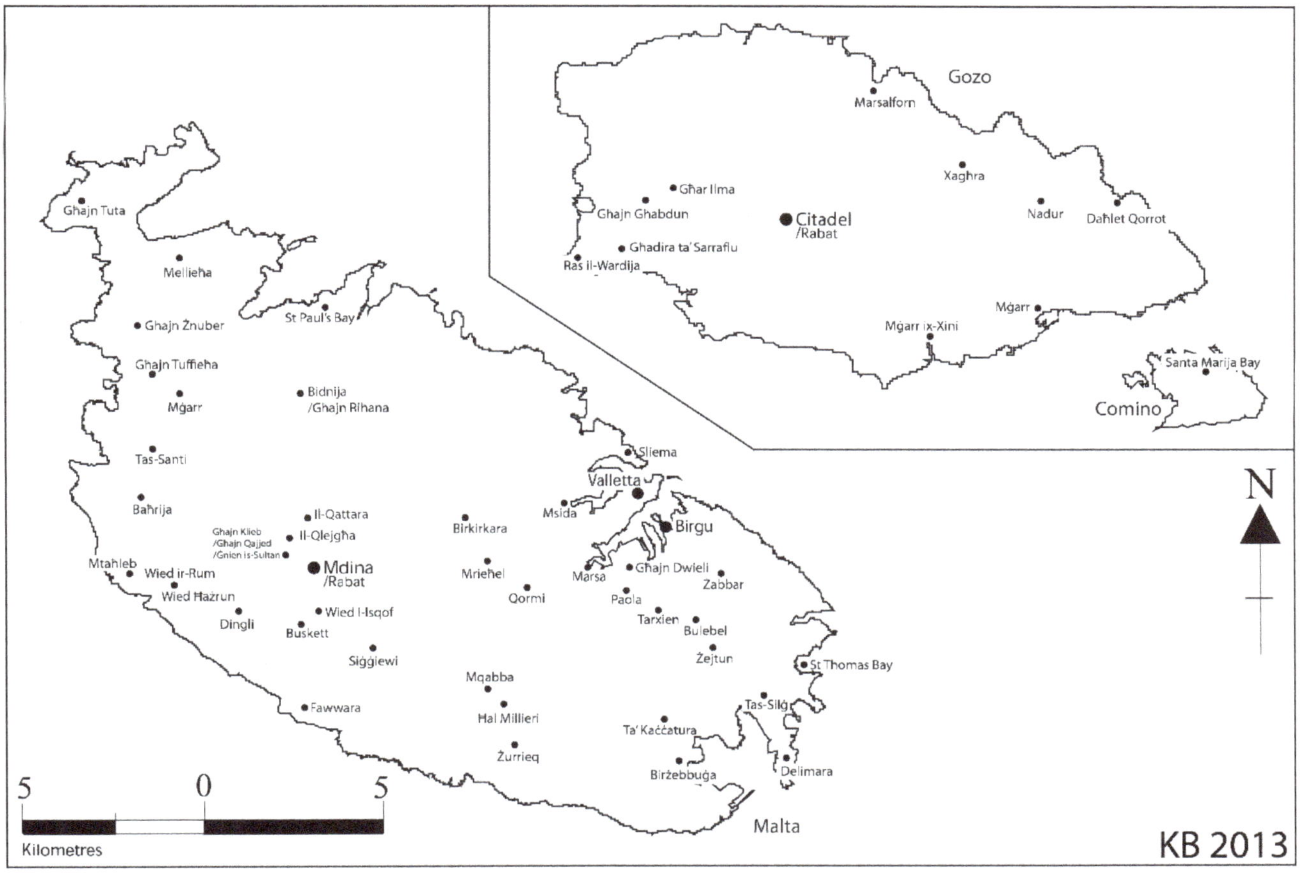

Fig. 2.12 Map of the Maltese archipelago showing the spatial distribution of the main place-names cited in this book.

useless and, if it were not for the mean-sea-level aquifer, no fresh water deposits would exist on the island.[13]

The Globigerina and the Lower Coralline formations located to the south of the Great Fault contain an estimated mean-sea-level aquifer which extends over 67.27 km² (Bowen-Jones *et al* 1962, 43-44). For the past 140 years, the mean-sea-level aquifer has largely been the archipelago's principal source of water supply (Morris 1952, 3-17). Even though occasionally tapped since at least the first half of the seventeenth century, prior to the 1870s, the dynamics behind the workings of the mean-sea-level aquifer remained practically unknown (Bosio 1604, 30; Abela 1647, 128; Morris 1952, 2, 3-17). The tapping of mean-sea-level aquiferous sources prior to the British period is discussed in chapters five, six and seven. A number of hand-excavated wells in the Żabbar district in south-east Malta were also observed to be yielding a fresh water source, but are difficult to date (Morris 1952, 66).

Fig. 2.13 Marly deposits present in Lower Globigerina Limestone horizons at Għajn Dwieli (PA2), Paola, overlooking French Creek. Source: the author.

2.6.2 The Globigerina Limestone perched aquifers

The archipelago's main aquiferous sources are currently derived from the various Upper Coralline Limestone perched aquifers and the mean-sea-level aquifer. However, these are not the only aquifers available locally. Section 2.4.2 has already partly examined the water retention potential of the Globigerina Limestone deposits. The formation of minor perched aquifers within Globigerina strata is largely unchartered territory and is currently being

[13] The rock stratigraphic sequence for the island of Comino, starting from top to bottom consists of: (a) Tal-Pitkal Member, located in most areas of the island; (b) Ġebel Imbark Member, exposed outcrops of which are present in the western, southern and eastern sections; and (c) Mtarfa Member, which is only exposed in a small area south of Santa Marija Bay. See Debono and Xerri 1993, Geological Map of Malta, Sheet 1. For a preliminary study which partially examines the hydrology of Comino see Buhagiar K. 2008A.

investigated by Geotechnical Engineer Adrian Mifsud as part of his doctoral research. There is nevertheless, sufficient geological and archaeological evidence to support the claim for the existence of highly localised minor perched aquifers in Globigerina Limestone formations. In the early 1950s, it had been noted that a substantial portion of the Globigerina deposits were composed of significant quantities of impervious marls (Hyde 1955, 97; Morris 1952, 55-56; Pedley *et al* 2002, 83).

One such localised aquifer is positioned in the syncline between Mqabba and Żurrieq in south-west Malta (Figs. 2.12, 2.14). This area is characterised by an exposed Lower Globigerina Limestone plain and holds a small perched aquifer which is probably the result of a marly bed located in the lower sections of this geological formation. This ground-water resource is too limited and of no commercial value to the local water company, but in the past it was certainly an indispensable asset to local farmers, who tapped this hydrological resource by means of wells (Morris 1952, 56). Besides increasing the commercial value of agricultural land, the tapping of such aquifer sources would have enabled the cultivation of a wider range of crops than normally made possible by arable land relying solely on annual precipitation. Malta Resources Authority data pertaining to the Survey of Private Groundwater Abstraction Sources compiled in 1943 indicate the location of most of these wells, the date of excavation of which is not known.[14]

Ħal Millieri, one of Malta's Late Medieval settlements, also located in the countryside area separating Żurrieq from the Qrendi and the Mqabba settlements, was likewise probably situated above another perched aquifer formed within a Middle Globigerina outcrop (Figs. 2.12, 2.14).

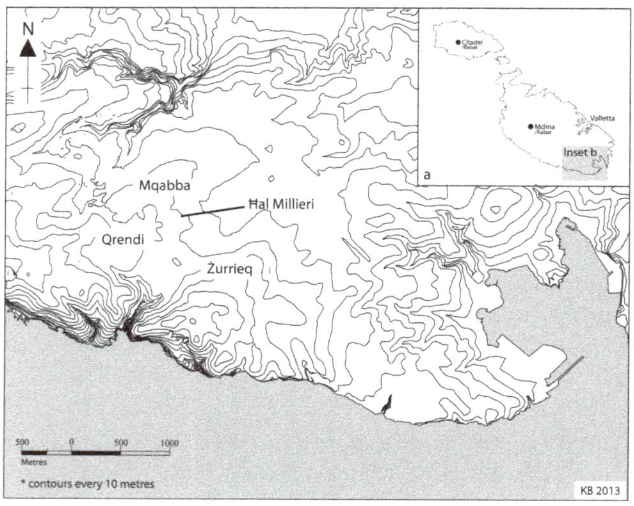

Fig. 2.14 Map of the Maltese islands (a), with inset (b) showing the location of Żurrieq and its neighbouring settlements. Base map data: MEPA.

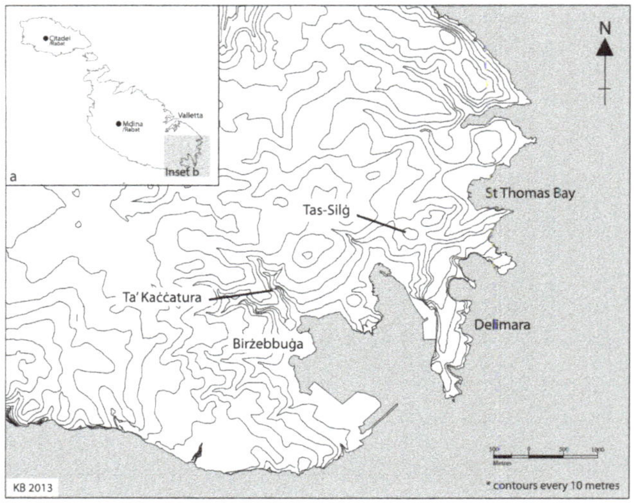

Fig. 2.15 Map of the Maltese islands (a), with inset (b) showing the location of the Birżebbuġa area and other place-names mentioned in the text. Base map data: MEPA.

The first human occupation of this area dates to ca. 1450-800 BC and corresponds to the Borġ in-Nadur phase of the Bronze Age period (Blagg *et al* 1990, 53-54). The remains of a trapetum and large ashlar masonry blocks at Ħal Millieri are likewise indicative of a Roman-period settlement in the vicinity (Blagg *et al* 1990, 43-44). To what extent past settlement of the Ħal Millieri area was conditioned by subterranean water availability still needs to be satisfactorily addressed. The fact that a field at Ħal Millieri was in 1523 called Il-Menqa – a place where water gathers – is indicative of poor rock porosity, thus allowing rain water accumulation at the surface, or close to it (Wettinger 2000, 373).

Fig. 2.16 The artificial pond known as L-Għadira ta' Sarraflu at Kerċem, Gozo, is of unknown antiquity and is located in a Middle Globigerina Limestone deposit. Source: the author.

[14] The water sources in question are in this survey classified as wells number 162, 163, 164, 165, 166, 167 and 2311. All are located in private property to the east and south-east of the Ħal Millieri *casale* in the territory of Żurrieq and flank either side of the road known as Tal-Marsa.

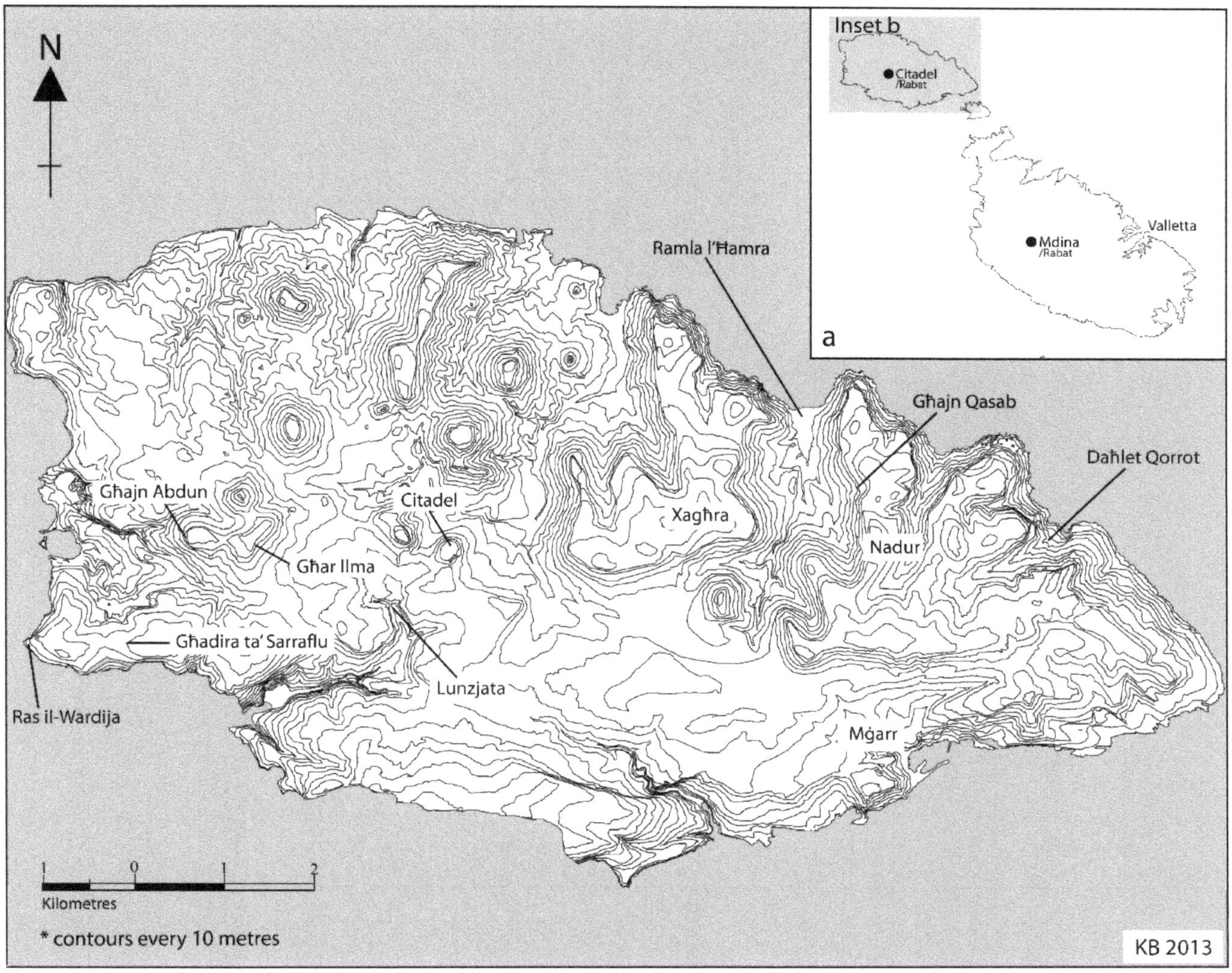

Fig. 2.17 Map of the Maltese islands (a), with inset (b) showing the location of the principal Gozitan place-names mentioned in the text. Base map data: MEPA.

Another localised perched aquifer detected within Middle Globigerina deposits is located in the Delimara and St Thomas Bay area of south-east Malta (Morris 1952, 56; Debono and Xerri 1993, Geological Map of Malta, Sheet 1) (Figs. 2.12, 2.15). As in the case of the Żurrieq Globigerina perched aquifer, this was in the 1950s not regarded worthy of consideration for the purpose of commercial extraction, but remains of vital importance to the local farming community. The presence of substantial quantities of marly deposits in the Delimara promontory was in the early 1990s further revealed by rock quarrying in preparation for the construction of the first phase of the Delimara power station compound. The Tas-Silġ multi-period sanctuary site in the vicinity is, likewise located within Middle Globigerina deposits. The exploitation of the hydrological resources at Tas-Silġ since antiquity is discussed separately in chapter three, section 3.2.3.

That all three Globigerina Limestone formations discussed in section 2.4.2 above contain intermittent but often substantial pockets of bluish-grey marly deposits is uncontestable (Morris 1952, 52-57; Hyde 1955, 34-42; Bowen-Jones *et al* 1962, 43).[15] However, it is the distribution and density of these deposits which determines the formation of localised aquifers within the Globigerina Limestone strata. These have so far not been given an official name, but are colloquially referred to by masons and building contractors as Il-Kaħla, a name reflecting the nature of the bluish coloured clay deposits embedded within the Globigerina Limestone strata. When naturally exposed, Kaħla deposits are whitish in colour, but the colour range can vary from dark beige to pale greyish-blue when freshly excavated (Figs. 2.9, 2.10).[16]

[15] Clay presence within tested samples of Middle and Lower Globigerina Limestone deposits is partially investigated in Cassar 2004, 19-20.

[16] Personal communication by Geotechnical Engineer A. Mifsud.

Fig. 2.18 The Ras il-Wardija Punic sanctuary at Kerċem, Gozo. General view of the Middle Globigerina rock outcrop within which the sanctuary is located. Source: the author.

Kaħla deposits are sometimes overlain by a yellow coloured deposit called Is-Safra which is different from the Lower Globigerina deposits which stone masons refer to as Franka. Both the Kaħla and the Safra deposits contain a larger percentage of clay and their permeability to water is much lower than that of the Franka deposit.[17] Rock excavations in Kaħla deposits do not drain easily when filled with water (Fig. 2.8). An exploratory borehole drilled at Santa Maria Estate, Mellieħa, in order to obtain geological samples for laboratory analysis led to the detection of a ca. 6 m thick Kaħla-type deposit at a depth of 18 m below surface level. The borehole was purposely filled to the brim with water, and after a fortnight-long interval during which no precipitation occurred, it only registered a mere drop of ca. 0.61 m.[18] These results confirm much earlier observations noting that boreholes driven down to mean-sea-level through Globigerina formations only yielded a minimal water supply, thus hinting at the largely impervious nature of this deposit (Morris 1952, 60).[19]

Personal field-research led to the detection of another Globigerina localised perched aquifer in the Għadira ta' Sarraflu area at Kerċem, Gozo. The Għadira ta' Sarraflu is a sizeable, artificially excavated basin having a circumference of ca. 128 m and is located at map reference 27904 88504 (Figs. 2.12, 2.17). It is likely that this basin pierces a substantially large impermeable deposit embedded in a Middle Globigerina layer, above which rainwater collects. The seemingly artificial hollow, known as l-Għadira, exposed a localised perched aquifer, which in this area is located just a few metres below surface level (Fig. 2.16). A similar situation was observed at the Punic sanctuary site of Ras il-Wardija close-by, excavated between 1964 and 1967 (Fig. 2.17).[20] Of the two cisterns located within the precincts of the sanctuary, the water retention capabilities of the Middle Globigerina deposit can be best appreciated from the one at the south-westernmost extremity of the site. Although not lined with a mortar rendering, this cistern still has excellent water retention capabilities (Figs. 2.18, 2.19, 2.20).[21]

Fig. 2.19 The Ras il-Wardija Punic sanctuary at Kerċem, Gozo. View of the rock-cut chamber containing niches, its fronting terrace and one of the two cisterns present within. Source: the author.

Fig. 2.20 The Ras il-Wardija Punic Sanctuary at Kerċem, Gozo. Detail of the square-shaped cistern with rock-cut steps leading down to it. Marly deposits present within this Middle Globigerina rock outcrop give this cistern excellent water retention capability. Source: the author.

[17] Personal communication by Geotechnical Engineer A. Mifsud.

[18] Personal communication by Geotechnical Engineer A. Mifsud.

[19] Another sizeable outcrop of Middle Globigerina deposit covers large tracts of the Attard, Mrieħel, Santa Venera, Ħamrun and Marsa settlements. A borehole excavated into the Santa Venera side of this deposit was originally intended as a recharge access point for the mean-sea-level aquifer. The poor permeability levels of this deposit, however, meant that only minimal quantities of rain water managed to infiltrate. Personal communications by Hydrologist M. Cremona and Geotechnical Engineer A. Mifsud.

[20] For a brief description of the archaeological remains at Ras il-Wardija see Bonanno 2005, 340-41.

[21] See chapter three, section 3.2.5. A similar situation was detected by Geotechnical Engineer A. Mifsud in a construction site at Mrieħel, Malta, where pile bores started filling up with water ca. 30 minutes following their excavation (Fig. 2.11).

Lower Globigerina strata containing marly deposits are colloquially referred to as soll. The soll is characteristically softer than the other Globigerina deposits, but is not easily hand sawn when in a damp state. Soll deposits weather rapidly when exposed to the elements. Due to their high clay content, they have very poor water absorption capabilities (Cachia 1988, 48-49; Cassar 2010, 14-15). On a freshly exposed surface, it is difficult to distinguish soll from any adjoining Lower Globigerina deposits. In due course, however, exposed soll deposits experience accelerated deterioration (Cassar 2004, 14; Cassar 2010, 13-14).

One or more localised perched aquifers were probably located in Lower Globigerina Limestone deposits at Corradino Hill in the territory of Paola, where substantial quantities of exposed kaħla-type deposits were noted by the present writer in the quarried rock-face to the immediate west of French Creek (Fig. 2.13). The presence of localised water deposits in this area is also hinted at by the toponym of the locality – Għajn Dwieli (PA2).

The principal drawback with aquifers located in Globigerina deposits is the sluggish rate of water percolation. This dissolves much larger quantities of salts and calcium carbonate than water percolating through Upper and Lower Coralline Limestone formations (Morris 1952, 59-60). On the other hand, this would probably have hardly mattered to past rural communities that managed to gain access to these ground-water resources. It is possible that even though highly localised in nature, a number of Globigerina Limestone perched aquifers provided small neighbouring communities with a perennial or quasi-perennial water supply. Within a semi-arid landscape context the presence of such natural hydrological resources must certainly not be overlooked.

Chapter Three

Water Resources Management Technology: An overview of ancient water resources management practices

3.0 Introduction

Limestone landscapes have since antiquity encouraged hydraulic experimentation, especially in circumstances of water shortage. Mediterranean hydrological resources are generally insufficient to allow dry-season irrigation, and spring-fed summer crop cultivation is only restricted to micro-regions possessing water-bearing geological deposits. In many instances, the main purpose of water technology was the successful cultivation and production of crops and, in a Mediterranean context, was mainly geared for intensive garden cultivation (Horden and Purcell 2000, 237-50). Settlement location and attempts at improving the hydrology of a given region were mainly conditioned by the prevailing geology and topography (Crouch 2004, 9-10).

The principal aim of this chapter is to provide the necessary archaeological and historical framework for the various water-harvesting typologies and techniques discussed in the subsequent chapters. Section 3.1 is an overview of principal elements forming part of hydraulic archaeology and deals with water capture and storage. The presented data document water-related strategies practised in arid and semi-arid zones, namely the Near Eastern and Mediterranean regions, which are worthwhile considering when assessing water management possibilities within a Maltese context. As specified in sections 1 and 1.1 water distribution and allocation fall beyond the scope this publication. These methods of water delivery and allotment, are as a consequence, not covered in this section. The various modes of ancient water management and techniques employed in the Maltese archipelago, within an urban and rural context since the Phoenician-Punic period until Late Antiquity, are investigated in section 3.2.

3.1 Water Capture and Storage

Water capture involves the interception of surface or subsurface water for its utilisation for agricultural or domestic purposes. Water storage involves the preservation of the captured water in either subterranean cisterns or surface reservoirs for later usage and in this section is touched upon through a brief discussion on cistern development in a central Mediterranean context.[1]

The water capture systems discussed in this section include field-terracing, flood-water irrigation, urban sewage diversion, wells, *qanat* water extraction systems and water galleries (adits), *cuniculi* and diversion dams (*azuds*).

3.1.1 Field-terracing as an aid to water conservation

Millennia of observations enabled ancient societies to become knowledgeable in many aspects of their surrounding environment. In the Near East, water management and irrigation can be traced back well into antiquity. It aided the diffusion of the practice to other areas of the known world, particularly North Africa and the Mediterranean region (Horden and Purcell 2000, 237). Ancient Near Eastern farmers encouraged water conservation systems through the construction of small terraced fields. The practice involved the building of a series of parallel surface stone walls along the contours of hills, with the resultant intermediate gaps being filled up with soil (Bazza 2006, 5). Terraces slowed down runoff precipitation movement downhill, enhancing water absorption by the field soil, whilst preventing water logging. Water movement over the field surface was enhanced through the adoption of contour ploughing. As a further technological development to this system, fields were at a later date watered by an intricate network of open canals and wooden aqueducts which channelled water from a nearby source (Bazza 2006, 5). Terrace landscape modifications can alter the topographical characteristics of entire regions and last for millennia. At a more micro-local level, by contrast, some landscape interventions last for only a season and are difficult to date with accuracy (Horden and Purcell 2000, 235).

3.1.2 Flood-water irrigation

In arid and semi-arid regions where ground-water deposits are either inexistent or have not been satisfactorily exploited, land cultivation strategies have for millennia centred upon the harvesting of surface rainwater runoff during and in the immediate aftermath of flash floods.[2] A series of walls in *Wadi Mansur* in Libya dating to the first few centuries AD appear to have acted as conduits, channelling runoff water from the *wadi* sides downstream.

[1] See section 3.1.8.
[2] Arid, or dryland, boundaries are commonly defined through rainfall measurement. A precipitation of 250 mm or lower defines dry-land boundaries. Hyper-arid zones can receive as little as a mean annual precipitation of 80 mm. See Spellman 2000, 19-21.

Several surviving tracts of wall still serve their original purpose (Gilbertson and Hunt 1996, 200; Gilbertson and Chisholm, 1996, 17-52).[3] Experimentation with this method of dry-farming apparently commenced ca. 4000 years ago in the *Wadi Faynan* district in southern Jordan (Barker and Gilbertson 2000, 7).

A similar system of flood-water irrigation in the Negev desert in Israel was discovered in recent years to allow the successful cultivation of a desert region during the rainy season. Experimental archaeology was successful in showing that agricultural yields from the Negev desert zone using such systems could in fact approximate those of the Mediterranean (Bruins 1986, 3; Rosen 2000, 45-47).[4] With an estimated runoff of 15 per cent, an average annual rainfall of 100 mm could here be transformed into the equivalent annual rainfall amounting to over 400 mm (Rosen 2000, 50-51).[5] In order to efficiently store surface runoff water during brief flash floods in the arid zones of North Africa, cisterns were occasionally built within the flood plain. Hot desert environments exert more pressure on settlements and their supporting agrarian systems than those located in wetter latitudes. Agricultural activity in these areas would have thus required more labour input and encompassed the construction and maintenance of a series of dams and terrace systems (Hodge 2002, 53; Rosen 2000, 57).

Archaeological field surveys held during recent years have revealed the concern of Classical Age Greek settlements, particularly geared towards the harnessing of runoff rain water generated by the early autumn flash floods (Crouch 1993, 35-36). Efficiently designed and built drainage canals in public areas and alongside residential roads transported flood water and waste water into specifically designated countryside areas for irrigation purposes. These works were considered to be of a sufficiently important nature as to warrant public expenditure (Crouch 1993, 37-38).

*3.1.3 Diversion dams (*azuds*)*

Another method of surface water capture involves the construction of dams either within the course of a stream to divert spring water or, in the case of semi-arid or arid areas with no perennial water availability, rain runoff flowing through a valley bed during, and in the immediate aftermath of, a rainfall event (Mattingly 1995, 148-52; Rynne 2000, 17; Holt 2000, 61; Magnusson and Squatriti 2000, 222-23; Glick and Kirchner 2000, 274; Kirchner 2008, 477-78; Wilkinson 2010, 147; Gerrard 2011, 7-8). Also referred to as *azuds*, such dams characteristically divert only a portion of the water flowing through the stream or valley system and unlike the "gravity dam", or *presa*, do not stretch along the whole width of the watercourse (Gerrard 2011, 8).

3.1.4 Urban sewage diversion into countryside areas

Another method of agriculture, which was possibly practised in a number of Mediterranean Classical Age settlements, consisted in the diversion of settlement sewage outflow into a countryside district. On reaching the area earmarked for cultivation, the sewage was then allowed to soak away, replenishing the soil with a mixture of nitrates and water. More appropriately known as the "soakage system", this method of agriculture was still practised at Aidone, a town close to the ancient settlement site of Morgantina in Sicily until a few decades ago.[6] Sewage was here spread out on what can be termed as "sewage farms" or "sanitary fields" (Crouch 1984, 362; Crouch 2004, 64-65). Given the relatively high temperatures characteristic of the Mediterranean region and the action of anaerobic bacteria present in human waste, any resulting offensive smell, quickly subsides (Crouch 1984, 362).

A similar system of agricultural intensification was employed in regions of Europe during the medieval period, and is still in use in several regions of present-day China, greatly boosting the annual agricultural yield (Crouch 1984, 362; Crouch 1993, 29).

3.1.5 Well-water capture technology

A well can be most aptly defined as the attempt of man "to obtain water from the earth, vertically below the spot where it is required, when it is not obviously present at the surface" (Forbes 1955, 146). Well-water availability and distribution depend on the available local geology and expertise. Wells tap subterranean springs or aquiferous sources through the excavation of circular, rectangular or square-shaped vertical shafts. Whenever wells were dug into unstable geological deposits, these were frequently lined by brick, stone or wood in order to prevent sectional collapse (Hodge 2002, 51-58; Crouch 2004, 17-24; Chapman Davies 2008, 29).

Well-shaft penetration was primarily conditioned by the depth of the tapped aquifer. Raising water from within

[3] For similar rainwater harvesting systems in north-east Marmairca (north-west Egypt) see Vetter *et al* 2009, 9-23.
[4] In the central Negev area, archaeological surveys have revealed that the existence of numerous large wine presses hint at the presence of agricultural industry geared towards grape and wine production. The presence of six towns, each having an estimated population of 20,000 people was also detected.
[5] The *wadis* surrounding the Ben Ulid region in the Tripolitanian pre-desert are one of the last areas where flood-water farming is still known to be active. Date palms, figs, plums and ancient olive trees are still to be found along four or five kilometres of the modern wadi floor, which periodically also yields a crop of barley. See Gilbertson *et al* 2000, 137.
[6] Personal communication by a farmer cultivating land in the Aidone countryside close to the Morgantina site.

deep wells to the surface was always problematic. It was not uncommon for the rope to break under its own weight and that of the bucket load of water whilst being hoisted to the surface (Hodge 2002, 54-56). At the Greco-Roman settlement of Morgantina in the territory of Enna, Sicily, there were areas where it appeared that the aquifer could not be tapped through the sinking of wells as this was too deeply located below surface level (Crouch 2004, 57). During the Roman period, wells were commonly provided with a well-head or low parapet wall (Hodge 2002, 52; Chapman Davies 2008, 30). In the island of Lampedusa, wells were located close to coastal areas at the mouth of several small valleys (Ercoli 2000, 63),[7] thus allowing relatively easy access to the mean-sea-level aquifer.

Vitruvius gives some information on methods used in antiquity for the tapping of underground water sources. Metal vases buried overnight were examined the following day for traces of water collecting within. Buried, unfired clay vessels, were also used as detectors of earth moisture (Morgan 1960, 227-28). Evidence as to the whereabouts of subterranean hydraulic deposits was achieved through the observation of surface vegetation. The ancient Greeks were quite familiar with Mediterranean flora, which favoured growth in copious water conditions and used the fig tree, the rosemary bush, the bramble as well as moss as markers for the availability of water (Crouch 1993, 23).

Vapours observed rising from the ground surface at dawn were considered as being another pointer to the availability of a ground-water source. In instances where a well shaft successfully tapped a water source, Vitruvius recommended digging at its bottom a series of tunnels fanning in all directions. This was intended to bring an additional water supply to the centre of the well (Rowland and Noble Howe 1999, 96-97).

3.1.6 Qanat *and adits*

An elaborate system of water extraction from underground aquifers, more commonly referred to as *qanat*, was achieved through a series of narrow galleries which channelled the tapped water to specifically designated areas (Horden and Purcell 2000, 251-52). *Qanat* originated from the Persian region and its neighbourhood over 2600 years BP and was a mode of water retrieval principally used as an aid to irrigation at a micro-regional level (Glick 1979, 217-18; Watson 1983, 107; Lightfoot 2000, 215-18).

The *qanat* system consisted in the tapping of groundwater sources through a series of wells, interconnected by tunnels, the gradient of which facilitated gravity-fed water extraction at a predetermined point at the base of a ridge or mountain (English 1968, 112; Beekman *et al* 1999, 441; Lightfoot 2000, 215-16) (Fig. 3.1). Only the first shaft to be excavated is in fact a "true" well. The other shafts merely serve the purpose of aeration pits and/or maintenance shafts, permitting the ease of entrance and exit of the personnel involved in the upkeep of these water systems (Judson and Kahane 1963, 78-79; Glick 1970, 182). Shafts were commonly spaced between 15 to 150 m apart. The wider spacing of shafts appears to reflect the superior technical and engineering skill of the workforce engaged in the project (Goblot 1979, 35-36, 39). The excavated debris was sometimes heaped up in a ring around the mouth of shafts in order to prevent runoff rainwater seepage during flash floods, which might result in gallery water contamination. The average gallery gradient of typical *qanat* systems is of 0.5 metres per kilometre. The tapped aquifer water is commonly conveyed through the gallery system by means of a channel carved into the gallery floor (Hodge 2002, 23).

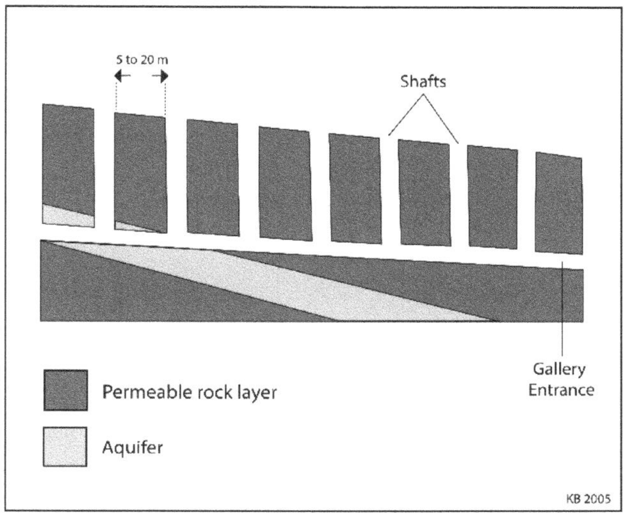

Fig. 3.1 Profile of a *Qanat* system. The gallery tunnels right through the permeable rock layer in order to tap the overlying aquifer. The slight gradient of the water tunnel channels the collected water to the gallery entrance. The shafts were commonly used both for the initial surveying works and for the maintenance of the *qanat* system.

Qanat tunnels are on average not more than a metre wide and one and a half metres high, but may extend underground for a considerable distance. A familiarity with mining techniques probably facilitated shaft and gallery excavation, especially in areas where the aquifer lies at a considerable depth (Forbes 1955, 150-51; Wertime 1968, 927-35; English 1997, 188-89; Hodge 2002, 22). This successful system of ground-water extraction expanded westward to the Mediterranean and parts of Egypt and eastward to Afghanistan, the oases settlements of central Asia and Chinese Turkistan (Glick 1979, 217-18; Watson 1983, 107).

The term *qanat* or *qanawat* is Arabic in origin and means "lance" or "conduit" (English 1968, 170-81). *Kanat*, *ghanats* and *quanate* are other common derivatives, and

[7] The dating of these wells, however, is not known.

the Biblical city of *Qanatha* located in present-day Syria, was named after its *qanat* systems (Hodge 2002, 21). In most areas of North Africa, *qanat* are generally known as *foggara*, *hattaras* in Morocco and *karez* or *kariz* in Iran. The Spanish linguistic equivalent for *qanat* is *madjira* and the Madrid settlement can be literally taken to mean "place of *qanat*" (Goblot 1979, 19-22, 137; Mattingly 1995, 34-35; Hodge 2002, 21; Wilson and Mattingly 2003, 235-78; Chapman Davies 2008, 27-28).[8]

Fig. 3.2 The Villa Conti spring at Aidone, Sicily. Detail of the water gallery entrance and the fronting water storage reservoir. Source: the author.

Qanat systems were dug into permeable rock strata in order to tap the aquiferous deposits present within. Galleries were left unlined and rock porosity was not interfered with. Tracts of galleries excavated into brittle and unstable rock strata which posed a danger of collapse were occasionally strengthened through the insertion of large terracotta rings known as *kaval*.[9] *Qanat* were commonly excavated in regions receiving between 100 and 300 mm of average annual precipitation. The volume of water generated by *qanat* systems fluctuates from gallery to gallery. In modern Iran, the overall average water-yield is of ca. 1,300 m³ per day (Goblot 1979, 41-42). Some of the lesser yielding *qanat* are located in the Saharan region, where the average yield of extracted water amounts to 432 m³ or less a day (Hodge 2002, 24).

Dating the stages of diffusion of *qanat* technology is extremely difficult (Lightfoot 2000, 216). It appears that *qanat* were already a common water-retrieval device in eighth-century BC Iran, and probably spread to Arabia by the sixth (English 1997, 188-89).[10] *Qanat* were likely introduced into Egypt by the Persians at around 500 BC (Goblot 1979, 53-59; English 1997, 188-89; Hodge 2002, 21). *Qanat* have recently been discovered in the western desert of Egypt, south of the Khargeh Oasis in the region of Dush, an area where Persian (Achaemenid) influence has long been known to have existed. The archaeological investigation of the area has correlated *qanat* systems to hydrological innovations introduced into this region during the Achaemenid occupation (Nesbit and O'Hara 2000, 112-13; Lightfoot 2000, 217-18; Wuttmann 2001, 109-35). *Qanat* diffusion reached Syria and Jordan during the Romano-Byzantine period (ca. 64 BC to AD 660). It is probable that in Egypt *qanat* were maintained and expanded by the Roman conquerors. The Romans also incorporated *qanat* technology in subterranean sections of aqueducts, examples of which survive in Tunisia and Turkey (Lightfoot 2000, 218-219; Hodge 2002, 21-22).

The westward diffusion of *qanat* technology to the Mediterranean coast of North Africa is more problematic, and possibly experienced an initial stage of limited diffusion during the classical era, especially in areas falling under Roman domain (Glick 1970, 175-76; Wilson and Mattingly 2003, 261-64). At Timgad in Algeria, a Roman colonial town founded by Emperor Trajan in ca. AD 100, aqueducts were apparently fed by *qanat* systems (Goblot 1979, 123). Henri Goblot, a leading authority on *qanat*, believes that further westward diffusion into Morocco and Spain, only took place around the eleventh century AD, and was brought about by the expansion of Islam (Goblot 1979, 136-58; Glick 1970, 175; Wilson 2004, 136-137; Wilson 2006, 212-14).[11]

Qanat introduction into Sicily and Cyprus has also been tentatively attributed to Muslim expansion in these territories (Bazza 2006, 5-6). However, it has been recently proposed that *qanat* introduction in Sicily might have experienced two separate waves of diffusion, the first dating to the Greek period and the second to the Muslim occupation of the island (Walthall 2008, unpublished

Fig. 3.3 The Villa Conti water gallery at Aidone, Sicily. Detail of a formerly unexplored section of the gallery interior. The darker rock-section on the left is a natural fault line, which was partly exploited by the gallery diggers, possibly to speed up the excavation process. Source: the author.

[8] See also Encyclopaedia Britannica Deluxe Millennium Edition, CD-ROM, (U.K., 1999).

[9] *Kaval* rings had an approximate diameter of 1.5 metres and were inserted one into the other to form a continuous tube. See Goblot 1979, 35.

[10] The Greek historian Polybius (ca. 200-118 BC), makes reference to a dense network of underground canals in Iran, which conveyed a water supply to places which were previously not irrigated. The galleries were already ancient in Polybius's time and are a possible reference to *qanat* technology. See Patton 1954, 169-71.

[11] For data on *qanat* and adit water capture systems in the Aragón district of Spain see Gerrard 2011, 8-11.

Fig. 3.4 The Contrada Merenda spring at Aidone, Sicily. Detail of the gallery entrance and the fronting water trough. Source: the author.

course paper).[12] Personal field-research at Morgantina in Sicily, in collaboration with A. Walthall revealed the probability that the perched aquifer gallery systems in question are not true *qanat* systems and are best classified as water galleries, or adits (Figs. 3.2, 3.3, 3.4, 3.5).[13] Even though they consist of a near-horizontal perched aquifer gallery, the water extraction systems which were partially investigated do not appear to contain shaft wells – a principal component of *qanat* systems.[14] Furthermore, Walthall's identification of adits tunnelling below the Morgantina settlement, must be juxtaposed not only with the apparent absence of shafts, but also with the challenging relief of the surrounding terrain which makes these only accessible from the settlement exterior. The perched aquifer galleries are positioned at an elevation of ca. 35 m lower than that of the Morgantina settlement and for these to be reached, a tract of country lanes ca.1 km long has to be tackled. If these adits were already extant in the Morgantina landscape in antiquity, water transportation between these and the overlying settlement would have been a difficult and challenging task.[15]

Other areas of Sicily are faced with a similar dating dilemma. Rock-excavated subterranean water systems in ancient Acragas,[16] the foundation of which dates to 580 BC, are comparable in dimension to the galleries of *qanat* systems. Their function is still unclear, but they were used either as aqueducts, or as sewers for waste water or as aquifer water extraction points (Ercoli 2000, 57-59; Crouch 2004, 27, 33). Whilst tentatively dating these systems to the Greek and Roman periods, D.P. Crouch points out that some of them, if not all, may be of a Muslim or Later Medieval date (Crouch 2001, 23-24).[17] Of the Sicilian galleries, the Palermitan ones are the best documented (Biancone and Tusa 1997; Ercoli 2000, 59-60; Wilson 2004, 137-38; Lofrano *et al* 2013). Commonly referred to as *ingruttati*, ca. twenty-five of these water capture and extraction systems have been identified in the Palermo plain and because these contain shaft wells, they can be classified as full-fledged *qanat* systems (Lofrano *et al* 2013, 1666-69).[18] In 1525, Palermo was still fed water

Fig. 3.5 The Contrada Merenda spring at Aidone, Sicily. Detail of the dry-stone lined tunnel interior and the masonry slabs capping the ceiling. Source: the author.

[12] This hypothesis was presented following the partial investigation by Walthall of perched aquifer galleries at the Greco-Roman settlement of Morgantina in the territory of Enna, Sicily. L. Ercoli similarly proposes that some of the *qanat*-type systems of Palermo might predate the Muslim period. See Ercoli 2000, 59-60.

[13] My familiarisation trip to Morgantina was made possible through the intervention of S. Lucore of the American School at Athens and the director of the archaeological excavations in the bathhouse area of this site.

[14] It is only the Cittadella-Gambero and the San Francesco perched aquifer springs which are possibly connected to a mother well. Walthall 2008.

[15] Toponomastic evidence shows that Morgantina and the neighbouring Aidone settlement started being associated with spring water availability during the Muslim period. As a matter of fact, the Aidone settlement was called *Ayndun* during this period, meaning "springs of water". See Lombardo 1999, 5; Trovato 1999, 27.

[16] This corresponds to the present-day settlement of Agrigento, Sicily.

[17] See also Burns 1974, 403, 408-9, who also proposes a Greek date for the Agrigento galleries. It is possible that the Greeks borrowed *qanat* technology from the Persians and introduced this within a Greek settlement framework in order to increase water supply. See Forbes 1955, 164; and Lightfoot 2000, 218.

[18] Professional diggers colloquially identified as *puzzari* excavated the Palermitan *ingruttati* galleries using only broad-bladed picks, shovels and oil lamps. See Lofrano *et al* 2013, 1668-69.

via the *Uscibene qanat* and the city paid fees for its maintenance (Lofrano *et al* 2013, 1667). The antiquity of the Palermitan *qanat* is still a matter of scholarly debate, but ceramic fragments recovered from within a secure archaeological context in a *qanat* system at Villa Raffo (Palermo) are of a twelfth or thirteenth-century date (Biancone and Tusa 1997, 375-89).[19] Several *qanat*-type water capture systems have also been identified in the Campania region of mainland Italy. Colloquially referred to as *qanate*, there is still no apparent consensus on the dating of these subterranean systems (De Feo *et al* 2010).[20]

Qanat are generally situated at a useful altitude in relation to the workable land and have the advantage of using lesser slope than above-surface canals. The underground tunnel forming a core part of the *qanat* system is instrumental in avoiding water loss through surface evaporation, and since *qanat* galleries are usually located in the upper sections of the terraced landscape, the use of water lifting devices, such as the *noria*, is unnecessary (Glick 1970, 175-82).

The digging of *qanat* was usually restricted to arid lands where water irrigation was extremely scarce and involved a major investment in agricultural intensification (Beekman *et al* 1999, 440). *Qanat* systems are not immune to variations in water flow, especially when seasonal or smaller localised aquifers of alluvial fans are tapped. Their construction is a highly technical endeavour, normally undertaken by a hereditary class of labourers known as *muqanni* (English 1997, 191; Lightfoot 2000, 216). In present-day Iran, *muqanni* are still responsible for the building and maintaining of *qanat* systems.[21]

Qanat construction is considered by many to be a costly and labour intensive enterprise, mainly conditioned by the amount of water desired and the skill of the *muqanni* entrusted with the excavation project (*English 1997, 193-94*). It has been noted, however, that the excavation of a *qanat* project can also be cost-effective in the short term. Whilst failing to comment on the approximate timeframe necessary for the completion of an average *qanat* system and the expenses involved, C.S. Beekman *et al* put forward their well-reasoned hypothesis that a team of qanat excavators would have seldom numbered more than ten members. Physical constraints posed by the nature of the project itself would have directly impacted the workforce size engaged in the *qanat* excavation project. The workforce would be usually divided into two equal parties, one working at ground surface level and the other within the *qanat* system (Beekman *et al* 1999, 441).

In order to ensure a perennial water supply, trial shafts, also referred to as "mother wells", were presumably excavated during the months of August or September – the time of the year when the water table levels are at their lowest in the Mediterranean, North African and Near Eastern regions (Glick 1979, 238). Following the digging of the "mother well", calculation was made as to where the gallery entrance was to be located. Only basic tools were employed in the digging of the vertical shafts and the underlying near-horizontal tunnels: picks, shovels, baskets and small castor oil lamps. In subterranean conditions, it was the castor oil lamps which indicated a lack of oxygen, thus making another vertical shaft necessary in order to enhance the airflow in the *qanat* interior (English 1997, 193).

Past literary sources give no evidence on the employed methods of excavation. Nonetheless, in present-day Iran *qanat* are still being constructed using traditional techniques (English 1997, 193; Hodge 2002, 20). A modern three-kilometre long *qanat* project excavated in mid-twentieth-century Iran, consisting of a bifurcated tunnel and two "mother wells", capable of supplying enough irrigation water for half an acre of land once every 24 hours, took 17 years to complete. A contemporaneous one-kilometre-long *qanat* with a 45 m deep mother well in the village of Kerman in Iran took 27 years to complete. The delay in project completion was largely due to changes in ownership of the subterranean water extraction system.[22] Likewise, it was estimated that a four-kilometre-long *foggara* in the Fazzān district of Libya having shafts spaced 10 m apart and an average depth of 12 m would have required 48,000 man days to construct (Wilson and Mattingly 2003, 273).

Prior to the publication of this study, *qanat*-type systems and adits in the Maltese archipelago were discussed in preliminary papers by E.P. Teuma and K. Buhagiar (Buhagiar K. 2002; Teuma 2005; Buhagiar K. 2007; Buhagiar K. 2008B; Buhagiar K. 2012A; Buhagiar K. 2012B). *Qanat* and water gallery availability within a local context will be expounded further in other sections of this book.[23]

3.1.7 Cuniculi *water drainage systems*

A dense concentration of subterranean galleries known as *cuniculi* discovered in the Latium and Veii regions of central Italy has long baffled archaeologists. Even though at face value these might appear to be very similar in

[19] Apart from Morgantina, personal field-research in Sicily has succeeded in identifying other galleries at the Castello di Lombardia (Enna), S. Lucia di Mendola and Ferla, both in the territory of Syracuse, and Scicli in the territory of Ragusa. More extensive investigation of these water systems is necessary, but their context is seemingly representative of the High or the Late Medieval periods. See chapter eight, section 8.1.

[20] Dates ranging from the Roman to the Norman period have been tentatively proposed for the *qanate* subterranean systems of Campania. See De Feo *et al* 2010, 647-55.

[21] The best known *muqanni*, who come from the desert city of Yazd in Iran, are paid high wages and command respect. Owing to the hazardous nature of their work, folk customs and beliefs have evolved around the *muqanni* profession. See English 1997, 191.

[22] The respective approximate construction costs of these *qanat* systems were of $10,000 to $11,000 per kilometre in the late 1950s. See English 1997, 194; Horden and Purcell 2000, 238-39.

[23] See chapters six to eight.

structure and technological principle to *qanat* systems, specialised research has shown these to be more related to the drainage of waterlogged deposits into which they are excavated, rather than the conveyance of a water supply (Judson and Kahane 1963, 78-79; Horden and Purcell 2000, 247; Hodge 2002, 45). *Cuniculi*, the literal meaning of which is "rabbit burrows", are composed of a near horizontal main tunnel, the roof of which is pierced by a series of vertical rectangular-shaped shafts. Livy reports that *cuniculi* were military devices aimed at facilitating access into besieged cities (Judson and Kahane 1963, 74).

Having similar dimensions to those of *qanat* systems, cuniculi are generally slightly over 0.5 m wide and around 1.75 m high, but cuniculi twice as wide have also been recorded (Judson and Kahane 1963, 78-79). Vertical shafts are generally spaced between 31 to 34 m apart and in areas where the *tufo* was recognised as being too brittle, they were re-enforced by means of a brick or masonry lining. On average, shaft depth is of 4 to 6 m, but when a *cuniculo* tunnels into a ridge, the shafts may be up to 30 m deep. The gradient of *cuniculo* galleries varies from 12 to 26 m per kilometre and contrasts sharply with the much gentler gradients employed by *qanat* systems and Roman aqueducts (Hodge 2002, 46). It is not known whether shafts were simply used as a device which facilitated *cuniculi* excavation, or were also a means through which the galleries could be accessed for the purpose of easier maintenance (Judson and Kahane 1963, 84-85).

Cuniculi systems hint at an Etruscan and Roman-period dating and remained in use till at least the fall of the Western Roman Empire (Judson and Kahane 1963, 84-85). Pliny the Elder's mention of underground tunnels used for carrying water furnished with vents every ca. 70 m, is a possible reference to *cuniculi*. Pliny further states that water flowing through a *tufo* deposit became sweet and light and that the *tufo* acted as a filter, keeping back the dirt (Pliny – Natural History, Lib. Xxxi, 57; Jones W.H., 1948, 407-9, 412-15). It is still unclear if *cuniculi* were inspired by the *qanat* water extraction systems, or whether they were an independent development. In case of the former instance, the proposed date for the *qanat* technology diffusion into mainland Italy is the 600 - 500 BC period, perhaps through contact with Carthage.[24]

It appears unlikely that *cuniculi* were also used as irrigation devices. The current interpretation of the functions of these subterranean systems centres on the waterlogged environment created by expanses of impermeable *tufo* strata in the Latium and Veii regions of Italy. A successful cultivation of these areas could not be possible before the land was drained of its surplus water resources. This was achieved through the burrowing of *cuniculi* in order to lower the water table level in the subsoil (Dalmiglio 2003, 38). The absence of water distribution canals and other water storage devices outside *cuniculi* systems further strengthens this hypothesis. It also appears that the use of *cuniculi* reduced surface erosion (Ward-Perkins 1962, 1643).

Another basic difference between *qanat* and *cuniculi* is that the former are able to supply specifically selected areas with a perennial water source. Most *cuniculi* on the other hand, carry water during the rainy season, when it is not really necessary for crop growth, but almost dry up during the summer months. Unlike *qanat*, most of the investigated *cuniculi* are neither dug at right angles to the valley side, nor located at a suitable elevation, that is, at a higher level than the land to be irrigated. Most cuniculi were, furthermore, found to move in a direction parallel to that of the valley and were located at a relatively low altitude, almost on a level with the valley bed. The fact that the majority of the *cuniculi* systems end in narrow segments of valleys where there is only limited availability of arable land is again indicative that these were not used for the purpose of field irrigation (Judson and Kahane 1963, 74-103). Apart from land drainage, it appears that *cuniculi* were at times also used for the purpose of lake drainage. *Cuniculi* were in some cases connected to aqueduct systems, or used as devices through which the necessary water momentum to drive mills and other water-powered machines was achieved (Judson and Kahane 1963, 89; Hodge, 2002, 47).

3.1.8 Cisterns: development and diffusion

A cistern can be more appropriately compared to a tank meant for rainwater storage. In antiquity, cisterns were either rock-excavated or built at surface level through the employment of brick or masonry. Cistern excavation, as opposed to construction, was in the past considered to be a more cost-benefit undertaking. It appears that expenditure constraints and maintenance-related issues limited the diffusion of above-ground cisterns, but when available, these were more commonly associated with the storage of an aqueduct water supply (Hodge 2002, 61, 279-80).

Cistern diffusion was aided by the fact that the involved technology was simple to harness, making it possible for cistern excavation to be undertaken without the need of outside technical expertise (Crouch 1993, 38). Cisterns were regularly used for the harvesting of rain runoff from roofs and courtyards. Occasionally, cisterns were also used as water settling tanks for the filtration of rain and aqueduct water (Hodge 2002, 273-75; Mays 2010, 123-26). In the case of the troglodytic settlement of Petra in Jordan, harvested rainwater originating from overlying cliff areas was diverted into domestic and municipal cisterns for later utilisation (Crouch 1993, 22). In antiquity, cisterns had various shapes and sizes, with the commonest

[24] It is not as yet known if the Phoenicians were familiar with *qanat* technology in pre-Roman days. See Goblot 1979, 120-21.

being square, rectangular or cruciform in shape. The barrel-vaulted type of roofing was the most common cistern covering method used in antiquity. Nonetheless, covered and uncovered cisterns appear to have been equally common (Hodge 2002, 58-59, 97-98).

One crucial factor ensuring the successful completion of any cistern-related project was the availability of waterproofing mortar, necessary for rendering the interior of cisterns watertight.[25] The wider regional context suggests that whilst various cisterns in Italy and Sicily are of a sixth-century BC date, cistern waterproofing technology appears to have become a feature common to the Mediterranean region as from the fourth century BC.[26] The scientific investigation of cisterns has revealed how the composition of the waterproofing mortar and the number of layers applied were probably also dependent on local geological and climatic circumstances (Hodge 2002, 97-98).

Accurate water storage and consumption data in a central Mediterranean context are so far only available for the Hellenistic period settlement of Morgantina in Sicily. The settlement's cisterns each had on average a capacity of 30 m^3. Considering that the annual precipitation of the areas is of 700 mm, cisterns were filled once a year through a roof collection area of 45 m^2. Cistern owners at Morgantina were in possession of large enough domestic water supplies to carry them through the dry summer season (Crouch 1993, 22-25).

3.1.9 Wells and cisterns: an aid to settlement in the Phoenician-Punic, Greek and Roman periods

In comparison to cisterns, wells appear to be the most ancient hydrological system and since antiquity these were generally preferred to cisterns as they provided a fresher and more abundant water supply (Hodge 2002, 51-52, 60). Only limited specialised literature on Phoenician and Punic water capture and storage systems has been published to date. Available evidence, mainly from Carthage in Tunisia suggests that this North African settlement primarily relied on a major spring or well-chamber called the "Fountain of a thousand amphorae". It was located at the foot of the Bordj Djedid cliff and was originally constructed in Punic times. Nonetheless, while many of the investigated cisterns at Carthage may be Punic in origin, most of them, including the great cisterns at La Malaga and Bordj Djedid, date to the Roman period (Harden 1980, 30-31, 125).[27] Wells were also discovered in domestic and industrial-related buildings in the Phoenician-Punic city of Kerkouane in Cape Bon area of Tunisia (Fantar 1988, 176; Bondì 1988, 261). In the close-by Punic Fort of Ras el-Drek rain water was harvested in "tub" cisterns (Bondì 1988, 265). It is also interesting to note how in a number of Phoenician-Punic settlements, rain runoff was not always channelled into any available underlying cisterns. Excavations at the Phoenician settlement of Sarafand (Sarepta), in Lebanon, revealed houses which directly disposed of their roof runoff water into streets, which apart from being walkways, also served as drainage systems (Pritchard 1978, 94).

For the Greek and Roman periods, the tapping of sub-surface sources by means of wells and the storing of harvested rainwater in cisterns were the principal factors which enabled the settlement of an area (Hodge 2002, 5, 48). This system is best illustrated at Morgantina, Sicily, where both water-harvesting systems supplemented one another (Crouch 1984, 360). Before their wide-spread appearance in Classical and Hellenistic period Greek sites, the use of cisterns appears to have been mainly restricted to Minoan and Mycenaean settlement sites. Archaeological fieldwork in several Greek and Roman settlements was pivotal in demonstrating how privately owned wells and cisterns provided the bulk of the necessary water supply. The construction of aqueducts appeared at a much later stage and mainly provided settlements with a surplus water supply, necessary in the running of bath complexes and water fountains (Hodge 2002, 5, 21-49; Crouch 2004, 30-33, 49-51).

3.1.10 Concluding remarks

Of the above-mentioned modes of water capture and storage, field-terracing, wells, cisterns, *qanat* and water galleries/adits are all present in a Maltese context. The absence of *tufo* deposits in Maltese geological stratification automatically disqualifies *cuniculi* availability locally. Nonetheless, an overview of *cuniculi* water capture systems is included in section 3.1.7 above, not only to showcase *cuniculi* structural similarity to *qanat* technology but also to indicate that their basic function was distinctly different from those of *qanat* systems. To date, no diversion dam water capture system of an archaeological or historical interest was observed in the Maltese archipelago, the principal reason for this being the almost complete absence of natural perennial springs and watercourses. One rare example of this method of water capture might have, in antiquity, been employed at Għajn Tuffieħa (MG15) in the Mġarr (Malta) territory in order to divert surface perennial spring water to the close-by

[25] Exceptions to this are instances where cisterns are excavated within largely impermeable rock strata. For a discussion on the Maltese situation on this matter see section 2.6.2.

[26] Personal communication by F. Schön, who specialises in cistern technology at the Institute for Classical Archaeology at the University of Tübingen.

[27] The hydrological setup of the Phoenician city of Tharros, located on the west coast of Sardinia was, as at Carthage, based on a series of private and public cisterns aimed at storing roof runoff. See Mazzuccato *et al* 1999, 117-33. Phoenician cistern technology was apparently based on the Early Iron Age Canaanite practice of lining storage cisterns with impermeable lime plaster, which made the long-term storage of any harvested water possible. See Harden 1980, 125-26.

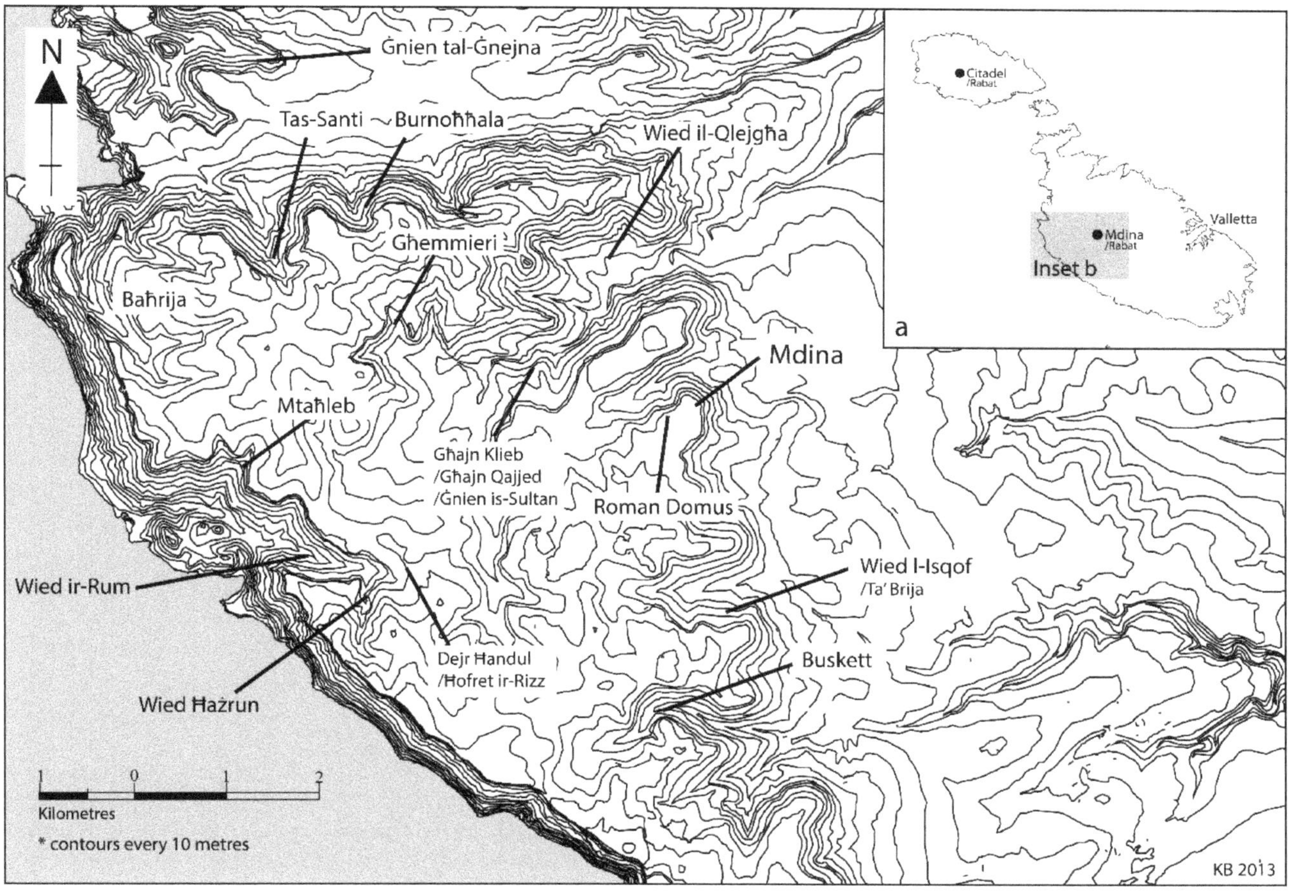

Fig. 3.6 Map of the Maltese islands (a), with inset (b) showing the spatial distribution of principal place-names in the Mdina area mentioned in the text. Base map data: MEPA.

Roman Baths site.[28] Similarly, there is no archaeological evidence for urban sewage diversion into countryside areas. If such a method of agricultural intensification was adopted in the Maltese archipelago in antiquity or the subsequent centuries, this would have presumably been confined to rural areas located in close proximity to the urban centres of the Melite and Gaulos. This, however, still awaits historical and archaeological justification.

3.2 The Maltese Context: The Phoenician, Punic and Roman periods (8th century BC to 6th century AD)

The emerging Maltese scenario shows how during the Phoenician, Punic and Roman periods, settlements were in the main located on either exposed Upper Coralline Limestone or Globigerina Limestone outcrops. An evaluation of the available archaeological repertoire shows the widespread use of cisterns as a means of harvesting runoff rain water. Ground-water sources were also tapped – clearly indicating the ability to detect and harness aquifer potential. Many of the sites discussed hereunder are multi-period ones and were utilised during the Punic, Roman and later phases. In the absence of a secure archaeological context, the dating of water-related features is difficult to ascertain (Bonanno 2005, 108).

3.2.1 Wells and cisterns

Two main types of ancient cistern typologies were reported for the Phoenician, Punic and Roman periods: the *bottle* or *bell-shaped*, and the *cuboid* (Bonanno 2005, 107). Both cistern typologies were rock excavated. Bell-shaped cisterns were accessed by means of a small opening at the surface and were circular in plan (Bonanno 2005, 107). Cuboid cisterns are by far the more elaborate and square and rectangular plan variants have been recorded. Cisterns conforming to the cuboid typology were roofed over by means of rows of large horizontal slabs of stone, resting on either horizontal arches, or thick stone beams set on vertical pillars. The best known example of such an instance comes from the multi-period site of Ta' Kaċċatura at Birżebbuġa (Ashby 1915, 63-64), and is discussed in more detail in section 3.2.4. Cisterns were made watertight through a mixture of finely crushed pottery and pozzolanic cement imported from the Campania region of mainland Italy (Bonanno 2005, 107).[29] Phoenician, Punic and

[28] The Għajn Tuffieħa Roman period baths are discussed separately in chapter six, section 6.1.9.

[29] Cistern waterproofing technology appears to have become a feature common to the Mediterranean region as from the fourth century BC. See

Roman period remains reported from Malta show that the majority of settlements and establishments were probably geared towards runoff rain water-harvesting, from both roof and ground-level catchment areas.

3.2.2 *The Roman Domus at Rabat (Malta) and other neighbouring minor sites*

An analysis of the hydrological systems adopted at the Roman Domus and its surrounding area sheds evidence on how the hydrological needs of the ancient town of *Melite* centred on the use of a combination of cisterns and wells (Fig. 3.6).[30] The area occupied by the Roman Domus site measures approximately 2800 m² and contains about ten cisterns and four wells. The 1920 excavation campaign at this site identified a total of six cisterns, the average depth of which was 4.6 m, and which were capable of storing a copious quantity of water. Two wells within the same site tapped the perched aquifer at an average depth of 12 m below surface level (MAR 1921-1922, 5), and provided the area with a perennial water supply. Rain runoff was collected and channelled towards cisterns by means of surface rock-hewn canals (MAR 1921-1922, 5).

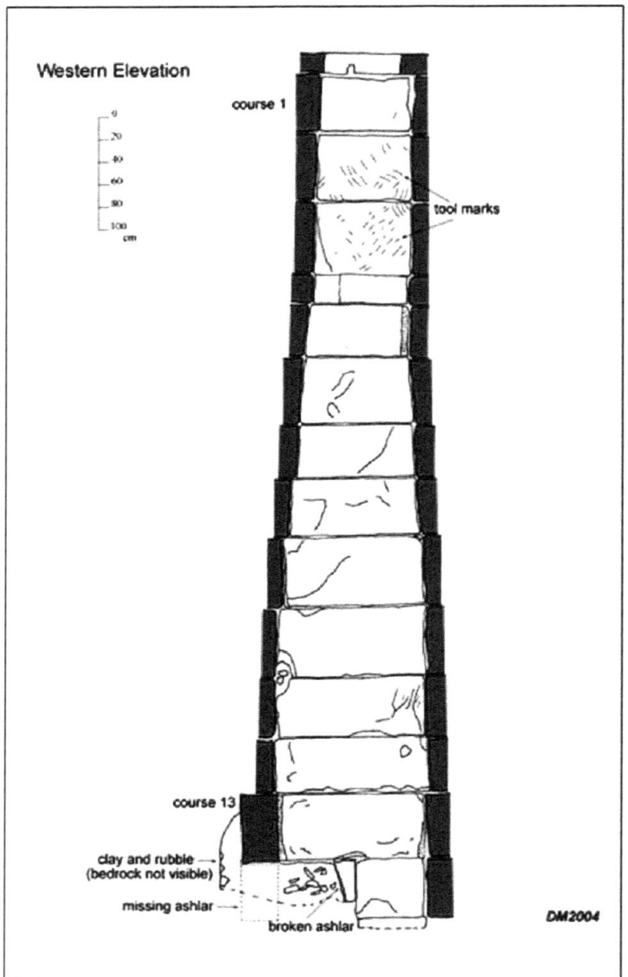

Fig. 3.7 Elevation plan of a shaft well at the sanctuary site of Tas-Silġ in the territory of Marsaxlokk, Malta (after Bonanno 2009, 28; drawn by D. Mizzi).

The 1923-1924 campaign at the same site revealed the presence of other water cisterns and two further wells. At a depth of 9.1 m below surface level, the wells were found to be in good condition, with water flowing over a layer of Blue Clay (MAR 1923-1924, 5). In the excavation campaign of the following year, another two large water cisterns were cleared. These were provisioned with proper mouths and terracotta pipes for the channelling of rain water collected from roof runoff (MAR 1924-1925, iv). The Roman Domus area was reutilised and converted into an extensive Muslim cemetery during the Later Medieval period (Dalli 2006, 274-75), but in all probability the above discussed water systems are a direct reflection of the water management strategies employed here during the Punic, Roman and Byzantine periods.

Ruins possibly belonging to the Roman period were discovered to the north-west of Mdina, in a field known as Ix-Xagħra in the Ġnien is-Sultan district, limits of Rabat (Malta). Amongst other remains, there is the mention of a large water tank, one side of which was constructed out of concrete (MAR 1909-1910, 7; Ashby 1915, 48). In another instance, lead pipes and a concrete vaulted chamber were reportedly identified in the garden of a certain Tomaso Zabbar and possibly dated back to the Roman period (Abela 1647, 33). The apparent vestiges of a hypocaust were discovered in the monastery of S. Pietro at Rabat (Malta) in 1720 (Ciantar 1772-1780, 135). The remains of baths are mentioned at Dejr Ħandul, to the north of the Dingli settlement. Lead pipes, marble and stone troughs, together with marble columns were all noted as being present in the area (Abela 1647, 36; Ciantar 1772-1780, 149).[31] Other wells were discovered elsewhere outside Rabat (Malta). Notable amongst these was a bell-shaped well at Notary Zarb Street, Attard, which yielded Roman period ceramic sherds. The bottom had a radius of 1.62 m and was cut into a perfect circle. A clayey deposit covered the well floor to an average depth of 0.30 m (MAR 1965, 2).

3.2.3 *The Tas-Silġ Sanctuary*

Several wells supplied the Tas-Silġ sanctuary in the territory of Marsaxlokk with a ground-water supply, the archaeological context of which suggests a Punic period

section 3.1.5. A. Bonanno personally communicated that in the case of Malta, this date can be probably pushed back to ca. the sixth century BC.

[30] The Roman Domus is an upper class Roman period dwelling containing exquisite mosaic floors at Rabat (Malta). The mosaics are typologically datable to ca. the last quarter of the second century and the first quarter of the first century BC. See Bonanno 2005, 306-17.

[31] For a discussion on the probable location of Dejr Ħandul see chapter five, section 5.2.4.

Fig. 3.8 Exterior view of the well located in the south area of the Tas-Silġ sanctuary site at Marsaxlokk, Malta. Source: N.C. Vella on behalf of the Department of Classics and Archaeology, University of Malta.

Fig. 3.10 The access point to one of the two vertical shafts tapping a localised aquifer in the north area of the Tas-Silġ site at Marsaxlokk, Malta. Source: Malta/Gozo Cistern Exploration Project.

date (Bonanno 2005, 106-8).[32] One of the wells, located in the southern area of the site (Bonanno 2009, 27), is over 13 m deep and was excavated in a Middle Globigerina deposit (Figs. 2.15, 3.7, 3.8, 3.9). This rock stratum contains substantial bluish-grey marly deposits which limit water percolation and which made possible the formation of a localised perched aquifer in the Delimara – St Thomas Bay area (Morris 1952, 56). Ashlars line the well sides and probably served as buttressing to prevent the sectional collapse of the marly deposits in which it is excavated. Due to the dislocation of several ashlar blocks and the resultant possibility of a cave-in, the investigation of the shaft was curtailed as a precautionary measure before the bottom was reached.

The investigation of the vertical walls of the shaft revealed a series of horizontal bands of reddish-coloured iron residue, which in the present writer's view represent the different water levels present within the shaft over the centuries.[33] Iron nodules are a common feature of Globigerina deposits and may dissolve in water to form a layer of iron oxide accumulation. Bluish-grey coloured clay deposits were also noted in the crevices between the ashlar masonry and are indicative of the use of the shaft as a well. No water deposits were detected during the excavation process of the well. This is probably either the result of a lowering of the water-table level due to over-

Fig. 3.9 Detail of well shaft interior located in the south area of the Tas-Silġ sanctuary site at Marsaxlokk, Malta. Source: the author.

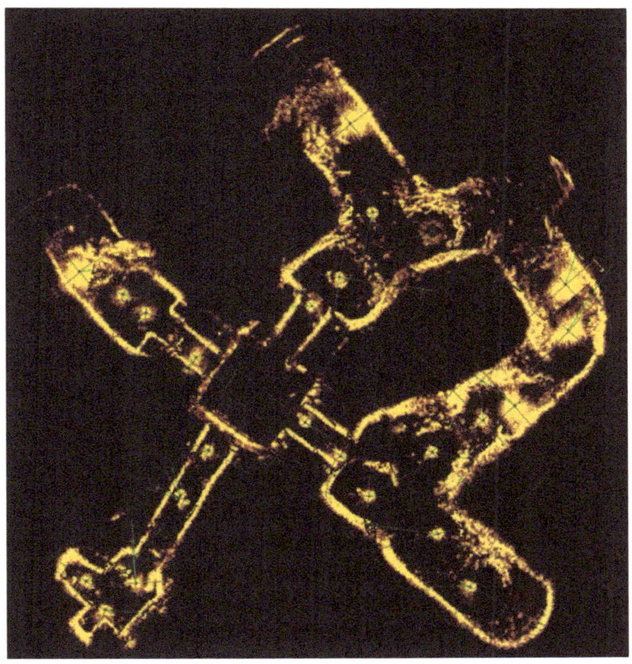

Fig. 3.11 Sonar data map image of two probable wells in the north area of the Tas-Silġ sanctuary at Marsaxlokk, Malta. Source: Malta/Gozo Cistern Exploration Project.

[32] Water availability at Tas-Silġ during the Late Medieval period is indicated by the Bir Riqqa place-name (also given as Bir Ricca), first documented in 1504. See Wettinger 2000, 49. This water source is located on the north side of the sanctuary site. Personal communication by A. Bugeja.

[33] These data are based on personal observations and data collected whilst excavating the shaft in Area D of Tas-Silġ South in July 2003.

extraction or past tectonic activity, and/or the formation of new fissures which encourage water percolation and dissipation.

Another two masonry-lined vertical shafts in the north side of the Tas-Silġ site are located in the peristyle area of the sanctuary (Fig. 3.10). These give access to a series of subterranean chambers connected by means of narrow corridors (Fig. 3.11).[34] The lack of any visible mortar applied to the interior of these subterranean chambers, leads the present writer to suspect that this setup functioned as a well. It is likely that this tapped the localised perched aquifer ground-water source accumulating in the largely impervious Middle Globigerina Limestone deposit prevalent in the area.

3.2.4 The Ta' Kaċċatura site

Some of the more significant remains regarding Phoenician, Punic and Roman hydrological techniques came to light from the multi-period site at Ta' Kaċċatura at Birżebbuġa (Fig. 2.15). The remains of a rural olive oil production edifice were investigated here nearly a century ago (Ashby 1915, 52), but have since been left in a state of neglect. The visible remains appear to have been constructed around the second century BC, on the foundations of earlier structures (Bonanno 2005, 300-1). A rectangular cistern beneath the peristyle area is roofed over by means of a slightly arched rough stone structure set in mortar (Fig. 3.12). Internally, the roof was lined with white cement, laid on a temporary frame of canes, in turn supported by seven beams. The cistern floor slopes from the north-west to the south-east and was connected to an earlier bell-shaped cistern located close-by (Fig. 3.14). It appears that the hydrological setup of the peristyle area consisted in the channelling of the harvested rain water into the bell-shaped cistern. This then overflowed into the adjoining rectangular water tank. The bell-shaped cistern possibly served the purpose of a water settling vat and is probably of an earlier date than that of the edifice in which it was incorporated (Ashby 1915, 56).[35]

The most noteworthy hydraulic-related work at Ta' Kaċċatura is located to the immediate south-west of this

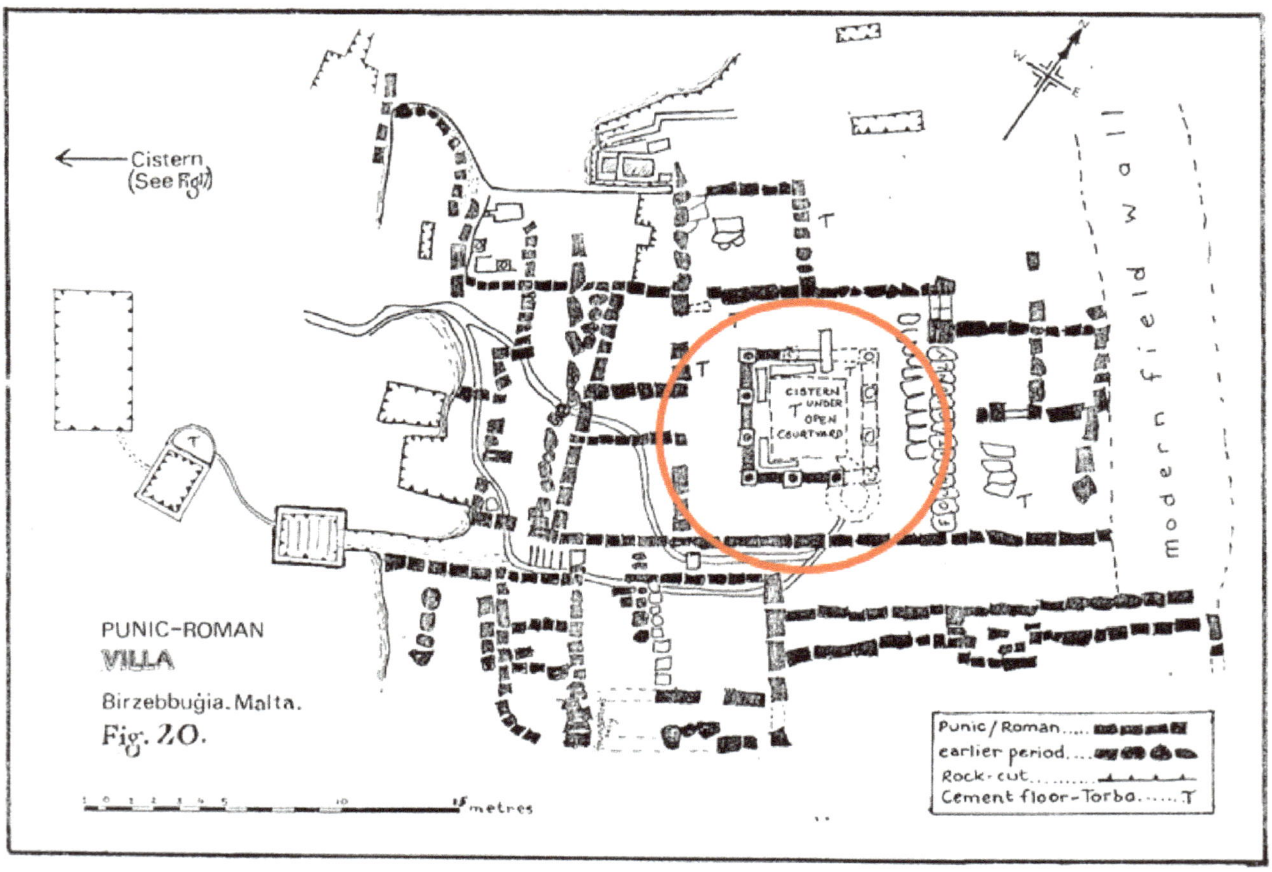

Fig. 3.12 Plan of the remains of the Punic-Roman olive oil production establishment at Ta' Kaċċatura in the territory of Birżebbuġa, (after Ashby 1915, 52).

[34] A note of appreciation goes to T. Gambin, C. Clark and C. Olstad of the Malta/Gozo Cistern Exploration project for making available all the collected data. See http://users.csc.calpoly.edu/~cmclark/MaltaMapping/site57.html and http://users.csc.calpoly.edu/~cmclark/MaltaMapping/site58.html

[35] A Roman amphora and other ceramic material recovered from within, securely date these cisterns to the Roman period.

Fig. 3.13 Interior view of the Ta' Kaċċatura cistern at Birżebbuġa, Malta. Source: the author.

agricultural establishment and is of a probable Punic date (Bonanno 2005, 107-8). This consists of a large rock-hewn cistern which perhaps served as the quarry site for the countryside villa and the adjoining olive oil production site (Fig. 3.13). Apart from a portion of the cistern which measures 2.7 by 2.6 m, projecting from the south-east side, its main area measures 10.3 by 10.25 m. The depth of the cistern varies between 3.9 and 4.2 m and it could hold approximately 422.3 m³ of water. Its interior is accessed by means of sixteen downward steps and was made watertight by a mortar coating. It is roofed over by four rows of flat slabs carried by cross beams, in turn supported by twelve pillars. Water channels supplied the villa structure with running water obtained from this subterranean reservoir. The Ta' Kaċċatura cistern was compared by Ashby to another found at Cisternae di Vitagliano in Puglia, which at the time had been wrongly attributed to the prehistoric period (Ashby 1915, 64). The Ta' Kaċċatura cistern also bears a superficial resemblance in its architectural design to Roman period cisterns at Eleftherna in Crete, which have much larger dimensions and which are completely excavated in calcareous limestone deposits (Fig. 3.15).[36]

3.2.5 *The Ras il-Wardija Sanctuary at Kerċem (Gozo)*

The Ras il-Wardija Punic sanctuary in the limits of Kerċem, Gozo, excavated between 1964-1967, provides interesting data on the harvesting of surface runoff water in cisterns excavated in a Middle Globigerina deposit (Figs. 2.17, 2.18, 2.19, 2.20). The two cisterns lie in close proximity to one another. The water retention properties of the Middle Globigerina layer, mentioned in sections 2.4.2

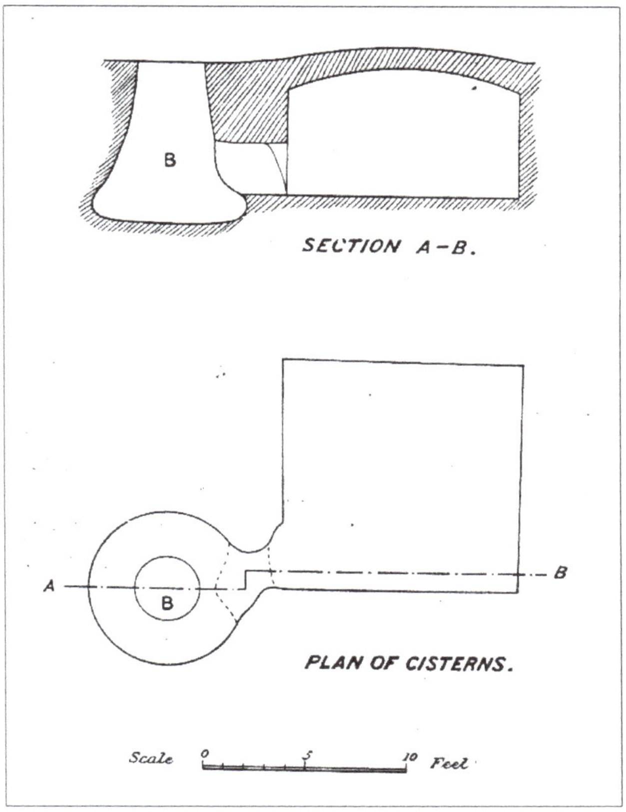

Fig. 3.14 Plan and section drawing of two interconnected cisterns hewn below the peristyle area at Ta' Kaċċatura, Birżebbuġa (after Ashby 1915, 55).

and 2.6.2 can be best appreciated from the most accessible water tank at the Ras il-Wardija sanctuary. Located to the south-west of the sanctuary area, this open-air reservoir is not lined with mortar, but still has excellent water preservation properties. This water tank, which the *Missione Archeologica Italiana a Malta* reports associate with a sacred pool for ritual bathing, is square shaped, measures around 3.7 by 3.6 m, is nearly 3 m deep and has a water storage capacity of around 37.5 m³ (Caprino 1966, 136) (Fig. 2.20). The bottom of the tank could be reached via a flight of ten now badly weathered steps cut into its western wall. Recesses were dug into the southern and eastern sides of the tank at a subsequent stage in order to increase its water storage capacity. The purpose of a neatly cut rebate which perimeters the four sides of the tank is possibly indicative of the presence of some sort of roofing arrangement aimed at covering the tank area (Caprino 1966, 137; Caprino 1967, 83-84; Buhagiar M. 1988, 75-76), possibly to prevent water evaporation during the arid summer months. A nearby water channel probably fed the water tank with surface runoff rain water.

The other cistern at the Ras il-Wardija site is bell-shaped and has a rectangular orifice which measures 1.52 by 0.67 and 0.42 m and which is enclosed in a rock-cut well head that is ca. 0.30 m high (Fig. 2.19). The cistern is currently ca. 3 m deep and has an estimated current water capacity

[36] Personal onsite field-research held in August 2008.

Fig. 3.15 Interior view of a massive rock-hewn cistern at Eleftherna, Crete. Source: the author.

of 14.5 m³. There are indications that this was originally of smaller proportions.[37] Access into the interior of the cistern was facilitated by means of a series of foot and hand holes carved into its south wall.

3.3 Interpreting the Evidence

The data presented in section 3.2 are only a representative sample of the documented remains for the Phoenician, Punic and Roman period for the Maltese archipelago. They do, however, give an indication of the local hydrological context and strategies resorted to during these periods. Most sites relied on a system of either wells or cisterns or, as in the case of the Roman Domus, on a combination of both. Particularly significant is the fact that a number of archaeological sites had during the periods in question already recognised the water retention potential of the Globigerina Limestone strata, as is the case with the sanctuaries at Tas-Silġ and Ras il-Wardija.

From a Mediterranean perspective, the question of the diffusion of *qanat*-type water extraction systems until Late Antiquity still needs to be more satisfactorily addressed. The available water-related evidence for Malta does not indicate the employment of *qanat* systems at this point, but personal research is on-going. The Għajn Tuffieħa Roman period baths (MG15) are discussed separately in chapter six, section 6.1.9. It is significant that the water setup of this site in antiquity appears to have relied exclusively on a natural perched aquifer spring and not a subterranean water-extraction gallery system. The same applies to the bathing system of the Villa at Ramla Bay, Gozo. Located at the valley's estuary close to the shore, the bathing establishment seems to have utilised a natural spring flowing through the Ramla l-Ħamra valley system.[38]

It is only at Bidnija in the territory of Mosta and at Ta' Baldu (RA43) in the Rabat (Malta) territory that personal field-research led to the identification of perched aquifer galleries which were located in close proximity to sites of a probable Roman and Byzantine date. The association of a water gallery at Bidnija with Roman period remains at Ġebel Għawżara is inconclusive and necessitates further investigation.[39] A water gallery which adjoins a circular bath at Ta' Baldu appears to date to a much later period. The utilisation of the Ta' Baldu site during the Roman period remains questionable, with such claims being mainly based on the apparent presence of a *trapetum* located in a nearby cave.[40] An important factor which cannot be overlooked when studying past water technology concerns the tendency to classify most modes of hydrological engineering works as "Roman" (Shaw 1991, 63-91; Hodge 2002, 250-53). This was the case with the bath remains at Ta' Baldu which, when first documented in 1869, were reported to belong to the Roman period (Caruana, 1882, 93). A re-interpretation of the surviving remains, however, strongly hints at a Late Medieval or Early Modern date.[41]

It is possible that several of the hydrological systems described above in section 3.2 remained in local use until the close of the Byzantine period in 870 AD. On the other hand, the historic and archaeological repertoire for Byzantine Malta is limited (Brown 1975; Magro Conti 1996, 52; Dalli 2006, 31-32), and so far fails to shed effective light on issues related to the hydrological setup and strategies adopted during this period.

[37] The original capacity of the cistern was estimated at 9.5 m³ of water. See Caprino 1967, 82-83.

[38] For a discussion on the Ramla l-Ħamra bath remains see Magro Conti 1996, 98-105.

[39] Docter *et al* 2012, 121-23. For a more detailed discussion on the water-related evidence for Bidnija refer to chapter seven, sections 7.5 and 7.6.1.1.

[40] For a discussion on the hydrological resources of Ta' Baldu see chapter five, section 5.2.1.1.

[41] See chapter five, section 5.2.1.1.

Chapter Four

The Changing Maltese Hydrological Landscape: 1800 till the present

4.0 Introduction

Water-related evidence for the British period of Maltese history is reasonably well documented, making it possible to discern changes taking place in the local hydrological framework since 1798 with relative ease. The island's primary source of water supply during this period was dependent on a series of perched aquifer springs dotting the countryside in north-west Malta which in the early seventeenth century were connected to the Wignacourt Aqueduct system.[1]

Changes in the utilisation and management of the Maltese hydrological resources were triggered off by interludes of extended drought which afflicted the archipelago for almost a decade starting from 1834. Extended dry spells caused the public and private cisterns and reservoirs to dry up – a situation which constrained a significant portion of Malta's population to rely upon direct access to perched aquifer resources for their survival. It was such exceptional circumstances that prompted a nation-wide change in Malta's hydrological management strategy from the later part of the 1840s. This eventually led to a progressively decreasing reliance on the perched aquifer systems and to the harnessing of increasing quantities of mean-sea-level aquifer sources. Changes to Malta's hydrological framework during this period were also motivated by a steady increase in water demand due to rapid demographic growth. The consolidation of dense urban centres to the south-east of Malta together with the boost in economic activity experienced in the areas adjoining the main harbours also influenced water infrastructural policies.

Water-related intervention in Malta and Gozo for most of the nineteenth century is best documented through data attained from an analysis of Government despatches currently deposited in the National Archives at Rabat (Malta).[2] The despatch documentation discussed hereunder consisted of an official channel of correspondence between the administrators of Malta and the British Government. Apart from being a valuable primary source of evidence, the presented data appear to have been hitherto untapped. Despatch manuscripts vividly describe the unfolding tense situation caused by periods of prolonged drought. They also provide an analysis on the shortcomings of the local water situation from 1840 onwards. More importantly, despatch documents for the latter part of the nineteenth century provide ample evidence accounting for the changing hydrological strategies being implemented locally during this period.

Due to Malta's importance as an island fortress, it was unthinkable that its hydrological deficit should be solved through the importation of water resources from other British-held territories. The distant location of these did not make water transportation a strategically and economically viable process. This meant that from the 1850s onwards, the only workable option had to revolve around the intensification and maximisation of Malta's limited hydrological resources. Successfully increasing the water supply potential of the Maltese archipelago was not the only milestone achieved during this period. As from 1912, public water sterilisation by means of chlorination was introduced (Zammit T. 1924, 43). During this same period, the open channel system still widely employed throughout Malta and Gozo was replaced by means of the closed pipe system (Zammit T. 1924, 33; Morris 1952, 57). This measure was undertaken in order to eliminate the unwarranted contamination of the canal water from surface seepage.

Attempts to intensify the water supply resources of Malta led to the excavation of a substantial number of new galleries. These tapped both the perched and the mean-sea-level resources of the archipelago. As discussed in more detail in section 4.2, these galleries, the excavation of which commenced in the latter part of the 1840s, are clearly distinct from the perched aquifer galleries which are the principal subject of discussion of this book. In many instances, nineteenth and early twentieth-century documentation refers to these as the "old galleries" (NAM. GOV 01.3/11, Despatch To, no. 5, 1871, ff. 502-7). The nineteenth and early twentieth-century water-yield of several of these older gallery systems is also available (Morris 1952, Table 9).[3]

4.1 The Drought of 1834-1841

The need for better water management and the harnessing of new resources was induced by a succession of dry years

[1] For a detailed analysis of the springs feeding the Wignacourt Aqueduct system see chapter five, section 5.2.4.
[2] Despatch documentation is classified as NAM. GOV 01.3/11.
[3] The water-yield data presented cover the period between 1893 and 1943. Despatch documentation dating to the first half of the nineteenth century does give water-yield data for some of the more important perched aquifer galleries discussed below. See NAM. GOV 01.3/6, Despatch To, no. 62, 1841, f. 55.

between 1834 and 1841 where the recorded precipitation was significantly below the average norm. Lack of rainfall caused crop failure in 1840 and 1841, necessitating the importation of wheat and grain (NAM. GOV 01.3/6, Despatch To, no. 65, 1841, f. 63).[4] The gravity of the situation can be assessed through a string of Despatches from Lord John Russell, all of which date to 1841. The first entry dated to 18 January 1841 outlines how the water storage infrastructure in the Birgu, Senglea and Cospicua settlements was based on rainwater harvesting alone (NAM. GOV 01.3/6, Despatch To, no. 4, 1841, f. 4).[5] The water management strategies employed in many of the other Maltese settlements were likewise geared towards the harvesting of rain water in cisterns (NAM. GOV 01.3/6, Despatch To, no. 81, 1841, ff. 81-82). Exceptions to this were: (a) Valletta and Floriana, together with a small string of settlements which adjoined the Wignacourt Aqueduct line; and (b) Mdina and Rabat (Malta), which, as a result of their location above the perched aquifer sources, had easy access to the water galleries. Despatch documentation

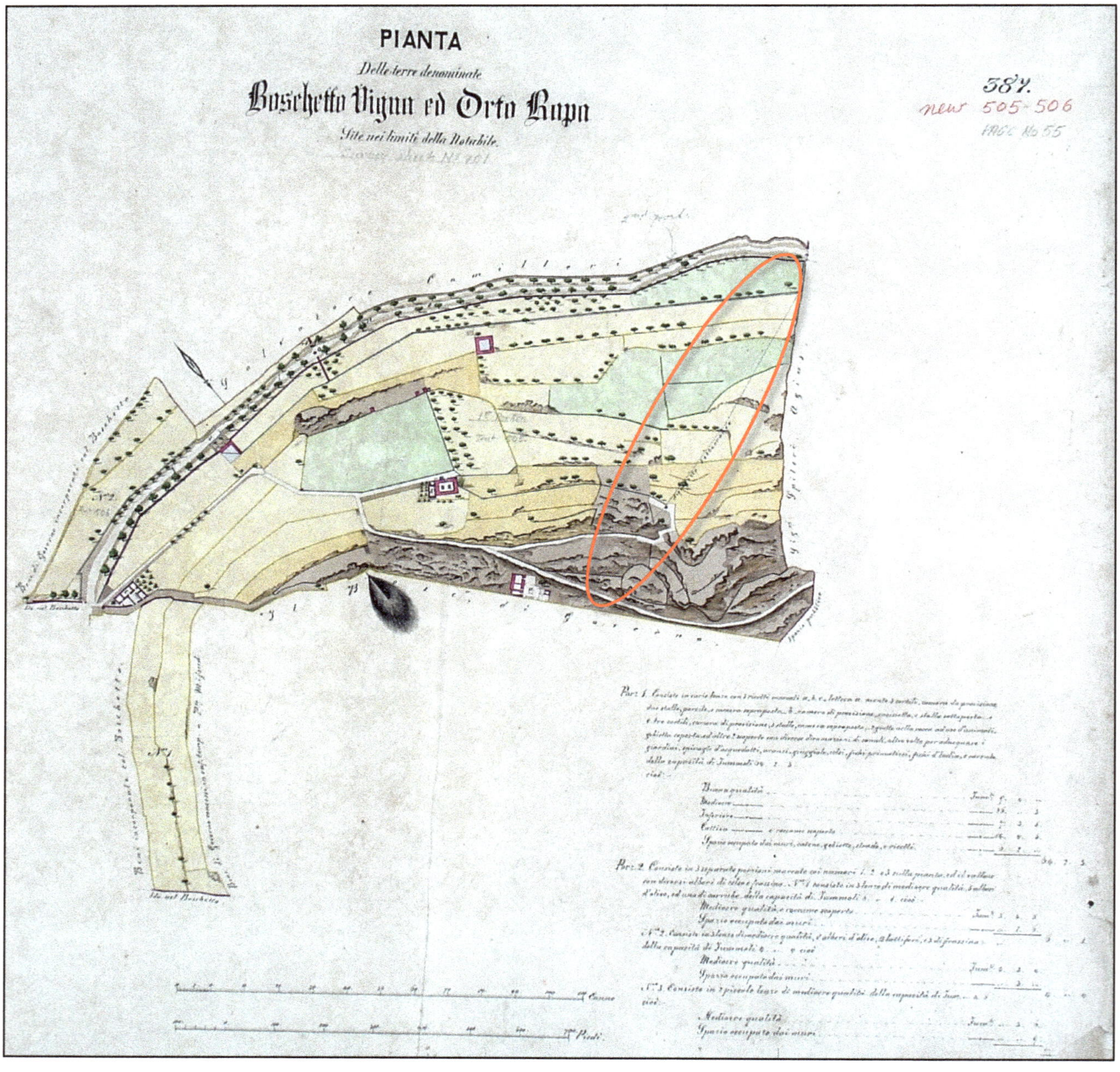

Fig. 4.1 The 1867 *cabreo* plan for the Ta' Rapa area (RA48) at Buskett indicating the whereabouts of the 1840s subterranean aqueduct (circled in red), which was connected to the Wignacourt Aqueduct system. Source: NAM. Descriptive Plans of the Crown Property in Malta, 1867, ff. 505-6.

[4] For rainfall observations in Malta in the late 1830s and the 1840s refer to Mitchell 1962, 59-61.

[5] Due to the prolonged nature of this drought, in August 1841 the public reservoirs and private cisterns were still empty. The public fountains were reportedly surrounded by inhabitants from morning to night and public order was only maintained through direct police intervention. See NAM. GOV 01.3/6, Despatch To, no. 57, 1841, f. 51.

never comments on the water situation in Gozo, making it likely that the water-related hardships being faced by Malta during this period were not being experienced there.[6] The Lord John Russell documentation exposes the fact that until the late 1830s, Malta's water supply was still directly reliant on the Knights' period hydrological framework.

Public and domestic cisterns in the Three Cities were empty by early 1841. Consequently, the need of water transportation became a priority, a task accomplished through the use of navy floating tanks filled with water provided by the Rabat (Malta) aqueduct sources.[7] The rural settlements of Malta were in a similar water-deficient situation and by February 1841 it was feared that their ca. 60,000 inhabitants would descend onto the urban centres, placing an increased demand on an already limited water supply.[8] In direct response to this unfolding water crisis, measures were taken to connect several springs at Buskett to the main canal of the Rabat (Malta) aqueduct system (Figs. 3.6, 4.1, 4.2, 4.4). This measure increased the water flowing through the Wignacourt aqueduct by 30,000 gallons a day (NAM. GOV 01.3/6, Despatch To, no. 8, 1841, f. 7).[9] By July 1841, this additional water supply brought much relief to the water-starved Grand Harbour region. Nonetheless, water was still a scarce commodity, but at least all inhabitants and troops stationed in the harbour side towns were supplied with an adequate daily provision of water (NAM. GOV 01.3/6, Despatch To, no. 57, 1841, f. 51). The course followed by the subterranean aqueduct in the Ta' Rapa tenement (RA48) at Buskett is also indicated on the Descriptive Plans of the Crown Property Cabreo for this property (NAM. Descriptive Plans of the Crown Property, 1867, ff. 505-6) (Fig. 4.1). This same source also mentions the existence of *spiragli d'acquidotti* or aqueduct shafts which could give access to the subterranean canals for maintenance purposes.

It is likewise possible that the two springs of Ta' Baldu (RA43) and Ħofret ir-Rizz (RA27), which the Malta Survey Sheet Links no. 85 shows as being connected to the

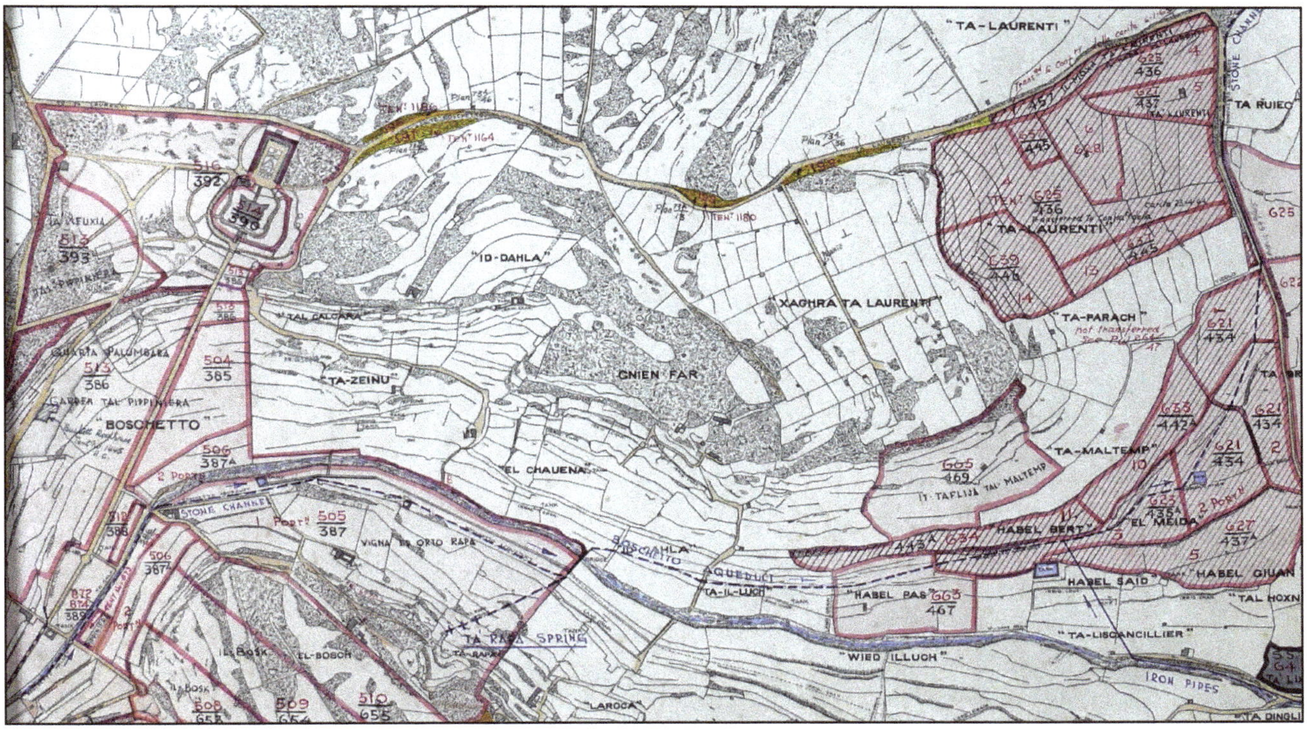

Fig. 4.2 Map excerpt from Malta Survey Sheet 101. The blue coloured intermittent lines show the course followed by the Buskett springs (RA6). Water from these was derived from formerly extant subterranean galleries. In 1841 it was diverted to the Wignacourt Aqueduct system. Source: NAM. Government Property Survey Sheet Links, Sheet 101.

[6] Gozo's higher percentage of Blue Clay deposits with respect to the available land ratio result in more copious perched aquifer resources than in Malta. The spatial distribution of Blue Clay deposits in Gozo also means that perched aquifer sources are within easier reach of a larger percentage of the population (Fig. 2.3).

[7] It appears that navy floating tanks consisted of barges onto which water storage tanks were fitted.

[8] By 19 February 1841 the cisterns of various villages were already empty, leading their inhabitants to commute a distance of several miles in order to obtain drinking water from the aqueduct fountains. NAM. GOV 01.3/6, Despatch To, no. 8, 1841, f. 7.

[9] This correspondence does not single out the Buskett Springs in question. Field-research has shown how the majority of the springs at Buskett originate from within already extant perched aquifer galleries – a matter also confirmed by T. Zammit in 1924. In 1924, the water supply from the Buskett branch of the Wignacourt aqueduct was described by Zammit as originating from the following springs: (a) Tas-Senia (RA53); (b) Tas-Sala (RA52); and (c) Ta' Rapa (RA48). Water at Tas-Senia originated from within a gallery, the ceiling of which appears to have been pierced by a number of shaft-wells. The springs at Tas-Sala and Ta' Rapa originated from within a hill known as *La Roca*, due south of Buskett valley. The Tas-Sala gallery was 44 m long, whilst the Ta' Rapa gallery is described as being ca. 81 m deep. The springs of Għajn il-Kbira at Girgenti and three sources originating from within Wied l-Isqof were eventually connected to the aqueduct system. See Zammit T. 1924, 35, 38-39. The location of the Tas-Sala and Ta' Rapa springs is also indicated in the Malta Survey Sheet 101 dating to 1929; and NAM. Government Property Survey Sheet Links, Sheet 101. The approximate map reference for the Tas-Sala spring is 45836 68589. The Ta' Rapa spring is located at map reference 46149 68474.

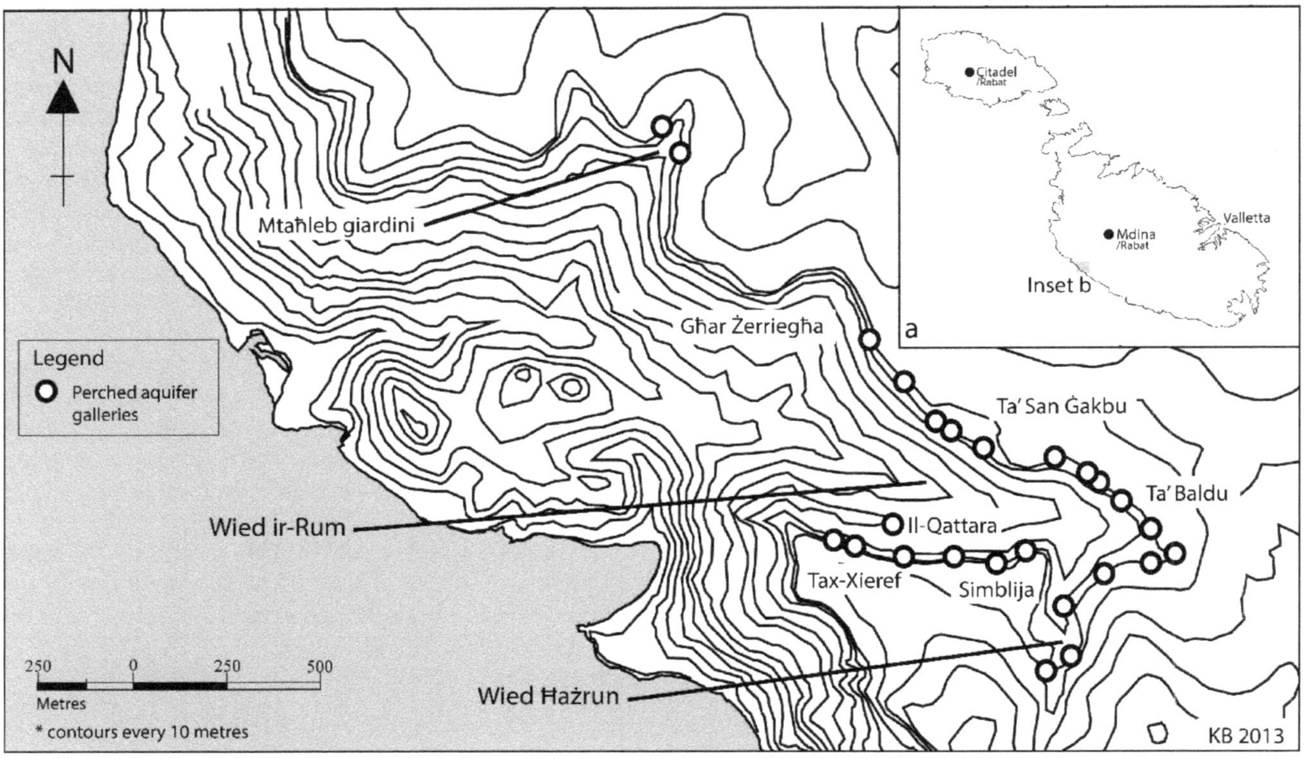

Fig. 4.3 Map of the Maltese islands (a), with inset (b) showing the spatial distribution of perched aquifer gallery entrances at Wied ir-Rum (RA56), Wied Ħażrun (RA55) and the Mtaħleb *giardini* (RA35) in the territory of Rabat (Malta). Base map data: MEPA.

Wignacourt system, also mark efforts undertaken during this period to increase the available water supply resources of Malta.[10] The Ta' Baldu spring in question – probably tapping the same aquiferous source as the spring at Ħofret ir-Rizz, – steers clear of a much older water extraction system located in the nearby Għar Ta' Baldu. This matter is discussed in more detail in chapter five (Fig. 5.47).

The water crises affected maritime activity negatively. Due to a shortfall of rain since the previous October of 1840, all tanks on both sides of the Valletta harbour were empty, leading the Admiral Commander-in-Chief to instruct navy vessels visiting Malta to arrive provided with water supplies. By February 1841, the fleet was instructed to proceed to Syracuse in Sicily in order to obtain its necessary water supplies from there (NAM. GOV 01.3/6, Despatch To, no. 8, 1841, f. 7).[11] By 22 May 1841 adequate provision was taken to divert navy-related shipping traffic to Mġarr (Gozo), where a watering place for ships and frigates had been constructed by channelling some of the springs there to the landing place (NAM. GOV 01.3/6, Despatch To, 1841, f. 30). Many of the Mġarr (Gozo) springs probably originated from a series of perched aquifer galleries. These are discussed in more detail in chapter six, section 6.1.11.4.[12]

The month of September 1841 again yielded no rainfall. A Despatch dated to 10 September expresses the sense of relief that a water source was discovered "near the head of the Great Harbour" (NAM. GOV 01.3/6, Despatch To, no. 63, 1841, ff. 56-58). The exact location of this spring remains unknown, but reports that this was located nearly 9.14 m below sea level makes it possible that this originated from within either the largely impermeable Middle Globigerina Limestone or the Lower Globigerina Limestone deposits, which in areas of the Grand Harbour area are positioned below mean-sea-level (Fig. 5.102). The yielded water was pumped to the surface by means of a watermill and conducted by pipes to both sides of the Grand Harbour.[13]

[10] Ta' Baldu and Ħofret ir-Rizz are both located within the territory of Rabat (Malta), close to the Dingli settlement (Figs. 3.6, 4.3, 4.10).

[11] It appears likely that in February of 1841 the Government authorities were still unfamiliar with the hydrological resources of the island of Gozo.

[12] An 1864 description of the water resources of Mġarr (Gozo) lists the availability of seven springs: (a) Għajn Barrani (GO5; GO16); (b) Għajn ta' fejn il-Barri (GO16); (c) Għajn Ħluk (GO16); (d) Għajn Sielem (GO16); (e) Għajn ta' Franġisk Spiteri (GO16); (f) Għajn Rajes (GO16); and (g) Għajn Fawwara (GO16). In 1889 water from these sources was pumped to Rabat (Gozo). See Zammit T. 1924, 31. This information is based on an 1863 report drawn-up by G. Schinas quoted by the 1924 Zammit publication. In Zammit's paper bibliography (page 48) this study by Schinas is simply referred to as "Malta-report" and could not be traced. See also the plan of the Mġarr (Gozo) *giardino* found in NAM. Descriptive Plans of the Crown Property, 1861, f. 54.

[13] The spring was yielding 150 tons or 150,000 litres of water a day. Had this spring not been intercepted, the local water situation would have been so serious that commercial shipping would have had to be driven away from Malta. NAM. GOV 01.3/6, Despatch To, no. 62, f. 55. By September 1841, 150 tons of water were being shipped to the Three Cities on a daily basis to

During this same period additional measures were undertaken to provide a number of villages with an adequate water supply. These involved conducting water from the lands of *Bulebel minore* (ZA1) to the public road near the Żejtun settlement, and the channelling of water from the L-Imgiered area of Qormi (MA10) to a public road close to Luqa, from the Buskett valley area to Siġġiewi and from *Casal Nuovo* (Paola) to a public road there (NAM. GOV 01.3/6, Despatch To, no. 63, 1841, ff. 56-58). The mention of water at Bulebel, L-Imgiered and Paola is particularly significant. These are areas dominated by the presence of Lower Globigerina Limestone deposits and strongly indicate the discovery of localised aquifer sources located in Lower Globigerina deposits.[14] The sinking of shafts probing for a water presence was also underway in the environs of the Three Cities (NAM. GOV 01.3/6, Despatch To, no. 63, 1841, ff. 56-58). Shaft boring was taking place within a context where the basic dynamics of the workings of the mean-sea-level aquifer still had not been understood.

The almost three-year-long drought was only brought to an end following abundant rainfall in the latter half of October 1841 (NAM. GOV 01.3/6, Despatch To, no. 70, 1841, f. 66). Malta's limited water resources, in particular during episodes of prolonged drought, were from this point onwards always considered to be an extremely serious matter. This was in all probability the factor triggering a drastic change in the hydrological landscape that Malta was to experience during the ensuing decades.

Gozo barely features in 1841 Despatch documentation. This indicates that the island was not as negatively influenced as Malta. Apart from being less densely populated than the larger island, Gozo has a much wider Blue Clay distribution base than Malta, thus providing easier access to perched aquifer deposits. It is also worth noting that the available 1841 documentation makes no mention of the failure of any of the Rabat (Malta) springs during this three-year rainless interlude. Considering that most of these originate from within perched aquifer galleries, the manner by which these subterranean water systems can cushion the adversity of dry spells, even ones extending for years at a stretch, is noteworthy. The reference to failed crops, including corn, cotton, fruits and green crops during this period (NAM. GOV 01.3/6,

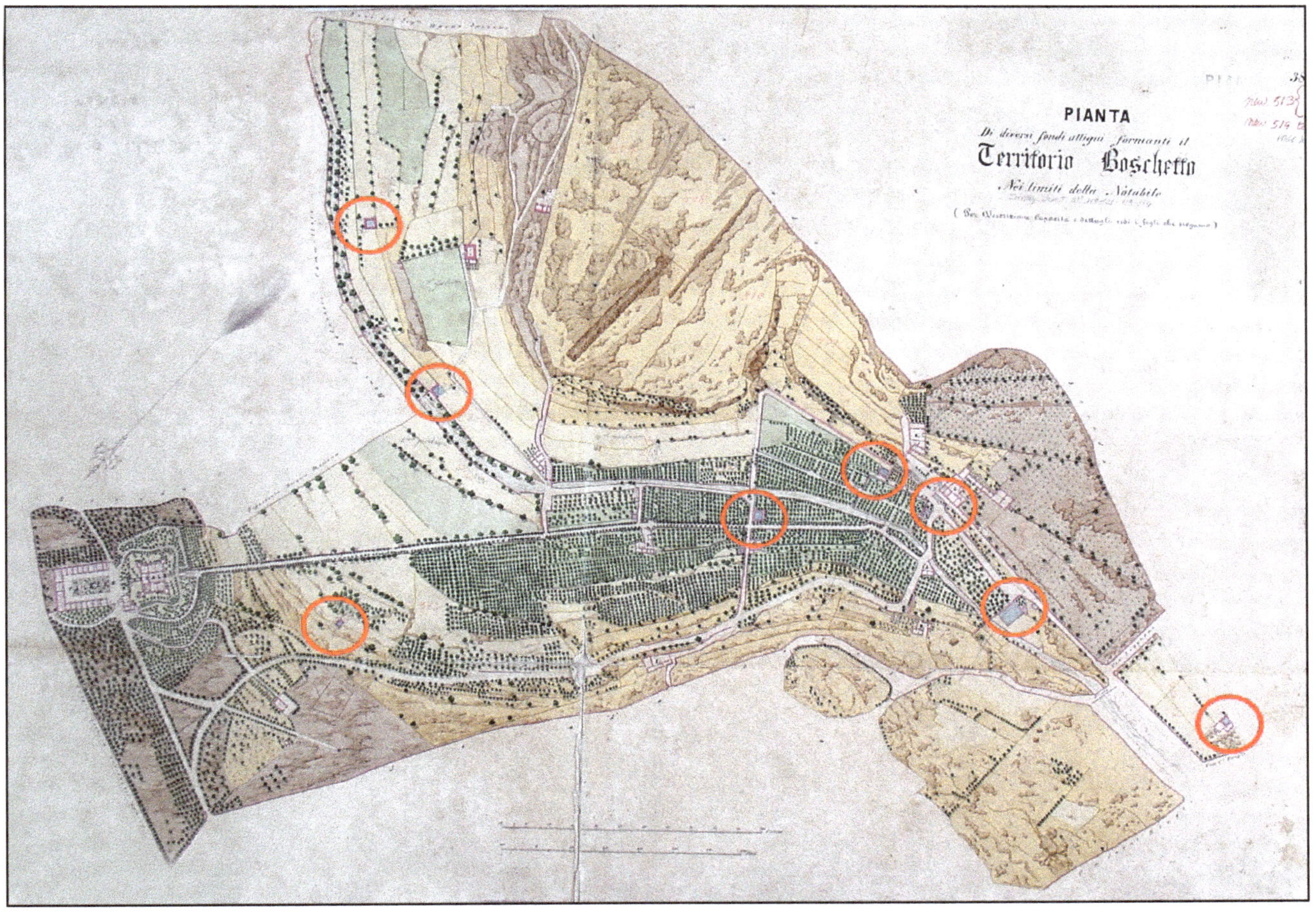

Fig. 4.4 The 1867 *cabreo* plan for the Buskett territory (RA6) in the limits of Rabat (Malta). Circled in red are a series of water collection reservoirs which also indicate approximate spring location. Source: NAM. Descriptive Plans of the Crown Property in Malta, 1867, ff. 513-6.

meet the hydrological needs of the over 20,000 inhabitants residing there. See NAM. GOV 01.3/6, Despatch To, no. 63, 1841, ff. 56-58.

[14] The localised perched aquifer at Bulebel covers an area of ca. 2 km². Personal communication by Engineer P. Micallef.

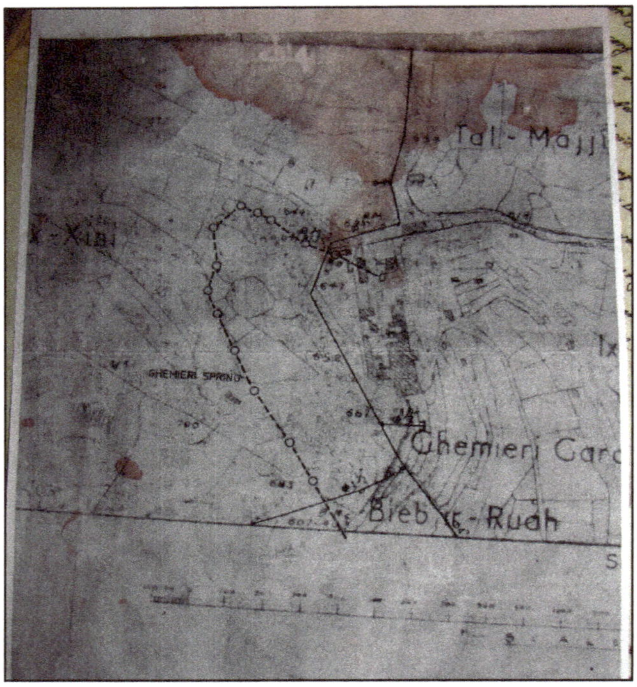

Fig. 4.5 A copy of the Survey Sheet for the Għemmieri area showing the west and north-facing boundaries of the Gomerino estate (RA15) together with the course followed by the subterranean perched aquifer gallery, which feeds Gomerino with a perennial water supply. Source: Personal archive of V. Tabone.

Despatch To, no. 81, 1841, ff. 81-82), must in this respect be a reference to arable land, the cultivation of which was either dependent on rainfall or on irrigation water obtained from adjoining cisterns or reservoirs supplied with harvested rainwater.

In an attempt to alleviate the water scarcity Malta was experiencing during this period, water from a spring in the Għemmieri (RA15) area was in 1842 also connected to the Wignacourt aqueduct supply (NAM. GOV 01.3/6, Despatch To, no. 22, 1844, f. 303) (Fig. 3.6).[15] It appears that this perennial water source originated from the perched aquifer gallery which feeds the Gomerino estate (RA15), close to the Fiddien territory (RA11), with a perennial supply (Fig. 4.5). The tapped water was channelled through the Bieb ir-Ruwa and the Fiddien areas where it was then joined to the Wignacourt Aqueduct system (NAM. Government Property Survey Sheet Links, Sheets 71-72) (Fig. 4.6).

4.2 Developments in the Maltese Hydrological Scenario: The mid-1840s till the end of the nineteenth century

Work on the Gozo aqueduct which provided Rabat (Gozo) with a perched aquifer supply was completed in 1843.[16] In 1844 work was well underway on the Fawwara aqueduct to supply the Three Cities with perched aquifer water (NAM. GOV 01.3/6, Despatch To, no. 16, 1844, ff. 290-92).[17] Both the Gozo and the Fawwara aqueducts were apparently connected to already extant water resources originating from perched aquifer galleries (Fig. 4.7).[18] The availability of pre-existing perched aquifer galleries is also supported by Despatch documentation. An 1845 entry makes reference to the identification of a spring located in a private garden in the Fawwara surroundings, which, if in Government possession, could be linked to the new aqueduct system feeding the Three Cities with water (NAM. GOV 01.3/7, Despatch To, no. 60, 1845, f. 66). Similarly, in 1866, as a measure to counteract the effects of the 1866-1867 drought, the government of the time had to purchase water flowing from two gardens belonging to the Bishop (Zammit T. 1924, 22). Personal archival and field-research has revealed that one of these springs originated from within a perched aquifer gallery known as L-Għar ta' l-Isqof (RA25) and for which there is documentation dating back to 1657 (Fig. 3.6).[19] In 1867, the government was likewise approached by Annunziato Schembri, who proposed the tapping of a spring in his property located between Ġnien il-Kbir and Buskett in the Rabat (Malta) district for the sum of £100 (Zammit T. 1924, 22). The spring may have originated from within a perched aquifer gallery.

Drought was again recorded in 1846-47 and 1866-67. The precarious water situation experienced by Malta in the first half of the 1850s was probably instrumental in motivating the Literary and Scientific Institute of Malta to hold a literary contest entitled *The Water Supply of Malta* in 1854. The chosen entry was that of Prof. Nicola Zammit, who in his essay tried to prove that spring water in Malta had its origin on the continent and was independent of the local rainfall. In order to increase the water supply of Malta Nicola Zammit proposed: (a) the enlargement of private cisterns to a size which was proportional to their collecting area; (b) the collection and storage of runoff water from roads and public squares in specifically built reservoirs; (c) the damming of valleys in order to turn these into large water reservoirs; and (d) the tapping of subterranean springs by means of artesian wells (Zammit T. 1924, 15-16; Micallef *et al* 2004, 201).

---

[15] For further information on the hydrological features present within the Gomerino estate see chapter five, section 5.2.1.3 and chapter seven, section 7.3.3.

[16] The springs tapped by the Gozo aqueduct were Għar Ilma (GO13) and Għajn Għabdun (GO8) which were connected together and water conducted on arches to Rabat (Gozo). See Zammit T. 1924, 31. Both the Għar Ilma and Għajn Għabdun springs originate from within perched aquifer galleries, but apart from the mention of these springs by T. Zammit in 1924, I did not come across any other documentation mentioning the spring(s) tapped by the Gozo aqueduct. G. Azzopardi, a professional archaeologist and curator of the Gozo Natural History Museum has communicated that popular tradition always pointed towards Għar Ilma, located in the Santa Luċija territory of Gozo, as being the source for the aqueduct water. At the turn of the twentieth century Għar Ilma had a daily summer yield of 9092 litres and a winter yield of 31822 litres per day. Għajn Għabdun had a daily summer yield of 27276 litres and a daily winter yield of 59099 litres. See Zammit T. 1924, 30-31.

[17] In the 1880s the Fawwara Aqueduct yielded only a third of the water-yield retrieved from the Wignacourt Aqueduct system. See Chadwick, 1884, i.

[18] The springs connected to the Fawwara aqueduct were: (a) Tal-Lunzjata (SI7); (b) Tal-Fawwara (SI6); and (c) Ta' San Ġorġ (SI5). All originated from the base of cliffs in the Ġebel Ċiantar area. See Zammit, T., 1924, 35; Morris, 1952, 3.

[19] See chapter five, section 5.3.

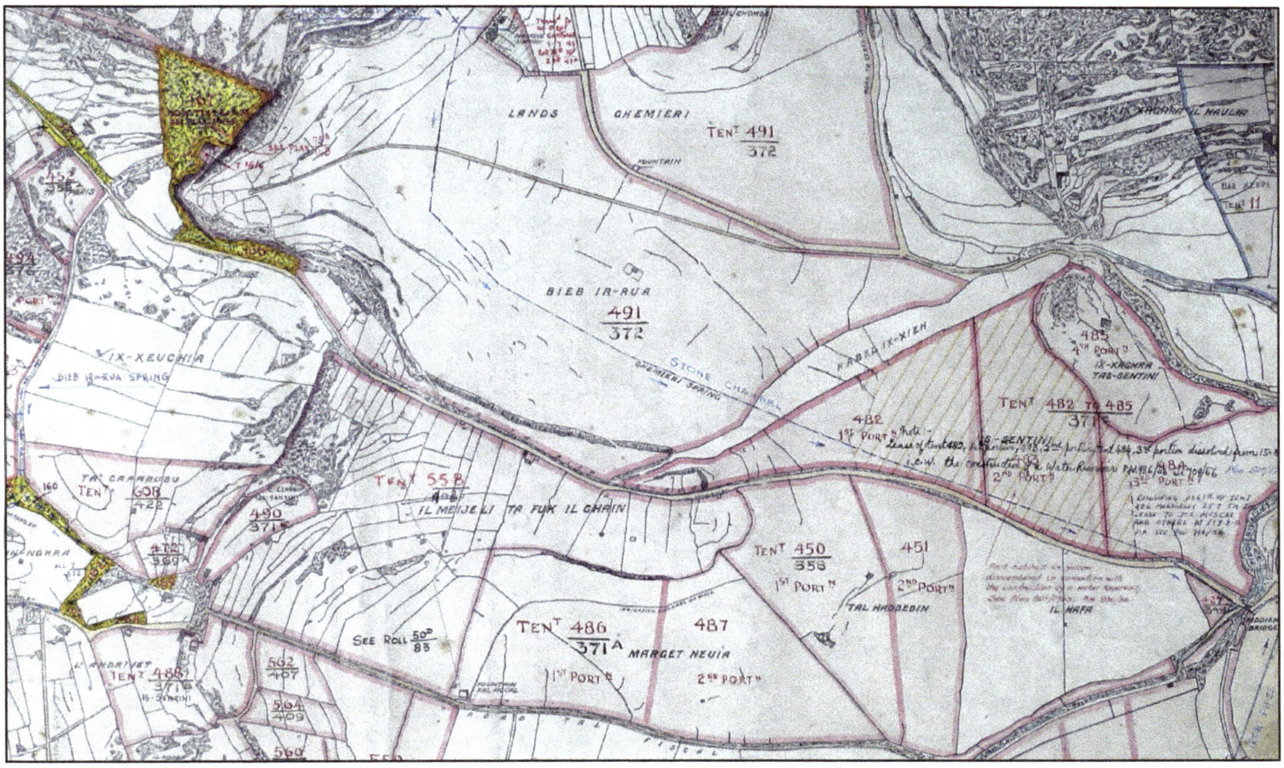

Fig. 4.6 Map excerpt from Malta Survey Sheet 71. The blue coloured intermittent line indicates the course followed by the Għemmieri spring (RA15) canal after a portion of its water-yield was diverted and joined to the Wignacourt Aqueduct system in 1842. Source: NAM. Government Property Survey Sheet Links, Sheet 71.

N. Zammit's proposal concerning the excavation of artesian wells was given some serious consideration a decade later during the drought of 1866-1867. As no expertise on artesian well systems was available locally, recourse was made to the Board of Trade in the United Kingdom and to any other department which could provide assistance in this matter (NAM. GOV 01.3/11, Despatch To, no. 18, 1867, ff. 299-301). Even the help of the Geographical Society, presumably a reference to the Royal Geographical Society of London founded in 1830, was sought in order to obtain more detailed information on Maltese geology. These contacts were made in an effort to determine if the boring of artesian wells could be implemented within a Maltese geological framework.[20] In many ways, this can be interpreted as being the first serious attempt leading to a proper understanding of the Maltese geological scenario, which eventually paved the way for the identification of the mean-sea-level aquifer. In one of the subsequent Despatches, the assistance of a person conversant with matters related to hydrology was specifically asked for. Rainfall data for the years 1864-1866 were also provided (NAM. GOV 01.3/11, Despatch To, no. 25, 1867, f. 311). 595 mm of rainfall were recorded for 1864, 556 mm for 1865 and 266 mm for 1866. Whereas rainfall levels for 1864-1865 conform to current average precipitation norms, those for 1866 characteristically pertain to an arid zone.

By June of 1867, Mr Bateman was appointed Government expert on matters related to hydrology.[21] Under his supervision, several water exploratory shafts were commissioned. It was the data and expertise accrued from this exercise that eventually led to the identification of the mean-sea-level aquifer. In July of the same year, a newly excavated 23.7 m deep shaft, located close to the Qormi (MA10) settlement, was found to yield good quality water at a depth of 1.82 m below sea level and yielded a discharge of ca. 250 litres of water per hour (NAM. GOV 01.3/11, Despatch To, no. 81, 1867, ff. 394-99).[22] It was planned from the outset that this was to be furnished with pumps, an engine and pipes for the distribution of the retrieved water. The Qormi shaft was eventually extended to a depth of 38 m and five lateral fissures yielding a total of ca. 509 litres of water per hour were discovered (NAM. GOV 01.3/11, Despatch To, no. 108, 1867, ff. 440-41).[23] Up to October of 1867 it had still not been comprehended that this shaft-well was tapping a mean-sea-level water source. As

[20] In order to furnish the Geographical Society with some information of Maltese geology, a copy of Captain Spratt's paper on the Geology of Malta and Gozo, published under the auspices of Sir William Reid, was dispatched. See NAM. GOV 01.3/11, Despatch To, no. 18, 1867, f. 301. An appropriate understanding of Maltese geology led T.O. Morris in 1952 to concede that no artesian water sources were to be found locally. It was also noted that shafts boring deep below sea level were only expected to yield saline water deposits, the salt concentration levels of which sometimes exceeded those of sea water. See Morris 1952, 119.

[21] Mr Bateman was a Civil Engineer who was asked by the then Duke of Buckingham to investigate the Maltese hydrological situation.

[22] Shaft excavation works commenced on 10 June 1867 and a water-yield was first noted on 24 July of the same year. See NAM. GOV 01.3/11, Despatch To, no. 101, 1867, ff. 423-30.

[23] See also NAM. GOV 01.3/11, Despatch To, no. 87, 1867, ff. 406-7; NAM. GOV 01.3/11, Despatch To, no. 83, 1867, ff. 400-1; NAM. GOV 01.3/11, Despatch To, no. 101, 1867, ff. 423-30. Only four men could be engaged in the shaft excavation works at any given time. A day's work only increased the shaft's depth by one foot.

no water was rising to the surface by internal hydrostatic pressure, it was realised that this could not be classified as an artesian well.

A despatch dated 30 December 1867 stated that the overall success of the Qormi shaft-well led to the excavation of other wells in various parts of Malta, all of which appear to have resulted in a good water-yield. Water from these was to be lifted to the surface by means of motor-driven pumps and a distribution network for the retrieved water installed (NAM. GOV 01.3/11, Despatch To, no. 137, 1867, ff. 495-97). By April 1868 more shaft-wells were excavated in the countryside around Qormi, whilst others excavated in the Rabat (Malta) countryside presumably tapped the perched aquifer resources of this area.[24] Water-related intervention during 1867 only involved the excavation of vertical shafts. The excavation of horizontal galleries only commenced in 1868 and only targeted the connection of a number of shaft-wells by means of horizontal galleries in order to increase the water catchment capacity of these systems (NAM. GOV 01.3/11, Despatch To, no. 40, 1868, ff. 554-60).

Despatch documentation indicates that the Rabat (Malta) shaft-wells were excavated in the countryside area between Dingli and the *Citta Vecchia* (Mdina) (NAM. GOV 01.3/12, Despatch To, no. 5, 1871, ff. 502-7).[25] The exact whereabouts of these is not specified, but their location is indicated on the Government Property Survey Sheet Links (Fig. 4.8), as well as the Government-owned pumping stations which dot the Rabat (Malta) countryside (NAM. Government Property Survey Sheet Links, Sheet 87). As these were positioned well away from the edges of ravines and cliff-faces, the perched aquifer galleries discussed below are typologically different and predate this period. This is also emphasised in the despatch documentation where a distinction is made between the "new sources" and the "old springs". It is also stated that the newly excavated wells and galleries did not diminish water from the older sources.[26]

Figure 4.7 Map excerpt from Malta Survey Sheet 125 showing the course followed by the Fawwara Aqueduct and the location of the tapped sources. The tapped water appears to be derived from formerly extant galleries. The springs in question are of San Giorgio (SI5), Fawwara (SI6) and Annunziata (SI7). Source: NAM. Government Property Survey Sheet Links, Sheet 125.

[24] One of the shaft-wells at Qormi excavated during this period was found to contain so much water, that a ten horse power engine containing pumps capable of lifting ca. 90 litres of water per minute could lower the water level within by only a few feet after hours of pumping. See NAM. GOV 01.3/11, Despatch To, no. 40, 1868, ff. 554-60.

[25] Other shafts and galleries were excavated in the countryside area between the Capuchin Convent at Rabat (Malta) and Dingli in 1885-1886. These tapped the perched aquifer deposits of the area. See Zammit T. 1924, 29, 39. The Rabat-Dingli galleries appear to mostly follow the course of the principal roads connecting both settlements. Soon after the excavation of these galleries it was noted that water percolation in Upper Coralline Limestone deposits increased greatly after rainfall but diminished rapidly during the dry season. See Mercieca 1931, 9, 12.

[26] In 1869 water quantities derived from the Rabat perched aquifer varied from 483,254 to 1,136,522 litres per day. See NAM. GOV 01.3/11, Despatch To, no. 5, 1871, ff. 502-7. Water exploratory campaigns held in the Upper Coralline Limestone plateaus during this period led to the sinking of 175 shafts and the excavation of 8.3 km of galleries. See Micallef *et al* 2004, 200. In order to curtail expenses and avoid the trouble of gaining access to privately owned land, most of these shafts and galleries were aligned along public roads. Furthermore, it transpires that several of the older springs which personal research has shown to originate from within perched aquifer galleries were partly surveyed in the first half of the 1950s. See Morris 1952, 82-83. Unfortunately, no further detail of this survey project was made available to the present writer at the time of finalising this book.

Fig. 4.8 Map excerpt from Malta Survey Sheet 87 showing the course followed by the newly excavated perched aquifer galleries at Rabat (Malta) in 1868. Due to land ownership issues, these were mainly excavated below the roads connecting Rabat to Buskett and Dingli. Source: NAM. Government Property Survey Sheet Links. Sheet 87.

In 1867 plans were submitted by Charles Andrews suggesting measures by which the water supply of Malta could be improved. Amongst the proposed measures were the damming of valleys and ravines and the storage of water in reservoirs constructed in different parts of the island (Zammit T. 1924, 19). By 1870, water from a number of newly discovered Rabat (Malta) springs was connected to the aqueduct system and measures were also undertaken to reduce water transportation losses within the Wignacourt Aqueduct (NAM. GOV 01.3/12, Despatch To, no. 38, 1870, ff. 357-59) (Fig. 4.9).[27] Works on this aqueduct system included the inspection of the whole line of canals, the cleaning of the stone canals, installation of iron pipes and replacement of worn out stone canals. Areas where the water canal passed close to the surface were paved over in order to prevent damage from any agriculture-related activities (NAM. PW 270, 11.01.1872, ff. 77-80). During this period, works were also ongoing in order to connect the Rabat (Malta) aqueduct system with a number of fountains providing formerly unserved settlements with a water supply. This entailed the construction of a number of reservoirs or, in the case of the Lija settlement, the emptying and restoration of a reservoir that had fallen in disuse and been backfilled with earth (NAM. PW 270, 18.09.1872, ff. 177-78; NAM. PW 270, 1.05.1873, ff. 293-94).

That the hydrological properties of the Maltese geological deposits were not properly understood in the latter half of the nineteenth century is again illustrated through a description of the water situation in Gozo. A despatch dated August 1876 shows how until this point, it was the perched aquifer resources of the island which were being utilised. Apart from the construction of the aqueduct three decades earlier,[28] and the diversion of the Mġarr *giardino fontane* (GO16) to the quay area of the Mġarr harbour in the 1840s (NAM. Descriptive Plans of the Crown Property, 1861, f. 54), very little other water-related interventions appear to have otherwise taken place. In 1876, evidence for the presence of ground-water deposits in Gozo was acknowledged and the intention of the British authorities to start the excavation of a series of shaft-wells was made known (NAM. GOV 01.3/14, Despatch To, no. 120, 1876, ff. 467-69). By October 1876 a water source had been identified at Xagħra, and plans were underway to construct a pipeline in order to convey this source to Rabat (Gozo) (NAM. GOV 01.3/14, Despatch To, no. 166, 1876, ff. 515-21). It appears that the priority of the central government during this period was to make a water supply available to a wider spectrum of the Gozitan population.

Figure 4.9 Map excerpt from Malta Survey Sheet 72 showing the course followed by the old perched aquifer galleries of the Ghajn Qajjed area (RA19) at Rabat (Malta), which in the opening decades of the seventeenth century were connected to the Wignacourt Aqueduct project. Circled in red is the Għajn Qajjed pumping station commissioned in 1903. Source: NAM. Government Property Survey Sheet Links, Sheet 72.

[27] The new springs increased the water-yield of the Wignacourt aqueduct system by 700 m^3 per day. See Zammit T. 1924, 17; Micallef *et al* 2004, 200.

[28] The aqueduct was in 1876 feeding 6000 of Gozo's 18,000 inhabitants with water. See NAM. GOV 01.3/14, Despatch To, no. 120, 1876, ff. 467-69.

In late 1876 plans were made public for the construction of a number of covered tanks in Malta having the collective capacity of nearly seventy million gallons, which in years of abundant rain could be filled with the increased water flow originating from the springs. The intention behind the building of these reservoirs was that of harvesting excess rainwater originating from areas possessing perched aquifer deposits, which would otherwise have gone to waste. A connection between the Wignacourt and the Fawwara aqueducts was in addition planned in order to allow the conveyance of water from one aqueduct system to the other, should it be necessary (NAM. GOV 01.3/14, Despatch To, no. 166, 1876, f. 515).[29] It was also during the 1870s that cast iron pipes started being installed in various localities of Malta. The installation of four miles of such piping was aimed at improving water pressure in the Valletta and Floriana localities in order to allow roof-level tanks to fill up with water (NAM. GOV 01.3/14, Despatch To, no. 166, 1876, f. 515).

Precipitation levels were again deficient in the 1878-1881 period.[30] This must have conditioned the government in the 1880s to make available the necessary financial resources for the upgrading of reservoir infrastructure at Valletta, Floriana, the Three Cities and elsewhere. Provision was made to increase their storage capacity to over a million cubic metres of rain harvested water (Micallef *et al* 2004, 200). During this period cistern water was already known to be inferior in quality to perched aquifer water, especially since it is more prone to accidental biological contamination, a matter well illustrated by the Sutherland report of 1867 (Chadwick 1884, iv; Mercieca 1931, 13; Micallef *et al* 2004, 200-1).[31]

1881 was another dry year and this must have compelled the British Administration of the Islands to explore further possibilities to increase the local fresh water-yield. Between 1883 and 1894, the sanitary Engineer Osbert Chadwick took stock of the local water situation. The publication of various reports on the water resources of Malta and a further understanding of the dynamics behind

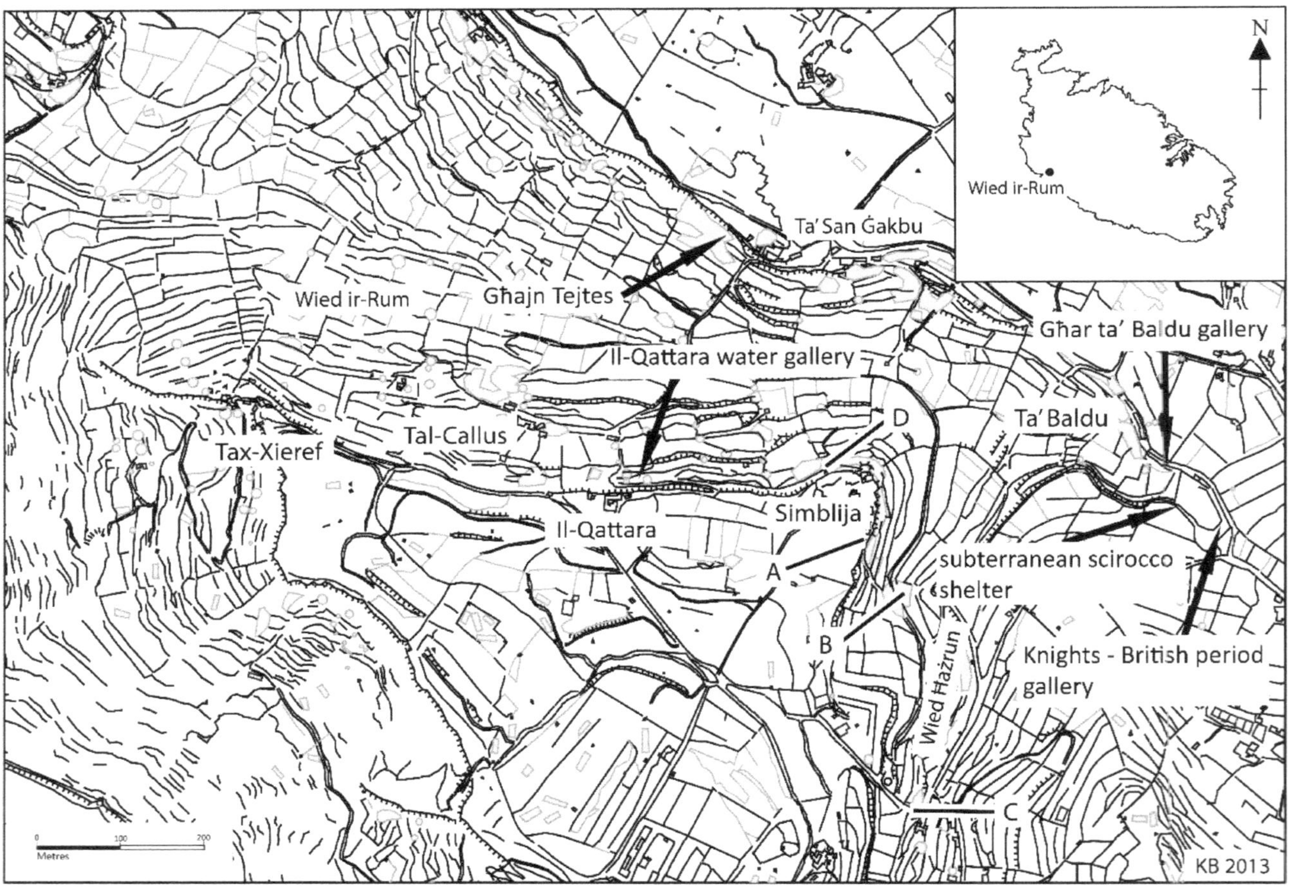

Fig. 4.10 Map of the Wied ir-Rum (RA56) and Wied Ħażrun (RA55) valleys showing the location of various water sources in these valleys. Il-Qattara (RA32), Ta' Baldu (RA43) and the Għajn Tejtes (RA20) springs originate from within perched aquifer galleries. "A" is a water cistern at Simblija (RA41), whilst "B", "C" and "D" indicate the location of other perched aquifer galleries. Topographical map image adapted from MEPA Map Server data.

[29] Writing in 1884, however, O. Chadwick claims that there still was no connection between the Fawwara and the Wignacourt aqueduct systems. See Chadwick 1884, i.

[30] The decade between 1880 to 1890 was again mostly rainfall deficient. See Morris 1952, 7.

[31] On water contamination during the 1880s see Caruana Scicluna 1888.

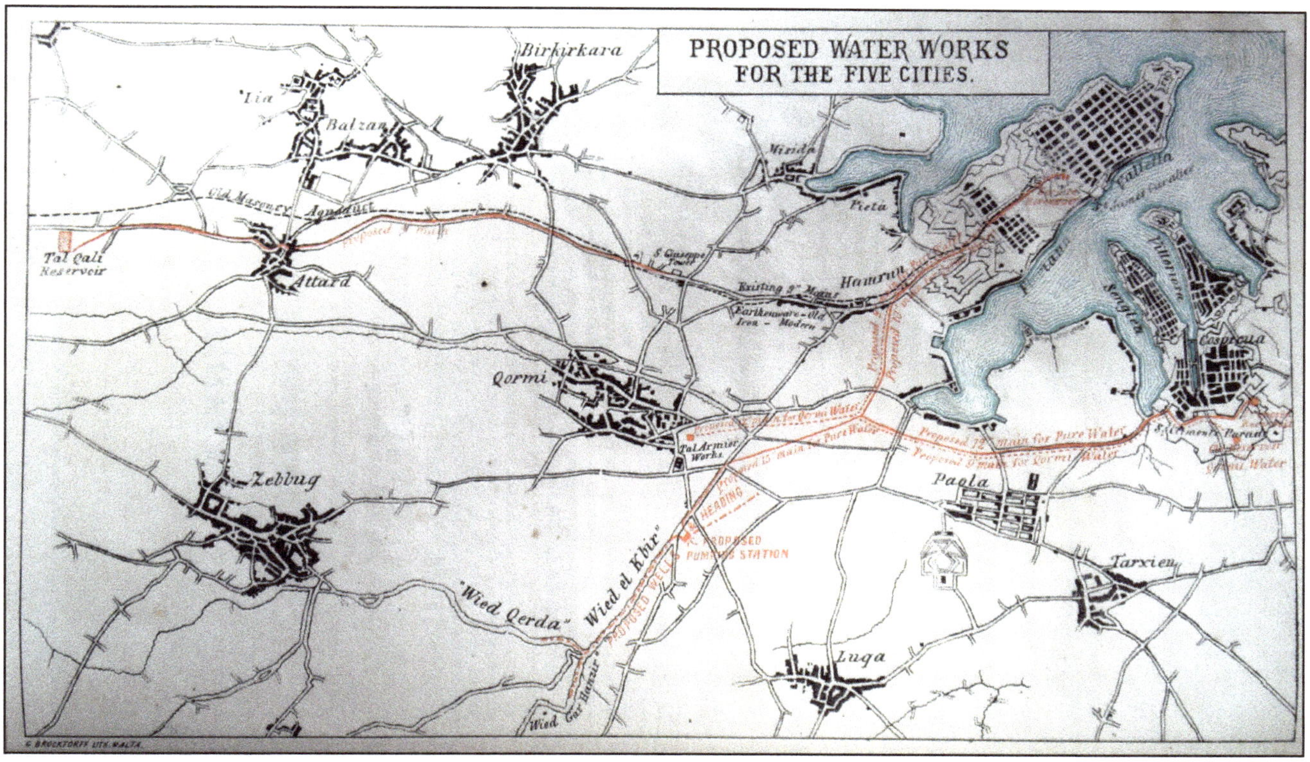

Fig. 4.11 Water works for the five cities as proposed by Osbert Chadwick in 1884. Source: Chadwick 1884, xi.

the workings of the mean-sea-level aquifer can be attributed to Chadwick's initial investigation (Zammit T. 1924, 26-27). A pressing need for a better water supply became evident when in 1883 it was established that most inhabitants of the Grand Harbour area received ca. 27 litres of water per capita per day while a large number of inhabitants living elsewhere did not even receive this amount (Fig. 4.11).[32] By 1894 there was the realisation that water pumped up from a mean-sea-level gallery at Wied il-Kbir (MA12) in the Qormi (MA10) district was of a high quality.[33] During this same year, measures by which water obtained from the Wignacourt Aqueduct springs could be purified through the use of sand filtration systems were also proposed (Chadwick 1894B, 3-4).

Efforts to increase the water supply resources of Malta and Gozo continued in earnest in the mid-1890s, especially considering the fact that the rainfall in the year ending 31st August 1896 was one of unusual scarcity (Chadwick 1897, 3). In 1896 attempts to increase the water supply of Gozo were well underway. These consisted in the excavation of galleries at Wied Żewwieqa (GO22) to the north of the Mġarr (Gozo) settlement. Trial shafts were also excavated in the Marsalforn hinterland in order to assess the water-yielding potential of the area. As the retrieved water was brackish in nature, the project was abandoned (Chadwick 1897, 12-13; Zammit T. 1924, 32*)*. On the other hand, shafts excavated at Wied Ħanżira close to Mġarr ix-Xini (GO17) provided a sufficient amount of potable water and prompted the excavation of galleries there (Zammit T. 1924, 32).

As a concluding comment, the risk posed by drought and population increase during the nineteenth century was initially averted through the diversion of spring water originating from already extant Upper Coralline Limestone perched aquifer galleries. At a subsequent stage, the excavation of wells tapping both perched and mean-sea-level aquiferous sources also buffered the effect of drought and increased fresh water availability. Throughout the nineteenth century, it was the tapping of ground water sources which provided resilience in an otherwise high risk situation.

4.3 The Twentieth Century

The turn of the century saw a more organised exploitation of the mean-sea-level aquifer as well as the commissioning of pumping stations for the conveyance of the intercepted water sources. The Għajn Qajjed pumping station was commissioned in 1903 specifically for the conveyance of water for consumption by troops stationed at Mtarfa (Mercieca 1931, 12) (Fig. 4.9). This spring water, mostly originated from the already extant perched aquifer galleries in the area including Għajn Qajjed (RA19) (Morris 1952,

[32] This average had regional differences. The inhabitants of the *Manderaggio* area in Valletta received less than fourteen litres *per capita* a day. See Chadwick 1884, ii.

[33] Chadwick described the subterranean aquifer of the Wied il-Kbir area as yielding, "almost the purest spring water which is to be found in Malta". See Chadwick 1894A, 6. The mean-sea-level galleries at Wied il-Kbir were to be extended further in the opening decades of the twentieth century. It is interesting to note that in 1894 Chadwick had already observed that after heavy pumping from the Wied il-Kbir galleries the salinity levels of the water increased. See Chadwick 1894A, 7.

17).[34] Water extraction levels from the Wignacourt springs for the second decade of the 1900s are given in Fig. 4.12. In 1909-1910 the Wied il-Kbir (MA12) and the Wied is-Sewda (MA13) wells were linked by means of a subterranean gallery, thereby increasing the local water production by 4550 m³ per day. During this same period, two serious outbreaks of typhoid fever were traced to an organic contamination of the perched aquifer sources. It was such instances which probably led to the commencement of water sterilisation via the process of chlorination as from 1909 (Zammit T. 1924, 43; Micallef *et al* 2004, 201). In order to further reduce the possibility of contamination of the water flowing through the stone canals of the Wignacourt Aqueduct, iron pipes were installed throughout the aqueduct system in 1910-1911.[35] As an excess of organic content was detected in the waters of the Għeriexem (RA26) and the Busugrilla (RA7) perched aquifer galleries, these were isolated from the Wignacourt Aqueduct system during this period (Fig. 4.13).[36] The contamination registered at Għeriexem and Busugrilla was probably due to organic seepage into the gallery system through shaft-wells.[37] As a measure to reduce the contamination of water originating from Għajn Klieb (RA18), Għajn Qajjed, Saqqajja (RA40) and a number of other old perched aquifer galleries, cast iron pipes were at times installed in the galleries' interior (Morris 1952, 57).[38] The use of the Ta' Dekozzu (RA45) and Ta' Liemu (RA47) springs was discontinued due to these being prone to bacteriological contamination (Zammit T. 1924, 41).

It is noteworthy that the already extant perched aquifer gallery of Burnoħħala (MG7) was extended by a few metres in 1908 (Fig. 3.6). The early twentieth-century intervention on this much older water system appears to have primarily concentrated upon the inner reaches of the gallery. This was done in order to assess if its water-yield, which was already supplying the Mġarr (Malta) and the neighbouring Binġemma (MG19) settlements, could be increased to also supply the St Paul's Bay and Mellieħa areas (Morris 1952, 17, 19). Between 1910 and 1914, two other galleries at Fawwara (SI6) and Annunziata (SI7), both

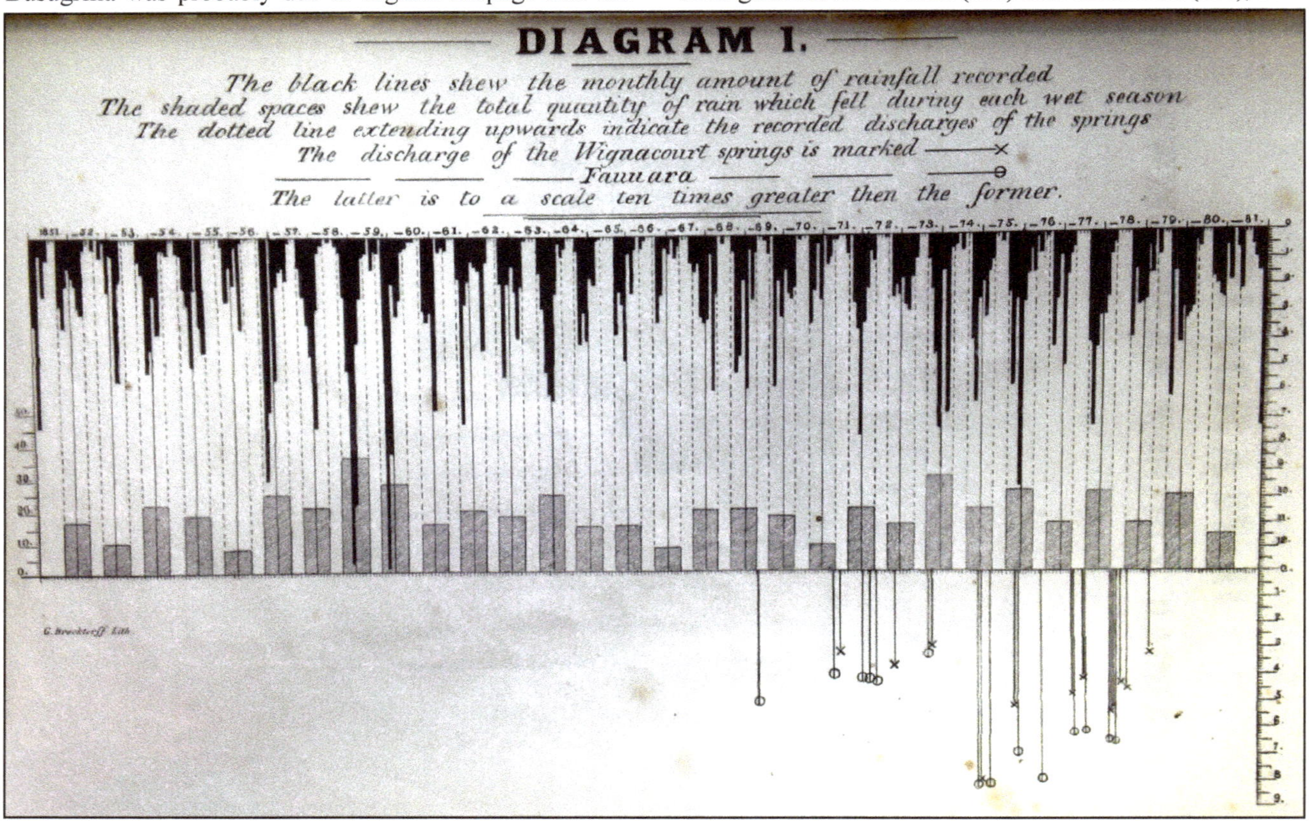

Fig. 4.12 Table showing the discharge levels for the Wignacourt springs in the early 1880s. Source: Chadwick, 1884.

[34] Plans involving the connection of the old Għajn Qajjed perched aquifer gallery with one or more of the new galleries located in between Rabat (Malta) and Dingli never materialised.

[35] The utilisation of sealed water pipes had already been suggested back in 1894 when it was realised that spring water was practically sterile when it left the rock but on its way to Valletta acquired ca. 200 bacteria per cubic centimetre. See Chadwick 1894A, 7; Chadwick 1894B, 6-7; Chadwick 1897, 9; Zammit T. 1924, 33-34. In 1930 the Wignacourt aqueduct was still in use, but only for the supply of second class water intended for use as irrigation and street cleansing purposes. See Mercieca 1931, 5, 19.

[36] The contaminated nature of the water yielded by the Għeriexem perched aquifer gallery had already been recognised in 1893. It appears that Għeriexem was for a while isolated from the Wignacourt springs during this period. See Morris 1952, 57.

[37] Writing in 1952, T.O. Morris identified shafts excavated into Upper Coralline deposits as a source of major aquifer contamination and proposed that the tops of all private shafts be walled-up as a measure to prevent access to surface washings. See Morris 1952, 84. Similar conclusions had already been reached by M. Blondel des Croisettes in the later seventeenth century when reporting on the water situation and quality of the Saqqajja and the Għeriexem perched aquifer galleries. See chapter five, section 5.3.

[38] The cast iron pipes installed within two perched aquifer galleries at the Għajn Tuffieħa *feudo* (MG15) also probably date to this period. See chapter six, section 6.1.9.2.

in the territory of Siġġiewi, were similarly extended by around 82 m in an attempt to increase their water-yield (Morris 1952, 19).[39]

The years from 1913-1928 were again mostly rainfall deficient. An insufficient availability of water compelled the authorities in charge to press on with further water-related exploratory campaigns. On completion of the Tal-Ħlas pumping station in 1924, also in the territory of Qormi, the Wied is-Sewda (MA13) pumping station was decommissioned.[40] Likewise, this tapped a mean-sea-level water source. An increasing demand for potable water during the course of World War I also led to the extension of existing mean-sea-level galleries. Works on the excavation of the Wied Dalam mean-sea-level gallery in the territory of Birżebbuġa dating 1920, however, only yielded a poor water output of an inferior quality (Mercieca 1931, 23; Morris 1952, 22; Micallef *et al* 2004, 201). In direct response to the "failed" Wied Dalam project, the Buskett springs were connected to the Fawwara ones by means of a pipeline (Morris 1952, 23).[41] A further extension of the Wied il-Kbir (MA12) and Wied is-Sewda galleries was undertaken between 1920 and 1939. However, this did not result in a marked improvement in the water-yield of this mean-sea-level system.

In 1935 plans for the commissioning of deep shafts boring below the mean-sea-level table were abandoned, only to be revived during the course of 1938. The prospect of a deep boring project was during this period overshadowed by the possibility of causing mean-sea-level aquifer contamination, and was eventually scrapped (Morris 1952,

Fig. 4.13 A 1919 map of Malta indicating which perched aquifer springs originating from galleries were by this period disconnected from the aqueduct system. Source: Robertson 1919, Appendix viii.

[39] For a 1919 plan showing the location of perched aquifer springs and galleries see Fig. 4.14.

[40] Works on the Tal-Ħlas pumping station commenced in 1916 and by August of the same year could already augment the already available water resources of the island with a limited, but much necessary water supply. See Mercieca 1931, 22.

[41] All these springs originate from old perched aquifer galleries. Plans to connect the perched aquifer galleries from the Ta' L-Isqof valley area to the Buskett – Fawwara system never materialised.

26, 29). This decision possibly influenced the commissioning of the Wied il-Għasel mean-sea-level gallery (MO6) at Mosta in 1938, an undertaking which increased Malta's ground-water-yield by 1900 m³ per day (Morris 1952, 62-63; Mangion *et al* 2004, 202).

The impermeable nature of the Upper and Lower Globigerina Limestone strata can be gleaned through the further water exploratory campaigns held during 1943. Boreholes were during this period drilled into Globigerina Limestone layers at a depth of fifteen metres below sea level. Twenty-three such shafts were excavated into Upper Globigerina deposits, whilst fourteen others were drilled into Lower Globigerina strata. The low water-yield of these galleries led to the abandonment of most and must have been instrumental in conditioning future exploration campaigns geared towards the exploitation of mean-sea-level aquifer sources (Mangion *et al* 2004, 202). As a matter of fact, 1944 saw the excavation of other mean-sea-level galleries at Ta' Qali (AT3) and the attempted extension of the Tal-Ħlas gallery system.[42] Between 1955 and 1965 other perched and mean-sea-level galleries were excavated with the principal works being concentrated upon the Ta' Kandja (MQ1) mean-sea-level galleries located in the Siġġiewi territory (Mangion *et al* 2004, 202).[43]

The "Survey of Private Groundwater Abstraction Sources" commissioned in 1943, contains data on privately-owned spring or well-type water sources found in the archipelago.[44] A number of datasheets pertaining to this survey make a distinction between the older wells and those excavated in the opening decades of the twentieth century, but provide little or no tangible evidence on the perched aquifer galleries, which are simply referred to as "old springs". They make no reference to excavated tunnels tapping the perched aquiferous sources.[45]

Since the excavation of the Ta' Kandja galleries which led to the extraction of a steady water supply from within, it has been commonly acknowledged by the Malta Resources Authority[46] and the Water Services Corporation,[47] that the maximum water extraction capacity from the mean-sea-level water sources had been reached and indeed surpassed on a number of occasions. This situation was further exacerbated in the 1970s with the excavation of another one-hundred and fifty boreholes in central and southern Malta in an attempt to increase water-yields.[48] The resulting increase in the salinity levels of this aquifer resource noted in the past decades has been counteracted through the application and development of reverse osmosis technology within a local context (Mangion *et al* 2004, 203-4). For the past half a century, water obtained from the archipelago's perched aquifer sources is no longer fit for domestic consumption. This is mainly due to high nitrate contamination originating from agricultural-related activities. Perched aquifer water is currently classified as a second-class source and is on the whole being used in agriculture-related activities.

4.4 The Agrarian Context

An overview of water utilisation as an aid to agriculture is best determined through the utilisation of the Descriptive Plans of the Crown Property *cabrei* dating to between 1861 and 1867. The emerging picture demonstrates that Upper Coralline Limestone perched and mean-sea-level aquiferous sources were both utilised as an aid to agriculture.[49] As indicated in sections 4.1 and 4.2, a sizeable number of the Upper Coralline Limestone springs originated from within subterranean galleries, the majority of which predate the British period.[50] Prior to the period under discussion, many of these springs were providing *giardini* areas with a perennial water supply and continued to do so throughout the British period.[51] Some perched aquifer gallery systems were during the course of the seventeenth, nineteenth and twentieth centuries also connected to aqueduct systems in order to supply water deficient settlements in south-east Malta with spring water (See chapter five, section 5.2.4).[52]

Mean-sea-level aquiferous sources were tapped both for agrarian and domestic utilisation in the Qormi and Marsa (MA10) basins as well as areas of the Burmarrad locality and Pwales, where the purpose was primarily agrarian (NAM. Descriptive Plans of the Crown Property, 1862, ff. 99, 102, 125-26, 145-46, 133; NAM. Descriptive Plans of

[42] The precarious water situation during the early 1940s is illustrated by the fact that continued heavy demand for water by shipping was straining Malta's water supply system to the utmost. The excavation of two boreholes at Ta' Qali (AT3) in 1944 led to the accidental interception of an "ancient system of underground channels" – a discovery of probable archaeological interest of which no further data are given. See Morris 1952, 34, 37.

[43] Since 1942, E.B. Baily had already implied that the mean-sea-level aquifer contained "extensive undersea-level untapped resources of usable water". See Morris 1952, 31.

[44] This database is now in the possession of the Malta Resources Authority.

[45] See also chapter one, section 1.9.

[46] This is the national entity in charge of managing Malta's water resources. See Mangion *et al* 2004, 206-7.

[47] This is the national entity in charge of Malta's public water supply and distribution.

[48] The newly drilled boreholes adversely affected mean-sea-level gallery production and drastically decreased the water recharge of the mean-sea-level aquifer system. See Mangion *et al* 2004, 203.

[49] This book focuses on water capture systems present within Malta's rural landscape. In sections 4.1 and 4.2, there is an apparent deviation from this norm because these take into consideration rural water capture systems which were already in existence during the Knights period, but which in the timeframe under discussion were utilised for domestic consumption.

[50] For water sources originating from within Upper Coralline Limestone perched aquifers utilised for agrarian purposes in Gozo and Malta see NAM. Descriptive Plans of the Crown Property, 1861, ff. 2-3, 14, 31, 34, 54, 57, 63, 80, 82-83, 109, 113, 149, 170, 179; NAM. Descriptive Plans of the Crown Property, 1862, ff. 162, 163-65; NAM. Descriptive Plans of the Crown Property, 1866, ff. 199, 202-4, 209, 215-16, 218, 220-22, 267-68, 278, 295-96, 375; NAM. Descriptive Plans of the Crown Property, 1867, ff. 459-63, 464-67, 468-78, 480-81, 486-87, 491, 496-97, 500-1, 505-6, 513, 523, 529-40, 541, 557-64, 566-67, 606-7, 620-45, 679, 681-82.

[51] For a definition of a *giardino* agricultural framework see chapter five, section 5.0. For a list of *giardini* and a description of their water sources, even during the British period see chapter six, sections 6.1.3 to 6.1.10.

[52] The early nineteenth-century Għajn Qajjed pumping station, similarly obtained its water supply from the nearby Għajn Qajjed (RA19) perched aquifer gallery. See Mercieca 1931, 12.

the Crown Property, 1866, ff. 204, 341-44). Crown property data for properties located in these areas frequently mention the availability of *spiere* or shaft wells and specify that water was being lifted to the surface by means of watermills (NAM. Descriptive Plans of the Crown Property, 1862, ff. 125-26, 145-46, 133). A number of agricultural estates in the Qormi and Marsa areas also contained reservoirs within which the retrieved mean-sea-level aquifer water was stored (NAM. Descriptive Plans of the Crown Property, 1862, f. 146). In some of these properties rain water was also harvested and stored in cisterns for later usage as irrigation water (NAM. Descriptive Plans of the Crown Property, 1862, f. 146).

Moreover, the Crown Property documentation also appears to make reference to attempts at harvesting small quantities of spring or rain water not within cisterns or reservoirs, but within either *ad hoc* shallow rock basins or small purposely built water collection recipients (NAM. Descriptive Plans of the Crown Property, 1861, ff. 26, 178; NAM. Descriptive Plans of the Crown Property, 1867, ff. 454, 459-63, 673). This applies to arable land located over both Coralline Limestone and Globigerina Limestone terrain. The captured water, even though in rather small quantities, appears to have been used as an aid to agriculture.

4.5 Concluding Remarks

The data gathered for the British phase of Maltese history unequivocally illustrate how until the middle of the nineteenth century, the archipelago's water framework was directly reliant on the hydrological infrastructure of the Knights' period. Until the third decade of the nineteenth century, no alteration works had been carried out on the Wignacourt aqueduct system. The aqueduct's water carrying capacity was only increased in 1841 when springs from the Buskett area of Rabat (Malta) were added to this water distribution system. During this period, the only harbour settlements connected to the aqueduct supply remained Valletta and Floriana. The Three Cities and most of the other settlements in southern Malta were still dependent on the harvesting of rainwater in cisterns.[53] Apart from Qormi, Marsa (MA10) and other minor areas, there is the possibility that since the sixteenth century, Birgu (BI1) had some access to mean-sea-level aquiferous sources.[54] The utilisation, during this period, of Globigerina Limestone perched aquifer resources, still has to be satisfactorily assessed. Nonetheless, evidence for the tapping of these aquifers at Bulebel and Paola is presented in section 4.1.

Fig. 4.14 A 1919 plan showing the location of perched aquifer springs and galleries. Source: Robertson 1919, Appendix vii.

[53] See chapter four, section 4.1.

[54] Bosio 1594-1602, 28-31.

Spells of periodic drought and rapid demographic growth further accentuated the water shortage Malta experienced from the 1840s onwards. The effect of water shortage on the smooth administration of an island fortress was a critical issue and led to the eventual upgrading of the archipelago's infrastructure. Previously untapped data in the form of Despatch communications between Malta and London gave a new insight into the water crises experienced throughout most of the nineteenth century. Despatch communication also provided me with a primary source of information on the sequence of events which eventually contributed to the detection of the mean-sea-level aquifer in the latter part of the century.[55] Even though Upper Coralline Limestone perched aquifer galleries were certainly in use during this period, primarily for agrarian-related usage, data presented in chapters six and seven amply demonstrate how this technology had been operative during the Knights' period. With the exception of objective (v) dealing with the context of the Maltese *giardino* agricultural framework, all the remaining objectives declared at the outset in chapter one, section 1.1, are touched upon in this chapter.[56]

[55] See chapter four, sections 4.2 and 4.3.

[56] For a reflection on the main objectives of this publication and an evaluation on whether these were successfully addressed see chapter eight, section 8.2.

Chapter Five

The Maltese Hydrological Landscape during the Knights' Period

5.0 Introduction

This chapter is the first of a two-part study investigating the hydrological practices and strategies adopted in the Maltese archipelago during the Knights' period between 1530 and 1798.[1] Section 5.1 takes stock of the known cartographic evidence for the Early Modern period, whilst an analysis of the hydrology-related situation as presented by Fra Gian Francesco Abela in 1647, is attempted in section 5.2. Section 5.3 deals with other water-related data for the Early Modern period. Particularly revealing is documentation, dating from the second decade of the seventeenth century onwards, on the Saqqajja (RA40) and Għeriexem (RA26) *fontane* and a water source at Wied l-Isqof (RA25), which provided the Ta' Brija estate (RA44) in the territory of Rabat (Malta) with a perennial water source.

Attempts were made by the author of this book to locate and investigate all water sources mentioned in this chapter. Many of the sources quoted below recurrently refer to *giardini* (sing. *giardino*) cultivations. The lack of an English equivalent term posed a problem of translation and *giardino* has therefore been kept in its original form throughout. Possible but inexact modern equivalents include orchards, plantations, gardens, market gardens and even small-holdings.

Water distribution and allocation for agricultural usage falls outside the scope of this study.[2] For this reason, these modes of water management are not represented in the case studies presented hereunder. As a concluding comment, in order to facilitate place-name readability, which due to Late Medieval and Early Modern unstandardised norms are in the original documentation frequently transcribed in differing ways, reference to the comparative table of toponyms (Appendix 2) is recommended.

5.1 Hydrology-Related Cartographic Evidence for the Early Knights' Period

The first known late medieval representation for the archipelago is the twelfth-century map by Al-Idrisi, the cartographer of Roger II. Malta is here included in a map of Sicily and other neighbouring lands (Agius-Vadalà and Ganado 1986, 2). Al-Idrisi describes Malta as a sizeable island with an eastward facing harbour. Other sea charts and portolans merely indicate the position of Malta together with its distance from other principal Mediterranean ports without including a more detailed map of the archipelago (Agius-Vadalà and Ganado 1986, 2; Gambin 2008, 127-28, 152-55). The first portolan known to contain a separate map of Malta is that by Piri Reis and dates to the 1520s (Agius-Vadalà and Ganado 1986, 2) (Fig. 5.1). This shows what appear to be two water drainage areas north of Mdina. It is possible that the larger of the water associated features represents the Fiddien (RA11) and Il-Qlejgħa (RA57) valleys water discharge towards Burmarrad, but due to the schematic nature of the map, this is difficult to determine with certainty.

Relevant water-related data from the map of Malta included in Quintin's description of Malta, published in 1536, are discussed in chapter seven, section 7.2. The next

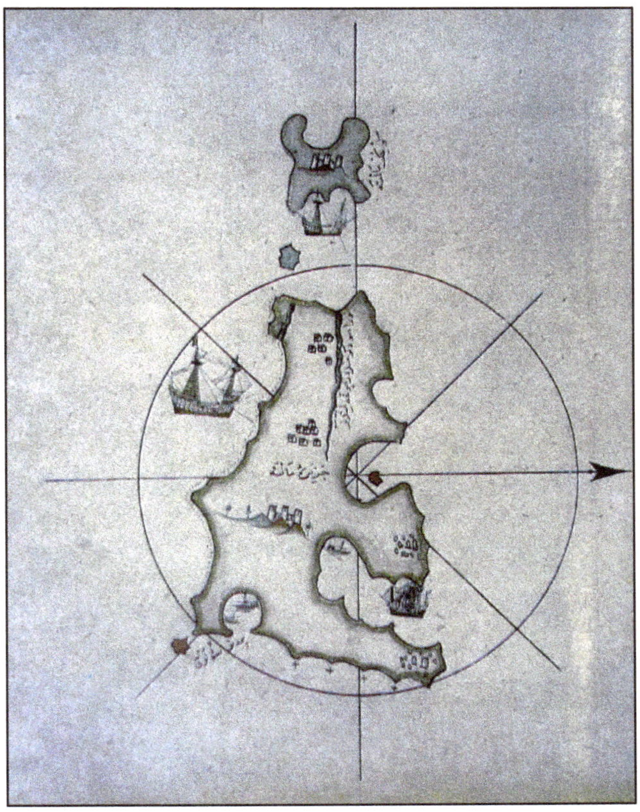

Fig. 5.1 The Piri Reis portolan of the 1520s. Source: Agius-Vadalà and Ganado 1986, 2.

[1] Hydrological data as revealed by the Knights' period *cabrei* manuscripts is discussed separately in chapter six.

[2] See chapter one, sections 1 and 1.1.

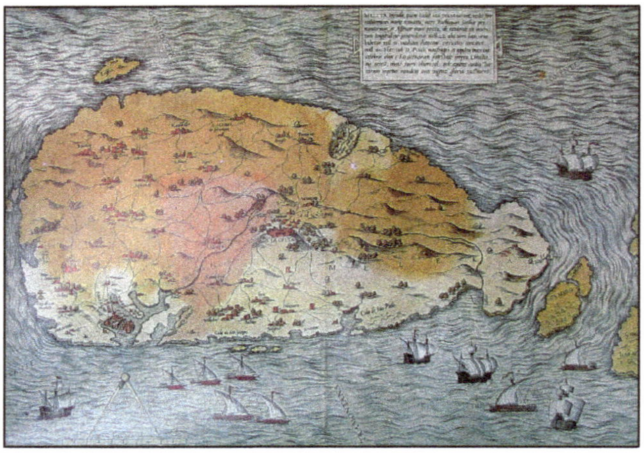

Fig. 5.2 Antonio Lafreri's map of Malta published in 1551. Source: Agius-Vadalà and Ganado 1986, 6.

relevant cartographic representation of Malta is a map by Antonio Lafreri printed in 1551. It depicts the islands of Malta, Comino and Cominotto and a small section of the south-facing Gozitan coast (Fig. 5.2). The map reproduces fifty-one toponyms, a rather extensive road network and several water-related features (Agius-Vadalà and Ganado 1986, 6).[3]

The principal significance of the map lies in the fact that it represents a well-defined valley system, located on the west coast of the island, wedged in-between the settlements of *Tartarni* and *Migiaro*. The latter probably demarcates the location of the present-day Mġarr settlement (Fig. 5.3). Marked as *vallone giardini*,[4] the spatial location of the only marked valley in this section of Malta, corresponds to that of Wied ir-Rum (RA56), an extensive valley system located west of Rabat (Malta) (Agius-Vadalà and Ganado 1994, 8).[5] Depicted on the map is what appears to be a watercourse passing through the valley bed. Trees are shown flanking the southern valley side. Wied ir-Rum is not mentioned by name, but the valley shape and its location on the western coast of Malta between the settlement of Tartarni and Mġarr leaves little doubt as to its identity (Figs. 3.6, 4.3).

To the immediate north of Wied ir-Rum is another water source marked as *Fonte* (spring). The schematic nature of the map makes it difficult to correlate with certainty to any specific water source known to exist in this part of Malta. The feature might, however, represent Ġnejna Valley (MG24), in which a number of perennial springs are known to exist. Another water source, similarly marked as *Fonte*, is shown to be approximately located at Marfa, close to Ċirkewwa (Fig. 5.4). This was identified as originating from the Wied Musa area close to Latmija (Agius-Vadalà and Ganado 1994, 7), and can possibly be associated with Għajn Tuta (ME2) – a major water-yielding source in the area. The approximate location of

Fig. 5.4 Detail from the Lafreri map of Malta of 1551. Apart from a water source to the north of the valley marked as "Vallone Giardini", another spring in the Ċirkewwa area of Malta can be tentatively identified with Għajn Tuta (ME2). Source: Agius-Vadalà and Ganado 1986, 6.

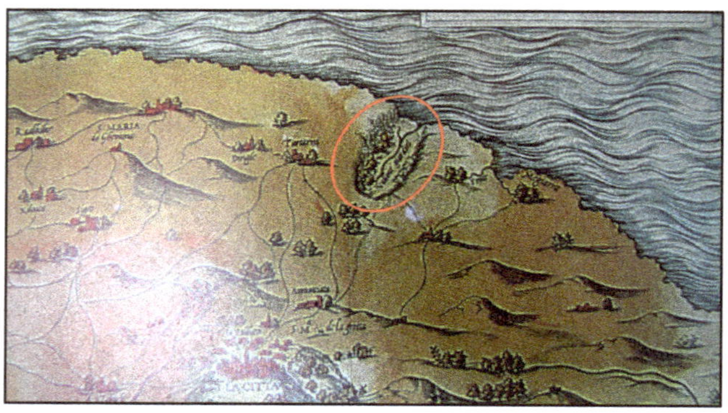

Fig. 5.3 Detail from the Lafreri map of Malta of 1551. The well-defined valley on the west coast of Malta is tentatively associated with Wied ir-Rum (RA56). Source: Agius-Vadalà and Ganado 1986, 6.

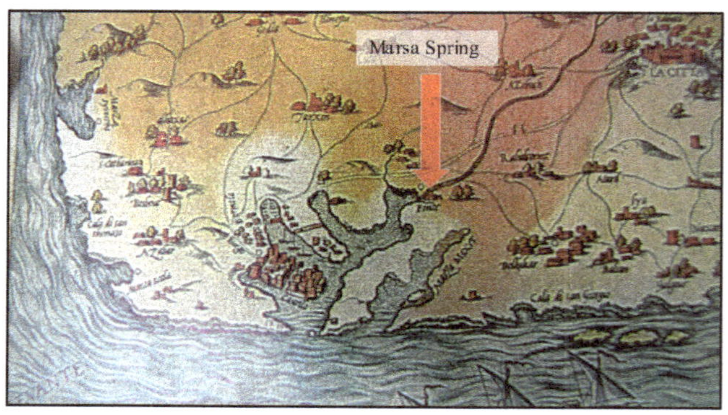

Fig. 5.5 Detail from the 1551 Lafreri map of Malta showing the approximate location of the Marsa spring (MA8). Source: Agius-Vadalà and Ganado 1986, 6.

[3] Agius-Vadalà and Ganado, 1986, 6.
[4] *Vallone* is the Italian equivalent for "valley".

[5] Agius-Vadalà and Ganado associate the *Vallone Giardini* location with the neighbouring Mtaħleb *giardini* (RA35).

Fig. 5.6 Anonymous map of Malta dating to the 1560s. Detail of the northern section of Malta. Source: Agius-Vadalà and Ganado 1986, Fig. 7.

the Marsa spring, in the inner reaches of the Grand Harbour, is also given and captioned as *Fonte* (Fig. 5.5). A stretch of land which approximately equates to the area covered by the Marsa watershed was in this map given a distinct coloration.

Fig. 5.7 Anonymous map of Malta dating to the 1560s. This shows the location of the Marsa spring (MA8), a water source close to Mtaħleb and the spring at Marfa (ME2). Source: Agius-Vadalà and Ganado 1986, Fig. 7.

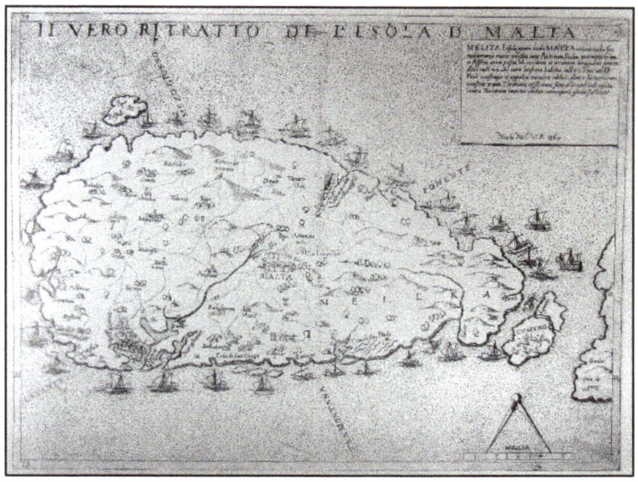

Fig. 5.8 Map of the Maltese archipelago by Nicolò Nelli dating to 1565. Source: Agius-Vadalà and Ganado 1994, Map 1.

The same water sources described above, including the *Vallone Giardini*, which the present writer's research tentatively associates with Wied ir-Rum, are also given in other cartographic representations of Malta. These consist of an anonymous map of Malta published around 1560 (Figs. 5.6, 5.7) and the Nicolò Nelli map dating to 1565 (Figs. 5.8, 5.9).[6] In another Great Siege map of 1565, possibly the work of Tommaso Barlacchi (Fig. 5.10), there are once again illustrated the valley and spring opening

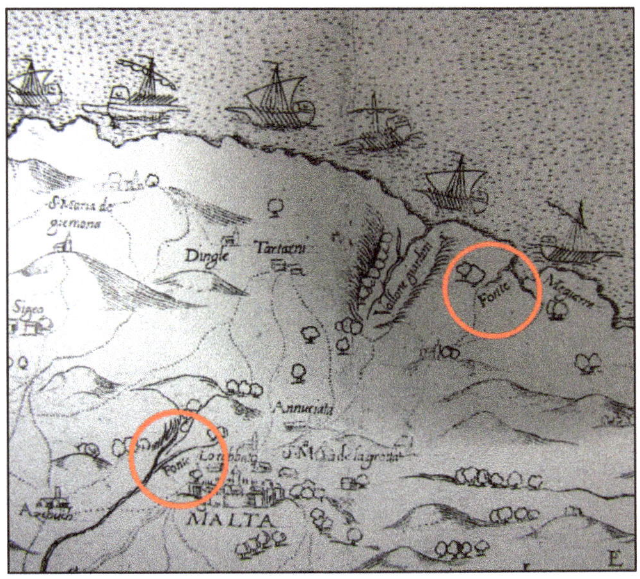

Fig. 5.9 The Nicolò Nelli map of 1565. Detail of central and western Malta showing the presence of water sources close to the fortified settlement of Mdina and *Megiarro* (Mġarr). Source: Agius-Vadalà and Ganado 1994, Map 1.

[6] The Nicolò Nelli map was modelled on the map of Malta by Lafreri, which dated to fourteen years earlier. See Agius-Vadalà and Ganado 1994, 3, 5.

Fig. 5.10 Map of Malta and Comino dating to 1565, attributed to Tommaso Barlacchi. Source: Agius-Vadalà and Ganado 1994, Map 13.

Fig. 5.11 Map detail of the western coast of Malta showing the location of an unnamed valley containing a water course. Tommaso Barlacchi (?) 1565. Source: Agius-Vadalà and Ganado 1994, Map 13.

Fig. 5.12 Map detail of the northern coast of Malta. Also shown is the location of the Marfa spring (ME2; circled in red), which on the map is marked as *fons*. Tommaso Barlacchi (?) 1565. Source: Agius-Vadalà and Ganado 1994, Map 13.

onto the west coast of Malta (Fig. 5.11). The spring close to Mġarr (Malta) is also illustrated, together with the springs at Marfa and Marsa (Fig. 5.12). Another anonymous cartographic representation of Malta, also dating to 1565, again includes the above-mentioned water sources (Agius-Vadalà and Ganado 1986, 45). The *Giardini Vallone* feature on the west coast of Malta are in this cartographic source not marked as a distinct valley system, but as an open area through which a water source flows (Fig. 5.13). The 1565 map of the Maltese archipelago by Domenico Zenoi (Fig. 5.14) is another illustration of water features given by the Lafreri map of 1551.

Another map of Malta produced by an anonymous cartographer, probably dating to 1551, indicates the whereabouts of the Marsa spring, the Marsa *giardino* (MA9) and a *giardino* in the landward side of the Petra Longa area (Agius-Vadalà and Ganado 1986, 8) (Figs. 5.15, 5.16). The Marsa *giardino* and Marsa *fonte* are again indicated in a coeval cartographic source by Giacomo Gastaldi (Agius-Vadalà and Ganado 1986, 10) (Figs. 5.17, 5.18). Of particular interest is the representation of another spring labelled *La fonte* (MO5), located to the east of Mdina (Agius-Vadalà and Ganado 1986, 10) (Fig. 5.17).

Dating to 1554 is a manuscript and nautical chart of the Maltese Islands by Battista Agnese (Agius-Vadalà and

Fig. 5.13 Anonymous map of Malta and Comino dating to 1565. Source: Agius-Vadalà and Ganado 1994, Map 44.

Ganado 1986, 2) (Fig. 5.19). This map not only marks the location of the Marsa spring but also two other water sources on the islands of Gozo and Comino respectively (Fig. 5.20). The origins of the Gozitan spring are marked as being close to the north shores of the island, with water being discharged into the sea on the south coast. Even though difficult to establish with certainty, this water feature might be tentatively associated with the Wied tal-Lunzjata valley system (GO21), where a perennial water source still flows through the valley bed. The Comino spring appears to be discharging water into the coastal area known as Ta' Bejn il-Kmiemen.[7] The schematic nature of the map makes it difficult to reconcile the exact location of the Comino water source with any known landscape features. As explained in chapter two, section 2.6.1, surface water availability on Comino is largely constrained. The spring marked in the Agnese map must, therefore, represent a seasonal surface runoff source, which would only flow over the island's surface for a limited amount of time following a heavy downpour.

A map of the Maltese Islands published by Henricus Petri in 1561, on the other hand, only makes reference to the Marsa *hortus* (Agius-Vadalà and Ganado 1986, 14) (Fig. 5.21). The presence of the Marsa spring is detailed in an anonymous map of 1563 depicting the Grand Harbour area and its surrounding localities (Agius-Vadalà and Ganado 1986, 16). It indicates the placing of the Marsa *Giardino* (MA9) and Għajn Filep (MA8), but these are uncaptioned (Figs. 5.22, 5.23). The 1563 anonymous cartographic source was also used as a prototype for later maps of the area (Agius-Vadalà and Ganado 1994, 11). Nicolò Nelli's 1565 map of the Grand Harbour region is one such instance (Fig. 5.24). Apart from the Marsa water source (Fig. 5.25), this map lists three different *giardini* (Figs. 5.24, 5.26). The first is the Marsa *giardino* (MA9), the second being the *Giardino del Gran Maestro* (the Grand Master's Garden –

Fig. 5.14 Map of the Maltese archipelago by Domenico Zenoi dating to 1565. Source: Agius-Vadalà and Ganado 1994, Map 36.

Fig. 5.15 Anonymous map of Malta probably dating to 1551. The map indicates the whereabouts of the Marsa spring (MA8), the Marsa *giardino* (MA9), and a *giardino* close to the landward side of Senglea. Source: Agius-Vadalà and Ganado 1986, 8.

Fig. 5.16 Detail from an anonymous map of Malta probably published in 1551 showing the approximate location of the Marsa spring (MA8) and the Marsa *giardino* (MA9). Source: Agius-Vadalà and Ganado 1986, 8.

Fig. 5.17 Image detail of Fig. 5.18. The Giacomo Gastaldi map of 1551 indicates the location of a *fonte* at Marsa (MA8), and the *jardin della Marsa* (MA9). Source: Agius-Vadalà and Ganado 1986, 10.

[7] This area is today more commonly referred to as the Blue Lagoon.

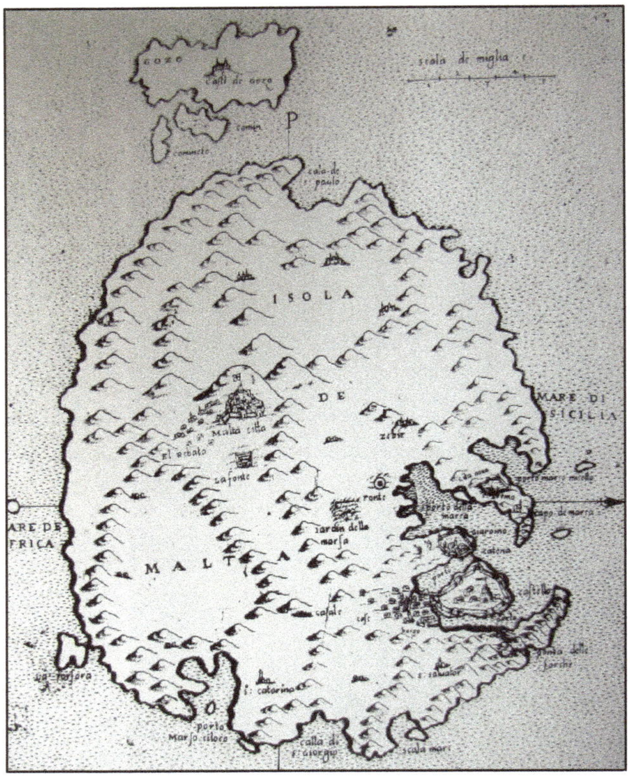

Fig. 5.18 Cartographic representation of the Maltese archipelago by Giacomo Gastaldi, probably dating to 1551. The map indicates the approximate location of the Marsa spring (MA8), the Marsa *giardino* (MA9) and another spring located to the east of the Mdina settlement. Source: Agius-Vadalà and Ganado 1986, 10.

Fig. 5.19 Cartographic representation of Malta by Battista Agnese dating to 1554. The only represented spring in Malta is that at Marsa (MA8). Source: Agius-Vadalà and Ganado 1986, 10.

PA1). The third is simply marked as *Giardino* (BI2) and is present in the area outside the landward bastions of Birgu. The *Giardino del Gran Maestro* is also listed as the *Giardino de la fontaina* in different map sources (Agius-Vadalà and Ganado 1994, 14). The fact that it appears to be located close to the Għajn Dwieli (PA2) or Wied Blandun areas raises the possibility that it was positioned close to a water spring originating from within a Globigerina Limestone catchment area – probably a localised perched aquiferous source.[8]

The Marsa spring is once more showcased in another Grand Harbour map by Nicolò Nelli (Figs. 5.27, 5.28). This principal water source at Marsa, referred to in historical sources as Għajn Filep (MA8), was included in most Great Siege Maps of Malta (Agius-Vadalà and Ganado 1994, 7).[9] According to Balbi da Correggio, water from Għajn Filep was continuously being drawn to supply the galleys, the forts of St Angelo and St Michael, as well as the cisterns of the Birgu (BI1) settlement (Balbi da Correggio 1965, 27, 44, 52). In the sixteenth century, Għajn Filep was already described as being somewhat brackish in nature. This conforms to data obtained from the Malta Resources Authority, which show that the springs of the Marsa area originated from mean-sea-level aquifer sources.[10] In a number of maps from this period Għajn Filep is showcased as consisting of a sizeable water trough or reservoir connected to a canal channelling water towards the nearby coast (Figs. 5.23, 5.26, 5.27, 5.28, 5.29, 5.30, 5.31).

Dating to between 1576 – 1581 are the Matteo Perez d'Aleccio frescoes which decorate the *Sala del Maggior Consiglio* at the Grand Master's Palace in Valletta (Agius-Vadalà and Ganado 1986, 466-69). These mention place-names formerly not included in previous cartographic representations of Malta.[11] The Perez d'Aleccio frescoes also provide some useful cartographic evidence on the location of several hydrological sources present in Malta.

The first fresco (Fig. 5.32) gives various details of the rural landscape. Even though the well-drawn valley which I tentatively associate with Wied ir-Rum is in the Perez d'Aleccio frescoes merely portrayed as an open space containing trees (Fig. 5.33), it is now clearly marked by the legend *Guiderum* (Wied ir-Rum). The fresco also includes the *Monte Alip* (Mtaħleb) and *Monte Pellegrino* (Ras il-Pellegrin) toponyms, but does not indicate the presence of water sources at Wied ir-Rum and Mtaħleb. The Marfa *fontana* is still listed and two other springs (*Le Dvi Fontane*) to the immediate north of Mdina are marked.[12] These latter springs appear to be closely related to a rectangular enclosure, marked as representing the *Giardino del Re*, which probably conforms to the Ġnien is-

[8] See chapter two, section 2.6.2.
[9] The first known recorded instance of Għajn Filep is 1543. See Wettinger 2000, 182.
[10] Personal communication by M. Sapiano.

[11] The Perez d'Aleccio frescoes appear to have been modelled on the Antonio Lafreri 1551 map of Malta, but list a much wider place-name compendium. See Agius-Vadalà and Ganado 1994, 5, 470.
[12] For a more detailed discussion on these water sources see sections 5.2.1.5 and 5.2.4.

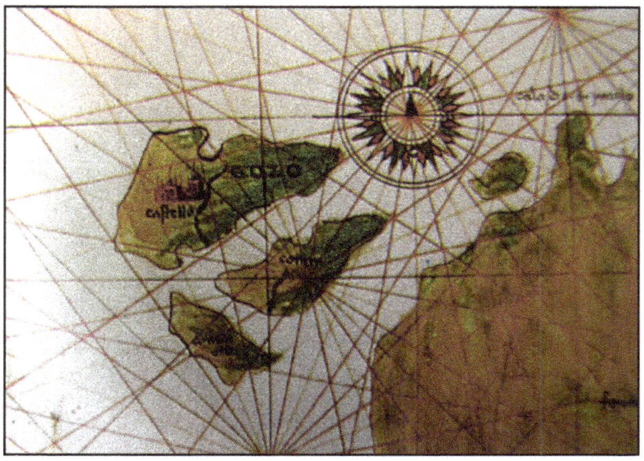

Fig. 5.20 Nautical chart of the Maltese Islands by Battista Agnese. This map portrays the existence of a spring on Gozo and another on Comino. Source: Agius-Vadalà and Ganado 1986, 10.

Fig. 5.21 Wood cut map of the Maltese archipelago published by Henricus Petri in 1561. Source: Agius-Vadalà and Ganado 1986, 14.

Sultan area (RA13; Fig. 5.33). It is also interesting to note that alongside the road from the *Citta Vechia* (Mdina) to Żebbuġ, there appears to be the artistic representation of a basin, above which is mounted a wooden frame, to which

Fig. 5.22 An anonymous map of the north-east coast of Malta dating to 1563. Source: Agius-Vadalà and Ganado 1986, 16.

Fig. 5.23 Detail of the anonymous 1563 map of the north-east coast of Malta. Marked on the map is the location of the Marsa spring (MA8). Source: Agius-Vadalà and Ganado 1986, 16.

a water lifting bucket was possibly attached. The presence of either a well or cistern might be indicated by such a depiction (Fig. 5.33).

The first Perez d'Aleccio fresco makes no mention of the Marsa water source. The Marsa *giardino* (MA9) and two other agricultural installations positioned close to the Cottonera area are, however, depicted. The location of the first of these corresponds to Corradino, close to the Għajn Dwieli area (PA2; Fig. 5.34). This same *giardino* is once more included in the second Perez d'Aleccio fresco showing the Turkish fleet's disembarkation at Marsaxlokk (Fig. 5.35). The location of the Għajn Dwieli agricultural installation is also shown on a slightly earlier *bozzetto*, also by Perez d'Aleccio, on which the frescoes were modelled. It has already been pointed out that this *giardino* is located in an area containing localised perched aquifer Globigerina Limestone water sources, the properties of which have already been discussed in chapter two, section 2.6.2.

Fig. 5.24 Map of Grand Harbour area by Nicolò Nelli dating to 1565. Source: Agius-Vadalà and Ganado 1994, Map 2.

Fig. 5.25 Detail of the Birgu and Senglea settlements and their two adjoining *giardini* (BI2). Nicolò Nelli, 1565. Source: Agius-Vadalà and Ganado 1994, Map 2.

The above discussed maps have to be interpreted with some caution. In most instances the maps were engraved by cartographers who, without visiting Malta, obtained their information from secondary sources. Furthermore, the given topographical details are often inaccurate, thus making it difficult to locate with certainty any of the marked landscape features (Agius-Vadalà and Ganado

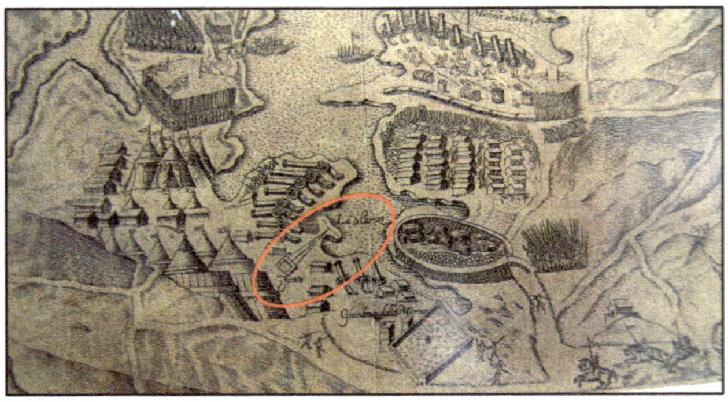

Fig. 5.26 Map detail showing the location of the Marsa spring (MA8) in the inner reaches of the Grand Harbour. Nicolò Nelli, 1565. Source: Agius-Vadalà and Ganado 1994, Map 2.

1994, 66, 73,113). Nonetheless, they do contain evidence for the presence of various water sources and *giardini* installations which dotted different areas of Malta.

It has been recently observed that the *giardini* located on plains in the south-east part of the island of Malta are in these cartographic representations almost always depicted as being surrounded by high walls (Saliba 2008, 58). Apart from demarcating the extent of these *giardini*, walls possibly also served the purpose of wind breaks and protected the cultivated crop from strong winds. Examples of such agricultural installations are the Marsa *giardino* (MA9), which is by far the most commonly represented. Several maps also show the Marsa *giardino* as being guarded by a series of fortified turrets.

To the north and north-west of Malta, *giardini* were located in valleys (Saliba 2008, 58), the principal of which appears to have been Wied ir-Rum. Mainly captioned as *vallone giardini* or *Guiderum*, this locality and its neighbouring areas appear to have been a major agricultural production site in sixteenth-century Malta. Furthermore, it appears that in most instances, both the harbour area *giardino* installations and those located to the north and north-west section of Malta were always positioned close to fresh water sources. The major difference between the two is that harbour agricultural installations tapped mean-sea-level or localised Globigerina Limestone perched aquiferous sources, whilst *giardini* to the north and north-west of Malta tapped Upper Coralline Limestone perched aquifer sources, possibly originating from water galleries.

5.2 Hydrology-Related Aspects as Presented by Fra Gian Francesco Abela in 1647

A more detailed analysis of the Maltese hydrological situation in the first half of the seventeenth century can be obtained through the examination of Gian Francesco

Fig. 5.27 Map of the Grand Harbour area by Nicolò Nelli dating to 1565. Source: Agius-Vadalà and Ganado 1994, Map 8.

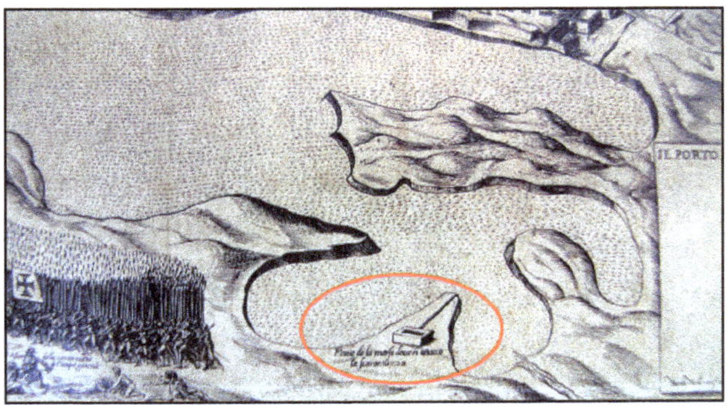

Fig. 5.28 Nicolò Nelli, 1565. Map detail of the Grand Harbour area showing the location of the Marsa spring (MA8). Source: Agius-Vadalà and Ganado 1994, Map 8.

Abela's *Della Descrittione di Malta* text. Of particular relevance to this field of study are the sections on the *Parte di Malta Inabitata* (Abela 1647, 64-75), the *Parte Habitata di Malta* (Abela 1647, 75-109), the section on *Dell'Acque della Fontana* (Abela 1647, 109-17), the *Descrittione dell'Isola del Gozo* (Abela 1647, 118-23), and that on *Della Fertilità di Malta* (Abela 1647, 125-38). In the first two sections, Abela diligently lists the main place-names of the uninhabited and the inhabited areas of Malta and often gives a brief description of the locality. The numerous water-related toponyms listed within give testimony to the importance these water-yielding areas were given in the seventeenth century. It is worth pointing out that the section on Gozo is much less informative. While several water-related features are listed in this section, it seems that Abela was relatively unfamiliar with the Gozitan territory.

The principal water-related terminology given by Abela consists of *fonte*, *fontana*, *aayn*, *bir*, *cisterna*, *pozzo*, *stagno*, *ghadira*, *ghar elma*, *gebbia* and *sorgente d'acqua*. Passing reference is also made to *falde* (aquifers) and *fiumicello* (stream). That Abela was acquainted with the basic dynamics of the perched aquifer water sources in the north and north-west Malta can be detected from his account of the Baħrija area: "*Il terreno alle falde è molto fertile, aprico, & ameno, & con molta copia d'acque*" (Abela 1647, 67). Conversely, very little consideration was made of the physical characteristics of the mentioned water sources. Terms such as *Aayn* and *Fontana* make it difficult to distinguish between a natural spring and an artificial water source. Towards the end of the section on the *Parte Habitata di Malta*, Abela does sparingly refer to the existence of numerous *mine*, *canali*, and *condotti di fontane* (Abela 1647, 101-2). Whilst in all probability, these are direct references to the perched aquifer galleries, there is lack of data to establish this with certainty. Elsewhere in the text, the nature of the majority of the named water-related features frequently remains unspecified.

Nevertheless, a close scrutiny of Abela's text, coupled with a sound geological, hydrological and archaeological knowledge of the area under examination, does make it

Fig. 5.29 Map of Malta and Comino dating to 1565, attributed to Tommaso Barlacchi. Source: Agius-Vadalà and Ganado 1994, Map 13.

Fig. 5.30 Tommaso Barlacchi (?), 1565. Map detail of the northern coast of Malta and the Grand Harbour area. The map indicates the location of the Marfa (ME2) and the Marsa (MA8) springs. Source: Agius-Vadalà and Ganado 1994, Map 13.

Fig. 5.31 Mario Cartaro, 1565. Map detail showing the location of the Marsa spring (MA8). Source: Agius-Vadalà and Ganado 1994, Map 24.

Fig. 5.32 Fresco painting by Matteo Perez d'Aleccio dating to 1576 -1581 depicting the arrival of the Turkish fleet. Source: Agius-Vadalà and Ganado 1994, Map 118.

possible to reach some preliminary conclusions. Toponomastic evidence given by Abela in the *Parte di Malta Inabitata* and the *Parte Habitata di Malta* differentiates between the hydrological strategies employed in the north and north-west sector of Malta and the remaining section of the island. Whereas the main water sources in the northern and north-western zones of Malta are of the *aayn* and *fontana* type, thus implying the utilisation of natural perched aquifer sources, *bir* and *cisterna* toponyms are widespread for the remaining section of Malta. It should be noted that there is a conspicuous absence of *bir*-related terminology in the *Parte di Malta Inabitata* section.[13]

When listing several valley water courses, Abela applies different terminology from that discussed above. The spring estuary for the coastal area of Mellieħa is described as *fiumicello* and *torrente*: *Il fiumicello, ò più tosto picciolo torrente, che s'incamina, e corre al mare nel feno*[...] (Abela 1647, 74). Reference to fresh water discharge into coastal areas, probably from the mean-sea-level aquifer, is made when describing the Msida and Marsa areas. *Vyed el Msida* is described as: [...] *stagno d'acqua, che vien' originato dà vn' fonte, il quale*

[13] Similar conclusions had already been reached by R. Grima. See Grima 2004, 341-42; Grima 2005, 68-70.

Fig. 5.33 Matteo Perez d'Aleccio (1576 - 1581). Fresco detail showing the western coast of Malta and the countryside area neighbouring the *Civitas*. Clearly indicated are the Wied ir-Rum (RA56), *Due Fontane* (RA14) and Ġnien is-Sultan (RA13) *giardini*. To the west of the Mdina fortified settlement there also appears to be depicted a water basin or well. Source: Agius-Vadalà and Ganado 1994, Map 118.

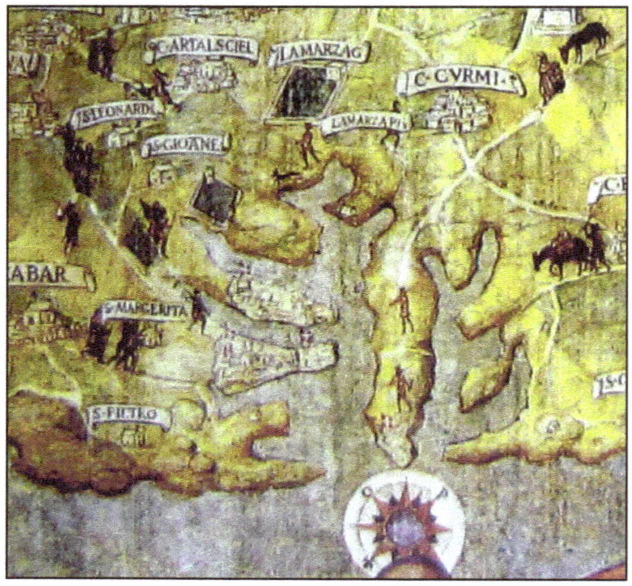

Fig. 5.34 Matteo Perez d'Aleccio (1576 - 1581). Fresco detail showing the Grand Harbour region. Indicated on the map are the Marsa *giardino* (MA9) and another located on Corradino hill close to Għajn Dwieli (PA2). Source: Agius-Vadalà and Ganado 1994, Map 118.

scaturische vicino al lido del mare, quasi bollendo [...] (Abela 1647, 91). *Dachlet el Kasab* is the place-name for two different areas at Marsamxett and Marsa, intimating the availability of a fresh water source in these areas (Abela 1647, 91, 93).

The same applies to the Qormi area (MA10). Whilst keeping in mind that the Qormi locality receives most of its autumn and winter water supply as surface rain runoff from the upstream areas of Żebbuġ and Siġġiewi, Abela

Fig. 5.35 Matteo Perez d'Aleccio (1576 - 1581). Fresco painting showing the disembarkation of the Turkish fleet at Marsaxlokk. Also indicated is a *giardino* on Corradino hill (PA2). Source: Agius-Vadalà and Ganado 1994, Map 119.

mentions the indirect tapping of mean-sea-level aquifer sources in this district. He describes the *Terra Cvrmi* area as: [...] *per l'acqua viua, che colà scorre in gran copia, e con molt'abbondanza si ritroua cauandosi il terreno pochi palmi* [...] (Abela 1647, 92). Abela's entry for Qormi also contains the only mentioned instance for the use of water lifting equipment: [...] *onde da per tutto & in ogni parte si sono fabricate molte machine, & ordigni da tirar in sù artificiosamente l'acqua per inaffiar quegl'horti, e giardini, da'quali si prouede d'herbaggi in buona parte la Città* [...] (Abela 1647, 92). Another entry on *Terra Cvrmi* is even more informative in showing that the organised large-scale exploitation of the mean-sea-level aquifer was a relatively new phenomenon and probably cannot be dated to prior the Knights' period (Abela 1647, 128).[14] It is here specifically mentioned that water was being lifted from the mean-sea-level aquifer by means of a water-lifting device known as *sienja*.

A subterranean underwater conduit in the *Bucana* area in the limits of Rabat (Malta) was also directly mentioned by Abela (Abela 1647, 82). This most likely formed part-and-parcel of the aqueduct system which channelled water towards Valletta. *Diar Handul* (Dejr Ħandul – RA42) is also listed as being the first spring to be channelled towards Valletta (Abela 1647, 66). Unfortunately, no evidence is given as to whether these were already extant springs which were diverted to Valletta or if water was specifically intercepted in part-fulfilment of the Valletta aqueduct project.

Whilst keeping in mind that a *giardino* place-name does not necessarily signify the availability of a water source, a close analysis of the *giardini* localities listed for the uninhabited and the inhabited sections of Malta, does reveal some notable disparities. *Giardini* listed in the uninhabited section of Malta frequently contain reference to spring water sources.[15] Various *giardini* listed under the inhabited part of Malta, on the other hand, do bring to light a different reality. The *giardini* of the *El Maatab* (NA1) locality overlooking the east coast of Malta contained olive plantations (Abela 1647, 85). The same may be said of various *giardini* at *Vyed el Dis* (GH1) in the limits of Għargħur, described as containing, [...] *alberi, d'oliueti, vigne, e di gran quantità d'amandorle, e carrobe* (Abela 1647, 85). The same situation applies to the close-by *El Sciüeki* (IK1) district, which is described as containing the same type of vegetation as that listed in the *Vyed el Dis* entry (Abela 1647, 85). The *giardini* of *Vyed Hal Màn* (LI1) are also described as containing, [...] *amandorle, carrobe, e fiche* [...] (Abela 1647, 85).

The *giardini* located in the north and north-west section of Malta, appear to have commonly contained orchard plantations geared towards fruit production. The mention of *buonissime frutta* grown in the Wied ir-Rum *giardini* (RA56) (Abela 1647, 65), and that of *regalatissime frutta* being cultivated at Ta' Baldu (RA43) (Abela 1647, 65), are all indications in this respect. It is possible that in the first half of the seventeenth century, these orchard installations were already geared towards the cultivation of fruit trees. Fruit tree cultivation requires a continual water supply and was probably concentrated in these areas due to the availability of perennial water sources. *Giardini* located in Globigerina Limestone areas close to the Għargħur and Lija localities, on the other hand, were used for the cultivation of olive trees, carob and almond trees. These are more drought-resistant crops requiring a far less copious water supply and much less irrigation.[16] Being located in valleys, *giardini* were naturally sheltered from the prevailing winds. Where this was not the case, high walls were built in order to act as wind breaks and shelter the tree plantations. Cases in point are the *giardino* Ta' Baldu at Wied ir-Rum and *Giardino Aayn Toffieha* in the territory of Mġarr, Malta.[17]

[14] Abela's entry on Qormi reads: [...] *oue per tutto quell tratto si sono rinouate da alcuni anni in qua molte machine da'nostri chiamate senie, per mezzo delle quali cauano & alzano l'acqua, che da per tutto si troua alquanto più bassa dell'altezza del mare, per inaffiare, & irrigare gl'orti, che vi si sono fatti.*

[15] This aspect is discussed in more detail in section 5.2.1.

[16] This matter was discussed with G. Knox, an expert geologist and hydrologist, with expert botanist, E. Lanfranco and with M. Vassallo. All agreed that from a hydrological perspective, carob and almond trees are more adapted to the Maltese climate and require a much smaller water supply than is necessary for the cultivation of fruit trees. A. Baldacchino, an expert environmentalist, likewise agreed that the cultivation of citrus trees requires water availability. A. Baldacchino personally communicated that citrus trees need to be irrigated on a weekly basis and each small to medium sized tree requires ca. 14 litres of water per week. See Blouet 1963, 76-77; Glick 1970, 29. The natural ground-water extraction rate for carob trees has been estimated at varying between 0.5 to 1.01 mm of water per tree per day. Morris 1952, 110.

[17] In the south and south-eastern part of Malta, *giardini* were less frequent. These were mainly located in the Marsa and Qormi area (MA9; MA10), and a number of which were possibly the result of an agricultural intensification process directly related to the building of Valletta. See Blouet 1963, 76-77.

Fig. 5.36 Exterior view of the Dejr il-Bniet (DI1) water gallery in the territory of Rabat (Malta). During the course of the seventeenth century the exterior section of this perched aquifer gallery was remodelled and given a monumental baroque-style appearance. Source: the author.

Fig. 5.37 Interior view of the Dejr il-Bniet (DI1) water gallery in the territory of Rabat (Malta). Circled in red is the square-headed doorway giving access to the original untouched section of the Dejr il-Bniet perched aquifer gallery. Source: the author.

5.2.1 The uninhabited section of Malta

The area covered in this section primarily encompasses the present-day territories of Mellieħa, Mġarr (Malta), Żebbiegħ, part of Mosta, Baħrija, Dingli, part of Girgenti and a section of the Rabat (Malta) countryside. It is an area in which exposed Upper Coralline Limestone and exposed Blue Clay deposits prevail. From a hydrological perspective, when compared to other regions of Malta, such a geological setup provides the ideal conditions for the formation of perched aquifers. A total of twenty-two *Aayn* toponyms are listed in this section. Apart from these, there is also mention of at least another seven *fonte* or *fontane* place-names. These are: *Deyr el Binet* (DI1), *S. Antonio tal Ghimeri* (RA15), *Ved tal Gineyna* (MG24), *Ta' Safsafa* (MG21), *Faüara* (MG11), *Giardino d'Angarao* (ME1), and *Giardino di S. Nicola* (ME7) (Abela 1647, 65-73).

Even though not directly specified by Abela, it is probable that many of the above-mentioned water sources originated from perched aquifer water galleries. The *Deyr el Binet* (Dejr il-Bniet) country estate may be used to illustrate the point. Abela describes the area as follows: *Case chiamate delle donzelle, perche quiui in una caverna, dala quale scaturisce un fonte, dicesi, che si smarrissero alcune fanciulle, che colà entrate non seppero poscia ritrouar l'vscita* (Abela 1647, 65). The mention of a cave from which a water source flows is almost certainly a reference to the perched aquifer water gallery present in the Dejr il-Bniet (DI1) tenement.[18] To this day, the gallery yields a copious water supply which irrigates the adjoining *giardino*. Its entrance was given a monumental baroque-style appearance during the course of the seventeenth century (Figs. 5.36, 5.37). Also significant is Abela's mention of the undated episode quoted above, where some girls lost their way through the water gallery network (Abela 1647, 65).

Although numerous *giardino* toponyms are listed by Abela in the *Parte di Malta Inabitata* section, in most instances no direct reference to the availability of water sources is made. However, taking into consideration the geological stratification of the north and north-west areas of Malta where such place-names are located, one gets the impression that the location of most *giardini* installations was probably determined by water availability.[19]

5.2.1.1 Wied ir-Rum

A concentration of water-related features is given by Abela as being found at Wied ir-Rum (RA56) and its immediate vicinity. The *Giardino la Kattara*, *Tal Callus*, *Ta' Scieref*,

[18] The map reference for the water gallery at Dejr il-Bniet in the territory of Dingli is 44530 69220.

[19] For Giardino locations listed in the Parte di Malta Inabitata see Abela 1647, 64-75. These are: Deyr el Binet, Giardino ta Cassia, Dachla, Vyed Casrun, Vyed el Rum, Giardino la Kattara, Tal Callus, Ta' Scieref, Di S. Giacomo, Di Baldu, Tal Nachla, Giardino del Mitahlep, Vyed el Bies, S. Antonio tal Ghimeri, Bahria, Giardino detto ta' Samti, Vyed Gherzuma, Gineyna tal Miggiar, Ved tal Gineyna, Ta Bengemma, Giardino tal Salib, Giardino tal Maaysciae(?), Targia ta' S. Zacchari, Wied Cannotta, S. Brancato, Bidenia, Vyed Bufula, Giardino del Puales, Giardino di S. Martino, Chazerna, Gnien tar-Ramla, Aayn Toffieha, Aayn Targia, Giardino di Sbetta, Vyed Santa Maria, Giardino ta' Rapa, Giardino d'Angarao and Giardino di S. Nicola.

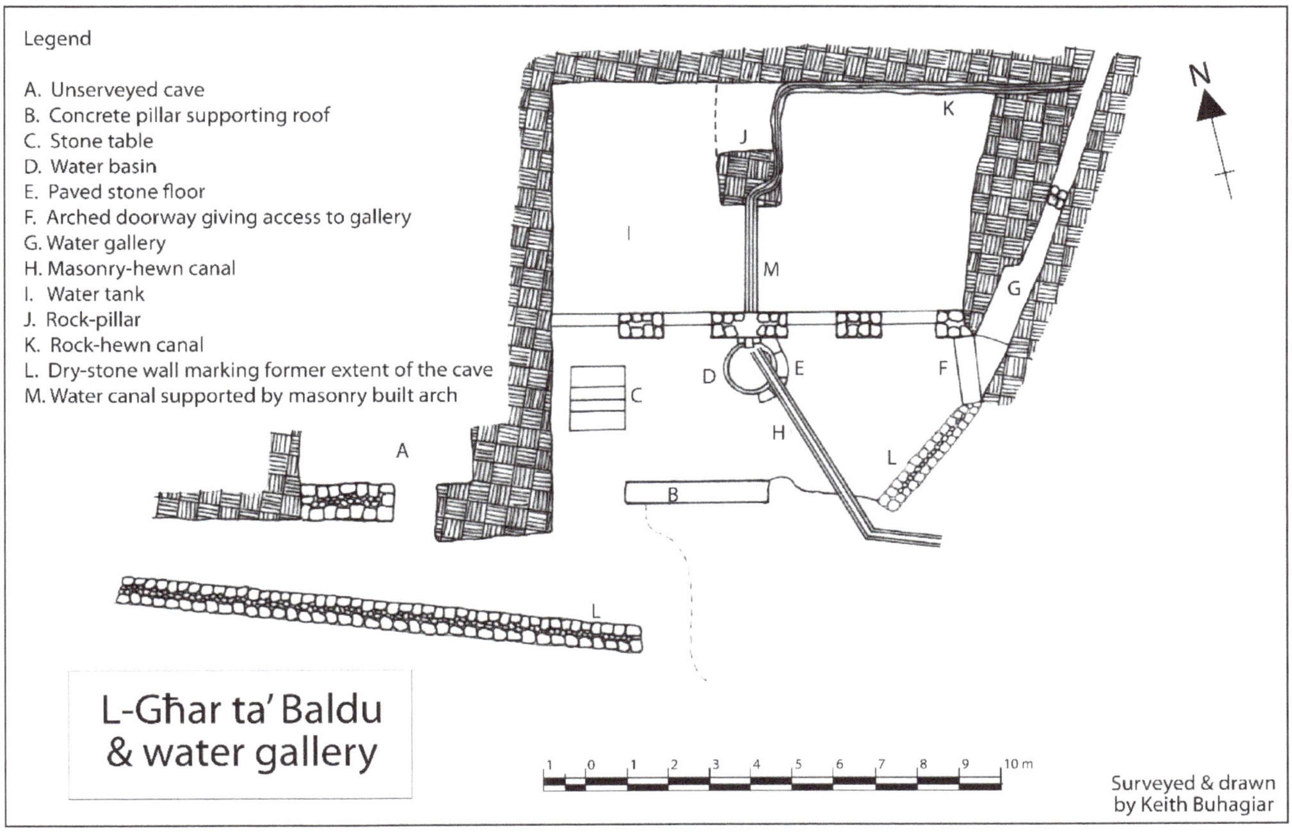

Fig. 5.38 The Għar ta' Baldu (RA43) perched aquifer gallery and its adjoining subterranean reservoir at Wied ir-Rum (RA56) in the territory of Rabat (Malta).

Di S. Giacomo and the *Di Baldu* areas are all separate *giardini* containing subterranean water features, in the Wied ir-Rum valley system (Abela 1647, 65) (Figs. 4.3, 5.39). Abela gives evidence for the presence of water in the Giardino la Kattara: [...] *nome originato dal gocciolare, essendoui una grotta, la cui parte superiore, continuamente distillandoui, forma un riuolo d'acqua, che inaffia il giardino* (Abela 1647, 65). Il-Qattara (RA32) is located on the southern flank of Wied ir-Rum, between Tax-Xieref (RA54) and Simblija (RA41; Fig. 4.10).[20] Five man-excavated caves and three water galleries are present in this area (Buhagiar K. 2002, 181-86). It is the largest and most imposing gallery that is associated with the Il-Qattara toponym (Fig. 5.40).

The Il-Qattara gallery is excavated into the side of a partly-collapsed natural cavern, in the ceiling and sides of which are stalagmites and other calcification features associated with water seepage. It has so far been difficult to determine if the Il-Qattara gallery was already a prominent hydrological feature in Abela's time, or whether this was excavated at a subsequent stage. On the other hand, Abela's description of Wied ir-Rum and its environs gives the impression of a series of long established *giardini* dependent on a series of reliable water sources. Bearing in mind that the Il-Qattara perched aquifer gallery is the primary water source for the *Giardino la Kattara* area, it is likely that its excavation was a critical element in the development of this *giardino*.

The situation at Il-Qattara is in many ways synonymous to that of the neighbouring *Ta' Baldu giardino* (RA43). The *Di Baldu* (Ta' Baldu) area was described by Abela as [...] *copioso d'acqua* (Abela 1647, 65). As in the case of the

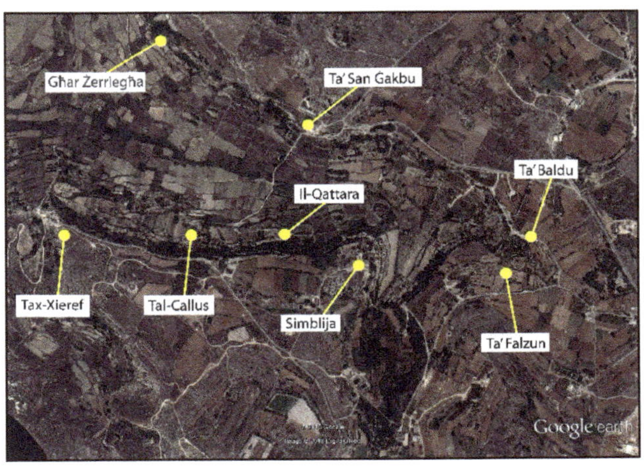

Fig. 5.39 Satellite view showing the location of *giardini* at Wied ir-Rum (RA56) and Wied Ħażrun (RA55) in the territory of Rabat (Malta).

[20] The map reference for the Il-Qattara gallery is 42620 69910.

Fig. 5.40 Exterior view of the Il-Qattara (RA32) water gallery excavated into the southern side of Wied ir-Rum (RA56) in the territory of Rabat (Malta). Source: the author.

Giardino la Kattara, Abela makes no mention of perched aquifer water gallery systems at *Ta' Baldu*. Onsite field surveys in the upper valley section of Wied ir-Rum, in which Ta' Baldu is located, have shown how the life source of this *giardino* is derived from three rock-cut perched aquifer galleries (Figs. 4.3, 4.10).

These water retrieval systems are of an unknown date, but there is enough evidence to show that at least one of the galleries was already in existence before 1647. This is located within a cave known colloquially as the *Għar ta' Baldu* (RA43) and is adjoined by a large subterranean water reservoir and a fronting space containing a large masonry water-settling vat and a stone paved floor (Buhagiar K. 2000, 44-45; Buhagiar K. 2002, 167-71) (Figs. 4.10, 5.38, 5.41). A carved inscription rendered in cocciopesto, located in the water basin area, records a 1629 date. This probably commemorates the occasion when the large subterranean reservoir was constructed and major changes brought about to an already extant perched aquifer water tapping system. It is likely that the original hydrological feature only consisted of a subterranean gallery. A subsequent re-organisation of the hydrological setup of the area would have involved the excavation of the subterranean water reservoir, into which the water of the gallery was diverted.

Another water gallery in the vicinity of *Ta'Baldu* supplied a circular rock-excavated space with a perennial water supply (Figs. 4.10, 5.43). The circular chamber was utilised as a cold water bath and, on stylistic grounds discussed hereunder, it has been tentatively dated to the Late Medieval period. Access to it is down a flight of four rock-cut steps. This subterranean chamber is, to the present writer's knowledge, typologically unique for the Maltese Islands and cannot be compared to other local examples. Its existence was first brought to the attention of A.A. Caruana in 1869, who subsequently included a description of it in his *Report on the Phoenician and*

Fig. 5.41 Interior view of the Għar ta' Baldu (RA43) in the territory of Rabat (Malta). Source: the author.

Fig. 5.42 Interior view of the subterranean circular-shaped bath located in the Ġnien ta' Baldu (RA43) at Wied ir-Rum (RA56) in the territory of Rabat (Malta). Source: the author.

Roman Antiquities in the Group of Islands of Malta, published in 1882.[21]

The chamber was again identified by the present writer during field-research held in the Ta' Baldu area in 1999 (Buhagiar K. 2000, 50-51). Its interior is fairly well preserved, and consists of rock-walls which are completely plastered over by means of a cocciopesto-type of grey-coloured mortar, and a circular bench which follows the circumference of the chamber (Figs. 5.42, 5.43, "B"). The chamber was provisioned with a perennial water source obtained from an adjoining water gallery, linked to it by means of a small rectangular-shaped water tank (Fig. 5.43, "C"). From this tank, water was gravity-fed through terracotta or lead pipes to a fountain which acted as a centre piece to the bath setup.

In 1999, only a few masonry fragments of the base of the fountain were found to have survived. Scale drawings of these were penned and archived at the University of Malta's Department of Classics and Archaeology.[22] The fountain appears to have been deliberately destroyed by one of the former land owners in the 1950s, in a search for lead and buried treasure. Until the early 1980s, access to the chamber was restricted by means of an iron gate,[23] the removal of which probably resulted in the substantial disfigurement of the rock-cut access doorway. Cut into the floor of the interior of the chamber, adjacent to the access doorway, is a rectangular-shaped shaft (Fig. 5.43, "E"). This contains three terracotta pipes still *in situ*, the function of which was to drain the surplus water elsewhere. Water level in the chamber was probably controlled by regulating the water flow through the pipes at this point. There is no evidence for water heating at Ta' Baldu.

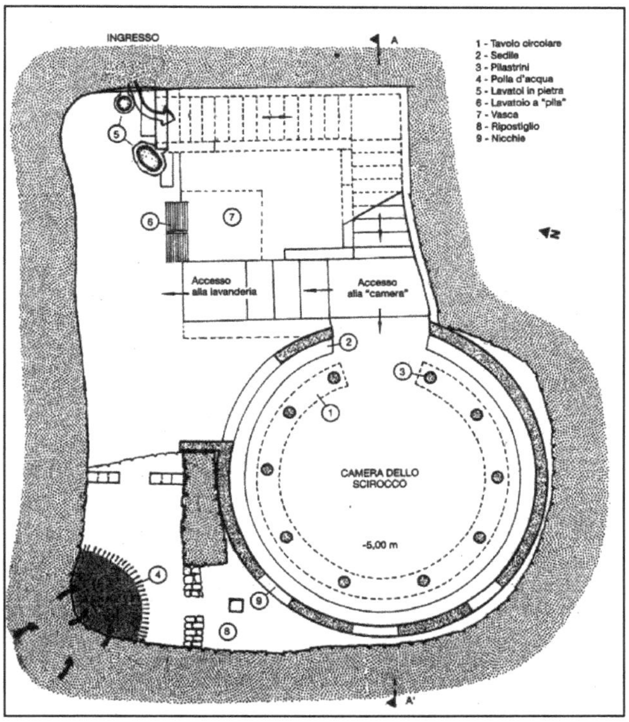

Fig. 5.44 Plan of the *camera dello scirocco* located in the premises of Villa Salerno, Palermo. The setup is tentatively dated to the seventeenth century AD. Source: Todaro 2002, 85.

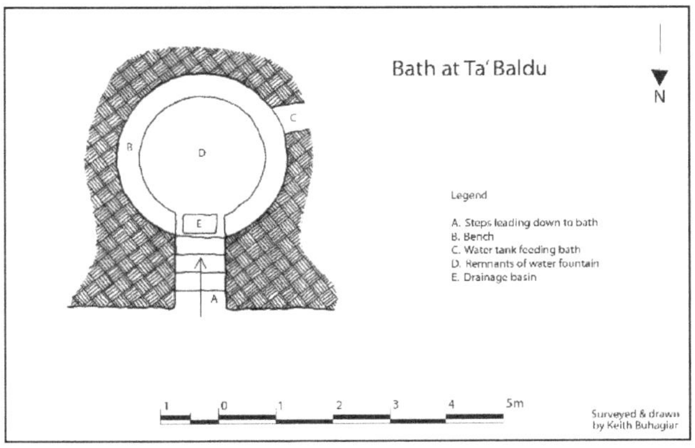

Fig. 5.43 The circular-shaped subterranean *scirocco* chamber located in the Ġnien ta' Baldu (RA43) at Wied ir-Rum (RA56) in the territory of Rabat (Malta).

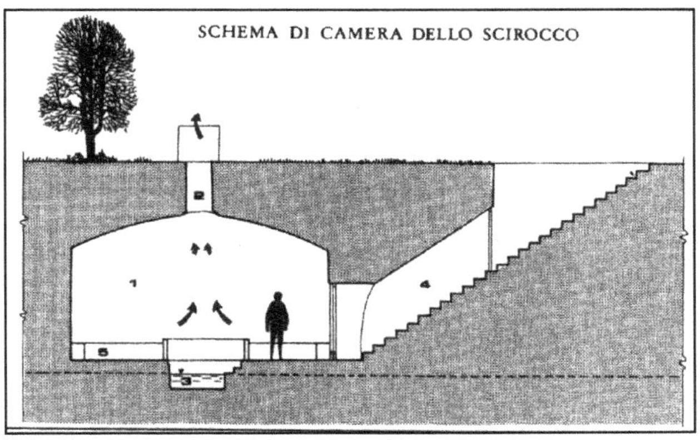

Fig. 5.45 Elevation plan showing the typical setup of a *camera dello scirocco*: (1) cooling chamber; (2) ventilation shaft; (3) spring; (4) access stairway; (5) seating accommodation. Source: Todaro 1989.

[21]Caruana's account of the remains is as follows: "I visited, on the 20th June last, one of these Roman baths, the property of Mr Giorgio de' Conti Sant Fournier, in the rustic tenement of Ta' Baldu, about one mile from the Casal Dingli. It was discovered in 1869, and is pretty well preserved. A circular room, excavated and vaulted in the rock plastered all over, about 44 ft. in circumference, with stone seats all round, forms the bath. There was a constant supply of water from a spring in the same tenement. Lead pipes conducted under-ground from a small tank, high up under the vault, on the right-hand side, fed with a water fountain in the centre of the room. Some remains of the basin, together with its support are still preserved. Below three or four steps at the entrance, there is another tank, 1 ft. 40 in. long by 1 ft. wide, with three holes furnished with lead pipes to carry the overflowing water into a large tank outside the bath." See Caruana 1882, 93. A bathing establishment at Ta' Baldu was first mentioned in 1772, but it is unclear if this refers to the subterranean bath investigated by Caruana in 1869. See Ciantar 1772-1780, 149.

[22] The drawings were completed with the help of N.C. Vella and M.E. Zammit.

[23] This information was passed on to the author by one of the former land owners of the Ta' Baldu tenement.

The interior is void of archaeologically-relevant deposits, but future scientific subsurface investigation of the fronting and adjoining fields will hopefully give a better indication as to the chamber's approximate date. A.A. Caruana, probably deceived by the presence of what might be a hollowed-out olive pipper basin in the *Għar ta' Baldu* close by, dated the bath to the Roman period (Caruana 1882, 93). Likewise, when publishing a brief report announcing the bath's re-discovery in 2000, the present writer tentatively assigned it to a Roman period date (Buhagiar K. 2000, 50). Based on previously unavailable evidence, a reinterpretation of the remains now suggests a Late Medieval or Early Modern date as being more likely. The chamber is typologically similar to other rock-excavated subterranean spaces in Palermo, Sicily, known as *Camere di Scirocco*.[24] The earliest documentary evidence for such chambers in Palermo comes from *Palazzo Marchesi*, and dates to the late fifteenth century (Todaro 1989, 122; Todaro 2002, 87). Likewise, notarial records recount how Baron Gerardo Agliata bought a countryside estate outside Palermo between 1542 and 1554 and took pleasure in enjoying the *scirocco* chamber available within with relatives and guests (Saeli and Saeli 2015, 345-46).[25]

Camere di Scirocco are in essence rectangular or circular-shaped rock-excavated chambers, the water supply of which is obtained by means of an underground spring or an adjoining subterranean water gallery (Figs. 5.44, 5.45). Their underground location, the flow of water and the presence of an aeration shaft, created a pleasant and refreshing microclimate during the warmer months of the year (Todaro 2002, 80-81; Saeli and Saeli 2015, 345-50). Commonly, *scirocco* chambers received water from a close-by *qanat* system (Saeli and Saeli 2015, 346). *Camere di Scirocco* became fashionable amongst upper class Palermo residents during the course of the seventeenth and eighteenth centuries, and were probably inspired by earlier medieval period chambers for which only indirect and limited evidence survives (Todaro 1989, 122). The principal difference between the Ta' Baldu bath and the Palermitan cooling chambers is the absence of an aeration shaft in the former.

The rural landscape context in which the *Ta' Baldu* chamber is located is of a probable Late Medieval date. More accurate dating can perhaps be established through the scientific analysis of mortar lining the bath interior. The reliability of mortar dating has been successfully tested on securely dated Punic period cisterns at Pantelleria.[26] The Ta' Baldu chamber and adjoining water gallery tunnel below arable land which nineteenth-century documentation identifies as *Ta Falzun* and Il-Għars (RA30) (Malta Survey Sheet 85, School of Military Engineering, Chatham, 1922; NAM. Descriptive Plans of the Crown Property, 1867, f. 497). For this property there is also the mention of *due spere d'acqua corrente* (NAM. Descriptive Plans of the Crown Property, 1867, f. 497),

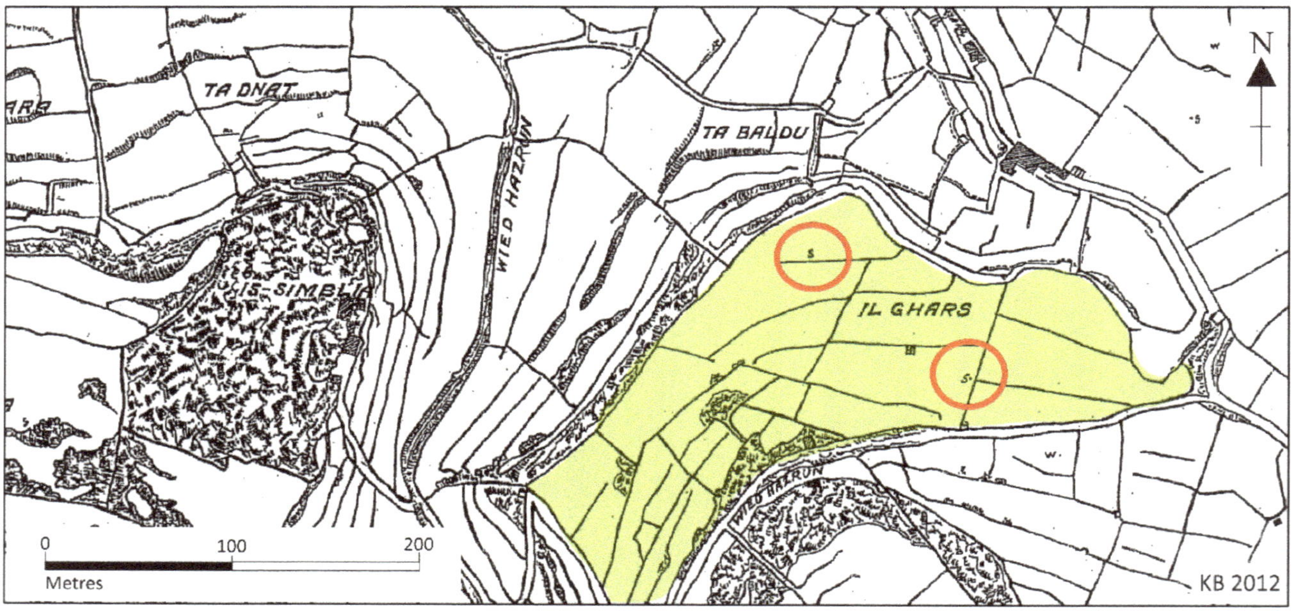

Fig. 5.46 Detail of Malta Survey Sheet 85 showing the land boundaries of the Ta' Falzun estate (RA30) at Wied Ħażrun (RA55) in the territory of Rabat (Malta). Circled in red are the two shaft wells mentioned in the 1867 *cabreo* plan for this area. Sources: Malta Survey Sheet 85, School of Military Engineering, Chatham, UK, 1922; NAM. Descriptive Plans of the Crown Property in Malta, 1867, f. 497; NLM. Treas. B. 292, f. 379.

[24] The existence of *Camere di Scirocco* was brought to my attention by N.C. Vella. *Scirocco* is the hot and humid wind that originates in the Sahara region and blows across the Mediterranean Sea into Southern Europe.

[25] This property is currently known as Villa Naselli-Ambleri. See Saeli and Saeli 2015.

[26] Data obtained from the "International Conference on Ancient and Modern Water Storage", organised at Pantelleria by the University of Tübingen, 11 to 14 May 2011. See also Schäfer *et al* 2014; Ercoli *et al* 2008, 323-30.

Fig. 5.47 Malta Survey Sheet 85 Map Links on which is marked the origin and the course followed by the Ta' Baldu Spring. This appears to originate from within a perched aquifer gallery, but is then conveyed to the Fiddien district in the territory of Rabat (Malta) by means of a subterranean stone canal. The course followed by the Misraħ Suffara, Ħofret ir-Rizz (RA27), Ta' Dekozzu (RA45), Iċ-Ċianti and Bonua springs is also indicated. The base map used dates to 1922.

which can tentatively be associated with shaft wells piercing the ceiling of the gallery feeding the Ta' Baldu bath with a perennial water supply and another located at map reference 43150 659920 (Fig. 5.46).[27] Another gallery or water tunnel in the vicinity appears to belong to either the Knights' or the British periods and is probably one of the sources which supplied the Valletta aqueduct system with water (Fig. 4.10). It is located in the upper reaches of the Ta' Baldu *giardino*, and on the *Map Links* volume, formerly archived at the Chief Draughtsman's Office in Floriana, it is labelled as "Ta Baldu Spring" and is shown to head towards the Sqaq Ħandu (RA42) and Fiddien (RA11) localities of Rabat (Fig. 5.47).

Also mentioned by Abela is the spring of *Aayn Teytes* (RA20) located in the Wied ir-Rum valley system (Abela 1647, 65) (Fig. 4.10).[28] A detailed map of the Mtaħleb area dating to 1665 has made it possible to approximate the boundaries of several of the *giardini* located in Wied ir-Rum (AIM Acta Civilia vol. 171, Case 12) (Fig. 5.48).[29] According to this map, *Aayn Teytes* borders the *giardini* bearing the same name and that of *Di S. Giacomo* (Ta' San Ġakbu – RA4). Personal field-research in the area has shown *Aayn Teytes* to originate from a perched aquifer water gallery excavated into the ravine below the Ta' San Ġakbu territory.

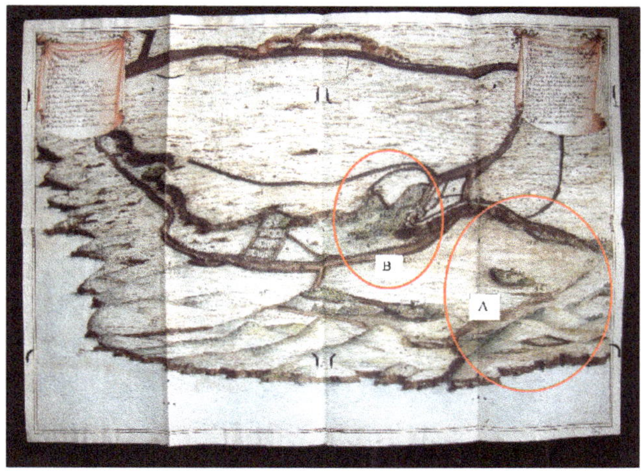

Fig. 5.48 Map of the Mtaħleb area drawn up in 1665. This indicates the location of the *Giardino di Ain Teites* (RA20), *Giardino di San Giacomo* (RA49), *Giardino ta Baldu* (RA43) and *Giardino e Terre tal Qattara* (RA32), all grouped under (A). The whereabouts of the Mtaħleb *giardino* (RA35) are indicated by (B). Source: AIM Acta Civilia vol. 171, Case 12.

5.2.1.2 The Mtaħleb giardino

Another example showing that *giardino* location in north and north-west Malta was often conditioned by perennial water sources is the *Territorio e giardino del Mitahlep* (Abela 1647, 65-66) (RA35; Fig. 4.3). Following a brief description of the Mtaħleb *giardino*, Abela lists six *Aayn* (spring) water sources. These are *Aayn il Kbira*, *Aayn San Giovan*, *Aayn Cior*, *Aayn Andria*, *Aayn Taiba* and *Aayn Ghrab* (Abela 1647, 66). In the case of *Aayn il Kbira* and *Aayn San Giouan* (RA24), it can be safely argued that these originate from within perched aquifer galleries. The

Fig. 5.49 An uncovered reservoir collecting runoff surface rain water in the Tal-Bruka (RA29) area in the limits of Mtaħleb. Source: the author.

[27] Only the location of one of the shaft wells is indicated on the 1867 *cabreo* plan of the *Ta Falzun* property and it appears to be flanked by a small water collection reservoir. See NAM. Descriptive Plans of the Crown Property, 1867, f. 497. The location of the other shaft is indicated on Survey Sheet 85 for the Il-Għars area. See Malta Survey Sheet 85, School of Military Engineering, Chatham, 1922.

[28] A map of the Mtaħleb area dating to 1665, lists Għajn Tejtes as a separate *giardino* from the adjoining Ġnien ta' San Ġakbu.

[29] This map was brought to my attention by M. Gauci of the Cathedral Museum Archives.

two springs are also located in the upper reaches of the Mtaħleb *giardino* (Fig. 5.51).

The exact location of *Aayn il Kbira* was established through the use of a detailed plan of the Mtaħleb area drawn up in 1665 by an anonymous cartographer in which it is listed as *Fontana di Giardino Grande* (AIM Acta Civilia vol. 171, Case 12) (Fig. 5.48). The hydrological setup of the water system depicted in this map has remained unchanged from at least the 1660s to the present. The spring originates from a subterranean water gallery, with the extracted water being channelled into a large open-air water reservoir located in an elevated location in the *Giardino Grande* area.[30] From here the water is gravity-fed via a series of open masonry canals to the various fields forming part of the *giardino del Mitahlep.*

Fig. 5.50 Uncovered reservoir at Tal-Bruka (RA29) in Mtaħleb. This unlined water storage basin is excavated into an impervious Middle Globigerina Limestone deposit. Source: the author.

Abela's brief description of *Aayn San Giovan* as *Fontana di S. Giovanni posta dentro una grotta* provides enough evidence to show that this water source also originates from within a water gallery. The practice of having a subterranean water reservoir excavated in the area immediately preceding the water gallery is a common occurrence in Wied ir-Rum and other locations of north and north-west Malta. *Aayn San Giovan* can be tentatively identified with a perched aquifer gallery which corresponds to this description and which is located on the other side of the valley from *Aayn il Kbira*. The gallery is presently known amongst the farming folk of the *Giardino Grande* (Ġnien il-Kbir) area as *l-Għar ta' l-għarwenin* (RA24), and it still yields small quantities of water.[31]

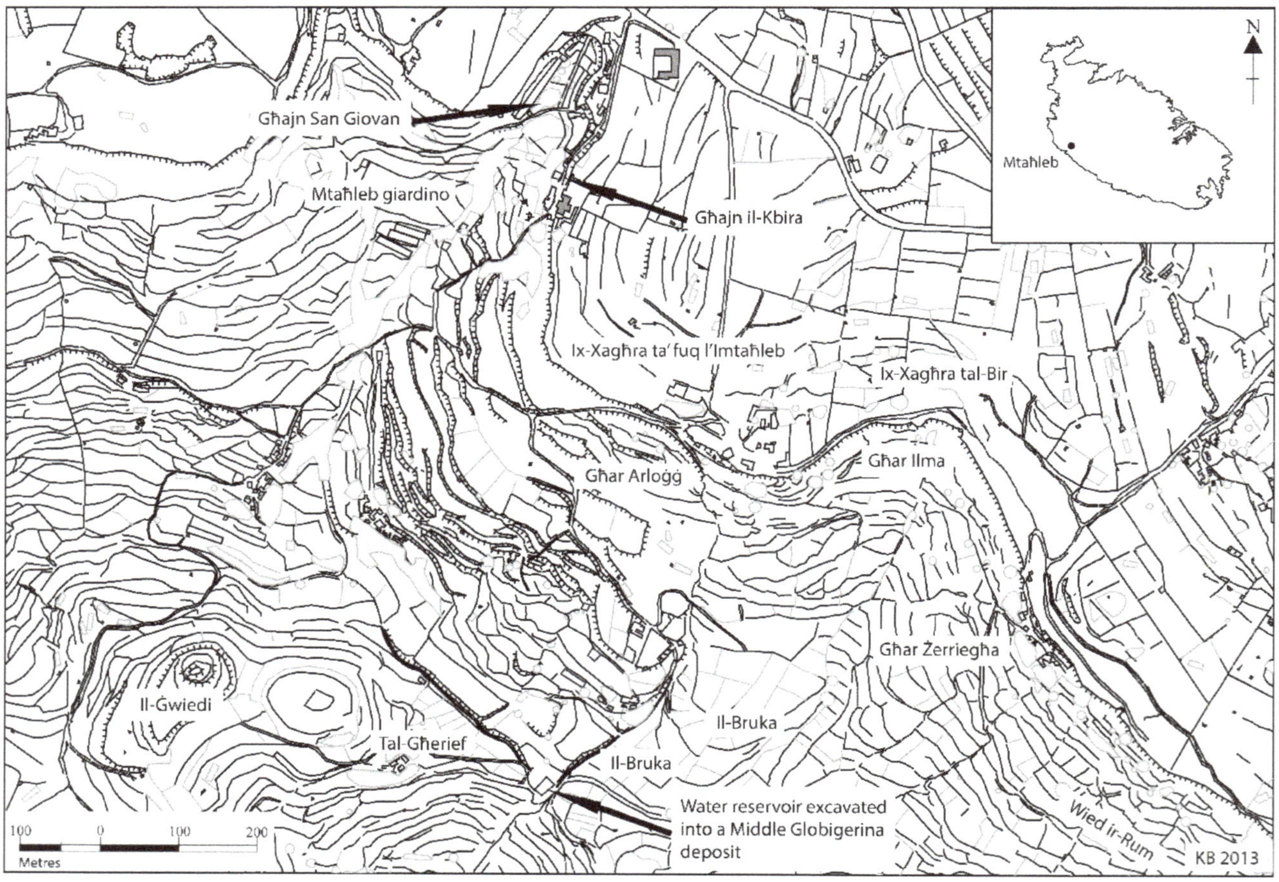

Fig. 5.51 Plan of the Mtaħleb *giardino* (RA35) located in the territory of Rabat (Malta), showing the location of Għajn il-Kbira (RA17) and Għajn San Giovan (RA24). Topographical map image adapted from MEPA Map Server data.

[30] The map reference for *Aayn il Kbira* is 41840 71100.

[31] The map reference for *l-Għar ta' l-għarwenin* is 41760 71160. The gallery name was coined during the 1940s after it became customary for British servicemen stationed in the Mtaħleb area to swim naked in the subterranean reservoir fronting the gallery. See Buhagiar K. 2002, 145, 150-1.

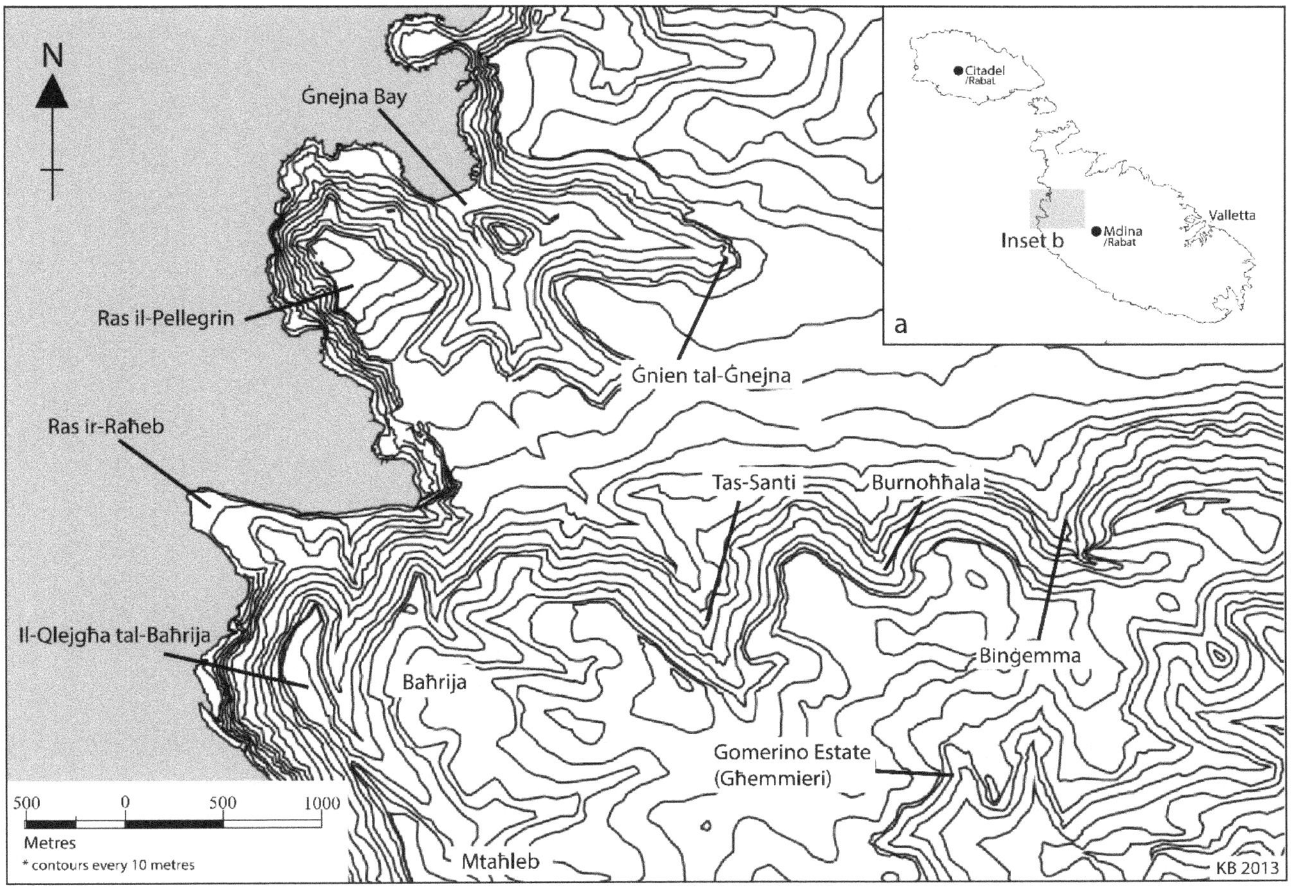

Fig. 5.52 Map of the Maltese islands (a), with inset (b) showing the spatial distribution of the principal place-names in north-west Malta mentioned in the text. Base map data: MEPA.

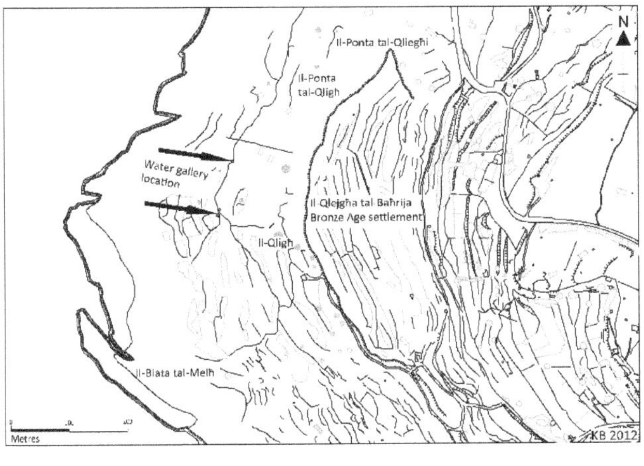

Fig. 5.53 Site plan of the Il-Qlejgħa tal-Baħrija area showing the location of two water galleries located below the west side of the Bronze Age settlement. Both are fronted by an uncovered water reservoir. Topographical map image adapted from MEPA Map Server data.

Abela's mention of *Aayn Taiba* gives no reference to its whereabouts. Its approximate location in the area of the *Giardino Grande* was nonetheless confirmed by the anonymous 1665 map of the Mtaħleb area (AIM Acta Civilia vol. 171, Case 12) (Fig. 5.48). The *Għar elma* toponym (Abela 1647, 66) which pertains to a location in the Mtaħleb area also betrays a water gallery type hydrological setup.[32] This toponym is synonymous with perched aquifer water collection either within the excavated gallery or in a fronting subterranean water reservoir (Fig. 5.51).

The presence of other water sources at Mtaħleb is attested by the *Uyed el Kasab helù* and *Bruca* place-names (Abela 1647, 66). The Bruca area is described as *Vaso, che contiene acqua, si dice anco BruKet elma, ò conserva d'acqua dalla somiglianza del luogo* (Abela 1647, 66).[33] Abela's description of the *Il-Bruka* (RA29) area confirms that an exposed Middle Globigerina deposit in the area had already been hollowed out and used for the purpose of water storage in the early seventeenth century. The *Il-Bruka* area is today synonymous with a large open-air, unlined water reservoir excavated into the Middle

[32] The first recorded instance for an *Ghar Ilma* place-name is 1470. See Wettinger 2000, 206.

[33] The map reference for the Il-Bruka area at Mtaħleb is 41810 70330.

Fig. 5.54 Interior view of a perched aquifer gallery in the Tas-Santi *giardino* (MG23). The gallery is excavated into an Mtarfa Member deposit and the rock-cut canal at its base facilitates water extraction to its exterior. Source: the author.

Fig. 5.55 Perched aquifer water collecting in the innermost section of the gallery (MG23) illustrated in Fig. 5.54. Source: the author.

Globigerina outcrop (Figs. 5.49, 5.50, 5.51). For a discussion on the water retention properties of Middle Globigerina deposits see chapter two, sections 2.4.4 and 2.6.2.

Fig. 5.56 A partly rock-cut, partly masonry constructed canal in the upper Tas-Santi (MG23) valley section channelling water out of the perched aquifer gallery (Figs. 5.54, 5.55) and which this study tentatively identifies with the *Ayn tal Carcara* (MG5) mentioned by G.F. Abela in 1647. Source: the author.

5.2.1.3 The Aayn Bierda and Aayn tal Carcara springs

Further *Aayn* toponyms which can be probably associated with perched aquifer galleries are *Aayn Bierda* (RA1) and *Aayn tal Carcara* (MG5). *Aayn Bierda* is listed by Abela as being in the area to the west of Il-Qlejgħa tal-Baħrija and Ras ir-Raħeb (Abela 1647, 67). Field surveys held in the steeply inclined terrain below the western side of Il-Qlejgħa tal-Baħrija have resulted in the detection of two water springs, both originating from water galleries. Located at map reference 39760 73100 and 39800 73190 respectively, both are of unknown antiquity and are fronted by open-air water reservoirs (Figs. 5.52, 5.53). The galleries are located in close proximity of troglodytic settlements and agricultural land; it may well be that one of these is the *Aayn Bierda* mentioned by Abela.

Aayn tal Carcara (MG5), on the other hand, is listed by Abela soon after the *Samti* (Santi) toponym and before the mention of the *Giardino detto ta' Samti* (Abela 1647, 68). In the upper section of the Tas-Santi (MG23) valley are two perched aquifer galleries hewn into an Mtarfa Member deposit in the north-facing ravine of the valley (Buhagiar K. 2002, 108-9; Buhagiar K. 2008B, 363-64).[34] The easternmost gallery is inaccessible and appears to be dry, but the other is relatively easy to access and still yields a perennial supply of water (Figs. 5.52, 5.54, 5.55, 5.59). A masonry canal channels water from this gallery towards an open-air water reservoir, from which a vast network of canals transports water from the reservoir to the lower portion of the Tas-Santi estate (Figs. 5.56, 5.57, 5.58, 5.60). One of these two perched aquifer galleries may be the *Aayn tal Carcara* mentioned by Abela in 1647.

Fig. 5.57 The uncovered reservoir fronting the perched aquifer gallery located in the upper reaches of the Tas-Santi valley (MG23; Figs. 5.54, 5.55). Masonry canals gravity feed water to agricultural territory located further downhill. Source: the author.

Fig. 5.58 The water trough adjoining the water reservoir (Fig. 5.57) in the upper reaches of the Tas-Santi *giardino* (MG23) in the territory of Mġarr (Malta). Source: the author.

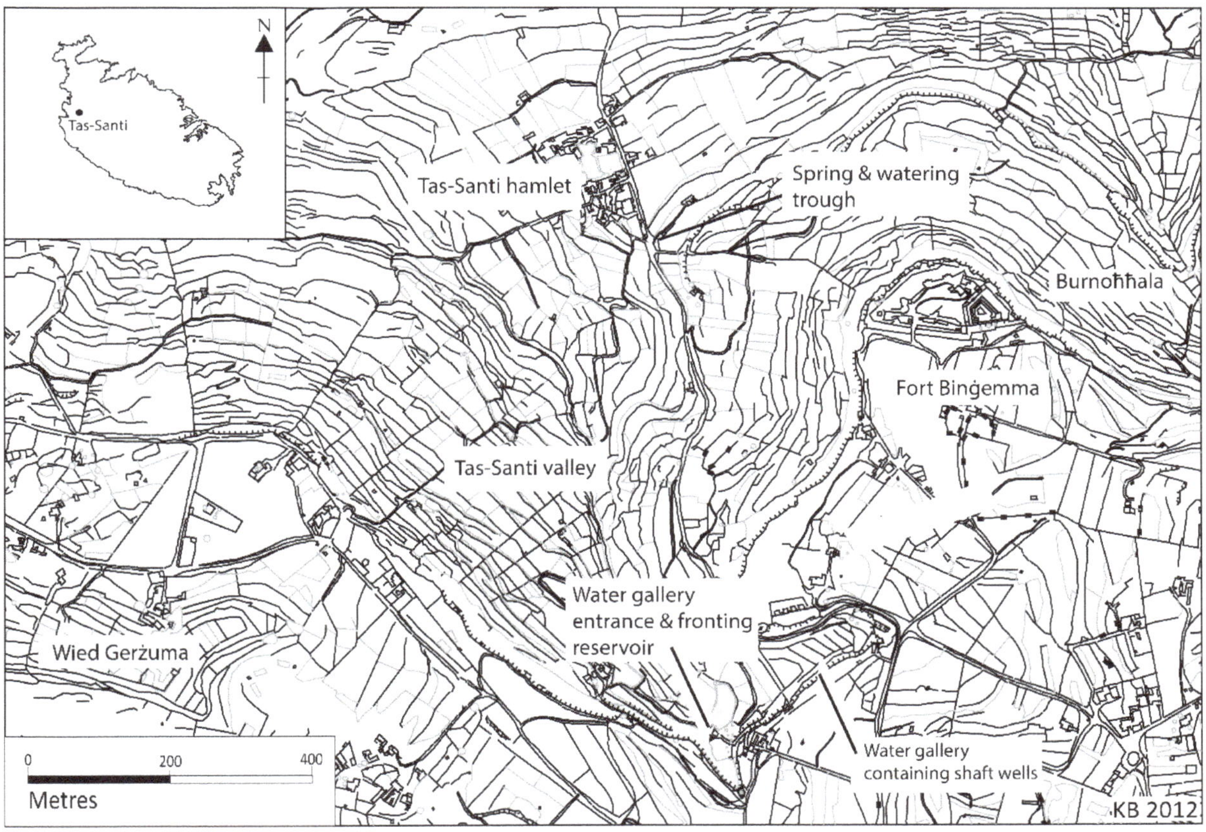

Fig. 5.59 Site plan of the Il-Qlejgħa tal-Baħrija area showing the location of two water galleries located below the west side of the Bronze Age settlement. Both are fronted by an uncovered water reservoir. Topographical map image adapted from MEPA Map Server data.

[34] The perched aquifer galleries are located at map reference 42160 72950.

Fig. 5.60 North-facing view of a section of the Tas-Santi estate (MG23). Dense reed growth is indicative of the abundant water deposits present in the upper section of the Tas-Santi valley. Visible in the background (circled in red) is the Tas-Santi hamlet. Source: the author.

Similarly, the *S. Antonio tal Ghimeri* area, more commonly identified today as Gomerino (RA15), is described as a *Giardino con fonte* (Abela 1647, 66). The spring originates from an expertly-excavated perched aquifer gallery, the roof of which is pierced by a series of shaft wells and is typologically similar to a *qanat* water extraction system (Figs. 4.5, 5.52, 5.62, 5.63, 5.64, 5.65, 5.66).[35]

5.2.1.4 Springs in the Mġarr (Malta) territory

A surprising concentration of water sources is listed by Abela in the Mġarr (Malta) area. Five of these appear to be located in the western section of the Mġarr territory and are named as *Aayn Sfuria* (MG20), *Aayn gefrà* (MG3), *Ciorciara* (MG8), *Ved tal Gineyna* (MG 24) and *Aayn* (MG1). *Aayn Sfuria*, which is described as *fontana del giallo, o da Safar, che suona partenza* (Abela 1647, 68), is located to the west of the Mġarr (Malta) settlement, to the south-east of Fomm ir-Riħ Bay (Figs. 5.67, 5.68).[36] In Survey Sheet 56, compiled and corrected in 1898, the area is marked as *Ta Ghasfuria*. In the current MEPA datasheets, the area toponym has been erroneously corrupted to Ta' Għasfura (MG20), meaning "of the bird".[37] Present in this area is an exposed deposit of Middle Globigerina Limestone. There is a strong probability that this rock stratum, characteristically impervious, functioned as the catchment area for this spring. It is possible that the name of the spring originally stemmed from the yellowish colour of the resultant Middle Globigerina deposit. The water retention properties of this geological deposit are demonstrated through the excavation, probably during the past century, of two large, un-mortared rectangular water reservoirs which retain water throughout the year.[38]

Fig. 5.61 A rectangular-shaped shaft, probably giving access to a perched aquifer gallery located in the upper valley section at Tas-Santi (MG23). A series of handholds and footholds facilitated access to the shaft bottom. Source: the author.

Fig. 5.62 General view of the Gomerino estate (RA15) showing the lush vegetation characteristic of the *giardino*-framework. The reeds in the top-right corner of the picture indicate the location of the water gallery entrance. Source: the author.

[35] The Gomerino estate perched aquifer gallery is located at map reference 43200 72670.

[36] Personal communication by M. Vassallo.

[37] Personal communication by M. Vassallo.

[38] These water reservoirs are located at map reference 41070 73960.

The location of *Aayn gefrà* (MG3), listed by Abela as *fonte cosi chiamato* (Abela 1647, 68), remains unknown. As this water source is listed soon after *Aayn Sfuria* and before *Redum pellegrin*, I suspect that this might be one of the springs originating from three perched aquifer water galleries excavated into the south side of Ras il-Pellegrin at map reference 40520 74730 (Fig. 5.68).[39] *Ciorciara* (MG8), described by Abela as *luogo per doue cala, e*

Fig. 5.63 The narrow square-headed doorway giving access to the perched aquifer gallery at the Gomerino estate (RA15) in the territory of Rabat (Malta). Source: the author.

Fig. 5.64 A circular shaft giving access into the perched aquifer gallery of the Gomerino estate (RA15). Source: the author.

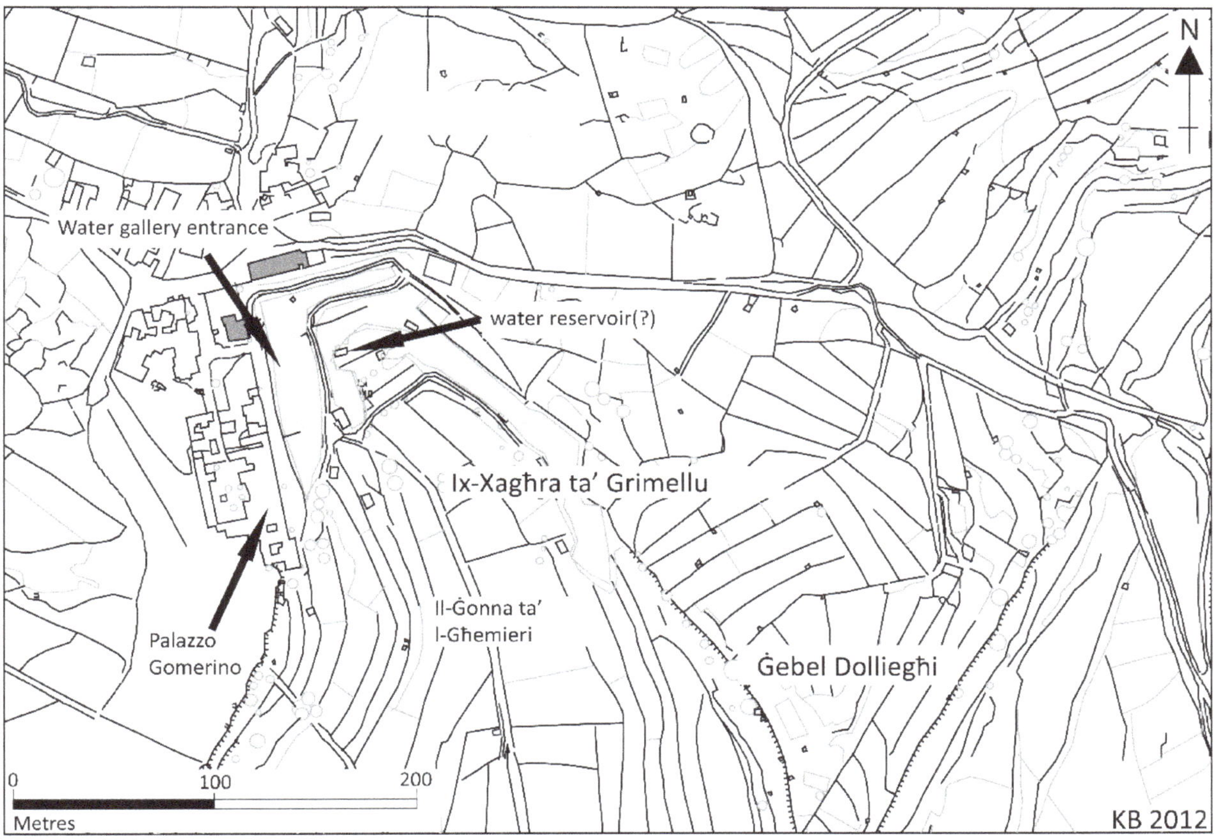

Fig. 5.65 Site plan of the Gomerino estate (RA15) in the territory of Rabat (Malta), showing the location of the perched aquifer gallery feeding this agricultural establishment with a perennial water supply. In the vicinity of the gallery entrance is a sizeable rectangular depression which possibly functioned as a water collection reservoir. Topographical map image adapted from MEPA Map Server data.

[39] These water sources were originally brought to my attention by A. Caruana, whose family cultivates land in the area. Arrangements to visit these perched aquifer galleries were undertaken by M. Vassallo.

scorre l'acqua denominate in questa guisa dal mormorio che fà cadendo is probably a reference to a seasonal spring which drained rain water into the Ġnejna valley system (Abela 1647, 68). The modern road giving access to Ġnejna passes through the *Ciorciara* area, approximately located at map reference 41810 75350 (Fig. 5.68, 5.71).[40]

Fig. 5.66 A rectangular-shaped shaft piercing the ceiling of the perched aquifer gallery at Gomerino estate (RA15) in the Rabat (Malta) territory. Rock-cut handholds and footholds (left) facilitate access to the gallery's interior. Source: the author.

Fig. 5.67 Former military entrenchment works in the Fomm ir-Riħ area of Mġarr (Malta), now function as rainwater collection reservoirs. The principal geological characteristic of this area is a sizeable, yellow-coloured Middle Globigerina Limestone outcrop with good impervious qualities, making it probable that the *Aayn Sfuria* spring (MG20) mentioned by Abela in 1647 formed within this rock deposit.

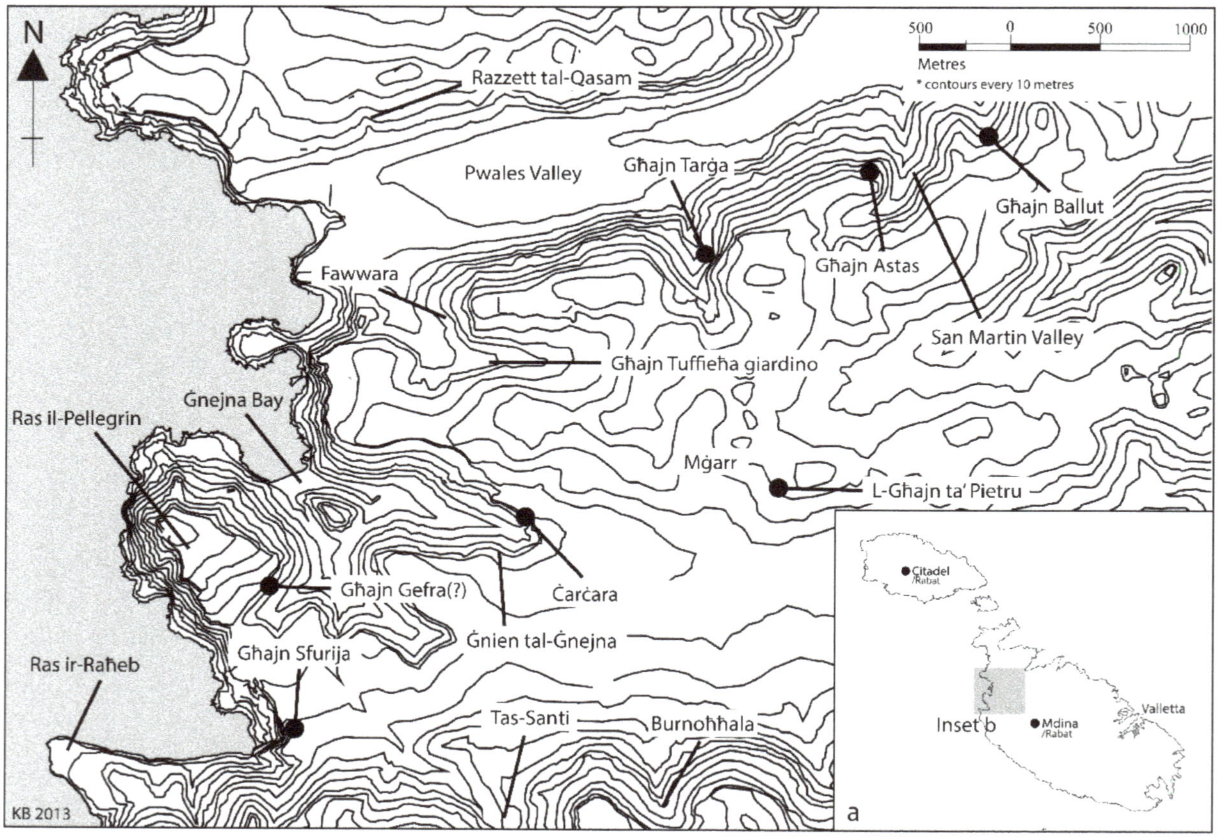

Fig. 5.68 Map of the Maltese islands (a), with inset (b) showing the spatial distribution of the springs in the Mġarr (Malta) territory mentioned by Abela in 1647 and the location of which was established by the present writer.

[40] Personal communication by M. Vassallo. The *Ic-ciarciark* toponym is also recorded on the 1903 version of the 1:2,500, Survey Sheet 35 for Malta.

Two other water sources were mentioned by Abela at *Ved tal Gineyna* (MG24), which he described as *Valle, ò torrente di esso giardino con due fontane* (Abela 1647, 69). The Ġnejna *giardino* (MG24) is located in the uppermost section of the Ġnejna valley system, at the approximate map reference of 42120 75160 (Figs. 5.68, 5.71). This fertile horticultural plantation is serviced by three water springs,[41] all of which appear to be perennial in nature and originate from perched aquifer galleries. One of the galleries is excavated into an Mtarfa Member deposit on the south side of the Ġnejna *giardino* at map reference 42010 75090. A collapse in the entrance section

Fig. 5.69 The collapsed entrance to a perched aquifer gallery excavated into the southern side of the Ġnien tal-Ġnejna (MG24) in the territory of Mġarr (Malta). Source: the author.

Fig. 5.70 The Ġnien tal-Ġnejna (MG24) in the territory of Mġarr (Malta). The uncovered water reservoir inside which the water extracted by the perched aquifer gallery illustrated in Fig. 5.69 was stored for later distribution. Source: the author.

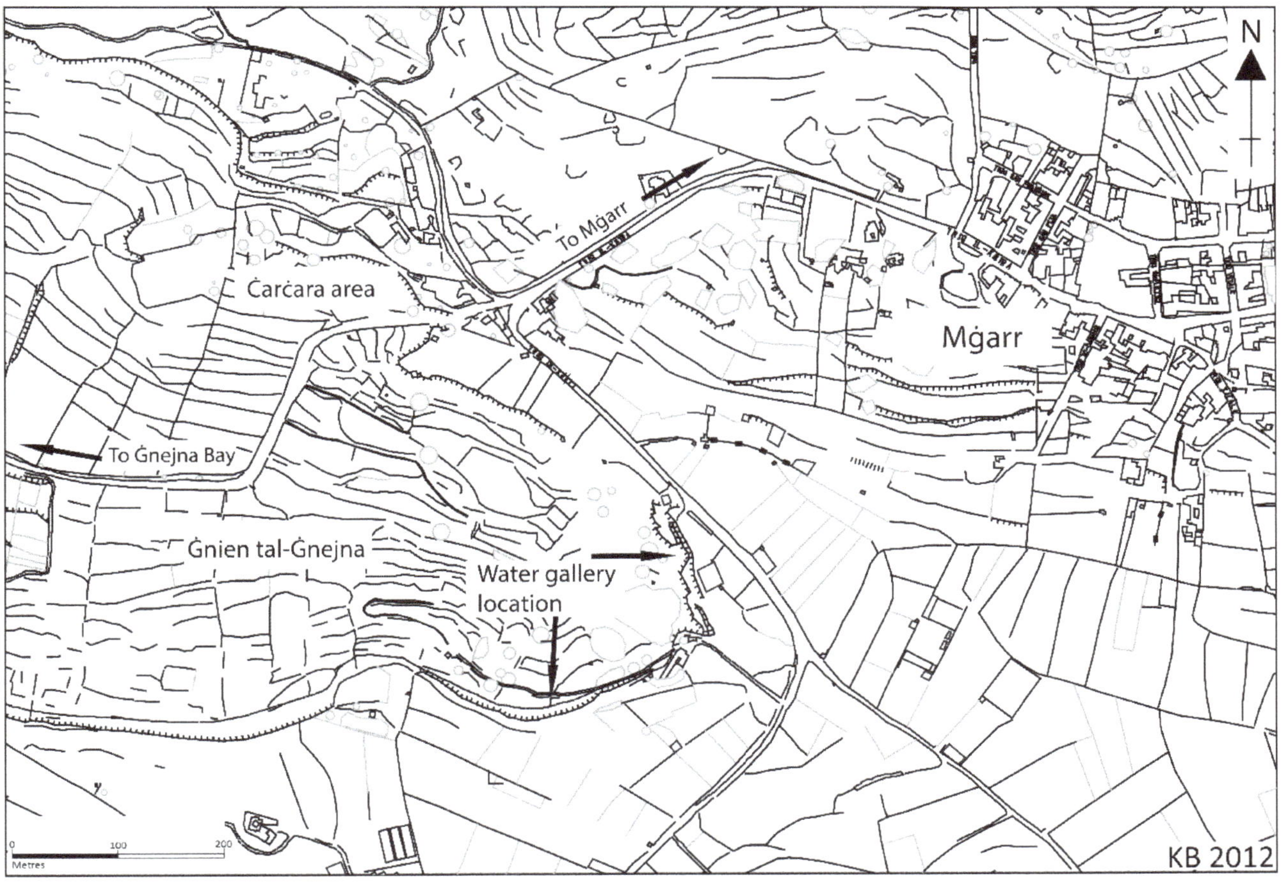

Fig. 5.71 Site plan showing the location of the Ġnien tal-Ġnejna (MG24) and the Ċarċara (MG8) area close by, in the territory of Mġarr (Malta). Two perched aquifer galleries are excavated into the north-east side of the Ġnien tal-Ġnejna. The other is located on the southern side of the valley. Topographical map image adapted from MEPA Map Server data.

[41] Personal communication by M. Vassallo.

Fig. 5.72 The Ġnien tal-Ġnejna (MG24) in the territory of Mġarr (Malta). Interior detail of the perched aquifer gallery tunnelling into the north-east valley side of this *giardino*. Source: the author.

Fig. 5.73 Calcite deposits covering the floor of the perched aquifer gallery illustrated in Fig. 5.72. These indicate that this water extraction system is centuries old. Source: the author.

of the gallery prevented its investigation (Fig. 5.69). The water retrieved by this gallery was stored in a fronting uncovered reservoir for later distribution (Fig. 5.70).

The other two perched aquifer galleries are excavated in close proximity to each other, into the north-eastern ravine at the Ġnejna *giardino* (Fig. 5.71).[42] The owner of the land in which both galleries are located claims that the water tapped by each of these water extraction systems is of a different quality from the other. It is indeed possible that these two galleries are tapping different aquifer sources.[43] Both galleries are excavated in Mtarfa Member deposits

Fig. 5.74 Detail of the calcite deposits lining the floor of the perched aquifer gallery illustrated in Figs. 5.72 and 5.73. Source: the author.

and were, at an unknown date, provided with a fronting subterranean water storage reservoir.[44] The galleries are difficult to reach and the investigation of only one of them was possible (Figs. 5.72, 5.75). Calcite formation on the gallery floor indicates that this is centuries old (Figs. 5.73, 5.74).

The location for another *Aayn* (MG1) water source described by Abela as *fonte con vaso, doue si dà à bere à gl'animali sotto il Migiarro* is not known (Abela 1647, 69). The same can be said of *Aayn Kasab* (MG4),

Fig. 5.75 Another gallery in the Ġnien tal-Ġnejna (MG24) is located ca. one and a half storeys above the valley floor and can only be accessed by ladder. Without the employment of a safety harness and climbing equipment its interior is difficult to access. Source: the author.

[42] The location of these perched aquifer galleries was indicated to me by M. Vassallo. He also made arrangements with the owner for the field-investigation of this *giardino*.
[43] Personal communication by M. Vassallo.
[44] This is an area of notable archaeological interest. A ca. 3.5 m high dry-rubble field wall built out of sizeable stone is visually similar to other known Bronze Age period structures. Furthermore, M. Vassallo personally communicated that flash flooding in this section of the Ġnejna valley had in the past temporarily exposed large sized ashlars typical of the Punic-Roman period.

described by Abela as *Fonte delle canne* (Abela 1647, 70). The *Aayn Kasab* toponym immediately precedes the *Torre Falca* place-name and it is possible that this water source was located in the Ta' Torri Falka area close to the north-eastern boundary of the Mġarr (Malta) territory.[45] The

Fig. 5.76 Exterior view of the Għajn ta' Pietru (MG1) at Mġarr (Malta). Source: the author.

Aayn (MG1) toponym can perhaps be associated with the locality known as Tal-Għajn, in which there is a water-yielding spot marked as "fountain" on the 1899-1900 1:2,500 Malta Survey Sheet 36 and which is known as L-Għajn ta' Pietru (Figs. 5.68, 5.76, 5.77, 5.78, 5.79, 5.80, 5.82).[46] It is also interesting to note that the nearby cemetery was located on the only exposed Blue Clay deposit in the area. During the rainy season, the cemetery would have been another possible water-yielding spot.

Aayn Ballut (MG2) is described by Abela as *Fonte delle quercie* (Abela 1647, 72). For this there are two possible alternatives, the place-name of both of which is Il-Ballut and which are located within relative close proximity. The first *Ballut* place-name refers to a site at Wardija on the outskirts of the Mġarr (Malta) territory and which overlooks Xemxija and St Paul's Bay (Fig. 5.81). The Il-Ballut place-name probably originates from a dense cluster of *Ballut*, or Holm Oak trees, which still characterise this area and are considered to be one of the few surviving forest remnants which might, possibly, date to the first millennium AD.[47] A water spring originating from within a perched aquifer gallery at map reference 44710 77530 feeds this area with a perennial water source and can be

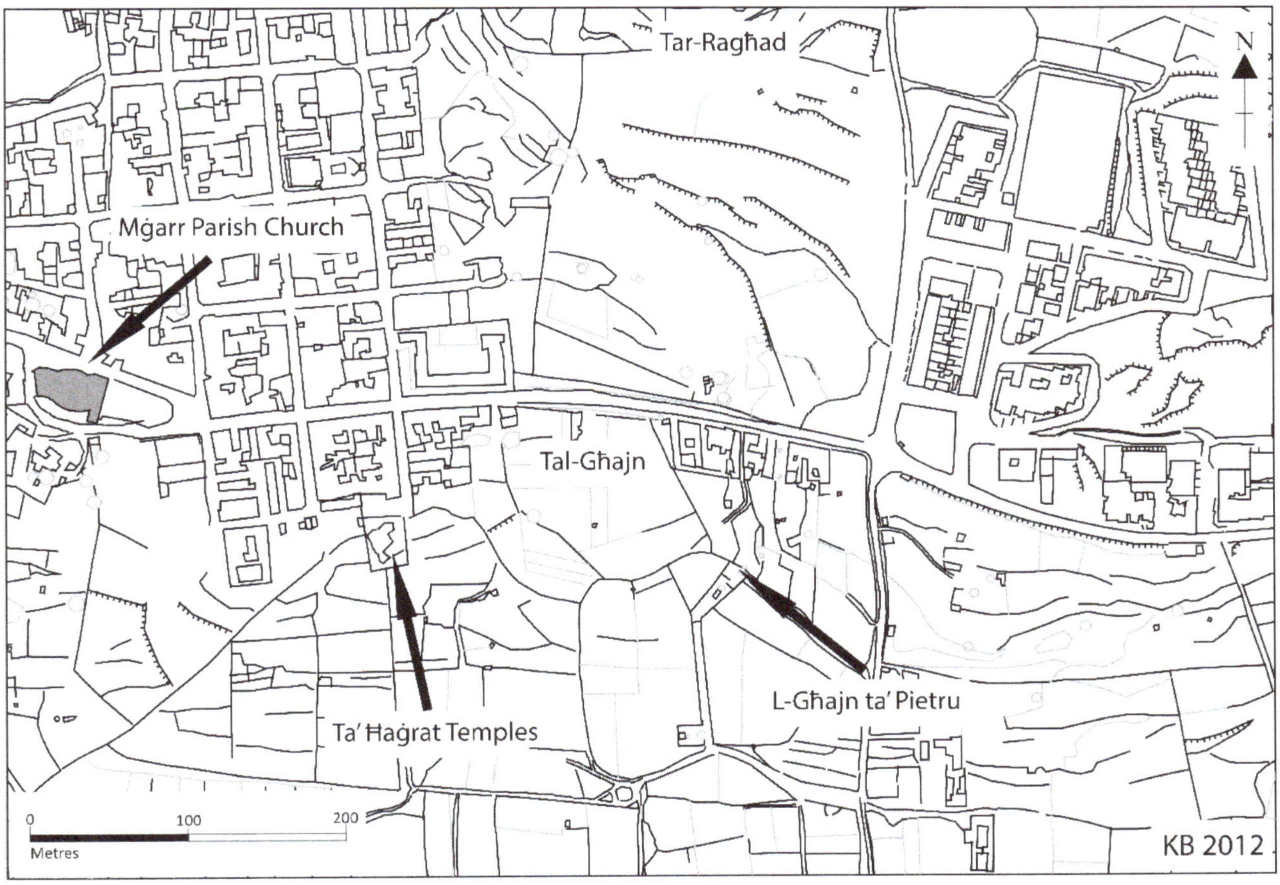

Fig. 5.77 Site plan showing the location of a water-yielding crevasse at Mġarr (Malta) known as L-Għajn ta' Pietru (MG1). Topographical map image adapted from MEPA Map Server data.

[45] The *Aayn Kasab* spring was discussed with M. Vassallo who suggested that this might have been located in the Tal-Falka area.

[46] Personal communication by M. Vassallo. Tal-Għajn is a perennial water source.

[47] Refer to http://www.mepa.org.mt/wardija.

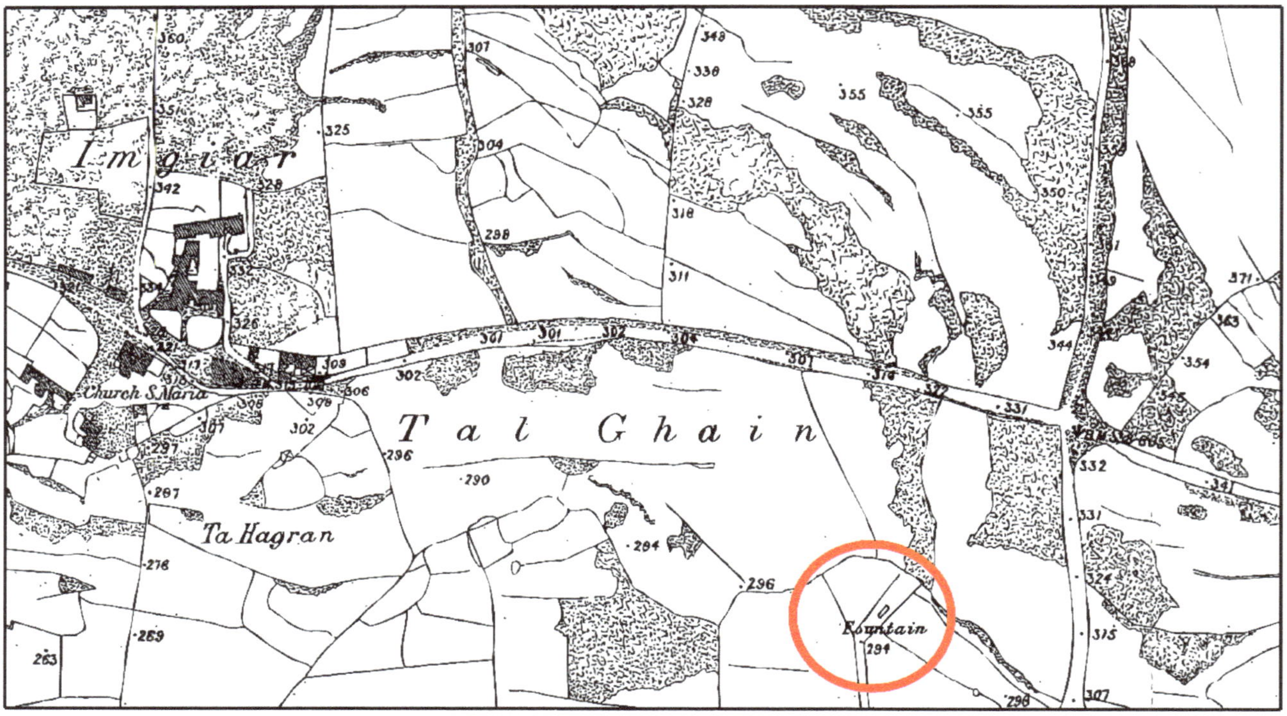

Fig. 5.78 Detail of Malta Survey Sheet 36 drawn-up in 1899 showing the location of the Għajn ta' Pietru (MG1) at Mġarr (Malta).

Fig. 5.79 Detail of the Għajn ta' Pietru (MG1) exterior at Mġarr (Malta). The short gallery is accessed through the visible narrow rock-cut opening. Source: the author.

Fig. 5.80 Detail of the narrow entrance giving access to the Għajn ta' Pietru spring (MG1) at Mġarr (Malta). The rock-excavated gallery is only ca. 1 m deep. A cross carving is present directly above the spring's entrance. Source: the author.

tentatively associated with the *Aayn Ballut* place-name mentioned by Abela (Figs. 5.83, 5.84, 5.85).[48]

[48] The first known mention of Għajn Ballut dates to 1500. See Wettinger 2000, 181.

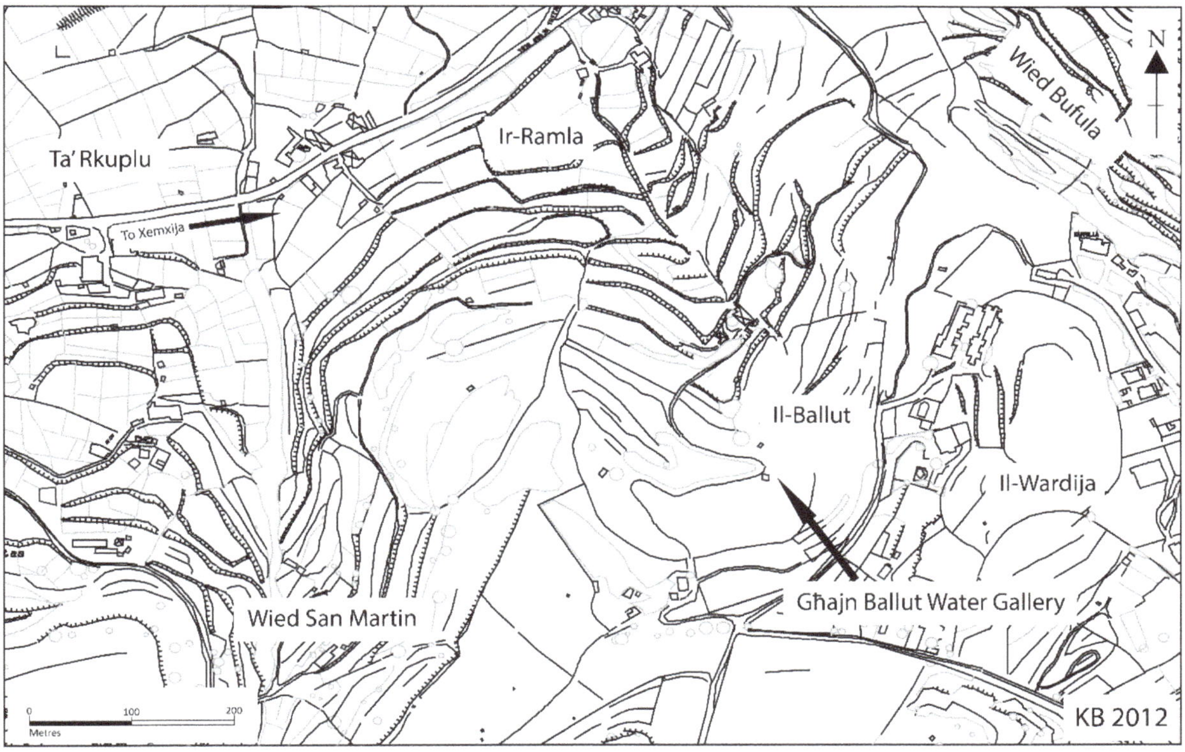

Fig. 5.81 Location plan of the Il-Wardija and the Wied Ta' San Martin areas showing the location of the Għajn Ballut perched aquifer gallery (MG2). Topographical map image adapted from MEPA Map Server data.

Fig. 5.82 Detail of the Għajn ta' Pietru (MG1) interior. The spring cannot be technically classified as a gallery, even though it is yielding a perched aquiferous source. Even though water flow is reduced to a small trickle during the summer months, it still yields a perennial water source. Source: the author.

Fig. 5.83 The entrance section of the Għajn Ballut perched aquifer gallery (MG2) at Wardija. Source: the author.

The other area also with an Il-Ballut place-name relates to a Government property overlooking Il-Pwales valley at map reference 42490 76850 (Figs. 5.86, 5.87). This tenement is known as Il-Ballut tal-Ħoxni (MG9) and is located in very close proximity to the Għajn Tuffieħa territory (MG15). To the present writer's knowledge, there does not appear to be Knights' period documentation on this property. On the other hand, Il-Ballut tal-Ħoxni is well documented in the Descriptive Plans of the Crown Property *cabreo* of 1866, in which mention is made of *tre sorgenti d'acqua munity con vasche e canali* (NAM. Descriptive Plans of the Crown Property, 1866, f. 216).[49] In two of the instances, water sources are clearly indicated on the site plan of this property as originating from the side of a ridge or ravine, making it likely that perched aquifer galleries are present within. The location of three water collection reservoirs is likewise indicated.[50]

[49] This property was formerly known as Il-Ballut ta' Għajn Tuffieħa.

[50] The presence of perched aquifer galleries at Il-Ballut tal-Ħoxni was indicated to me by both C. Grech and M. Vassallo. An interview with F.

Fig. 5.84 The Għajn Ballut perched aquifer gallery (MG2) at Wardija. Interior detail of the rock-excavated channel. Source: the author.

Fig. 5.85 The Għajn Ballut perched aquifer gallery at Wardija (MG2). The water channel carved into the gallery floor was widened at an unknown date. Source: the author.

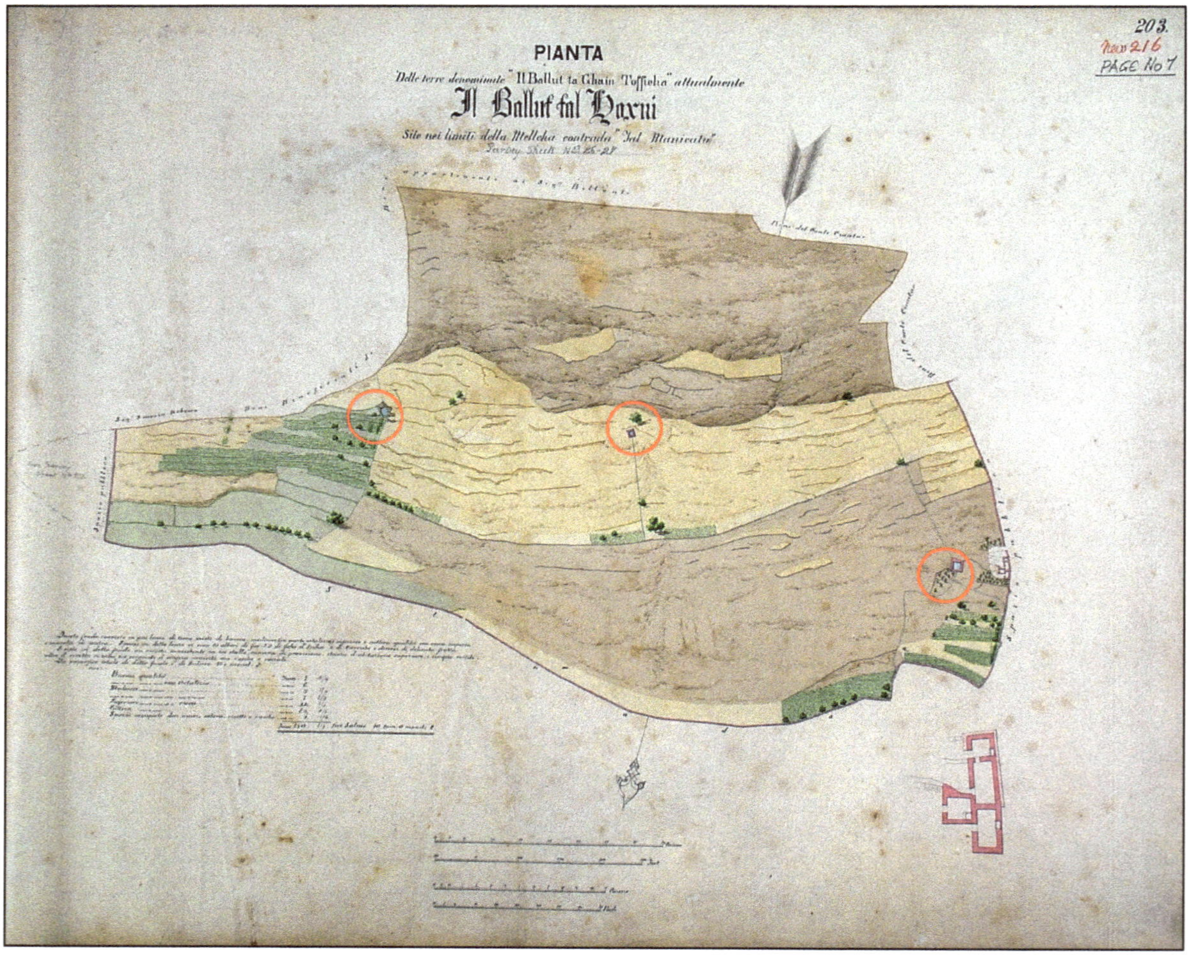

Fig. 5.86 A nineteenth-century plan of the Il-Ballut tal-Ħoxni (MG9) estate overlooking Pwales Valley in the territory of Mġarr (Malta). Indicated in blue are reservoirs which probably harvest water yielded by two perched aquifer galleries known to be present in this area. Source: NAM. Descriptive Plans of the Crown Property, 1866, f. 216.

Vella, an octogenarian born and brought up in the Il-Ballut tal-Ħoxni area confirmed that these springs originate from within perched aquifer galleries, but according to him, none can be identified with Għajn Ballut. He is inclined to associate Għajn Ballut with the Il-Ballut tal-Wardija area nearby. F. Vella communicated that the names for the Il-Ballut tal-Ħoxni springs are Tal-Ġuż and Tal-Gandott ta' Għajn Ħamiem.

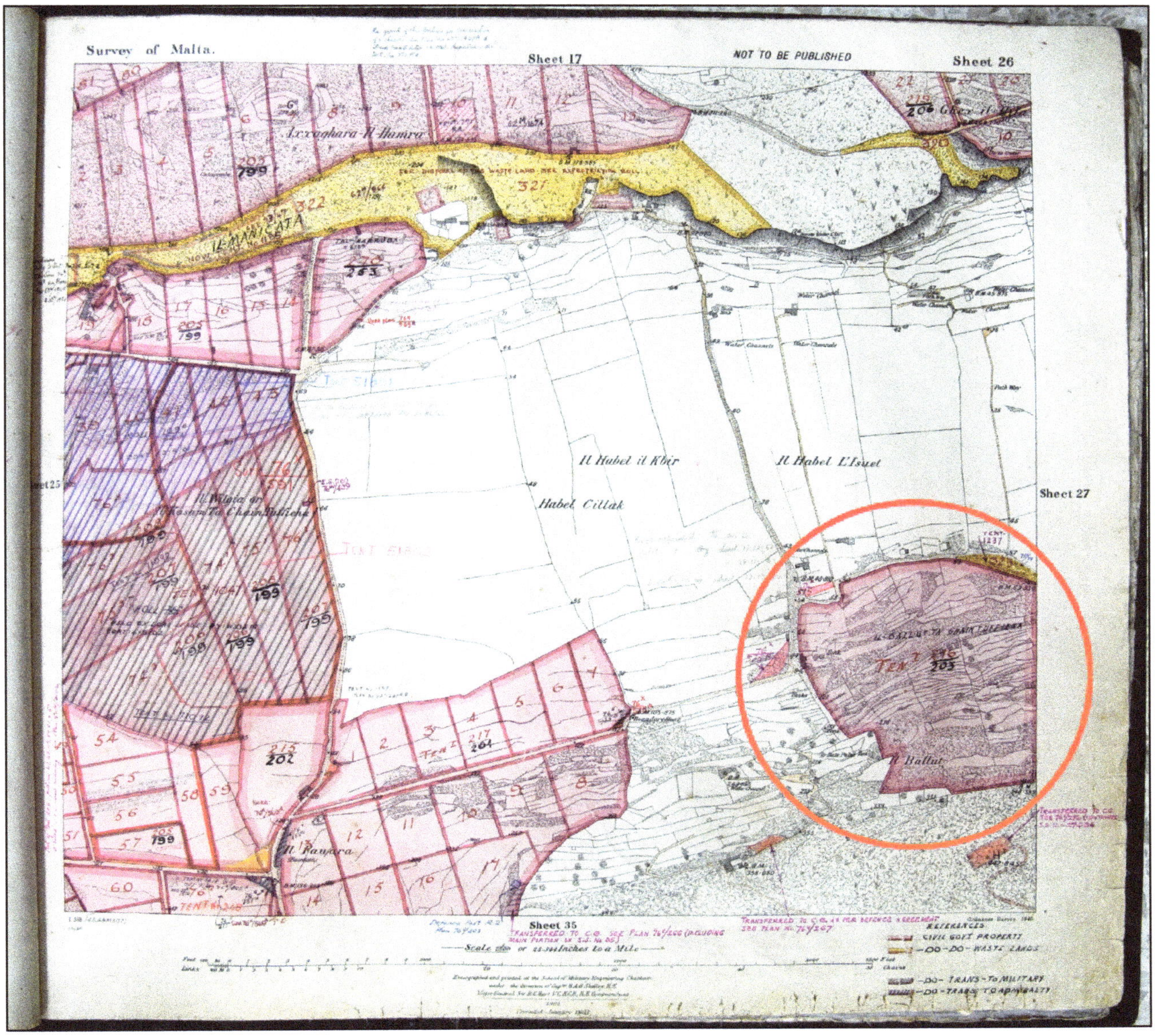

Fig. 5.87 Government Property Survey Sheet Links showing Government-owned property in the Għajn Tuffieħa (MG15) and Pwales valley area. The circle indicates the Il-Ballut tal-Ħoxni estate (MG9). Source: NAM. Government Property Survey Sheet Links, Sheet 26.

The location of two other close-by sources, *Il-Mina l-Kbira* (MG12) and *Il-Mina ta' Maġemba* (MG13), is not known, but they both appear to originate from within perched aquifer galleries. Field-research has again helped establish that the nearby *Aayn Estas* (Għajn Astas – MG14), simply listed by Abela as a *Fontana*, similarly originates from a perched aquifer gallery located at map reference 44022 77365 (Abela 1647, 72) (Figs. 5.68, 5.88, 5.89, 5.90, 5.91, 5.92). Another spring with the name of *Aayn Targia* is described as *Fonte dello scaglione, giardino* (Abela 1647, 72). The nature of this water source located in the Tal-Palma area on the outskirts of Mġarr at map reference 43140 76810 remains unknown (Figs. 5.93, 5.98).[51]

The *Ta' Safsafa* (MG21) locality is described by Abela as: [...] *del salce albero, quiui sono fontane, e ruscelli d'acque, dove per avventura erano anticamente piantati salci, che ricercano qualità di terreno acquoso* (Abela 1647, 72). *Ta' Safsafa* is located in a fertile agricultural territory and adjoins the Ta' San Martin valley in the outskirts of Mġarr (Malta),[52] where various perennial water sources originate from perched aquifer galleries. Abela's mention of both *fontane* and *ruscelli* in the *Ta' Safsafa* area is a possible indication that at least some of the hydrological sources in its neighbourhood originated from perched aquifer galleries.

The agricultural territory of *Aayn Toffieha* (MG15) is described as *Fonte del pomo, territorio, e giardino del sig.*

[51] Personal communication by M. Vassallo.

[52] The approximate map reference for the Ta' Safsafa area is 45120 78020.

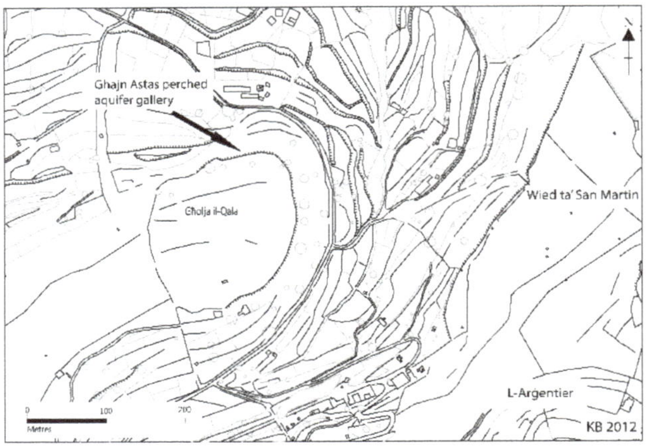

Fig. 5.88 Plan of the Wied ta' San Martin area in the limits of Mġarr (Malta) showing the location of the Għajn Astas (or Estas) perched aquifer gallery (MG14). Topographical map image adapted from MEPA Map Server data.

Fig. 5.90 Rubble walls flanking the entrance to the Għajn Astas perched aquifer gallery (MG14) in the territory of Mġarr (Malta). Source: the author.

Fig. 5.89 Water from the Għajn Astas perched aquifer gallery (MG14) in the territory of Mġarr (Malta) is channelled into this old uncovered reservoir. Source: the author.

Fig. 5.91 Interior detail of the Għajn Astas perched aquifer gallery entrance (MG14). The masonry wall (foreground) functions as a dam and increases the water level within the gallery system. Source: the author.

Gran Maestro Principe, questo fù già feudo (Abela 1647, 72). The *fonte* mentioned by Abela is unknown, but in the lower section of this horticultural production area is a perched aquifer water gallery, known as *Mina t'Isfel*,[53] which provides the area with a perennial water source (Fig. 6.72).[54] Further down the valley is another water-receiving spot known as *Fauara* (Abela 1647, 72). A derelict open-air water storage reservoir located at map reference 41450 76650 forms part of this water system.[55] The hydrological setup of both the Għajn Tuffieħa *giardino* and the neighbouring Il-Fawwara area (MG11) is discussed in more detail in chapter six, section 6.1.9.

5.2.1.5 The Mellieħa territory

Aayn Znuber (ME4), located just outside the boundaries of the territory encompassed by the recently designated Majjistral Nature Park, is simply listed as *Fonte de'pini* (Abela 1647, 72). This water source was located at map reference 41580 78649 and also originates from a perched aquifer gallery, the entrance section of which has collapsed (Figs. 5.93, 5.97).[56]

[53] Personal communication by M. Vassallo.
[54] Personal communication by M. Vassallo. The map reference for this perched aquifer gallery is 41680 76140.
[55] Personal communication by M. Vassallo.
[56] Għajn Żnuber was located through the valuable assistance of T. Gambin, who provided me with directions indicating the whereabouts of this *fontana*. The Għajn Żnuber gallery is approximately 148 m long and in the mid-twentieth century yielded over 4,500,000 litres of water per year. In summer its water-yield amounted to ca. 5460 litres per day. See Morris 1952, 101.

Aayn Zeituna (ME3) in the territory of Mellieħa, is listed as *fonte dell'oliuo, territorio della Chiesa Catedrale di S. Paolo* (Abela 1647, 73) (Fig. 5.93). The source of this water spring is unknown, but at least one perched aquifer gallery tunnels into the east side of Għajn Żejtuna valley at map reference 43651 80755. On the other hand, the *Giardino d'Angarao* (Ġnien Ingraw – ME1), described as: [...] *(nome proprio) cò vna picciola fontana* (Abela 1647, 73), and *Giardino di S. Nicola* (Ġnien San Niklaw – ME7) (Abela 1647, 73), listed as having a *fonte*, both originate from perched aquifer water galleries. An investigation of the innermost section of the Ġnien Ingraw gallery revealed a deeply incised Latin cross, flanked on either side by two carved dates. These read 1854 and 1849 respectively and probably record instances when maintenance works were carried out on the gallery interior (Figs. 5.93, 5.94, 5.99).

Fig. 5.92 Interior view of the entrance section of the Għajn Astas perched aquifer gallery (MG14). Source: the author.

The *Aayn Tuta* area (ME2), listed by Abela as containing a *fonte* [...] *posto quasi in mezzo alla Penisola* (Abela 1647, 75), is located at an approximate distance of 560 metres south-south-east of the Latmija subsidence structure in the territory of Mellieħa (Buhagiar K. 1997, 64-70) (Figs. 5.93, 5.100). There are actually two water-yielding spots in this area located at a linear distance of ca. 100 m from each other. The principal source appears to be the one located at map reference 40054 81621 and originates from a perched aquifer gallery, the outer section

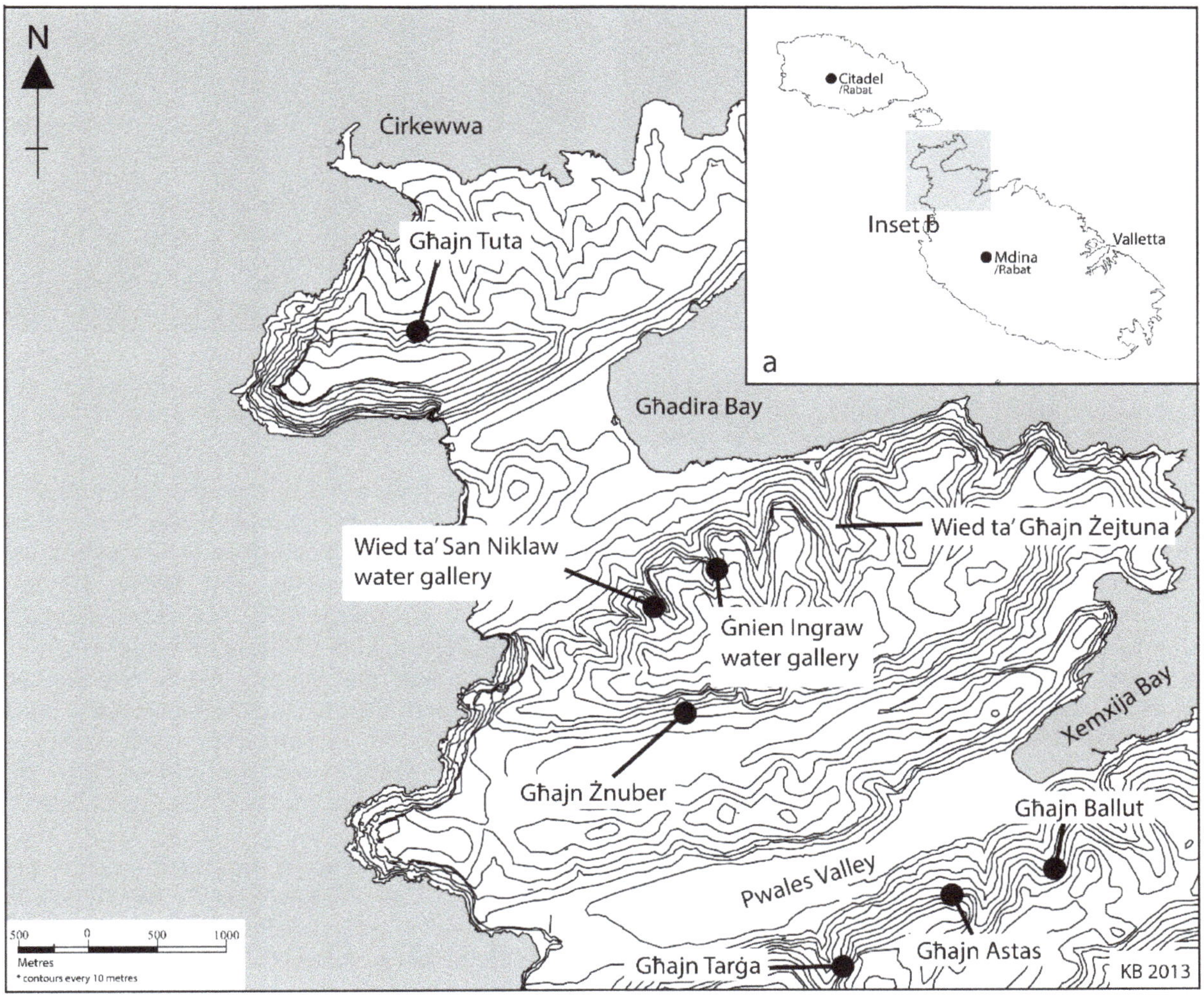

Fig. 5.93 Map of the Maltese islands (a), with inset (b) showing the spatial distribution of the springs in the Pwales and Mellieħa territories mentioned by Abela in 1647 and the location of which was established by the present writer.

Fig. 5.94 The innermost section of the Ġnien Ingraw perched aquifer gallery (ME1). Two dates "1854" and "1849" flank either side of a deeply incised Latin cross and probably mark past maintenance works carried out within the gallery. Source: the author.

of which is partly masonry-built (Figs. 5.95, 5.96). A shaft in one of the overlying terraced fields gives direct access to this water system. Lining the gallery floor is a finely carved masonry canal through which the tapped aquifer water is carried to the gallery's exterior. The other water source is located at map reference 39941 81610 and likewise appears to originate from a small perched aquifer gallery. Water retrieved by both galleries is harvested in fronting reservoirs for later distribution and use.

5.2.1.6 Other springs and water-related place-names

The *El Kattara* place-name (Abela 1647, 70), located in the Buqana area close to Wied il-Qlejgħa (RA57),[57] is

Fig. 5.95 General view showing the entrance to the Għajn Tuta perched aquifer gallery (ME2) and its fronting reservoir at Ċirkewwa, Malta. Source: the author.

Fig. 5.96 The Għajn Tuta perched aquifer gallery (ME2). Interior view of the initial gallery section. Source: the author.

described by Abela as *colatoio terreno*. The interpretation of this toponym cannot be easily determined, but it does seem to be water-related. Il Qattara is located in an area of exposed Upper Globigerina and Middle Globigerina deposits at map reference 46751 73466 (Debono and Xerri 1993, Geological Map of Malta, Sheet 1) (Fig. 5.101). Keeping in mind that flanking the western side of the Buqana territory there is an Upper Coralline Limestone perched aquifer, it is more than likely that water seepage from this area would be preserved and possibly stored in the largely impermeable Middle Globigerina plains.

Aayn Rihana, located in the Burmarrad area is described as *Fonte della Mortella* (Abela 1647, 70). Well-informed sources have actually confirmed the present writer's suspicion that Għajn Riħana (MO4) originates from within a perched aquifer gallery, one of the entrances to which

[57] The map reference for the *Il Kattara* area is 46780 73420.

must still be available at the approximate map reference of 46680 75877.[58]

The location for the *Aayn tal Calcara* (MG6), described as *fonte della furnace di calce* is likewise not known (Abela 1647, 73). However, a *Tal Calcara* place-name is recorded in the 1942 Ordnance Survey Sheet 101 for Malta. *Tal Calcara* is the area in which Verdala Palace is located and a number of *giardino*-type orchards are found in its neighbourhood. Ordnance Survey Sheet 101 shows the area to abound in water-related features. In the adjoining area of Id-Daħla (map reference 46147 68831), an open-air water reservoir, irrigation channels and a subterranean gallery are indicated on the map.

5.2.2 The inhabited section of Malta

The place-names listed in this section encompass the remaining, more accessible areas of Malta. An analysis of the given water-related toponyms reveals a largely different and more diverse hydrological situation from that through an analysis of Abela's description of the urbanised section of Malta. The bulk of the included land area consists of exposed Upper, Middle and Lower Globigerina plains, but areas containing exposed Upper Coralline and Lower Coralline deposits are also included. The water-related toponyms listed in this section are indeed a reflection of this largely varied geological reality.

Until the second half of the nineteenth century, the inhabitants of Malta only had limited access to the mean-sea-level aquifer. The absence of perched aquifers in most areas of central and southern Malta is attested by the presence of *Bir* place-names, which shows agricultural and settlement areas geared towards the harvesting of rain runoff in cisterns and reservoirs. Cases in point are place-names such as *Biar Giabbar* (Abela 1647, 100), *Birmiftuh* (Abela 1647, 105) and *Bir e taüyl* (Abela 1647, 105). At *Hal Caprat* (BIR1), Abela refers to the presence of cisterns associated with this abandoned settlement: [...] *v'è rimasto vestigio solo d'alcune cisterne* (Abela 1647, 91).[59]

Significantly, several of the *Bir* toponyms listed by Abela in the inhabited section of Malta also make reference to a

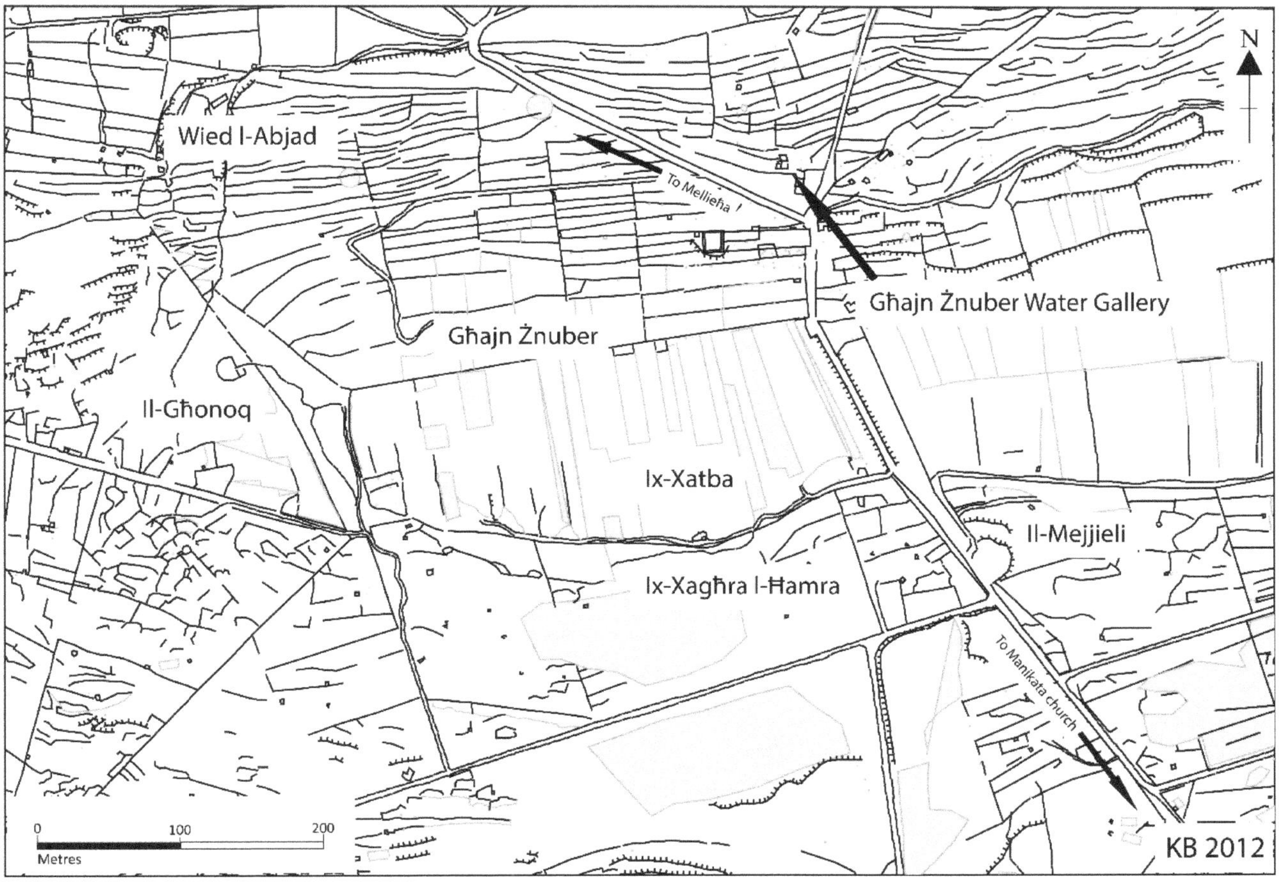

Fig. 5.97 Plan of the Manikata area in the territory of Mellieħa showing the location of the Għajn Żnuber perched aquifer gallery (ME4). Topographical map image adapted from MEPA Map Server data.

[58] The Għajn Riħana spring was located thanks to the assistance of C. Buhagiar who accompanied me on site and provided me with a detailed description of the Għajn Riħana gallery. See chapter seven, sections 7.5 and 7.6.1.1.

[59] As *Hal Caprat* was located on the eastern edge of a sizeable Middle Globigerina Limestone outcrop, however, there is the possibility that cisterns were tapping into localised perched aquiferous deposits, but this still has to be satisfactorily assessed.

vertical shafts sunk into the ground. The principal localities within which the mean-sea-level aquifer was tapped in this manner are Qormi and Marsa (MA10) (Abela 1647, 92). The tapping of the mean-sea-level aquifer is attested to by the descriptions provided for *Bir yahlef* (MA5),[60] *Bir Eebeyer* (MA2),[61] *Bir el Meru* (MA3)[62] and *Bir Buhagiar* (MA1)[63] place-names, all located in the Qormi-Marsa district (MA10).

The *Bir Szegrella* (RA7) place-name (Abela 1647, 80), by contrast, also described by Abela as *pozzo d'acqua viua*, probably tapped a perched aquiferous source. This is listed by Wettinger as being located at Rabat (Malta) between Ta' l-Abbatija tad-Dejr and Għeriexem (RA26)[64] – localities in which Upper Coralline Limestone deposits prevail. The same applies to the Ras il-Għajn toponym (RA38), a water-related feature near Għar Barka, described by Abela as: [...] *vn pozzo d'acqua viua molto profondo chiamato Ras el Eeyun, cioè à dire capo delle fontane, impercioche si tiene, che colà sia la scaturigione principale, e capo de'fonti di quelle contrade non molto lungi dalla Città, e che in qualche tempo essendosi fatta certa sperienza (come si suole) con olio, e morca sia poscia stato riconosciuto il segno in altre fontane anco discoste da questo luogo; inditio ben chiaro della dipendenza è corrispondenza con quello* (Abela 1647, 80). This description is extremely significant and appears to make reference to a vertical shaft located in the whereabouts of Għar Barka, giving access into the innermost section of a perched aquifer gallery. Being the innermost shaft-well, it is likely that this can be identified as the mother well – the shaft through which the existence of this water source was detected and which determined the feasibility of the perched aquifer gallery excavation project.[65] *Bir Giabrùn* situated in an unidentified locality close to Qrendi is similarly described as: [...] *contiene in fe acqua viua*, and possibly refers to the tapping of a localised aquifer forming within Globigerina Limestone deposits (Abela 1647, 100).

The majority of the *Aayn* place-names listed by Abela in the inhabited part of Malta tap perched aquifer water sources in the Rabat (Malta), Fawwara, Girgenti and Ġebel Ċiantar areas. *Aayn Hammem* (Abela 1647, 80), below the west-facing bastions of Mdina, *Aayn Clieb* (Abela 1647,

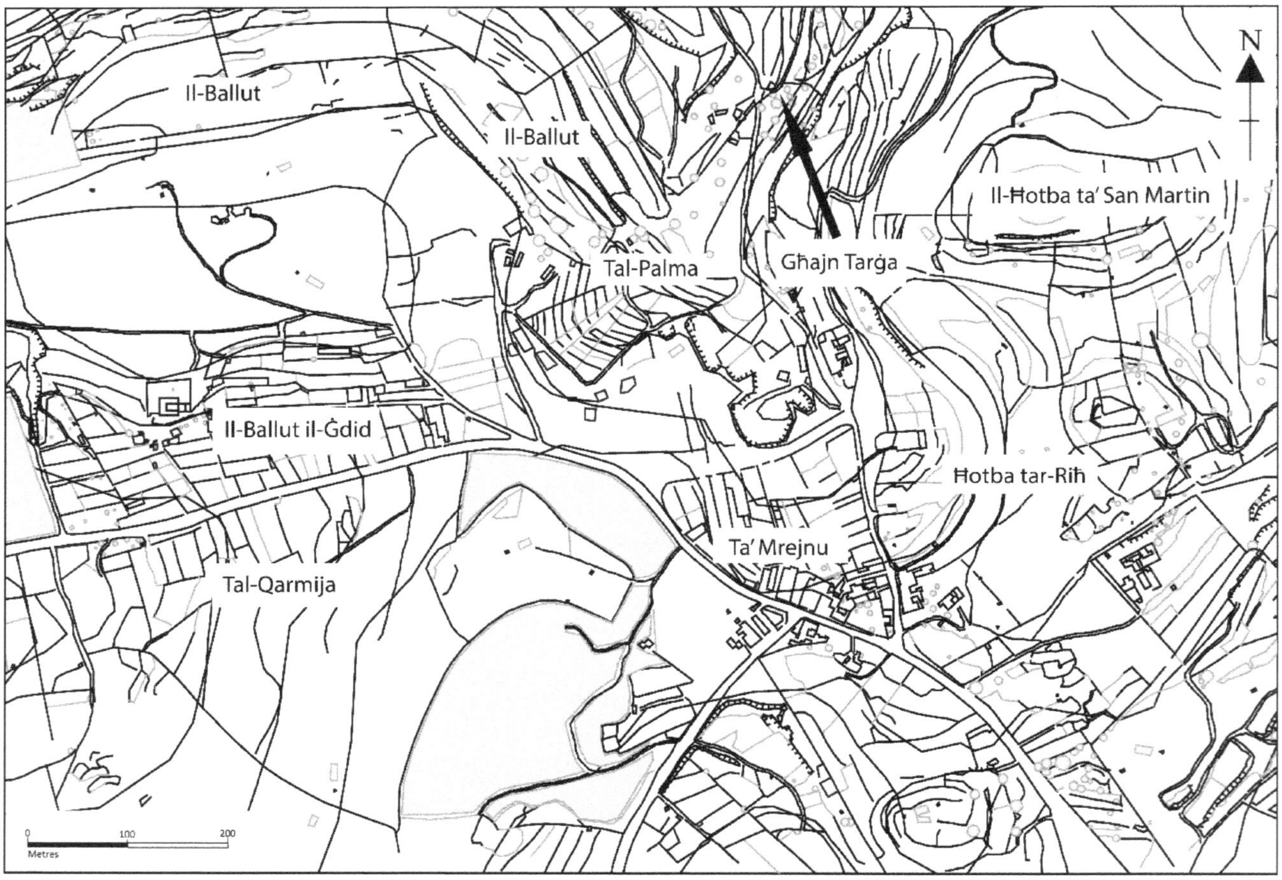

Fig. 5.98 Plan of the Tal-Palma area in the territory of Mġarr (Malta), showing the location of Għajn Tarġa (MG16). Topographical map image adapted from MEPA Map Server data.

[60] *Bir yahlef, pozzo d'acqua viua*. Abela 1647, 92.
[61] *Bir Ebeyer, cisterna d'acqua viua* [...]. Abela 1647, 93.
[62] *Bir el Meru, pozzo pure d'acqua viua*. Abela 1647, 93.
[63] *Bir Buhagiar, pozzo così nomato per le molte pietre* [...] *l'acqua è perenne, e continua, che non vien mai meno*. Abela 1647, 93.
[64] See the *Bir Żigrilla* place-name in Wettinger 2000, 53.
[65] See chapter three, section 3.1.6.

80) and *Aayn Cayed* (Abela 1647, 80) are such instances. *Aayn Pietro* (RA4) (Abela 1647, 81), *Alferez* (RA5) (Abela 1647, 81) and *Aayn Colia* (RA3) (Abela 1647, 81) are other perched aquifer springs, probably located in the Rabat (Malta) whereabouts, but their location is not known. Other perched aquifer water springs located in the Girgenti and Ġebel Ċiantar whereabouts are: *Aayn el Kibira* (SI2) (Abela 1647, 95), *Aaym Ghliem Alla* (SI1) (Abela 1647, 96), and *Noscieaa* (SI4) (Abela 1647, 96). *Fontana de' Laurenti* (SI3) (Abela 1647, 95) appears to have been located in close proximity to the Buskett woodland area.

The *Aayn Filep* (MA8) (Abela 1647, 93) and the *Torre di Aayn duyeli* (PA2) (Abela 1647, 94) place-names are located within Middle Globigerina Limestone formations. The former represents a mean-sea-level aquifer source, whilst the latter spring was quite probably derived from a localised perched aquifer in the Corradino area (Fig. 2.13). Abela remarks that the Għajn Filep spring was of an exceptionally good quality: [...] *si faceuano da questa l'acquate delle galere, e d'altri vasselli, perche è dotata di si rara, & eccellente qualità di conseruarsi più ch'alcun altra in lunga nauigatione senza putridire* (Abela 1647, 93). Għajn Filep probably formed as a result of the mean-sea-level aquifer pressuring against the largely impermeable Middle and Lower Globigerina deposits which perimeter the shoreline of the Grand Harbour region (Fig. 5.102). Aquifer percolation is in such instances limited by the impervious nature of the Globigerina deposits. The sheer mass of the mean-sea-level aquifer pressing against these geological deposits forces water upwards in select areas, and probably accounts for the existence of these surface springs. Water percolation within Globigerina formations is also conditioned by the presence of a series of rock fissures which encourage water percolation and which determine the surface spring location.[66] The *Stagno detto ta' Sirina* (MA11) (Abela 1647, 87) was also located in the Qormi area (MA10) and was probably formed by rain-water runoff and the poor porosity of Globigerina Limestone deposits.

Other mean-sea-level water sources found within Globigerina deposits are listed at *Vyed el Aayn*, described as *Valle, ò torrente della fontana, stagnone d'acqua, ò peschiera con torre* [...] (Abela 1647, 106). Apart from the reference to a spring at Marsaskala (WI2), there is the mention of a *stagnone* – a feature synonymous with water

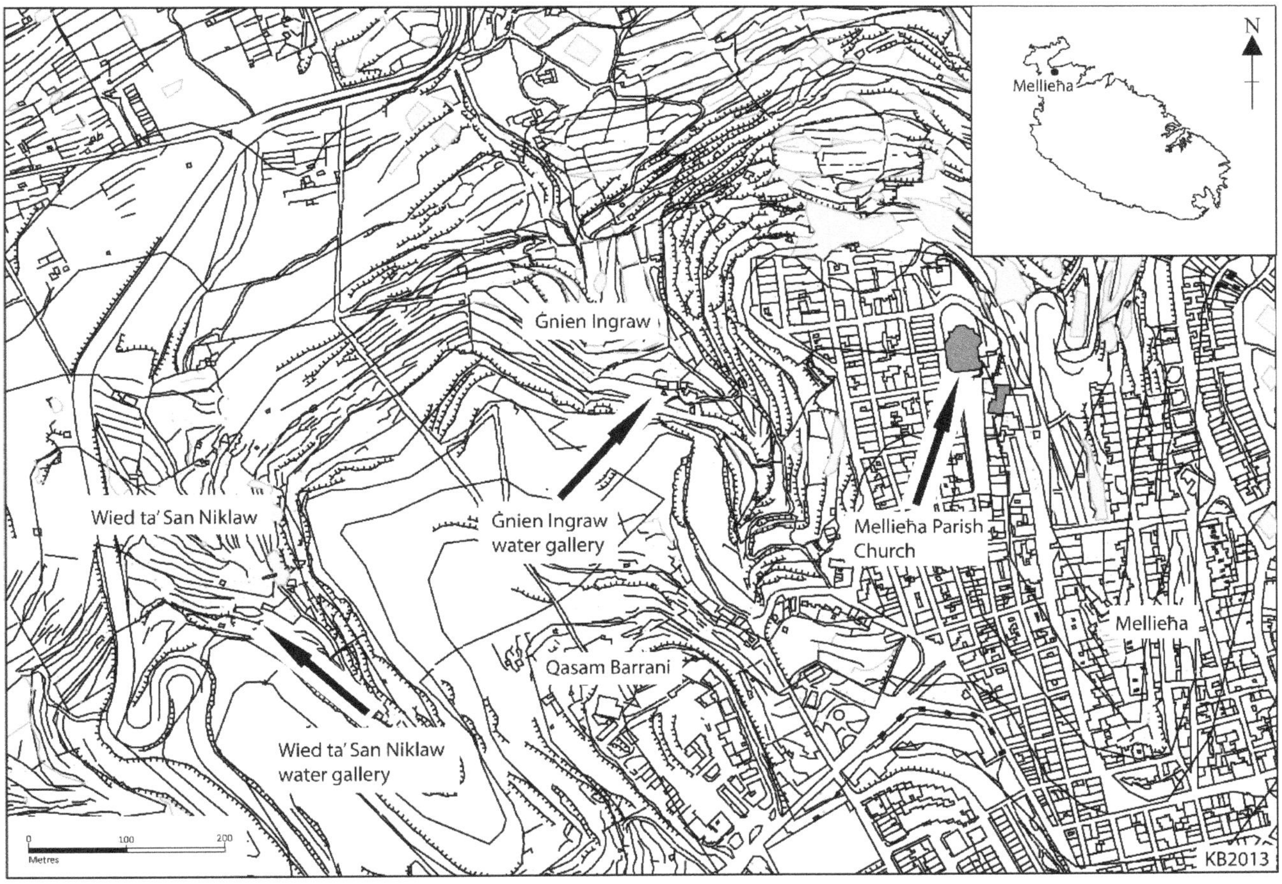

Fig. 5.99 Plan of the Mellieħa territory showing the location of the Ġnien Ingraw (ME1) and Wied ta' San Niklaw (ME7) perched aquifer galleries. Topographical map image adapted from MEPA Map Server data.

[66] Personal communication by M. Sapiano. See also chapter two, section 2.6.2.

collecting in an area of low rock porosity. The *Ghadira ta Rez* place-name (ZA1) (Abela 1647, 106), in the vicinity of the Żabbar settlement, also infers a similar situation and hints at the collection of water within an area of Middle or Lower Globigerina deposits. Although difficult to locate with certainty, *Ghadira di Torbet Gharfagna* (AT1) (Abela 1647, 86) and *Ghadira tal Bordi* (AT2) (Abela 1647, 86), both in the vicinity of Attard, are likewise probably located within such geological deposits. There are extensive surface Middle or Lower Globigerina limestone formations in areas south and east of Attard (Debono and Xerri 1993, Geological Map of Malta, Sheet 1).

5.2.3 *Gozo and Comino*

Containing a much higher percentage of Blue Clay than Malta, Gozo was always better provided with perched aquifer water sources. For Gozo Abela states: [...] *scaturiscono in essa molte fonti di perfettissime acque dolci, e fresche à marauiglia. Vi sono alcuni giardini, che producono eccellenti frutta* [...] (Abela 1647, 122). Abela's knowledge of Malta's sister island seems to have been limited. Only three spring water sources are listed: *Aayn Tili* (GO1) (Abela 1647, 121), *Aayn Mghelghel* (Abela 1647, 122) and *Aayn Barrani* (Abela 1647, 122). *Dachlet Korrot* (Abela 1647, 122) place-names. No water sources are mentioned for Comino (GO2). This is probably a direct reflection of the lack of perched aquifer water sources on the island, due to an absence of above sea level Blue Clay deposits.

5.2.4 *The Valletta aqueduct project*

The section entitled *Dell'Acqve Della Fontana* documents the early seventeenth-century sources tapped for the Valletta aqueduct project (Abela 1647, 109-17). These were *Diar Chandul* (RA42), *Hofret el Rez* (RA27), *Aayn Kaied* or *Kayed* (RA19), *Aayn Cirani* (RA2), *Aayn Teuzien* (RA21) or *Birzegrella* (RA7), *Nescieaa tal borgi* (RA36), *Tal Mitarfa* (RA51) and *el Megira tà Mesrah Mnusa* (RA9) (Abela 1647, 110-11). There is a lack of detailed description on the nature of these springs. Nevertheless, the emergent picture gives the impression that in most instances, a series of already extant springs were altered and modified in order to be utilised as the principal water sources feeding the Valletta aqueduct.

Abela states that *Diar Chandul* or *Chandut* (Dejr Ħandul) was the principal spring feeding the aqueduct project. It is described as *Il capo principale dell'acqua, che fù presa*

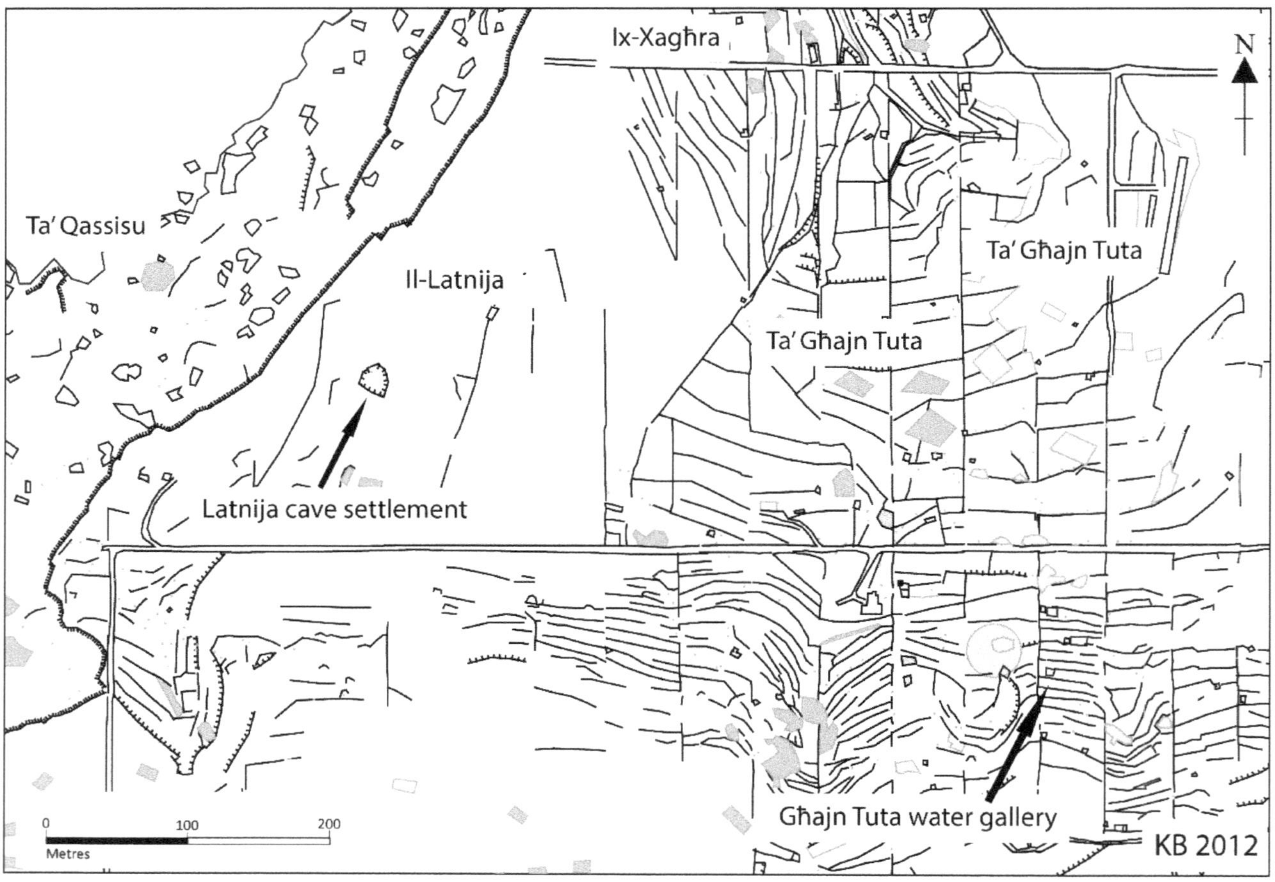

Fig. 5.100 Plan detail of the Ċirkewwa locality showing the location of the Latmija cave-settlement and of the Għajn Tuta (ME2) perched aquifer gallery in the vicinity. Topographical map image adapted from MEPA Map Server data.

[...] (Abela 1647, 110-11).[67] This demonstrates that a water source intercepted in this area was extended or remodelled through the excavation of a *mina*, or gallery, probably in order to amplify its water-yield. The gallery is approximately 60 m long,[68] and the extracted water exited the subterranean canal at a point called *El Sciaayra*. From this point onwards water was channelled towards the Ħofret ir-Rizz area (RA27), which reportedly contained the first *spiraglio*.[69] The meaning of *spiraglio* is open to interpretation, but in this context probably signifies the presence of a small opening or shaft.[70] Another *mina*, or subterranean tunnel, having a length of one hundred *canne*, was also excavated at this point.[71]

The first known mention of the Dejr Ħandul place-name is in 1399.[72] In 1542, the Dejr Ħandul area was in use as a vineyard (Wettinger 2000, 109, 113). Prior to research carried out by the author of this book, it appears that the *Diar Chandul* site had not been satisfactorily located. Themistocles Zammit lists Dejr Ħandul as being located on the high plateau to the north of Dingli (Zammit T. 1924, 9), but gives no specific point of reference. The British period survey sheets for the area surrounding Ħofret ir-Rizz were closely inspected for clues to the approximate location of the Dejr Ħandul toponym. Survey Sheet 85, which dates to 1922, does contain the mention of a *Skak Handu* (Sqaq Ħandu – RA42),[73] located 170 m from the northernmost perimeter of the Ħofret ir-Rizz karst subsidence structure (Figs. 3.6, 5.103). In Abela's

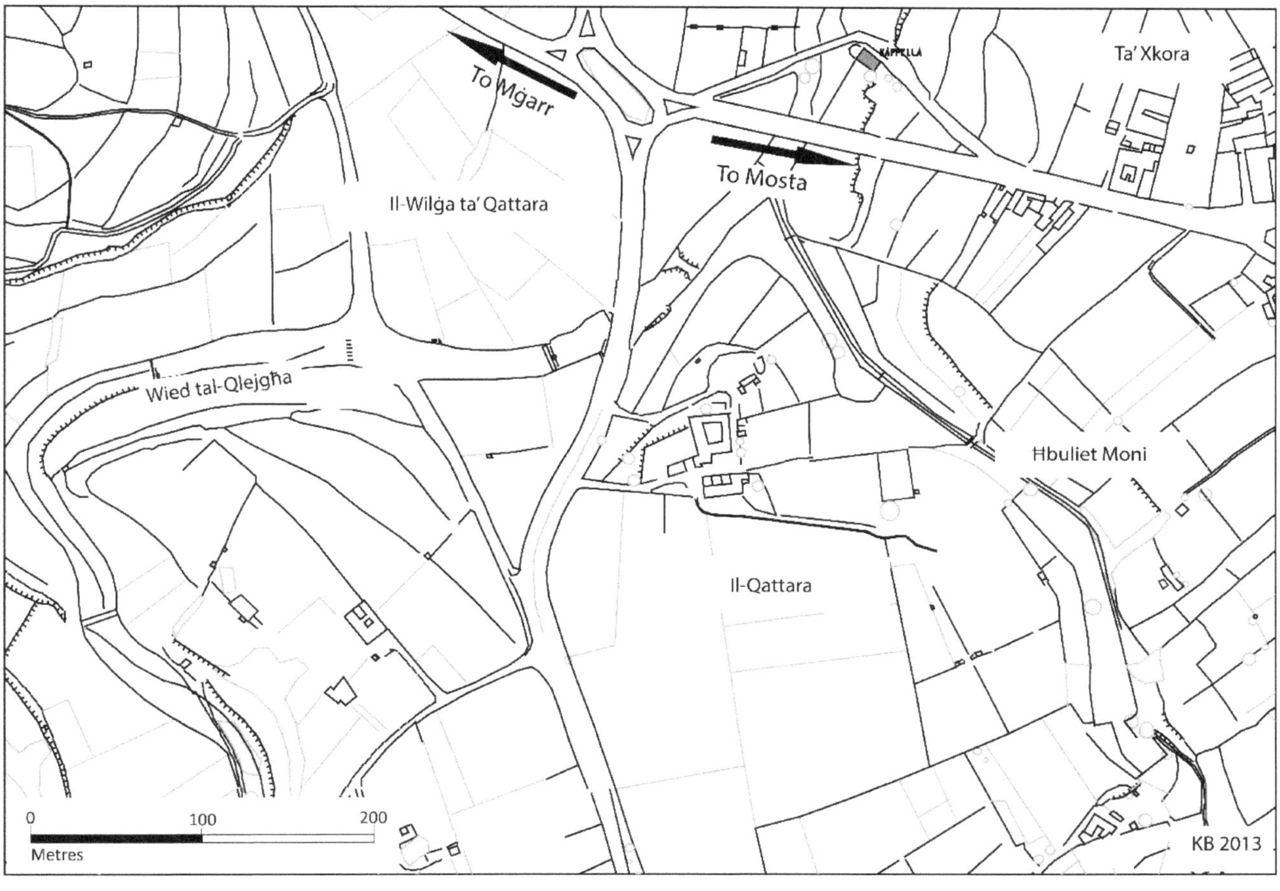

Fig. 5.101 Site plan showing the location of the Il-Qattara (MO1) area in the territory of Mosta. The Upper and Middle Globigerina Limestone deposits in this area limit sub-surface water percolation. Topographical map image adapted from MEPA Map Server data.

[67] The evidence given by Abela conflicts with T. Zammit's claim that a perched aquifer water source originating from within a subterranean gallery on the Ta' Baldu plateau was also tapped by the Wignacourt aqueduct project. As there is no mention in Zammit's account of the *Diar Chandul* area in the vicinity, I suspect that he was confusing this with the adjoining Ta' Baldu plateau. Zammit was actually not aware of the exact whereabouts of Dejr Ħandul. See Zammit T. 1924, 9, 36.

[68] The *canna* (sing. form) or *canne* (pl. form) are the traditional units of measurement utilised in Spain, Italy, Sicily and southern France. The *canna* varied in size, but was generally equivalent to two metres. See http://www.onlineunitconversion.com/meter_to_cana.canna.canne.html.

[69] According to T. Zammit, the Ħofret ir-Rizz spring originates from within a ca. 305 m long gallery excavated into the eastern edge of the Ħofret ir-Rizz karst subsidence structure. Zammit T. 1924, 36. The approximate location of this spring is marked on Malta Survey Sheet 85 dating to 1922.

[70] The matter was discussed with M. Pace Asciak, who occupies the post of Head of Department for Italian at the Secretariat for Catholic Education. Considering the context in which *spiraglio* is used, she associates this with the presence of a small opening. For another entry concerning the *Dair Handul* canal see Treas. B. 289, f. 73. This passed through a field known as *Tal Hueg*, in the Fiddien district. The canal contained a *spiraglio* – possibly a shaft allowing easy access to it.

[71] This would be roughly equivalent to a 200 m long subterranean tunnel.

[72] See the *Dejr Handun* entry in Wettinger 2000, 109.

[73] *Skak Handu* is located at map reference 43518 70472.

uninhabited section of Malta, the *Diar Chandul* toponym is preceded by the *Tà l'Estbal* place-name. The *Hofret el Rez* place-name immediately follows (Abela 1647, 66).

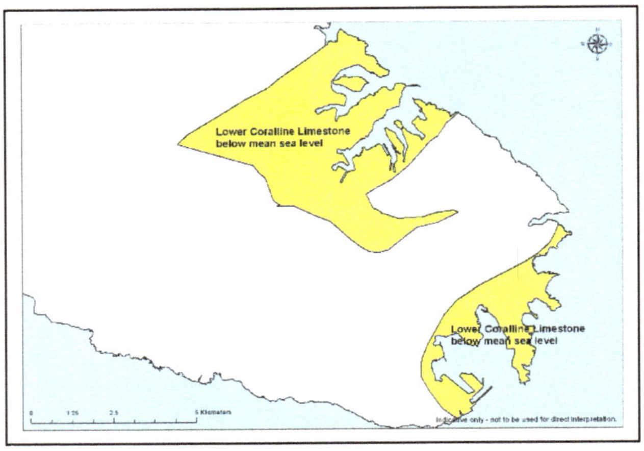

Fig. 5.102 Detail of the south-eastern section of Malta showing the location of submerged Lower Coralline Limestone deposits. Image courtesy of M. Sapiano, Malta Resources Authority.

The *Tà l'Estbal* toponym can be probably associated with the Li Sball place-name (RA33),[74] which on Survey Sheet 85 is indicated as being located ca. 300 m to the south-west of the *Skak Handu* place-name (Survey Sheet 85, School of Military Engineering, Chatham, 1922) (Fig. 5.103).

The Sqaq Ħandu area is located in an Upper Coralline Limestone deposit of the Mtarfa Member type, at an altitude of ca. 183 m above sea level. Springs originating from perched aquifer galleries at the neighbouring Wied ir-Rum (RA56; Fig. 4.3), and which are probably intercepting the same perched aquifer water source, are approximately located at the same height above sea level. The perched aquifer galleries at Ta' Baldu, at a distance of a little more than half a kilometre from Sqaq Ħandu, are actually located at a slightly higher elevation, at a height of ca. 187 m above sea level. This makes it probable that in the early seventeenth century, the perched aquifer was in this area located fairly close to ground surface level,[75] making it fairly easy to intercept. Easy access to a perched aquifer ground-water source provides a tentative explanation for the excavation at this point of the gallery tapping the principal water source of the aqueduct project.

Fig. 5.103 The *Skak Handu* (RA42) area north of Ħofret ir-Rizz (RA27) is tentatively associated with the *Diar Chandul* place-name listed by Abela in 1647. Source: Survey Sheet 85, Scale 1:2,500, School of Military Engineering, Chatham, 1922.

[74] The map reference for the *Li Sball* area is 43208 70279.
[75] The Geological Map of Malta shows the upper section of Wied ir-Rum, Ħofret ir-Rizz and Wied Ħażrun areas to be free from fault plains, the presence of which might result in the disruption of ground-water percolation. See Debono and Xerri 1993, Geological Map of Malta, Sheet 1.

Fig. 5.104 Reed growth in the Sqaq Ħandu (RA42) area in the territory of Rabat (Malta). This follows the course of a subterranean aqueduct canal channelling the Dejr Ħandul spring water in the direction of Fiddien. Source: the author.

Fig. 5.105 An exposed tract of the Dejr Ħandul canal transporting water towards Fiddien at map reference 43504 70686. Source: the author.

Fig. 5.106 An exposed section of the Valletta aqueduct canal in the Ta' Koronja (RA46) territory of Rabat (Malta). Source: the author.

Fig. 5.107 Detail of an uncovered canal section at Ta' Koronja (RA46) in the territory of Rabat, Malta. Source: the author.

At present, the perched aquifer is located at a depth of between 20 to 45 ft below surface level.[76] Various wells have been dug in this area since the mid-nineteenth century and supply a remarkable concentration of open-air reservoirs with water.[77]

Furthermore, it is significant to note that the principal water source feeding the Valletta aqueduct does not appear to originate from within perched aquifer water galleries dug into the vertical face of a ravine. Channelling water from any of the Wied ir-Rum or Wied Ħażrun (RA55) water galleries would have been problematic, because of the presence of higher terrain in the *Li Sball* area.[78] Such a scenario would necessitate the use of water-lifting machinery. Excavating a gallery in the Sqaq Ħandu area, on the other hand, presented the engineers in charge of the aqueduct project with a gentle downward incline leading towards the Fiddien valley area. This would have enabled the channelling of water towards Fiddien through the use of a gravity-fed water distribution system. The section of the partly-excavated, partly-constructed masonry canal channelling water in the direction of Fiddien is still visible to the immediate north-west of the Sqaq Ħandu at map reference 43481 70531. The presence of water still flowing through the canal is attested to by dense reed growth (Figs. 5.104, 5.105). The aqueduct canal carrying water from the Dejr Ħandul is again exposed at an above-ground level in the Ta' Koronja area (RA46) at map references 44196 71982 and 44369 72112 (Figs. 5.106, 5.107).

A point worthy of consideration is that *Diar Chandul* is not a water-related place-name. A scrutiny of Survey Sheet 85

[76] Personal communication by P. Micallef, a seasoned farmer who owns extensive tracts of land in the Sqaq Ħandu and Li Sball areas in the territory of Rabat (Malta).

[77] Personal communication by P. Micallef.

[78] It appears that British period documents had the tendency to refer to the L-Iżball area as the *Ta'Baldu plateau*. See Zammit T. 1924, 36.

also confirms that in the area to the immediate north of the Ħofret ir-Rizz karst subsidence structure, there are no water-related toponyms. Abela does mention that when the excavation works commenced at *Diar Chandul*, *tréta pezzi di canali di terra cotta* [...] were unearthed (Abela 1647, 111). The dimensions of these water-related features are given as follows: [...] *di diametro d'vn palmo, e di due, e mezzo di lunghezza*" (Abela 1647, 111).[79] The antiquity of this find is not known but Abela's description indicates that water was already intercepted in the area at an earlier date. Other finds mentioned by Abela as having been discovered at Dejr Ħandul are lead pipes, tiled floor sections and marble vases and columns (Abela 1647, 36), which are all suggestive of an earlier occupation of this site, possibly due to the availability of water.

Fig. 5.108 The Għeriexem perched aquifer gallery interior (RA26). Areas in the gallery system containing shaft-wells are sometimes bypassed through the excavation of a side passageway (left). In some instances, these appear to be later additions to the gallery system. Source: the author.

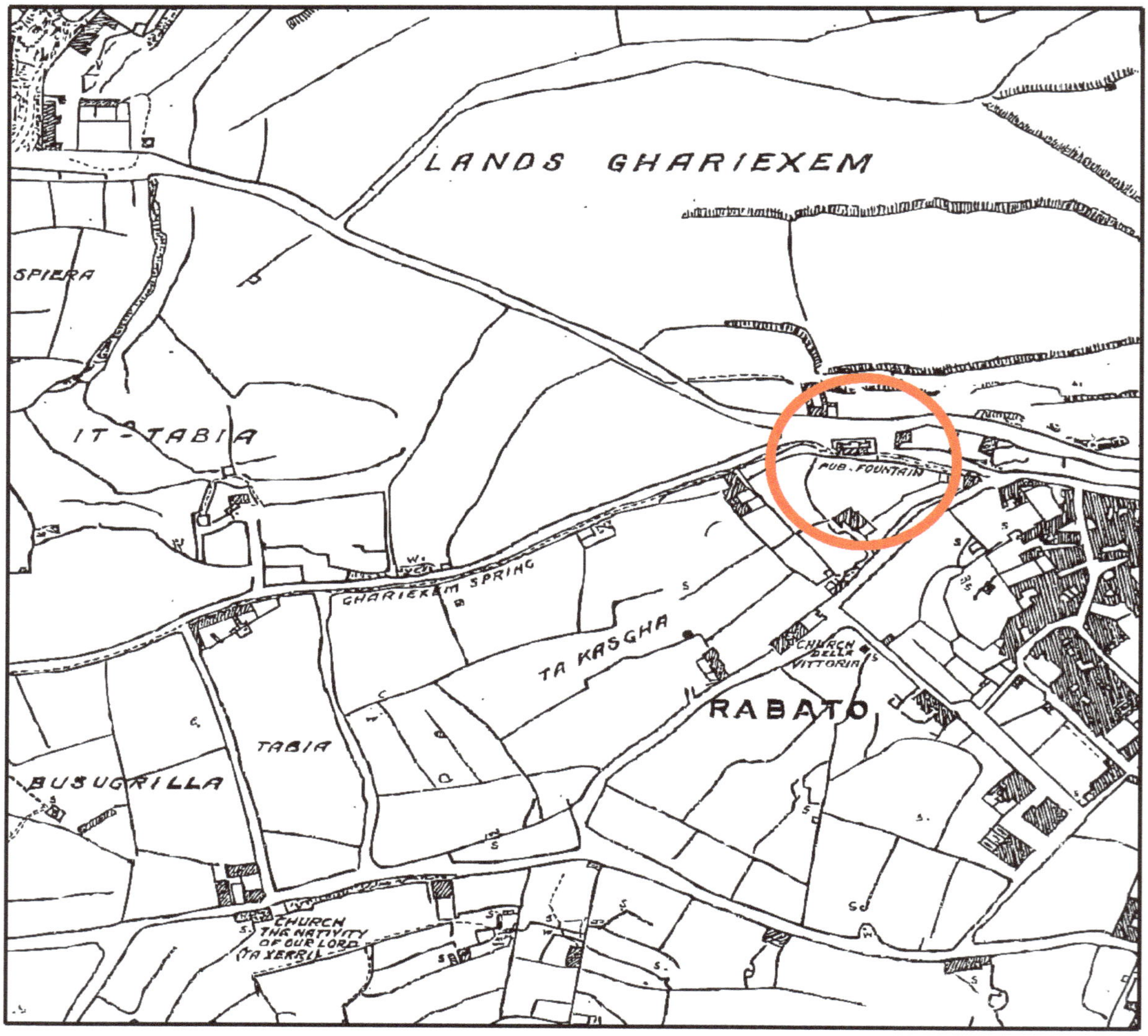

Fig. 5.109 1922 map detail showing the location of the Għeriexem fountain (RA26) and giardino. The fountain's water originates from within a perched aquifer gallery. Source: Survey Sheet 72, Scale 1:2,500, School of Military Engineering, Chatham, 1922.

[79] The earthenware pipe dimensions would be equivalent to ca. 1 foot in diameter and 2 feet long.

Fig. 5.110 The Għeriexem perched aquifer gallery interior (RA26). Water loss from within the gallery system is prevented through the occasional use of a masonry-lined water canal fitted into the gallery floor. Source: the author.

Fig. 5.111 Interior view of the Għeriexem water gallery (RA26). Rectangular-shaped shafts pierce the ceiling of this water system at intervals. Rock-excavated footholds and handholds (right) facilitate access to the gallery's interior or exterior. Source: the author.

Another three springs connected to the Valletta aqueduct project were located in the *giardini* of Għajn Qajjed, Ġnien is-Sultan (RA13) and their whereabouts and were given as *Aayn Kaied* (Għajn Qajjed – RA19), *Aayn Teuzien* (Għajn Tewżien – RA21), and *Aayn Cirani* (RA2) (Abela 1647, 111). Another spring which Abela refers to as *Birzegrella* (Busugrilla – RA7) (Abela 1647, 111) was either an alternative name for Għajn Tewżien or a minor spring supplementing the yield of the Għajn Tewżien source. There is evidence to show that both Għajn Qajjed and Għajn Tewżien originated from within perched aquifer galleries. This subject is discussed separately in chapter six, section 6.1.3.

Of all the springs tapped for the aqueduct project, Dejr Ħandul was the purest and better source (Abela 1647, 111). The least preferred was the Għajn Tewżien or Busugrilla spring. When water flowed through its canal it generated sediment, referred to by Abela as *brama* (Abela 1647, 111). That this was considered a nuisance is apparent by the exclusion of Għajn Tewżien as one of the aqueduct sources. Abela associated the sediments generated by the spring as being reflected in its name, *Vzien*, which he defines as an Arabic idiom describing the heavy or hard nature of this water source (Abela 1647, 111). Whatever the case, this reveals that *Aayn Teuzien* was present in the Rabat landscape well before its inclusion as one of the principal water sources feeding the aqueduct project.

Another spring tapped at *Vyed el Besbies* (Wied il-Busbies – RA59) channelled water towards the *Marget Meuya* (Marget Neuia – RA34) area close to Fiddien valley

(Abela 1647, 112).[80] Field-research has revealed the presence of a perched aquifer gallery at Marget Neuia known as the Tal-Fiskal spring.[81] This is discussed in more detail in chapter six, section 6.1.4.

5.3 Other Seventeenth and Eighteenth-Century Water-Related Documentary Sources of Evidence

In 1647 Abela only makes sparing mention of *mine, canali* and *condotti di fontane*, yet points out that these were a common feature of the early seventeenth-century Maltese landscape (Abela 1647, 101-2). On numerous occasions

Fig. 5.112 The Għeriexem perched aquifer gallery interior (RA26). Detail of one of the shafts which connects this subterranean gallery system to the exterior. Access is facilitated through the use of rock-excavated handholds and footholds. Source: the author.

Abela gives indirect and vague evidence that accounts for the existence of perched aquifer systems in the first half of the seventeenth century.

In this respect, documentation on the Saqqajja *fontana* (RA40) at Rabat (Malta), dating from 1626 onwards, is particularly revealing. A *bando* dating to September 1626 explicitly mentions that the public waters originating from the Saqqajja *fontana* were being illegally diverted into private residential cisterns.[82] Even though the content of this document refrains from mentioning the existence of subterranean water galleries and any associated shafts, the description of the Saqqajja *fontana* setup indicates the existence of a *qanat*-type water extraction system. Surface access to a number of shafts was from within private properties – a circumstance which prompted property owners to easily access the underlying gallery system and divert its water flow. The context in which the verb *serrare* is used is an apparent reference to the practice of blocking access to shafts piercing the ceiling of a subterranean gallery resorted to by government officials during this period.[83]

Even more informative is a first-hand report on the Saqqajja and the Għeriexem (RA26) springs by Mederico Blondel dating to 23 April 1695 (Univ. 187, ff. 2-3).[84] Blondel personally investigated the Saqqajja fountain and

Fig. 5.113 The Għeriexem perched aquifer gallery interior (RA26). Detail of the natural water-yielding veins tapped by the Għeriexem gallery. Source: the author.

[80] The spring at Wied il-Busbies mentioned by Abela, can be tentatively associated with a section of the present-day Wied il-Busbies spring which currently flows through a nineteenth-century man-excavated gallery close to the Iċ-Ċianti area of Rabat (Malta). One of the access points of this gallery system is located at the approximate map reference of 42734 71086. Entry to this perched aquifer gallery was obtained through the kind intervention of M. Sapiano who also accompanied me on site.

[81] The map reference for the Marget Neuia area is 43460 71660.

[82] This *bando* dating to 28 September 1626 is attached to the Univ. 187 manuscript (ff. 1-14). The bulk of the documentation contained in this manuscript dates to 1695-1741.

[83] The most important section of this *bando* reads: *Considerando la pubblica e commodita' commune del Popolo quale viene ammancare a causa di alcuni li quali posponendo il publico a loro utile e privato temeriarmente ardiscono di occupare parte dell'acque della Fontana della Saccaya con aprire molti condotti nelle proprie cisterne di case loro diminuendo la quantita dell'acque di detta fontana in grave scommodita del publico e della commodita'. Percio' ordiniamo et espressamente comandiamo che da mo' inanzi nessuna persona di qualsi voglia, grado, stato sesso e conditione si sia non ardisca ne presuma aprire simili condutti nei propri cisterni di casi loro, ordinando parimente che tutti gli li quali al* [?] *si trovino havere simili condotti* [?] *spandenti di detta acqua in case loro che fra il termine di giorni oro da contrasti da hoggi avanti li debbano molto bene serrare in modo tale che non possa passare acqua e contravendo in uno dei sudetti ordini incorreranno nella pena se sara homo di quattro tratti di corda, e se donna di* [?] *tre d'applicarsi alla publica fabrica e di altre pene riservate a S.A.S. e cio sensa permissione alcuna haura solamente relatione de nostri officiali.* See Univ. 187, f. 4.

[84] Mederico Blondel des Croisettes was a French architect and military engineer and in 1695 occupied the post of *Ingengnere della Sacra Religione e dello Stato e Sopra Intendente delle Fontane*.

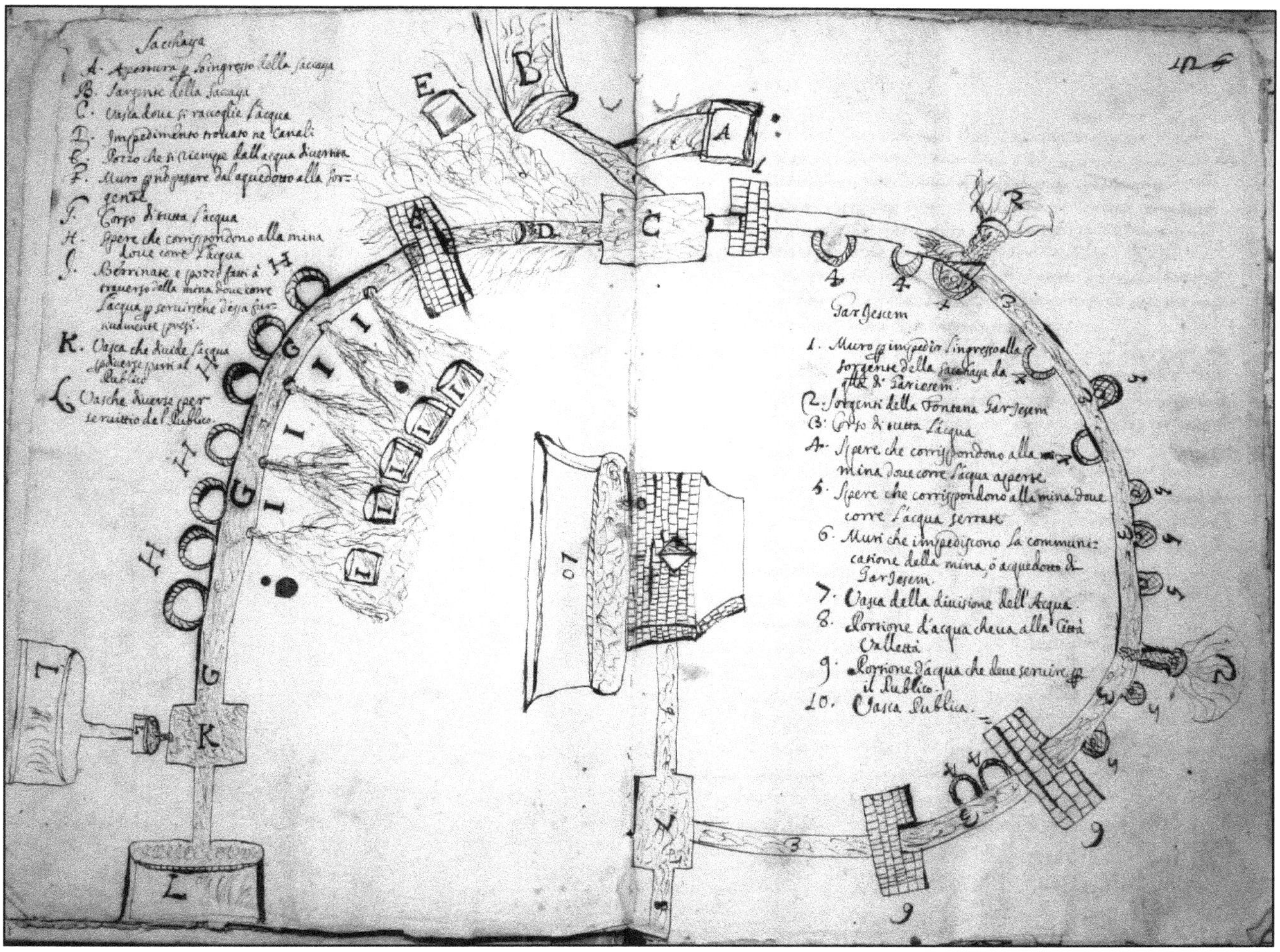

Fig. 5.114 Sketch plan showing the Saqqajja (RA40) and the Għeriexem (RA26) perched aquifer galleries. Indicated are a number of shaft-wells (labelled "H" and "4"), piercing the ceiling of the gallery systems. Source: NLM, Univ. 187.

commences his account with: *Siamo scesi nella mina stessa di detta Fontana che dal fosso della fronte della detta Citta' ove c'e l'ingresso, si distende sotto il Rabbato per buon tratto cioe' di circa cinque cento passi havemo con molto scomodo e stento interamente visitato a scorso per riconoscere ocularmente lo stato dell'acqua, dello suo condotto sotteraneo, dal capo ove la scaturisce viva ne sale immediatamente dale rocca, scorrendo parte sul suolo naturale del fondo della mina e parte in canali ordinarij di pietra, sin alla Fontana pubblica in sui si scende con molti scalini, sita nel principio della Piazza inanzi la citta'poco distante dal Monastero di S. Agostino* [...] (Univ. 187, f. 2).

From the above it is clear that Blondel is describing a subterranean perched aquifer gallery. The water flow within the gallery system was facilitated by a combination of rock-excavated and masonry canals installed in the gallery floor. Extremely significant is Blondel's mention and description of the *spiragli* or shafts, which he attributed as belonging to the ancient gallery framework: *Che vi sono alcuni spiragli che anticamente servirono allo scavar e'farla e de quali poi abusarono coloro che nelle cui case, cortile o giardini si ritrovano valendosene e convertendoli in uso loro private con alzarne l'acqua per via di corde* [?] *li quasi che da pozzi sorgenti* (Univ. 187, f. 2).[85]

The present writer's investigation of a ca. 200 m section of the Għeriexem perched aquifer gallery provided ample insight into the situation described by Blondel. At Għeriexem, a number of shafts likewise connect the water gallery with the surface and contain handholds and footholds in order to facilitate the entry or exit from this subterranean system (Figs. 5.108, 5.110, 5.111, 5.112, 5.113). Masonry walls were constructed beneath various shafts in order to partially dam the water flowing through the gallery system and make it possible to retrieve water from the surface by means of a bucket or any other type of water-lifting device.

[85] In another section of this report it is hinted by Blondel that the origins of this water system might date back to the Phoenician period. See Univ. 187, f. 2.

The illegal diversion by private individuals of the Saqqajja water system was a matter of concern to Blondel: *Che vi sono mine particolari che arrivano alla subetta per mezzo delle quali si deriva parte dell'acqua della Fontana, derivandola nelle cisterne d'alcune case del Rabbato* [...] (Univ. 187, f. 2). Private individuals had also illicitly accessed the gallery system, dammed sections of it and lined this with water-proofing mortar in order to store water within: *Quale mina private si vede e conosce essere stata altre volte serrate e murate con calce e puzzolana e poi riaperte con diversi artifici.*[86] Additionally, Blondel observed that substantial quantities of aquifer water tapped by the gallery system were being lost either through breaches in the masonry canal lining the gallery floor, or in the case of rock-cut canals, due to the natural porosity of the rock. The latter instance was actually singled out as being the principal factor accounting for water loss from within the Saqqajja gallery system. In view of this, it was suggested that sizeable lengths of the gallery floor should be fitted with well-mortared masonry canals in order to ensure a more efficient water extraction system (Univ. 187, f. 3).

Documentary sources dating to 1718 show that the blocking of shaft-wells giving access into the Saqqajja gallery system in 1626 and 1695 did not deter Rabat residents from continuing to tap this water source illegally. A 1718 report points out that walls were once again erected in the gallery system in order to function as dams and increase the water level within. Water was also being diverted to nearby private cisterns (Univ. 187, ff. 9-10). The 1718 report also pointed out that the shafts piercing the ceiling of the Saqqajja gallery were a potential source of a water contamination. The shafts, if left unsealed, could provide an entry point to potentially water polluting material and debris which could limit the water flow in the gallery system. The sealing-off of all *spiragli* or shafts was once again proposed (Univ. 187, ff. 9-10).

Documentation for the Għeriexem *fontana* dates to the early eighteenth century and as is the case with the Saqqajja water system, it reveals the presence of a perched aquifer gallery, the ceiling of which was breached by means of a series of shaft-wells (Figs. 5.108, 5.109, 5.111, 5.112). The base of the Għeriexem gallery contained a canal which channelled the retrieved aquifer water to the gallery exterior (Fig. 5.110).[87] Specific mention of shafts piercing the gallery at Għeriexem is made in an entry dating to December 1718 (Univ. 187, ff. 9-10).[88] As was the case with the Saqqajja gallery system, the major problem faced by Għeriexem *fontana* was water loss from within the gallery system resulting from the porous nature of the rock within which it flowed. The water canals were also being obstructed by mud, clay and other infiltration residue. Walls erected in different tracts of the gallery system, especially in areas where the ceiling was pierced by shafts, were also obstructing the water flow. In the case of the Għeriexem *fontana* it was again advised to block all shafts giving direct access to this water system (Univ. 187, ff. 9-10).

In the early eighteenth century, the water yielded by the Għeriexem *fontana* was synonymous with illness and disease. As a matter of fact, 1709 documentation attributes

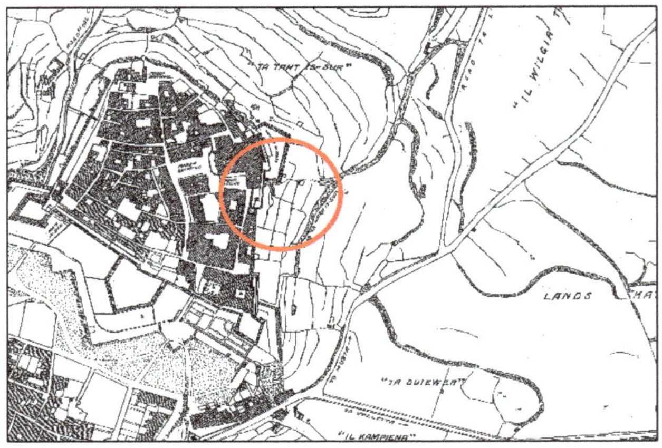

Fig. 5.115 1921 map detail of Mdina and the underlying fief of Il-Ħemsija (RA58). Clearly indicated in the area circled in red is a narrow country lane and footpath leading to a perched aquifer gallery tunnelling below Mdina. Source: Survey Sheet 73, Scale 1:2,500, School of Military Engineering, Chatham, 1921.

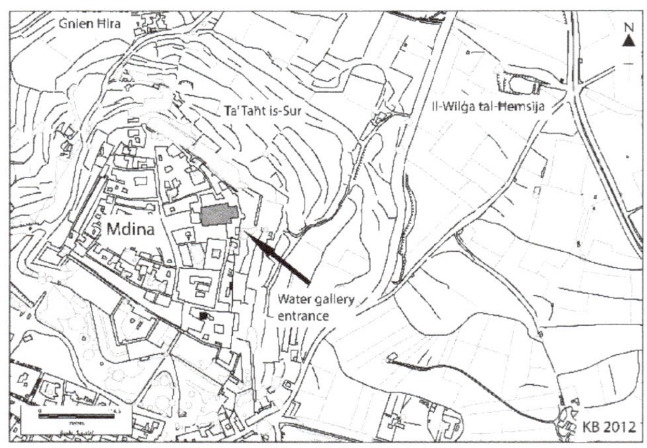

Fig. 5.116 Plan of the Mdina settlement (MD4) showing the location of the perched aquifer gallery tunnelling into an Mtarfa Member deposit below the Ta' De Redin bastions area (MD2). Topographical map image adapted from MEPA Map Server data.

[86] Blondel strongly recommended that all modifications carried out to the Saqqajja subterranean water system by private individuals should be reversed. There is mention of works aimed at blocking surface access to shafts being carried out in 1658. The expenses for this shaft-blocking exercise are also given. See Univ. 187, f. 2.

[87] [...] *li canali delle acque di Hariexem si ritrovano brutti in maniera che non correvano secondo il bisognio, gli fu fatta nel medesimo tempo l'istanza perche' fossero nettati.* See Univ. 187, f. 7 dating to 1718. An entry in Univ. 187, f. 8 dating to 1714 states that maintenance works at Għeriexem are to be addressed in an identical manner as the Saqqajja *Fontana*.

[88] Further reference to the Għeriexem perched aquifer gallery is found in Univ. 187, 101.

Fig. 5.117 Detail of the entrance section to the perched aquifer gallery which tunnels into an Mtarfa Member deposit below the De Redin bastions (MD2) at Mdina. Source: the author.

documented evidence of mosquito infestation in the Late Middle Ages.[89]

Bound to the Univ. 187 manuscript is a sketch plan illustrating the various components forming part of the Saqqajja and the Għeriexem *fontane* (Fig. 5.114).[90] This appears to date to the mid-seventeenth century (Univ. 187, ff. 1-14).[91] The galleries, shafts and cisterns forming part of both the Saqqajja and the Għeriexem water systems are in this plan clearly represented. It ought to be stated that this drawing is merely a schematic exemplification of both water systems. The fact that the Saqqajja and Għeriexem galleries are shown to be physically interlinked is probably incorrect. The partial investigation of the Għeriexem perched aquifer gallery in November 2012 clearly demonstrated that the innermost tract of this water extraction system does not connect to the Saqqajja *fontana*.[92]

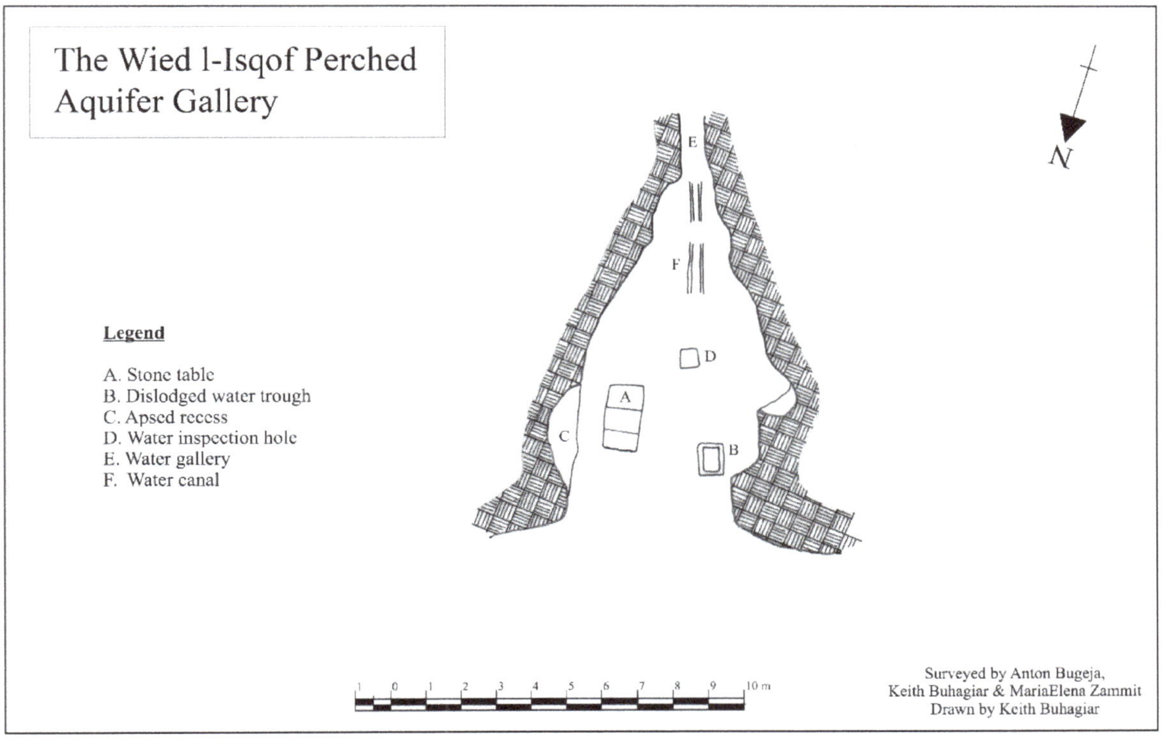

Fig. 5.118 Plan of the Wied l-Isqof perched aquifer gallery (RA25) located in the precincts of Wied l-Isqof in the territory of Rabat (Malta).

this to be the cause of death of a number of people living in the Għeriexem locality. The area was mosquito infested and measures were to be taken to reduce the copious quantities of water yielded by this *fontana* (Univ. 187, ff. 95, 97-98). For this and other sites, there is no known

Documentation dating to 3 October 1725 reveals that stagnant or contaminated water in the area immediately flanking the outermost curtain of the Mdina bastions was again the cause of illness and disease (Univ. 187, f. 95; Univ. 187, f. 102).[93] Of particular interest is the mention

[89] Personal communication by S. Fiorini.

[90] See Univ. 187. The *foglio* of the sketch plan for the Saqqajja and the Għeriexem *fontane* is not numbered but was inserted between ff. 8 and 9 of the Univ. 187 manuscript.

[91] Basing himself on calligraphic-related evidence, S. Fiorini proposed a mid-seventeenth-century date for this plan. See Abdilla 2003, 30.

[92] One of the current access points to the Għeriexem perched aquifer gallery is through the rock-excavated church of St Mary Magdalene located beneath the parvis of the St Paul's Collegiate Church. Access to Għeriexem was obtained through the kind assistance of E.P. Teuma.

[93] The mention of *acque incorrote e puzzolanti* in the case of the latter reference indicates the contamination of the Mdina perched aquifer sources during the early eighteenth century, probably due to sewage infiltration. Documentation dating to 1709 also proposed the covering of open water reservoirs located in the vicinity of the *Civitas*, and the areas

of *diverse mine e forami* tunnelling into the rock face beneath the Mdina Cathedral area.[94] The blame for land subsidence which in the early eighteenth century was already being registered in the De Redin bastion area was laid on such excavations: [...] *quali communicate per dentro una con l'altra han fatto restar in aria tutta la rocca, che sostiene la contina che e' contiguo con la S. Cat suddetta, quale mancando il terreno si smosse dal suo luogo poco meno di mezzo palma, e la medesima Cat. Ebbe pure il suo danno, che si vede crescere d'anno in anno nel suo corso* (Univ. 187, f. 95).

The described location of the *mina* mentioned in this entry conforms with that of a perched aquifer gallery which tunnels into an exposed Mtarfa Member deposit located below the De Redin bastion walls in the Bishop's palace area of Mdina at map reference 46307 71681 (Figs. 5.115, 5.116, 5.117). This gallery could have possibly supplied the Ħemsija estate (RA58) with a perennial water source. Another entry again makes unequivocal reference to this subterranean water system, the ceiling of which was pierced by shafts: *Fu similmente ben riconosciuta una mina manufatta scavata nello roccame sul piano della terra e si stende sotto il Bastione della Citta piena ed aque puzzolente d'altezza pali 3 ed sue gebietta fabricata e bitumata uicina all imboccatura di detta mina* [...] *e detta imboccatura con sua gebietta uengono a terminare sotto la radice del Bastione giust' à punto oue uiene fabricato il Palazzo Vescovile* (*Univ. 187, f. 100*).

The interior of this gallery still needs to be adequately investigated, but is reported as being twenty-seven *canne* long – thus having a length of ca. 54 m (*Univ. 187, f. 100*).[95] The presence of shafts in the gallery system would have allowed water retrieval from within the fortification walls of the *Civitas*. Another water-yielding canal was

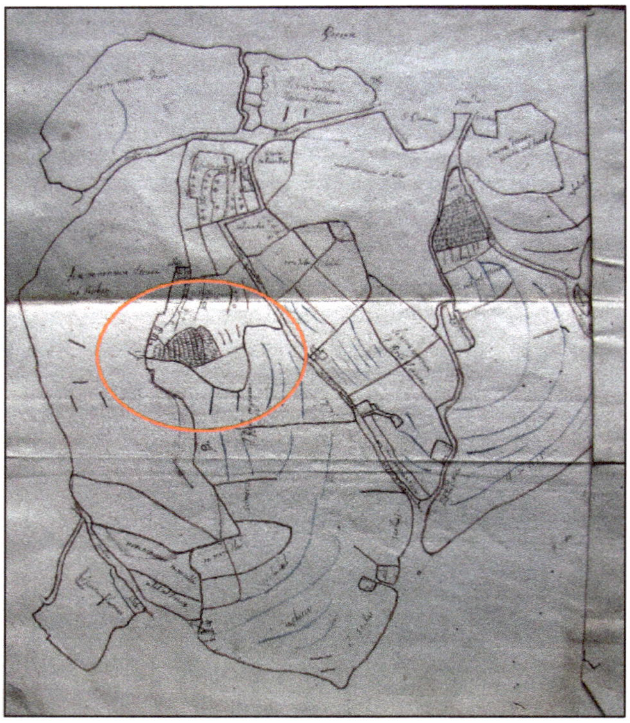

Fig. 5.119 Mid-seventeenth-century plan showing the Ta' Brija (RA44), Ta' Lawrenti (SI3) estates and the Church owned property in the territory of Rabat (Malta). The red circle indicates the location of the Wied l-Isqof perched aquifer gallery (RA25) and its fronting De Robertis portion of the Ta' Brija estate (RA44). Source: CEM, 114/7.

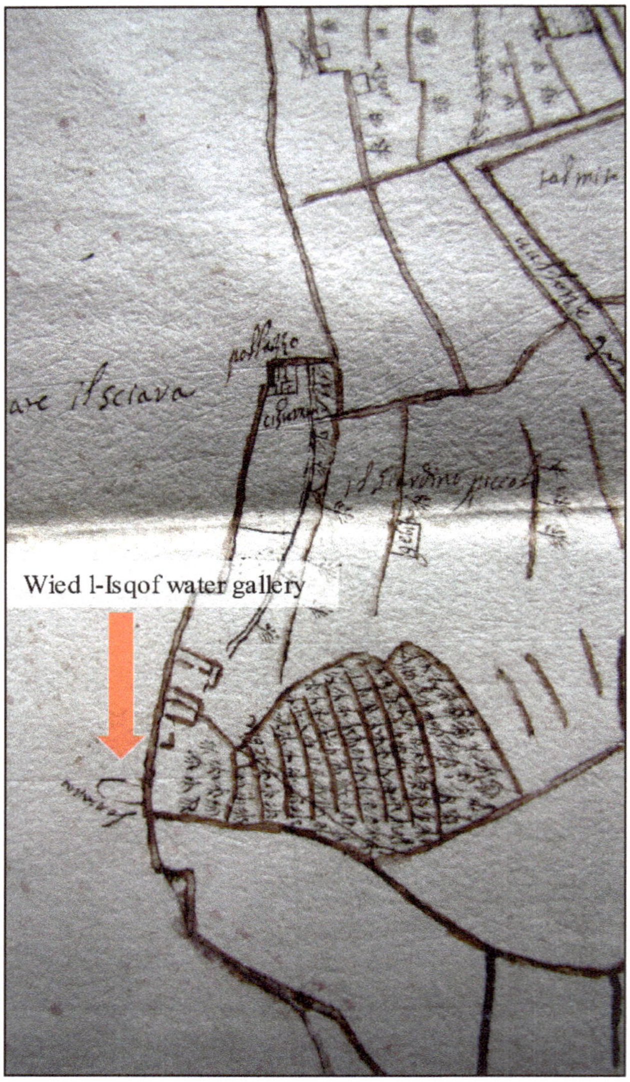

Fig. 5.120 Plan detail of the water gallery excavated into the side of the ravine at Ix-Xagħra tal-Buskett (Wied l-Isqof – RA25), its fronting reservoir and the *giardino* forming part of the Ta' Brija estate (RA44).

of *Tafflia, Gnien Chira, la Grotta detta uolgarmente la conzavia, Saccaya, Tabia* and *Bocana*. See Univ. 187, f. 99.

[94] This entry gives the impression that some of the rock-hewn features below the Mdina bastions might date to the late seventeenth or early eighteenth century. See Univ. 187, ff. 95, 99. There is also a general comment mentioning the presence of *aque, mine e spiragli* as being present at Mdina and its neighbouring area. See Univ. 187, f. 95. Another reference to *spere, gebie et altre mine sotterranee* located in close proximity to the Mdina bastions dates to 2 October 1725. See Univ. 187, f. 100.

[95] On the conversion of *canna* to metres refer to footnote 67.

located below the Magisterial Palace (MD3) at Mdina (*Univ. 187, f. 100*). A number of shafts, some of which yielded *acqua corrente* or spring water, were also documented within the Mdina settlement (MD4).[96]

A manuscript in the Curia Archives at Floriana, classified as CEM 114/7 and entitled *Acque (Seminario)*, dates to the mid-seventeenth century and gives a reasonably detailed picture of the workings of a perched aquifer gallery located in a rural context (*CEM 114/7*).[97] This documentation, dating to 1658-1659, forms part of a fact-finding report drawn up in view of a litigation over rights of water possession and access, between Giacomo Testaferrata de Robertis and the Church authorities. Central to this litigation were claims by Giacomo Testaferrata de Robertis that the Church authorities denied him access to the water sources necessary to irrigate a portion of the *Tabria* (Ta' Brija – RA44) estate known as the *viridarium* de Robertis located in the territory of Rabat (Malta) (*CEM 114/7, ff. 10-11*).

Fig. 5.121 The cobbled pathway which connects the Cagliares residence to the underlying perched aquifer gallery at Wied l-Isqof (RA25) in the territory of Rabat (Malta). Source: the author.

Fig. 5.122 Exterior view of the Wied l-Isqof cave and perched aquifer gallery (RA25) at Rabat (Malta). Source: the author.

The water source in question originated from a *mina*, or water gallery, located in a cave at Wied l-Isqof (RA25): [...] *realmente l'acqua che si litiga nasche dale Terre del Vescovato da una mina, la cui bocca riguarda verso tramontana, parte della quale aqua nasce primariamente da un pertuggio ò sia forame nella detta Rocca naturalmente situate da libeccio distante dalla detta bocca in dentro sessanta* [?] *canne, e mezzo ed il resto dell'acqua, ch'e in pusea quantita deriva dalla parte più interiore della medesima mina che tira verso al mezzo giorno de la lunghezza canne ottanta in circa, il tutto però nelle dette terre Vescovile e quest acque cosi unite si raccogliono nella beveratura fatta da Mons. Vescovo Cagliares tranta due anni sono, e più distante dalla detta bocca canne quindici le dette acque* [?] *ma che ci fosse la detta beveratura cascavano dentro alcune pile poste in un pezzo di spacio publico contiguo con le terre Vescovali, e del feudo della Tabria che sta di sotto, e molto vicino alla bocca di detta mina* [...] (CEM 114/7, f. 12).[98] This perched aquifer gallery yielded substantial amounts of water and in fact, the *viridarium de Robertis* portion of the *Tabria* estate was equipped with canals in order to channel this water to the various areas forming part of this *giardino* (CEM 114/7, ff. 12-13) (Fig. 3.6).

The above entry is accompanied by a detailed plan of the area under scrutiny, thus making it possible to identify the exact location of the mentioned *mina* (Figs. 5.119, 5.120). The perched aquifer gallery, several water reservoirs close by and cisterns are clearly indicated on the plan. The water gallery is shown to originate from within a subterranean feature excavated into the face of a ravine and is marked as a *fontana*. The territory in question complies with the *Teritorii del Vescovado* listed by Abela in 1647, who, nonetheless, made no reference to the existence of this subterranean perched aquifer extraction system (Abela 1647, 65). The water gallery is hewn into the inner end of an irregular shaped cave at map reference 46210 69451

[96] One such shaft was located below the Bishop's Palace (MD1) at Mdina, but was in 1725 reported as being completely dry. See Univ. 187, f. 100.

[97] The original manuscript title is almost completely worn out and only *dell acque* could be deciphered.

[98] According to this entry, the Wied l-Isqof gallery has a length of 160 m.

and is located beneath the early seventeenth-century country residence of Bishop Baldassare Cagliares in the Xagħra tal-Buskett whereabouts (Figs. 5.118, 5.122, 5.123).[99]

A cobbled path connects the old Cagliares residence with the underlying gallery and meanders past two small caves containing dry-rubble external and internal wall features (Fig. 5.121). Both path and caves are indicated on the plan for this area. The *giardino* underlying the perched aquifer gallery exterior is in this plan depicted as containing trees cultivated over ten different strips of terraced land (Fig. 5.120).

5.4 Concluding Remarks

The adopted methodology for the carrying out of this research, together with its limitations, have already been discussed in chapter one, sections 1.6 and 1.7. This chapter also touches upon all the objectives declared at the outset in chapter one, section 1.1, but data on these will be corroborated upon further in chapters six and seven.

A major drawback when interpreting the cartographic evidence presented in section 5.1 above involves: (a) the limited nature of cartographic data dating to the latter half of the sixteenth century; (b) the fact that in most instances, maps were engraved by cartographers who obtained their information from secondary sources; and (c) the fact that map topographical details are often inaccurate, making it difficult to locate with certainty any of the marked landscape features. On the other hand, source material dating to 1551 already makes a good argument for the presence of a well-defined valley system, the spatial location of which corresponds to Wied ir-Rum (RA56). Marked as *vallone giardini* this indeed indicates a prime agricultural site or area, but an interpretation of the term *giardino* based on this evidence is, at this stage, still unclear. Furthermore, it is significant to note that the term *fonte* or *fontana* can be associated with *għajn*-type springs which when located in north and north-west Malta, commonly originate from within man-excavated perched aquifer galleries in the Upper Coralline Limestone layer. An exception to this was the Marsa *fontana* which tapped a mean-sea-level aquiferous source.[100]

More evidence in this respect was obtained through an analysis of place-names given by G.F. Abela in 1647 in section 5.2 above. Whereas the principal water sources in the north and north-west zones of Malta are of the *għajn* and the *fontana* type, thus implying the utilisation of natural perched aquifer sources, *bir* and *cisterna* place-names are widespread for the remaining section of Malta. Unspecified *bir* place-names and listings can, in the absence of further evidence, represent both wells and cisterns, but where there is mention of a *cisterna d'acqua sorgente* this indicates the tapping of an aquiferous source, originating from either the perched or the mean-sea-level aquifer.[101] A *giardino* place-name does not necessarily signify water availability, but *giardini* references in areas located to the north and north-west of Malta frequently refer to the presence of spring water. Personal field-research did make it possible to correlate many of these water sources, commonly referred to by Abela as a *fontana*, with perched aquifer galleries excavated in Upper Coralline Limestone deposits. Furthermore, in the case of the *Deyr el Binet* (DI1) and the *Aayn San Giovan* (RA24), Abela does mention *fontane* placed in caves, which is another indirect reference to perched aquifer galleries.[102]

Fig. 5.123 Detail of the entrance section of the Wied l-Isqof perched aquifer gallery (RA25). Source: the author.

[99] Baldassare Cagliares was Bishop of Malta from 1615 to 1635. The cave fronting the Wied l-Isqof gallery bears a superficial resemblance to the l-Għar ta' Baldu at Wied ir-Rum, and is similarly equipped with a stone table. See section 5.2.1.1. No dry-stone wall screens the Wied l-Isqof cave entrance. The water gallery tunnels into the cliff-face and a narrow channel cut into the gallery and cave floor transports water to a masonry-built storage tank positioned to the east of the cave entrance which is also clearly indicated in the mid-seventeenth-century plan of the site.

[100] See chapter five, section 5.1. For a philological appreciation of *għajn*, *fonte* and *fontana* terminology refer to Appendices 3 and 4. An in-depth analysis of *għajn* place-name evidence during the earlier medieval period is attempted in chapter seven, section 7.6.1.

[101] See chapter five, section 5.2.

[102] See chapter five, section 5.2.1.

This analysis of the hydrological-related content of Abela's 1647 text does not come without its limitations. Very little consideration was taken by Abela of the physical characteristics of the mentioned water sources, commonly making it difficult to distinguish between a natural spring and an artificial water source. Even though *għajn* and *fontana/fonte* water sources are listed by Abela according to their locality or area, the exact location of these sources is not specified. Had it not been for personal field and archival research as well as Mario Vassallo's expert knowledge of the Mġarr (Malta) locality, the identification of the water springs/sources mentioned by Abela would have been difficult. On the other hand, this is an exercise which does not come without any related pitfalls, which, in this instance can involve the wrong association of a water spring with that mentioned by Abela due to changing place-name evidence or "failing" popular knowledge.

On the other hand, section 5.3 provides solid evidence for the existence of Upper Coralline Limestone galleries within a pre-Knights Maltese landscape. Furthermore, an analysis of documentation entitled *Acque (Seminario)* dating to the mid-seventeenth century shows how in the case of the Ta' Brija estate (RA44) at Rabat (Malta), a *mina*, or water gallery, was specifically marked as a *fontana* on the accompanying plan (CEM 114/7).

Chapter Six

Hydrology-Related *Cabreo* Documentation for Malta and Gozo during the Knights' Period

6.0 Introduction

A *cabreo* (pl. *cabrei*) is a collection of maps, plans of buildings and landscape views representing public or privately owned property. These are normally bound together in a single volume in which there also is a descriptive text section which corresponds to the illustrations presented (Ginori Lisci 1978, 20).[1] The maps and the corresponding text entries also include data on the locality in which the land in question is located, as well as any neighbouring properties (Ginori Lisci 1978, 20). It is possible that the words *cabreo* and *cabrero* were initially coined by the Knights of the Order of St John towards the end of the sixteenth century (Ginori Lisci 1978, 15-16). Available evidence indicates that the first *cabreo* was initially made use of in the Tuscan settlement of Lucca in ca. 1540 and dealt with a governmental practice aimed at controlling the land and its tenants.[2] Cabrei dimensions vary from volume to volume, but it has been noted that the earlier manuscripts tend to be smaller in size than the later ones, such as eighteenth-century volumes (Zammit Lupi 2002, 59).

The Treasury Manuscripts which form part of the sixteenth section of the Archives of the Order of Malta (A.O.M.) include a number of *cabrei*.[3] Several of the A.O.M. *cabrei* are leather bound, but most either have simple-limp or semi-limp parchment bindings.[4] These belonged to several of the Order's foundations,[5] most of which were established during the course of the seventeenth century and which commissioned the compilation of the *cabrei* manuscripts. It was during the first half of the seventeenth century that the Order recognised the need to employ the use of *cabrei* documentation in a local context, probably in a bid to record the extent of its possessions in Malta and administer these more efficiently (Blouet 1963, 72; Vella and Spiteri 2008, 22). The minutes of the Council of the Order for 1643 report that rent on a number of fiefs which became the Order's property with the annexation of Malta in 1530 had never been collected.[6] This shortcoming was addressed by the drawing up of the *Cabreo Magistrale*,[7] compiled in the years between 1643 and 1654, hence marking an effort to develop these assets further (Blouet 1963, 71).[8]

With mentioned circumstances in mind, it is significant to note that the first *cabreo* in Malta was only commissioned slightly over a century after the Maltese territory was passed to the Hospitaller Knights of Malta. This might be indicative of the fact that between 1530 and the first half of the seventeenth-century rural government territories received scant attention. It is possible that no concerted effort to improve the hydrological and agricultural potential of many of these territories was made during this period.[9] Should this be the case, it is likely that the agricultural framework discussed in the *Cabreo Magistrale* betrays a pre-1530 situation.[10]

It was common practice for *cabrei* manuscripts to be drawn through the direct engagement of architects and draughtsmen, technically known as *periti agrimensori* (Zammit Lupi 2002, 59).[11] These were responsible for surveying and drawing up plans of the property or

---

[1] *Cabrero* is the Spanish equivalent for, and is derived from, the Latin *capi brevium*. See Zammit Lupi 2002, 59. Zammit Lupi derives most of her information from the text by Ginori Lisci.

[2] Such manuscripts were originally referred to as *Martilogio*. Evidence for the first *cabrei* at Pistoia and Florence dates back to 1556 and 1565 respectively, and they were primarily referred to as *Campione*. Another term denoting a *Cabreo* was *Terrilogio*. See Ginori Lisci 1978, 13; Zammit Lupi 2002, 59.

[3] The Archives of the Order of the Malta (A.O.M.) *cabrei* manuscripts are housed at the National Library, Valletta. The Treasury manuscripts are divided into "A" and "B". Treasury "A" documents or *compendia* contain accounts books related to the sale, lease and maintenance of several of the Order's holdings. It is within the Treasury "B" manuscripts that all the Order's *cabrei* are included.

[4] For information on *cabrei* binding see Zammit Lupi 2002, 59.

[5] These recorded the movable and immovable property entrusted in the care of these foundations.

[6] With the acquisition of the Maltese archipelago as a fief from King Charles V, the Order automatically took possession of the crown lands in the Maltese territory. See Blouet 1963, 71.

[7] This *cabreo* consists of one volume and with the exception of a few missing plans, has survived in its entirety. See Blouet 1963, 73.

[8] The incomes derived from the establishment and growth of these foundations started to make a significant contribution to the Order's Treasury. See Blouet 1963, 72.

[9] By the end of the sixteenth century, economic problems stemming from a rapid local demographic increase and fluctuating international food prices placed the Order in a dire financial situation. During the early seventeenth century this situation was partly counteracted through the implementation of new patterns of administration and control and the regulation of local economic life. See Blouet 1963, 69.

[10] This would also seem to apply to later centuries. In a despatch dating to 27 November 1841, Lord John Russell discloses his views on the Maltese agricultural framework, which he considered to be largely archaic: "The general state of agriculture in these islands appears to a casual observer, or to those who have been used to see it carried on with all the modern improvements, to be extremely defective. In fact, the implements in use are the same as they have been in the memory of the oldest men, and probably no improvement in them has taken place from the time of the first occupation of the islands." See GOV. 01.3/6, Despatch To no. 81, 1841, f. 81.

[11] On the role played by the *periti* under the Knights of the Order of St John see Zammit A. 2009, 22-26.

tenement together with an accompanying description of it. It has to be noted that most of the earlier *cabrei* compiled by the Order of St John lack plans and illustrations and solely consist of a written survey listing the immoveable property (Vella and Spiteri 2008, 22). Where illustration is available, this generally involves single-line drawings in a monochromatic, dark blackish-brown ink (Fig. 6.1). Ink markings are mainly used to represent property boundaries. Shading, where present, is primarily used to provide contrast.[12] A particular case in point is the *Cabreo Magistrale* completed in 1654 (NLM. Treas. B. 289).

The eighteenth-century *cabrei* forming part of the A.O.M. collection of the National Library of Malta contain more elaborately decorated plans to which various colours are applied (Fig. 6.2). The principal section of a typical eighteenth-century *cabreo* consists of descriptions and illustrations of the urban and rural immovable property. Depicted in the survey are various features, including farmsteads, animal pens, sties and agricultural storage rooms, uncovered water tanks, cisterns and conduits amongst other features (Vella and Spiteri 2008, 23). The various tree species planted in a territory are also frequently included.[13] Several of the *cabrei* dating to this period also distinguish between the irrigated and the non-irrigated land. Most of the plans presented are accompanied by a scale-bar and a north-pointer, and in some instances, the name of the surveyor is also included. In some rare instances *cabrei* dating to this period also contain a watercolour rendering representing an artist's impression of the property in question and its general landscape setting.[14]

6.1 Interpreting the Data Provided by the Seventeenth and Eighteenth-Century Cabrei *Manuscripts*

For the purpose of this study, various Treasury B cabrei manuscripts belonging to the A.O.M. archives were consulted. The *cabrei del Magistero* 289, 290 and 291, amongst others, were found to be the most informative. The fact that *cabrei* were produced by *periti agrimensori* signifies that the plans contained within are usually a

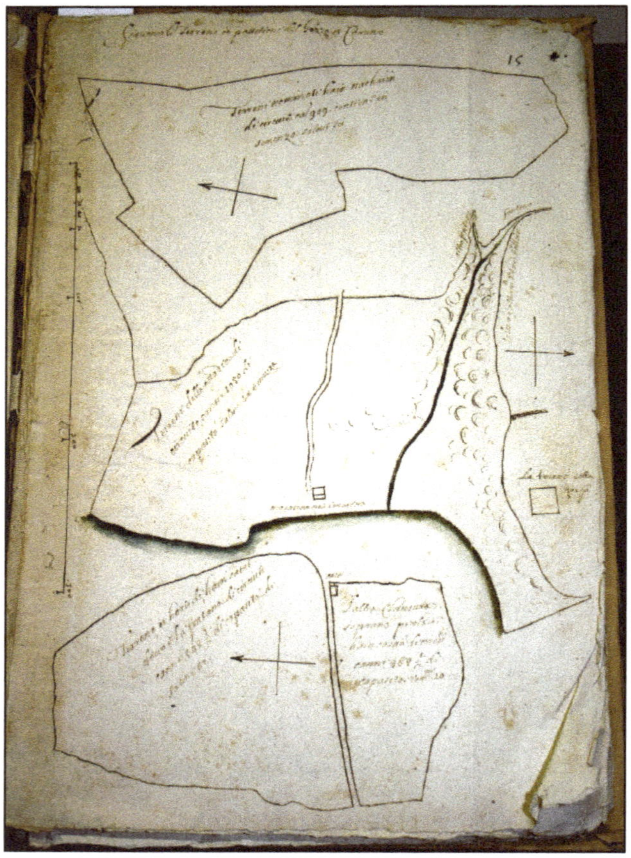

Fig. 6.1 NLM. Treas. B. 289, f. 15. Plan showing three separate parcels of land at (top to bottom) *hain nacharin* (GO11), *Madia* (GO16) and *hain casab* (GO12) in Gozo. In the instance of the *Terreni della Madia*, what appears to be a water course passing through a valley bed is clearly illustrated.

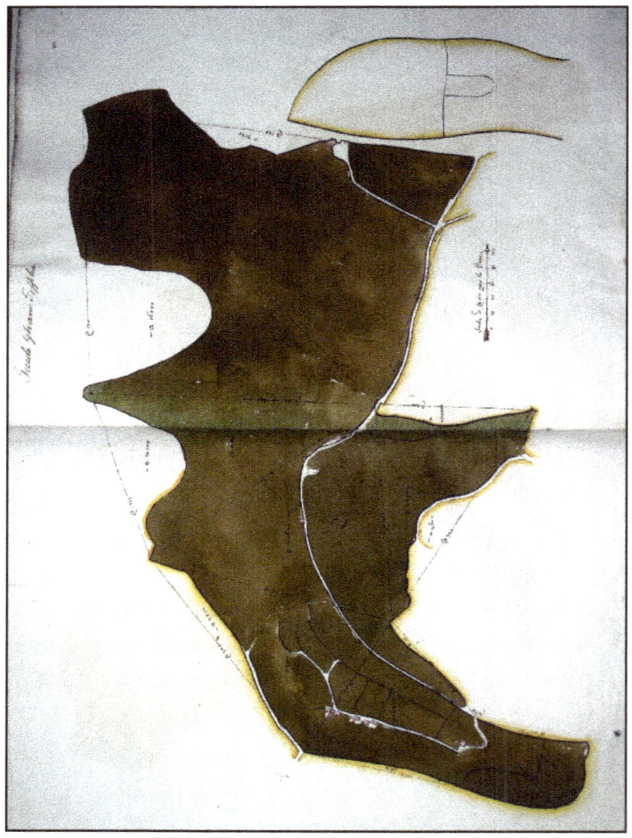

Fig. 6.2 NLM. Treas. B. 290, f. 100. Plan of the *Feudo Ghain Tuffiha* (Għajn Tuffieħa – MG15). The greenish-brown colouring shows the territory forming part of the Għajn Tuffieħa *feudo*. Farmsteads are shaded in pink, but water-related features are not illustrated on this plan.

[12] This situation is closely paralleled in the late sixteenth-century *cabrei* of northern Italy. See Ginori Lisci 1978, 165.

[13] This data were examined by C. Grech. See Grech 2001.

[14] The *Cabreo del Magistero* 290 is a particular case in point.

scaled-down version of the territory or tenement they represent. Through the use of land boundaries and a place-name analysis, as reflected in early twentieth-century survey sheets for Malta having a scale of 1:2,500, it is usually still possible to accurately locate rural tenements mentioned in these *cabrei*. Personal experience has also shown how a familiarity with the geology, topography, archaeological and historical remains in the areas described by the *cabrei* greatly facilitates the interpretation of these manuscript sources. For practical reasons, the various *cabrei* data referring to the same tenements or properties were merged and are here described as single entries according to the territory in which they are located. There are instances where the entries described below were supplemented with evidence obtained from the nineteenth-century Descriptive Plans of the Crown Property *cabrei*. The inclusion of this data made it possible to gain an understanding of changes to the fabric of *giardini* taking place in the eighteenth and the nineteenth centuries.

The first volume of the *cabreo del Magistero* is the most important primary source of evidence when trying to understand the agricultural and hydrological setup of rural tenements pertaining to the Order of St John in the first half of the seventeenth century. Compiled between 1643 and 1654, this source includes descriptions of over 150 plots of land. In various instances, *cabrei* 290 and 291 entitled *Cabreo del Magistero* Vol. 1 and *Cabreo del Magistero "Secrezia"*, detail further the water-related data given in *cabreo* 289. *Cabreo* 290 contains the plans. The textual description of the landholdings presented in manuscript 290 is to be found in *cabreo* 291.

From an illustrative point of view, the plans of *cabreo* 290 are of a much better quality than those presented in *cabreo* 289. Several of the water-related entries in *cabrei* 290 and 291 also make it possible to distinguish between the irrigated and the non-irrigated landscape. It is also through the *cabreo del Magistero* 290 that it can be established that grains, cotton and cumin were commonly cultivated in good quality agricultural land. Grains of the barley, *tuminia* and *mischiato* types, were regularly sown in medium quality land. Poor quality land could only support the cultivation of stronger grain, known as *mischiato*. Poorer quality land generally covered the Upper Coralline and Lower Coralline Garigue plains, commonly used for rough grazing. It is likely that medium and poor quality land only contained shallow soil deposits and had no access to perennial water sources for irrigation purposes. Mid-seventeenth-century accounts describing the cultivation of the small island on Cominotto are a typical case in point. The lack of a perched aquifer on Cominotto made the cultivation of small tracts of arable land directly dependent on seasonal rainfall. As a consequence, only *mischiato*-type grains were grown on this barren and rocky island.

Hydrology-related features were in most *cabrei* entries generically classified as either *cisterna* (cistern), *cisterna d'acqua sorgiua* (cistern of spring water), *pozzo* (well), *fontana d'acqua* (water fountain or spring), *fontanella* (little fountain or spring) and *gebia* (water reservoir). In *cabreo* 290 cisterns and wells are commonly represented by means of blue-coloured circles (Fig. 6.3). Unless the plan data are corroborated further by an accompanying text or a description of the property in question, making any distinction between a well or a cistern is unfortunately not possible (Fig. 6.4). Some subterranean water-related features are represented by means of an intermittent line, which, in a few instances, can be safely associated with a subterranean water tunnel or gallery. No technical distinction is made between perched aquifer and mean-sea-level aquifer sources. Some *cabrei* entries do make a direct reference to the existence of a *mina* – probably a reference to perched aquifer galleries. In the

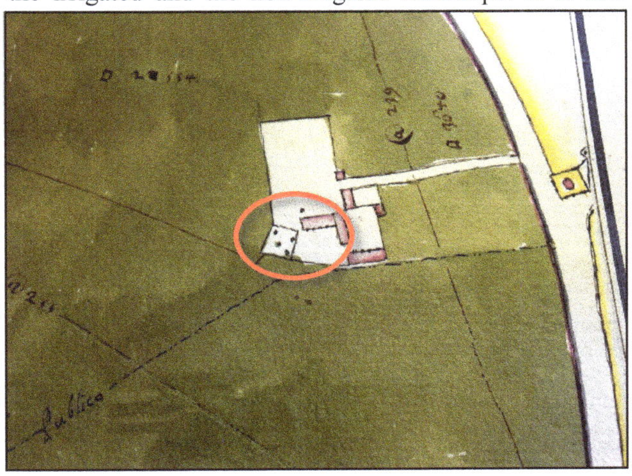

Fig. 6.3 NLM. Treas. B. 290, f. 26. Plan detail of the Government landholding at *Bulebel il piccolo* (ZA1). Encircled are five greenish-blue coloured spots which possibly represent water cisterns.

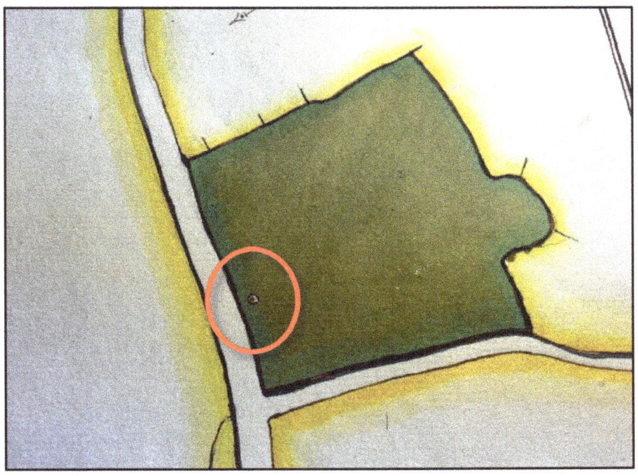

Fig. 6.4 NLM. Treas. B. 290, f. 31. Plan detail of the Government landholding at *Bulebel il Grande*. Encircled is a line drawing of a circle containing a dot in its centre. In *cabreo* 290 this convention was often used to denote water cisterns.

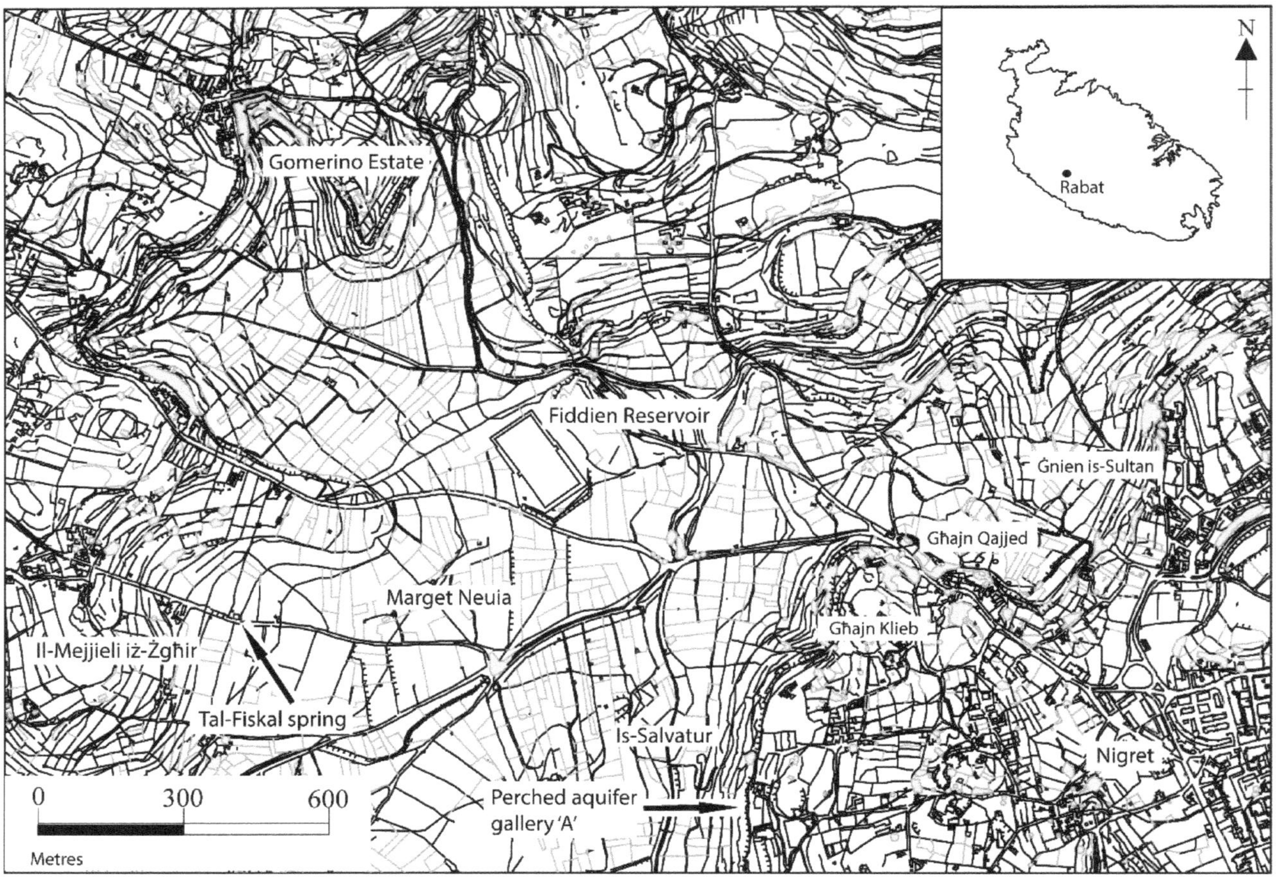

Fig. 6.5 Plan of the Rabat (Malta) territory showing the location of the Ġnien is-Sultan (RA13), Għajn Qajjed (RA19), Għajn Klieb (RA18) and the Tal-Fiskal (RA34) areas. Topographical map image adapted from MEPA Map Server data.

instance of *Cabreo* 290 plans, it was noted by the author of this book that water sources, seemingly originating from perched aquifer water galleries, were represented by means of a thin green-coloured line.

In conclusion, the reader of this book is to be reminded that water distribution and allocation for agricultural usage falls outside the scope of this study. For this reason, these modes of water management are not represented in the case studies presented hereunder. Moreover, in order to facilitate place-name readability, which due to Late Medieval and Early Modern unstandardised norms are in the original documentation frequently transcribed in differing ways, reference to the comparative table of toponyms (Appendix 2) is recommended.

6.1.1 *The central and southern plains*

Water-related entries in the *cabrei del Magistero* make it possible to distinguish between the irrigated and the non-irrigated landscape. Lands to the south and the south-eastern section of Malta are frequently listed as containing a *cisterna* (water cistern), or *pozzo* (well). Cisterns commonly imply the availability of non-irrigated land, thus marking water conservation efforts involving the harvesting of rain water, probably in subterranean water storage tanks. *Cisterne* were located at the *Fegò di Buleben Minore*, situated between the settlements of Tarxien and Żejtun. This contained [...] *quattro cisterne due bone, et l'altri due sono rempiti di Pietre, e terra* (Fig. 6.3).[15] Another cistern was recorded in the *Clausura Ta xiahra*, located between the settlements of Żabbar and Żejtun.[16]

The tapping of mean-sea-aquifer sources is detectible through the mention of a *cisterna d'aqua sorgente*[17] having the name of *bir il vuassa* (MA4),[18] to the north of *Clausura bir uasa* at Marsa (NLM. Treas. B. 289, f. 17). Both *Bir Buhagiar* (MA1; NLM. Treas. B. 289, f. 20), located in the Qormi-Marsa area (MA10), and *Bir hialef*

[15] "[...] four cisterns, two of which are good, the other two being filled with stones and earth." See NLM. Treas. B. 289, f. 5. These same water features are again mentioned in NLM. Treas. B. 290, f. 26; NLM. Treas. B. 290, f. 31.

[16] [...] *in detta Clausura una cisterna da circa dodeci palmi di fondo* [...]. See NLM. Treas. B. 289, ff. 5-6.

[17] This translates into a "cistern of spring or running water". The term *cisterna* can in this instance be taken to represent a well tapping an aquiferous source.

[18] *Bir il Vuassa* appears to have been located within the *Giardino della Marsa*. See NLM. Treas. B. 289, f. 18; NLM. Treas. B. 289, f. 26.

(MA5; NLM. Treas. B. 289, f. 30), in Marsa, are also described as a *cisterna d'acquà sorgente.*

Cabreo 289 also makes mention in the Marsa area of *canali grandi per dove corre l'acqua che viene dal Casal cormi et il gran vallone e va al mare* (NLM. Treas. B. 289, f. 25). The availability of surface fresh water or brackish water in the Qormi-Marsa area can perhaps be detected from place-name evidence such as *Clausura ta Siamar* (NLM. Treas. B. 289, f. 22), possibly derived from the Maltese word *Simar* - the Hollow-leaved Rush. This is a plant species that grows in valleys and damp areas (Sultana and Falzon 2002, 80-81). The *Clausura tal gadira* (MA6; NLM. Treas. B. 289, f. 22) place-name, also from the Qormi-Marsa area, has possible derivation from the Maltese word *għadira*, and might indicate the presence of a shallow water pool.

Another water-related feature representing either a well or a cistern, which the relevant *cabreo* 290 plans indicate to have been located in the middle of the road at Marsa, is represented by a circle enclosed by a rectangle (Fig. 6.6). This features prominently in a number of plans showing Government-owned landholdings in the Marsa area (NLM. Treas. B. 290, ff. 44-45, 48, 50, 52-54). Another two wells or cisterns are indicated in a clearing on the side of the road, adjoining the *Clausura tal Hammieri* (MA7), also at Marsa (NLM. Treas. B. 290, ff. 48, 50, 52-54; Fig. 6.7).

The existence of a brackish water source of no use to agricultural cultivation is recorded for either Birżebbuġa or the neighbouring Marsaxlokk settlement. This water source is described as *fù Gabellata con la bevratura ò sia fontana convense d'acqua salmastra (?) S. Giorgio di Marsa Scirocco oggi posseduta dal Publico* (MAR1; NLM. Treas. B. f. 13). Referred to as *Ghajn detta Andar ta Langarù* [...], this can be tentatively identified with either: (a) the public fountain located on the coastal edge fronting Triq iż-Żejtun at Marsaxlokk; or (b) the whereabouts of St George's Bay in Birżebbuġa.

6.1.2 The north and north-west districts of Malta

It is in the areas yielding water from the perched aquifer of north and north-west Malta that the majority of the *hayn* (spring) place-names mentioned in the *cabrei* manuscripts are located. Areas containing the availability of spring water were frequently developed into *giardini* or *ġnien* areas of cultivation.[19] These were typically located in sheltered valleys, where water could be obtained from either a perennial or seasonal water course flowing through the valley bed, or from spring water originating from the overflow of the perched aquifer located further up the valley sides. Properties in the island of Malta which subscribe to this model are: *Clausura sotto hayn Richana* (NLM. Treas. B. 289, ff. 45-46), *Giardino nominato hayn Cayet* (NLM. Treas. B. 289, ff. 53-54), *Giardino di due fontane* (RA14; NLM. Treas. B. 289, ff. 56-57), *Giardino di Hued iron tal Callus* (NLM. Treas. B. 289, ff. 84), *Fego del Santi* (MG23; NLM. Treas. B. 289, ff. 84-88), *Giardino e terreni scapuli di hayn toffeha* (NLM. Treas. B. 289, ff. 89-91), and *Gnien haggel* (NLM. Treas. B. 289, ff. 106-12).

Properties in Gozo in which *giardini*-type territories are located include: *Giardino della gran Fontana* (GO21; NLM. Treas. B. 289, ff. 143-45),[20] and the *Giardino di hain habdun* (Għajn Għabdun – GO8; NLM. Treas. B. 289, ff. 148-49). All the above-mentioned *giardini* are

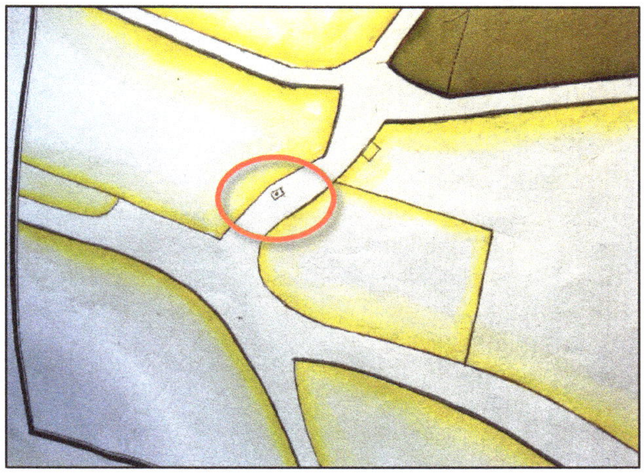

Fig. 6.6 NLM. Treas. B. 290, f. 45. Plan detail of the Government-owned landholding of *Due Clausure ta Sultan o ta Suimajna* at Marsa. Encircled is a probable well or cistern located in the roadside.

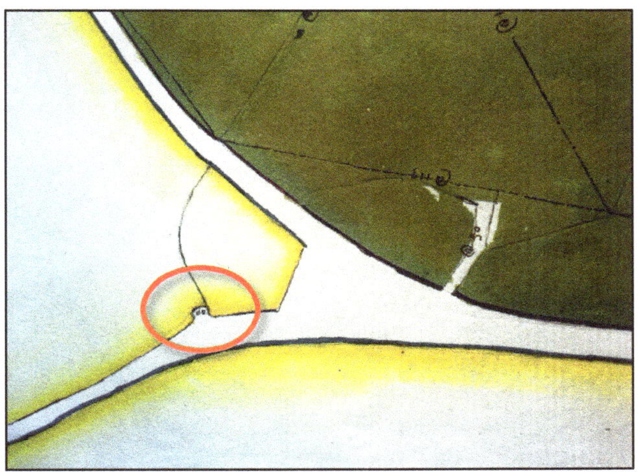

Fig. 6.7 NLM. Treas. B. 290, f. 48. Plan detail of the Government-owned estate of *Clausura ta Ciappuna* at Marsa, showing the location of two probable wells or cisterns.

[19] For a discussion on *giardini*-type agricultural sites see chapter five, sections 5 and 5.2.

[20] This territory mostly encompasses the area which is presently known as Lunzjata in the outskirts of Rabat, Gozo.

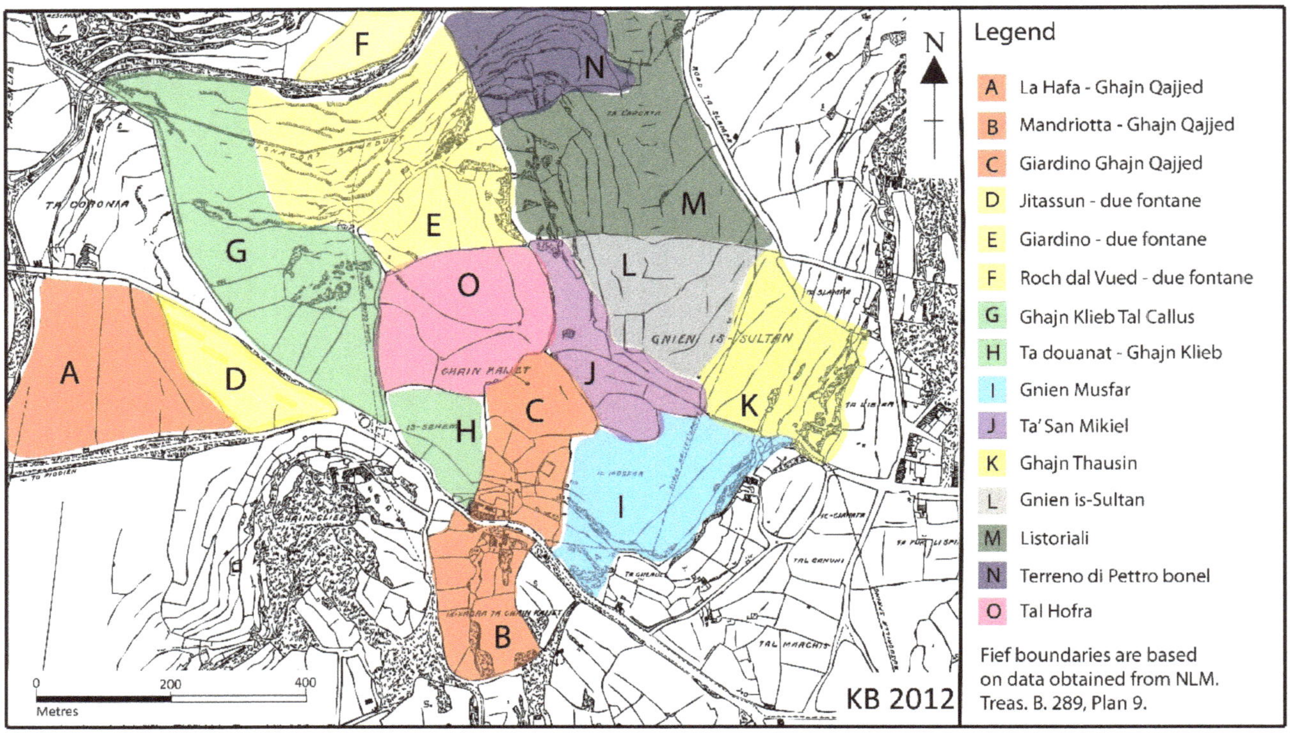

Fig. 6.8 1922 map detail of the Għajn Qajjed (RA19) and Ġnien is-Sultan (RA13) areas in the Rabat (Malta) territory, overlying which is property data showing the location of the various landholdings belonging to the Order of St John. Source: Survey Sheet 72, Scale 1:2,500, School of Military Engineering, Chatham, 1922.

located in areas where Upper Coralline deposits of the Mtarfa Member type prevail, and most contain perched aquifer water galleries which supply these agricultural tenements with a perennial water source. Even though the *Clausura sotto hayn Richana* and *Fego del Santi* are not listed as *giardini* in the *cabreo del Magistero* 289, personal field-research has shown these to contain a *giardino*-type setup which also consists of one or more perched aquifer water galleries furnishing these properties with a perennial water supply. Furthermore, the presence of water at Tas-Santi (MG23) is also indicated in the *cabreo* entry for this tenement (NLM. Treas. B. 289, ff. 84-88).

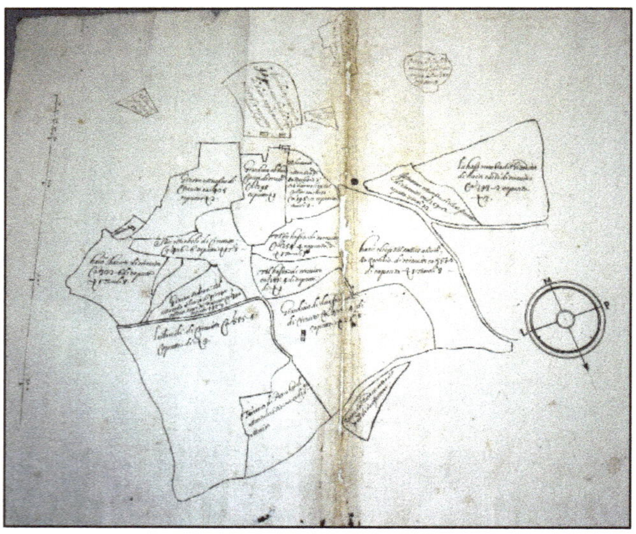

Fig. 6.9 NLM. Treas. B. 289, Plan 9. Seventeenth-century plan of the Ghajn Qajjed (RA19) and Ġnien is-Sultan (RA13) territories in the limits of Rabat (Malta).

6.1.3 *The Għajn Qajjed, Ġnien is-Sultan, Għajn Klieb and San Mikiel* giardini *of Rabat (Malta)*

Hayn Cayet (Għajn Qajjed – RA19), to the immediate west of Rabat (Malta), appears to have been one of the principal *giardini* in the possession of the Order of St John (Figs. 3.6, 6.5). The importance of this territory can perhaps be attested by the fact that this *giardino* is described in more detail than most of the other landholdings listed in *cabreo* 289. The entry for the Għajn Qajjed area is also accompanied by a comprehensive plan showing the location of the various mentioned landholdings (NLM. Treas. B. 289, Plan 9). Although land in this territory has since the seventeenth century been fragmented into smaller fields, in most instances it was still possible to work out the location of each parcel of land mentioned in the *cabreo* text entry (Figs. 6.8, 6.9).

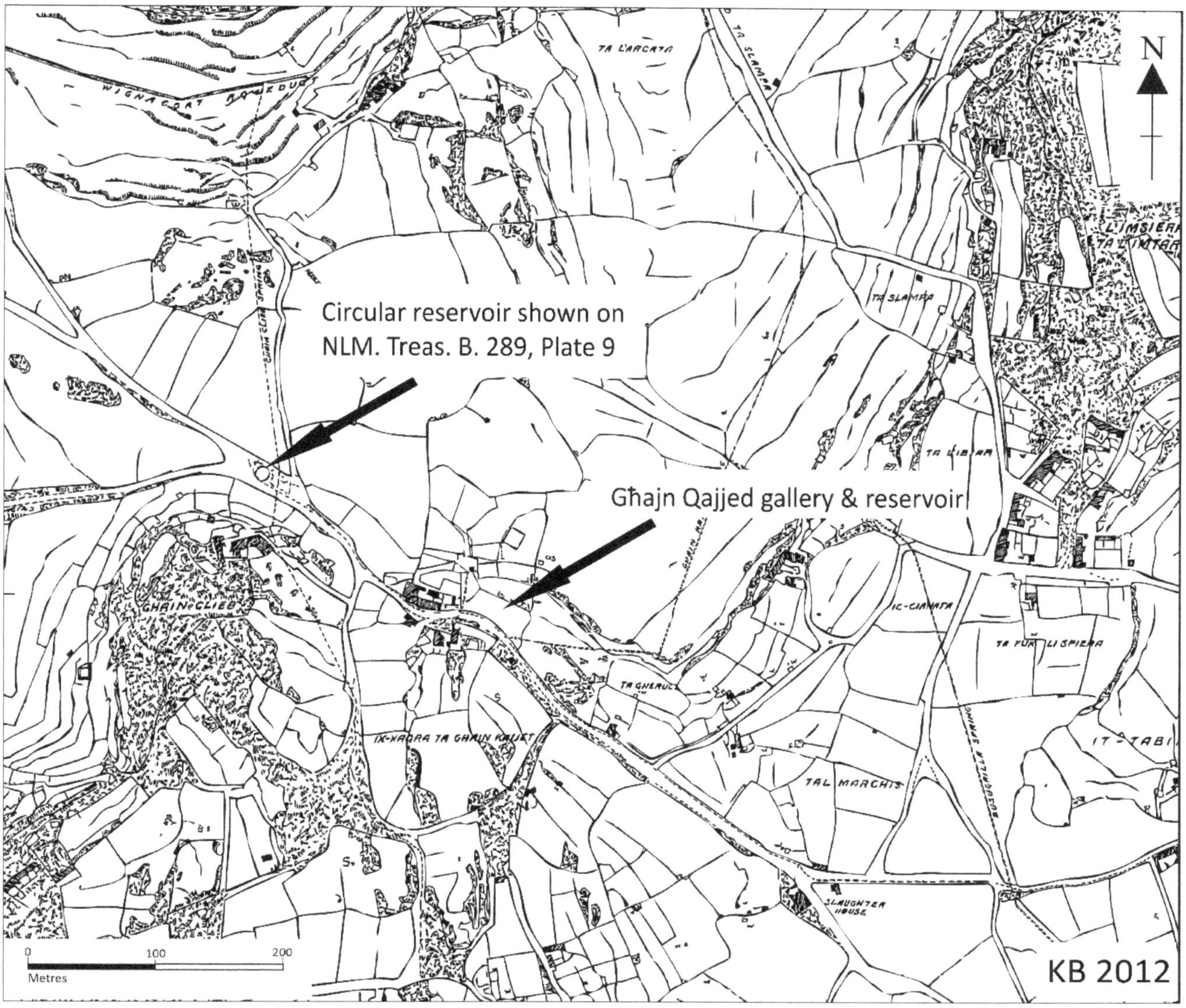

Fig. 6.10 Site Plan of the Għajn Qajjed area showing the location of the Għajn Qajjed (RA19) water gallery and its fronting reservoir. Source: Survey Sheet 72, Scale 1:2,500, School of Military Engineering, Chatham, 1922.

The principal feature of the Għajn Qajjed *giardino* (Fig. 6.8 "C"), was a large reservoir, described as the *gran gebia* (NLM. Treas. B. 289, f. 54). There is also the mention of canals, which transported water from the *gran gebia* to the *Giardino delle due fontane* (RA14) in the vicinity (Fig. 6.8 "E"), and the presence of a *lauatorio* – probably a washing basin.[21] Field-research successfully located the Għajn Qajjed spring at map reference 44739 71685, at the exact point indicated by the 1922 Survey Sheet 72 (Fig. 6.10).[22] Għajn Qajjed originates from within a perched aquifer gallery excavated into an Mtarfa Member deposit.[23] It is fronted by what appears to be a Knights' period reservoir which can be perhaps associated with the *gran gebia* mentioned in the *cabreo* documentation. It measures roughly 8 by 21 m, and the extracted water is stored within it for later distribution and usage. The gallery's entrance is visible, but because of dense vegetation growth, is difficult to access. The reservoir contains masonry pillars at every 1.5 m interval. Bridging the gap between each pillar there are still a fair number of masonry slabs which appear to have originally covered the whole extent of the reservoir (Figs. 6.11, 6.12).

[21] [...] *li canali che portano l'acqua dalla gebbia di hayn cayet al giardino delle due fontane per longhezza di can 49*. See NLM. Treas. B. 289, f. 53. The *giardino delle due fontane* is located further downstream from that of Għajn Qajjed, in an area in which no perched aquifer deposit is available. It is probable that the water yielded by the Għajn Qajjed perched aquifer gallery was also utilised to irrigate the adjoining territories of *Tal Hofra* and of *San Micheli*.

[22] See Survey Sheet 72, School of Military Engineering, Chatham, 1922.

[23] Personal communication by farming folk cultivating Government leased property at Għajn Qajjed. Writing in 1924, T. Zammit stated that the Għajn Qajjed gallery is ca. 50 m long. See Zammit T. 1924, 37.

Fig. 6.11 The Knights' period reservoir at Għajn Qajjed (RA19) in the territory of Rabat (Malta) which can perhaps be associated with the *gran gebia* mentioned in the *Cabreo* 289 documentation. The Għajn Qajjed spring which originates from within a perched aquifer gallery feeds this reservoir with a perennial water supply. Source: the author.

Fig. 6.12 The Għajn Qajjed (RA19) water reservoir in the territory of Rabat (Malta). This was probably covered over by means of loose masonry slabs in order to minimise water evaporation during the warmer summer months. Source: the author.

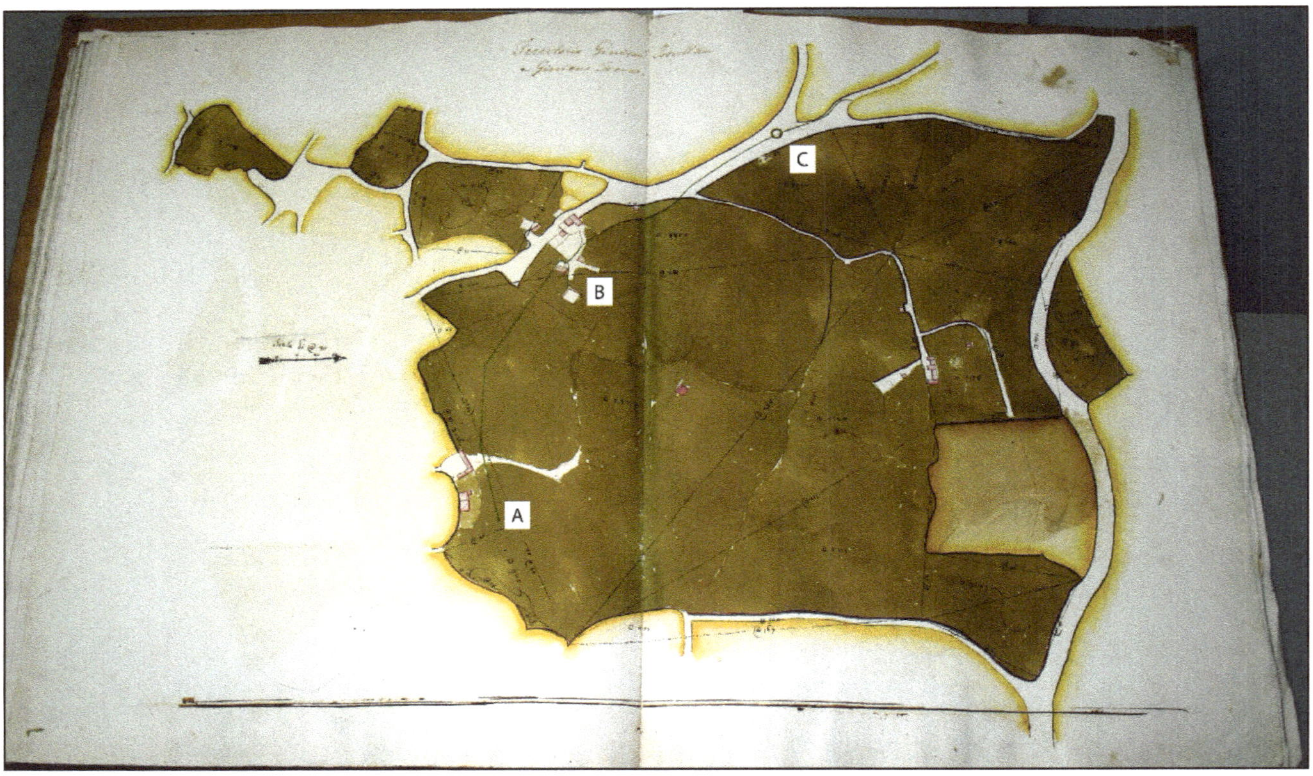

Fig. 6.13 NLM. Treas. B. 290, f. 85. Plan of the Ġnien is-Sultan estate (RA13) and its adjoining territories located in the Rabat (Malta) territory. "A" marks the subterranean canal which from the Għajn Tewżien spring (RA21) channels water towards the *Gebia Grande* "B" and another water reservoir "C".

Also significant is the description of the *Ginen Mosfar*[24] area (RA31; Fig. 46 "I"), where there is mention of excellent quality agricultural land in which are found *tre speraglie della' fontana di hayn Cayed chi va' alla' vatta* (NLM. Treas. B. 289, f. 62). Considering that the Għajn Qajjed spring originates from the Ix-Xagħra ta' Għajn Qajjed area (Fig. 6.8 "B"), the mention of *speraglie* in this particular case probably refers to vertical shafts piercing the ceiling of the subterranean canal connecting this spring to the Knights' period aqueduct system.

The *cabreo* description of Għajn Qajjed also gives a detailed insight into the early seventeenth-century water-sharing arrangement practised between the Għajn Qajjed *giardino* and its neighbouring *giardino* known as *delle due fontane* (RA14). It is significant to note that the perched aquifer gallery supplying the Għajn Qajjed *giardino* with a perennial supply is only referred to as

[24] The current Maltese place-name for this area is Ġnien Mosfar.

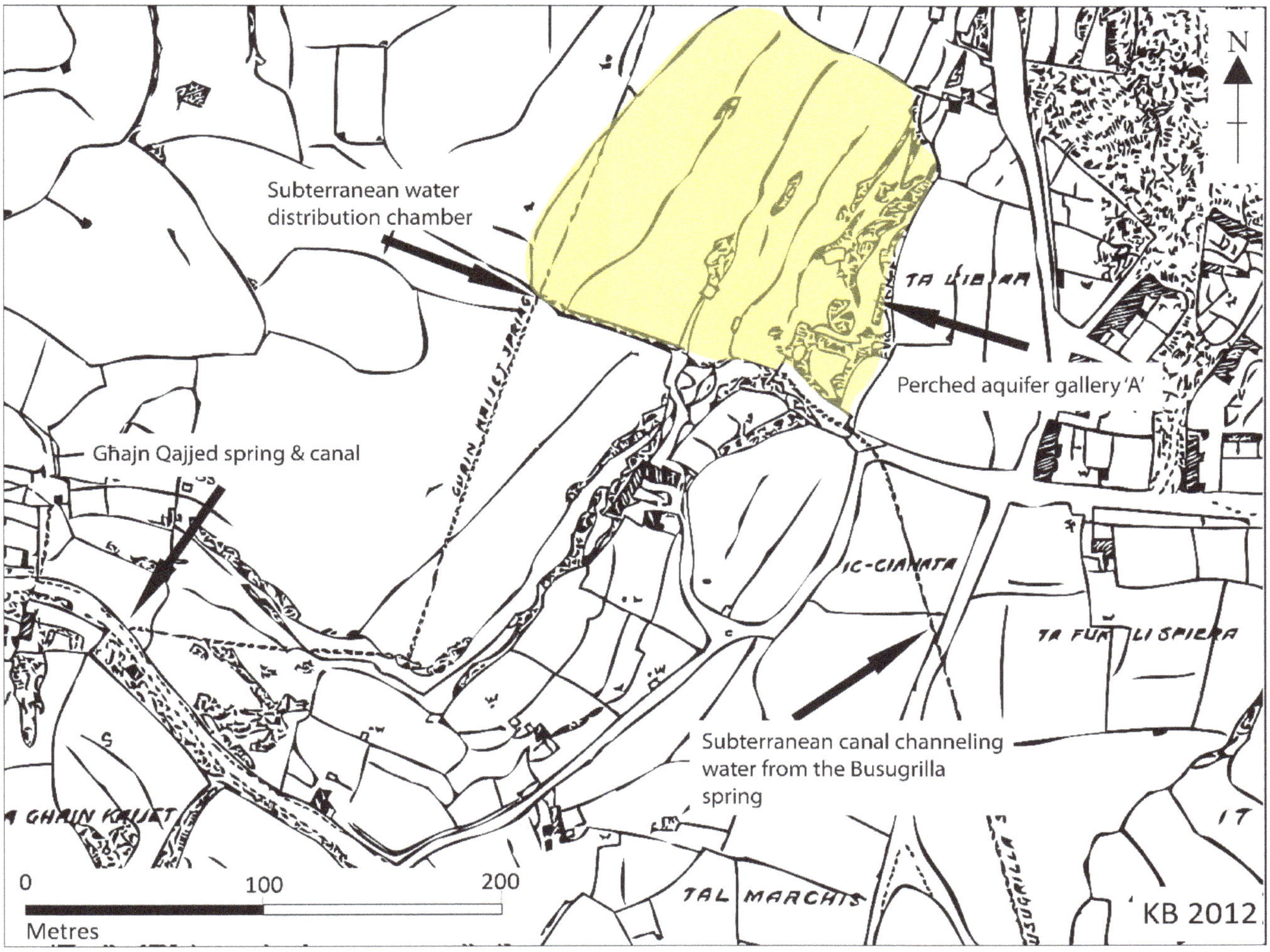

Fig. 6.14 Shaded in yellow is the area covered by the Għajn Tauzen (or Tewżien – RA21) tenement in the territory of Rabat (Malta). Source: Survey Sheet 72, Scale 1:2,500, School of Military Engineering, Chatham, 1922.

fontana or spring (NLM. Treas. B. 289, f. 54). In the 289, 290 and 291 *cabrei* manuscripts, no reference is made to the subterranean gallery tapping a perched aquifer source. That the Għajn Qajjed spring formed an integral part of the framework of the *giardini* of Għajn Qajjed and that of *delle due fontane* before its inclusion as one of the primary water sources feeding the Valletta aqueduct[25] can perhaps be detected through the documentation relating to a litigation between the *gabelloti*, or farmers, cultivating the land in this *giardino* and the administrative entity in charge of its management (NLM. Treas. B. 289, f. 54). It is likely that prior to the commencement of the aqueduct project, a water distribution agreement between farmers cultivating land in Għajn Qajjed and the adjoining *Giardino delle due fontane* was already in place.

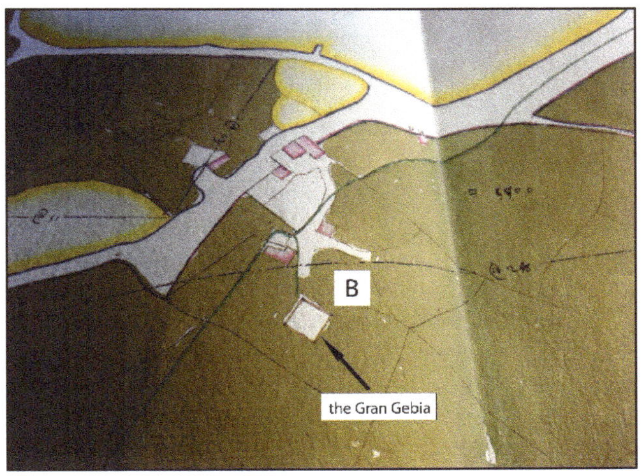

Fig. 6.15 NLM. Treas. B. f. 85. Plan detail of the *Territorio Gnien Issultan* (RA13). The green line shows the water being diverted from the Għajn Tewżien spring (RA21) to the *Gran Gebia* at Għajn Qajjet (RA19).

[25] Works on the Valletta aqueduct project commenced in 1610 and were completed in 1615. See Zammit T. 1924, 9-13. See also chapter five, section 5.2.4.

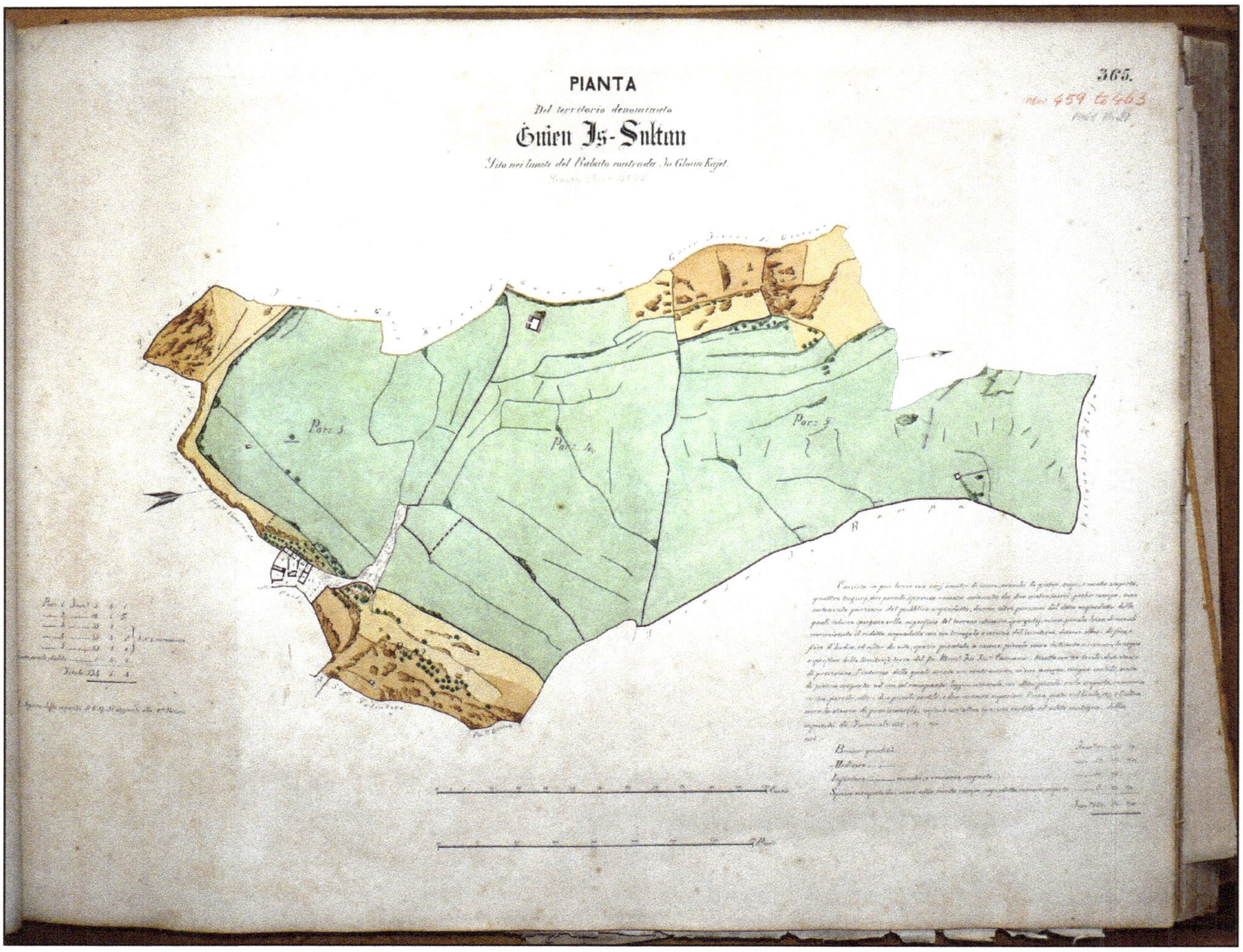

Fig. 6.16 NAM. Descriptive Plans of the Crown Property in Malta, 1867, ff. 459-63. Plan for the Ġnien is-Sultan tenement (RA13) in the territory of Rabat (Malta).

With the inclusion of Għajn Qajjed as one of the Valletta aqueduct's principal sources, only half of the total water yielded by the spring was kept for use in the above-mentioned *giardini*. Of the water available for irrigation purposes, three days of the week were dedicated to the watering of Għajn Qajjed. In the remaining four days, the water was channelled towards the *giardino* of the *due fontane* (NLM. Treas. B. 289, f. 57).[26] That this water distribution arrangement was in 1644 negatively impacting the irrigation of both *giardini* can be seen from the fact that farmers working land in both territories were lamenting with the authorities that the volume of water irrigating their land was not adequate to offset the inclemently hot and dry month of August. The matter appears to have been resolved by diverting part of the water-yield of the nearby *Hayn Tausen* to Għajn Qajjed.[27] The canal transporting water from *hayn Tausen* into the *gebia grande* at Għajn Qajjed passed through the agricultural territory known as *Ginen Mosfar* (Fig. 6.8 "I"), located to the east of Għajn Qajjed (NLM. Treas. B. 289, f. 62). This water management system is again mentioned in *cabreo* 290. What appears to be a partly

[26] Writing almost contemporaneously, Abela states that the Għajn Qajjed water source was only channelled towards Valletta by day. By night, it fed the *Giardino del Principato* (probably corresponding to *delle due fontane*), and that of *Aayn Kayed*. See Abela 1647, 111.

[27] A section of the entry for the *Giardino noiato* [*sic*] *hayn Cayet* reads: *Dentro il detto giardino di hayn cayet ui è una gran gebia, e il detto Giardino soleua pigliando la 'meta' integra di tutta l'acqua della' fontana di hayn cayet e poi la detta acqua si soleua spartire in due parti, cioe tre giorni della settimana repigliaua il giardino di hayn cayet e l'altri quattro il giardino delle due fontane, che e sotto francà alhorà dalla uicenda, che bisogna donare a detto ogni 15 giorni al Giardino del fu com. Perdicomati allhora dal quam(?) medico Corogna, conforme la concessione fatta dall' em(?) Mons. Vignacourt di bona memoria tre anni sono in circa ad instantia del sig cau. De Ponset si sono fatti li canali che pigliano l'acqua della fontana di hayn Teuzen che ua d scolare dentro la detta gebia e (?) che la detta acqua di hayn Teusen ne' é della medessima' quantita della meta' dell' acqua della detta fontana di hayn cayet il detto sig. Ponset per il mancamento hà uoluto chi fossero dati hore, e mezzo di continuo della suà acqua ualuta' per un quinto conforme il tutto allhora fu' uisto, e prouato in prisentia delli ss. Riceuitori e secreto. Dopoi lamentandosi li gabelloti di hayn cayet e dil giardino di due fontane che la detta quantita' d'acqua non si gli daua' al mese d'agosto 1644 il detto s. Cau. Ponset impresentia dil sig. cau. di carros et dil secreto Mazallet hà fatto misurare l'acqua della detta fontana di hayn cayet et trouatalà conforme lui dissi esser di 335 filetti di 32 l'unza n2s(?): equa della' fontana di hayn teusen 314 filetti 29 si vistô di accordo tra loro stabilito, chi oltri lacqua dellà fontana di hayn teuzen fosse data' alli detti giardini sei unzi dell'acqua di hayn cayet obligando pero alli gabelloti di detto dio giardini di daro una' uicenda ogni 15 giorni al detto Giardino dil fu Perdicomati* [...]. See NLM. Treas. B. 289, f. 54.

Fig. 6.17 The square-headed doorway giving access to the water distribution chamber for the Għajn Qajjed (RA19) and Busugrilla (RA7) springs at Ġnien is-Sultan (RA13) in the territory of Rabat (Malta). Source: the author.

Fig. 6.18 Interior view of the water distribution chamber at Ġnien is-Sultan (RA13) in the territory of Rabat (Malta). Source: the author.

subterranean canal system starting within close proximity to the border between Għajn Tewżien and Ġnien Musfar (RA31) properties heads towards the Għajn Qajjed area where it forks into two distinct channels (Figs. 6.13 "A", 6.15). A section of the canal leads to a square-shaped structure which probably conforms with the *gran gebia* mentioned above (Figs. 6.13 "B", 6.15 "B"). The other section of the canal heads towards the public road in order to supply neighbouring *giardini* with water (NLM. Treas. B. 290, f. 85).[28]

Of importance is the entry describing the *terreno di S. Michaele* (Fig. 6.8 "J"). This borders the agricultural estates of *Tal Hofra, delle due fontane* (RA14), *Listoriali, Gnien Sultan* (RA13), *Hayn Tausen, Gnien Musfar* and *Għajn Qajjed*. The *cabreo* description of the *terreno di S. Michaele* indicates that it was fed with a perennial water supply originating from the neighbouring *Hayn Teusen* territory located at map reference 45053 71769 (Fig. 6.8 "K"). Reference within the *terreno di S. Michaele* to *la bocca della' mina della fontana di hayn Tausen, che ua' alla vatta longa' canne 700 et fonda in alcune parte 45 palmi in alto 60* (NLM. Treas. B. 289, f. 63),[29] is significant in this respect. If the given measurement of 700 *canne*[30] is correct, this implies that the near-horizontal rock-excavated water-collection tunnel forming part of this gallery was ca. 1.4 km long. The gallery appears to have had an approximate internal height that varied between 3 and 4.5 m. The height of the gallery was probably the result of the lowering of its floor level. Personal experience has shown that this practice was frequently adhered to in instances where a receding perched aquifer level was being experienced, probably due to water over-extraction.

In 1647, *hayn Tausen* was also referred to by Abela as *Birzegrella* (Busugrilla – RA7) who likewise described it as a *pozzo d'acqua viua* (Abela 1647, 80, 111). The Busugrilla spring was fed by two other minor sources: *nescieaa tal borgi tal mitarfe* (RA36) and *Megira tà Mesrah Mnusa* (RA9; Abela 1647, 111). The location of these is not known, but as hinted at by the *nescieaa tal borgi tal mitarfe* place-name, they were presumably located further downstream, in close proximity to Mtarfa. The *Hayn Teusen* toponym could not be traced on the twentieth-century survey sheets for the area, but that of Busugrilla is clearly indicated (Fig. 6.14).

Personal field-research did not help to resolve this matter. It did, however, lead to the detection of another perched aquifer gallery at the approximate map reference of 45053 71769 (Fig. 6.14, perched aquifer gallery "A"). The current place-name of the area is that of *Ta l-Ibiar*, meaning "of the wells", and points towards the existence of a series of shaft wells piercing the ceiling of a perched aquifer water gallery. The presence of one or more water galleries containing shafts is also implied by the place-

[28] It is interesting to note that an 1867 description of Għajn Qajjed does make reference to a *mina* or perched aquifer gallery as well as the existence of water-related infrastructure geared towards the collection of surface runoff water originating from a neighbouring territory (Fig. 6.21). There is also the mention of a tract of the Wignacourt Aqueduct system which passes from within the Ġnien is-Sultan portion. See NAM. Descriptive Plans of the Crown Property, 1867, ff. 459-63. Apart from canals for the conveyance of water, in the nineteenth century, the Għajn Qajjed *giardino* contained a *vastissima gebia*, and a *piccola gebia coperta con soli capitelli*. Whilst the location of the larger water reservoir is indicated on the plan of the Għajn Qajjed *giardino*, the reservoir covered by unmortared masonry slabs placed alongside each other is not. As the *vastissima gebia* is never indicated in the Knights' period seventeenth and eighteenth-century documentation, this is probably a later addition. See NAM. Descriptive Plans of the Crown Property, 1867, ff. 468-70.

[29] This water source is listed by Abela as *Aayn Teuzien*. See Abela 1647, 111.

[30] On the conversion of *canna* to metres refer to chapter five, footnote 67.

name of another nearby field known as *Ta' Fuq l-Ispiera*.[31]

Another water-related entry located in the *chayn taosen ta naclia* territory is the mention of *tre boche della fontana di chain cayed che ua' ala Valletta* (NLM. Treas. B. 289, f. 64). The meaning of *boche* is open to interpretation. In this context it possibly refers to the presence of water inspection chambers which allowed easy access to the aqueduct canal conveying water to Valletta.

Fig. 6.19 The water distribution chamber at Ġnien is-Sultan (RA13) in the territory of Rabat (Malta). Interior detail showing the water canal which is channelling the water retrieved from the Għajn Qajjed (RA19) and Busugrilla (RA7) springs further downhill towards Fiddien valley. Source: the author.

Fig. 6.20 The water distribution chamber at Ġnien is-Sultan (RA13) in the territory of Rabat (Malta). Interior detail showing the outlet for the Busugrilla (RA7) spring which now only yields a small water trickle. Source: the author.

[31] See map reference 45147 71604.

Personal field-research has revealed that at the point where the subterranean canal transporting water retrieved from the Għajn Qajjed spring intersects with that of Busugrilla, there is what appears to be a Knights' period water distribution chamber (Fig. 6.14). Located in the Ġnien is-Sultan (RA13) *giardino* at map reference 44918 71775, the subterranean reservoir is accessed through a square-headed doorway and down a flight of steps (Fig. 6.17). It is covered by a barrel-vaulted ceiling and measures some 4 by 1.8 m. Visible at the far end of the reservoir is the canal through which the Għajn Qajjed water flows and which is still yielding a fair water discharge (Fig. 6.18). To the immediate left of the water distribution chamber entrance is the Busugrilla canal which, on the other hand, is only yielding a small trickle of water (Fig. 6.19). Another canal close to the entrance of the subterranean chamber channels the collected water towards Fiddien valley (RA11), where it joins with other springs feeding the aqueduct network (Fig. 6.20).

In the *Listoriali* area (Fig. 6.8 "M"), a large territory located downhill from Għajn Qajjed, there appears to have been another water distribution chamber (NLM. Treas. B. 289, ff. 58-59), which has so far not been located. From this point onwards the Għajn Qajjed, Busugrilla, Diar Ħandul (Dejr Ħandul) and *hued il bisbes* springs all proceeded towards the *Giardino di santo Antonio* as a single entity.[32]

A *gebia in mezzo strada* (a water reservoir in the middle of the road) is mentioned in the entry for *Hayn Clep tal Callus* (RA28),[33] but no further details are given. This probably corresponds to a shaded circle shown to lie by a roadside in plan 9 of *cabreo* 289 (Figs. 6.10, 6.21). This feature was also recorded on Malta Survey Sheet 72 for the Għajn Qajjed territory dating to 1922 and was probably used to harvest water obtained from the Għajn Klieb spring (RA18). Located at map reference 44542 71773 on the north-bound side of Għajn Qajjed Road, a section of the reservoir probably still survives concealed beneath debris.

It is worth noting that NLM. Treas. B. 289, 290 and 291 *cabrei* are unexpectedly silent about the adjoining Għajn Klieb *giardino* for which neither plan nor description appears to be available. The present writer's knowledge of the hydrological setup of Għajn Klieb is mainly derived from onsite field-research and the 1867 Descriptive Plans of the Crown Property *cabreo* data (Figs. 6.22, 6.25). The lack of Knights' period data on Għajn Klieb is possibly due to a name change experienced by the *Aayn Cirani* (RA2) spring sometime between the seventeenth and eighteenth centuries. *Aayn Cirani* is described by Abela as being an offshoot of

[32] See Survey Sheet 71, School of Military Engineering, Chatham, 1922,

[33] The eastern side of this property bordered the *Giardino delle due fontane*. See NLM. Treas. B. 289, f. 66; NLM. Treas. B. 289, Plan 9; NLM. Treas. B. 290, f. 85.

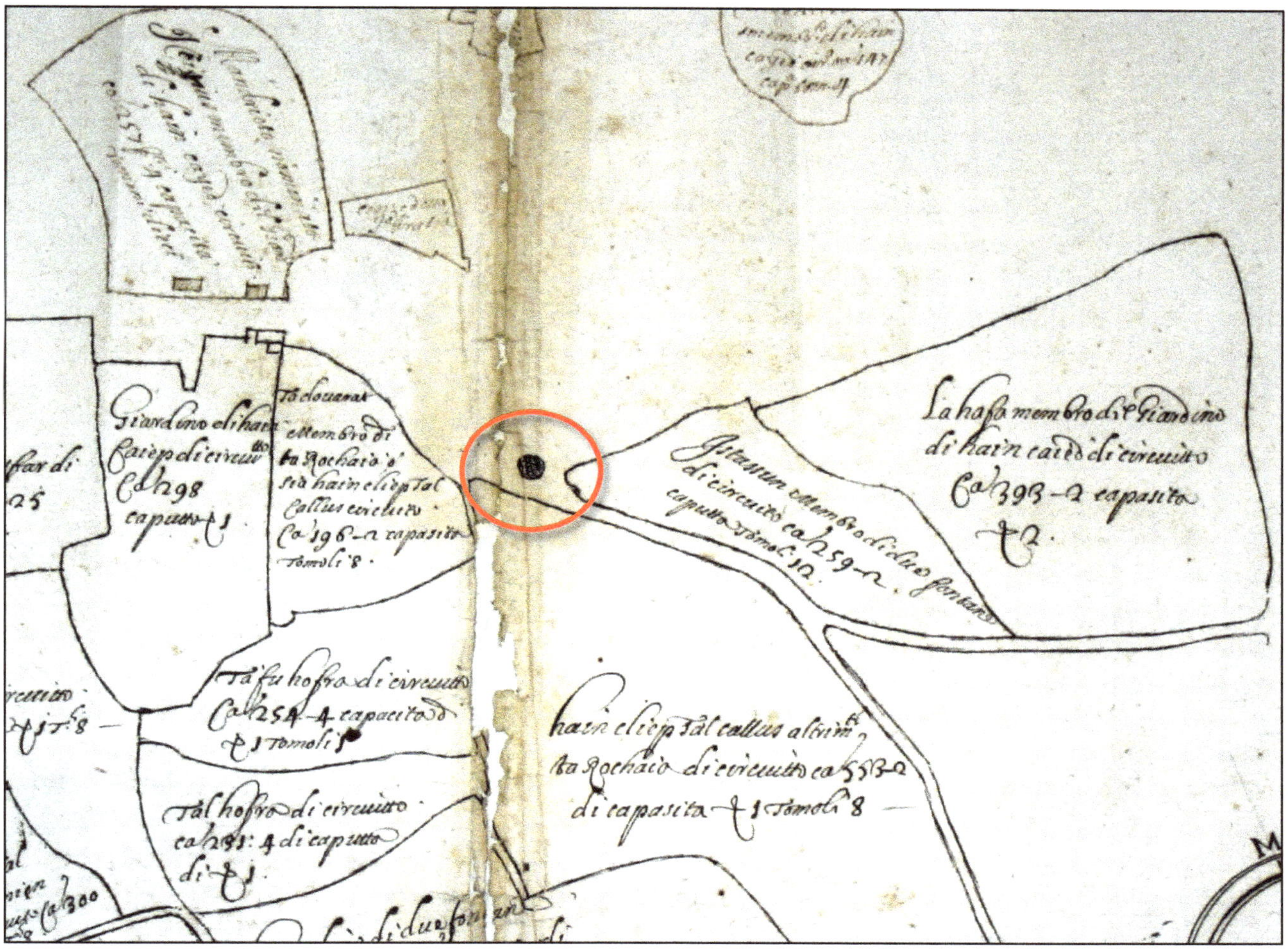

Fig. 6.21 NLM. Treas. B. 289, Plan 9. Detail of a circular feature located in the middle of the present-day Għajn Qajjed road and which was probably utilised as a water reservoir.

Għajn Qajjed (Abela 1647, 111). Its exact whereabouts remain unknown. By 1647, it appears that only a small quantity of surplus water was channelled into the aqueduct system, with most of the water being utilised to irrigate the Għajn Klieb *giardino* (Abela 1647, 112). It is claimed by Themistocles Zammit that *Aayn Cirani* later became known as Għajn Klieb.[34] Should this be the case, it explains why Abela only seldom mentions the Għajn Klieb place-name (Abela 1647, 80).

The Għajn Klieb spring is clearly marked on the British period Survey Sheet 72 of 1922. This appears to originate from within a subterranean gallery, the starting point for which is a field in the Nigret locality of Rabat at map reference 44661 71346 (Fig. 6.27). An investigation of the site did not, however, yield evidence of any water-related features. Located ca. 40 m away from this area at map reference 44688 71352 are the remains of a shaft well which possibly formed part of a perched aquifer water gallery system, and which this study tentatively associates with the Għajn Klieb spring (Fig. 6.23).

The field survey of the Għajn Klieb area also revealed another perched aquifer water gallery at map reference 44349 71486. The gallery is excavated into the west-facing escarpment in the Is-Salvatur area, close to Għajn Klieb. Its presence is indicated by dense reed growth, but due to the process of cliff-face detachment at this particular point, it is no longer accessible (Figs. 6.5, perched aquifer gallery "A", 6.24). In the above promontory, in close proximity to the gallery entrance, is a large rectangular-shaped rock-cut feature which measures approximately 11 by 4 m (Fig. 6.26). It appears to be over three and a half metres deep, and because of a rock-excavated canal leading to it, it can be conjectured that it was used as an open-air water storage reservoir. It is still unclear if the gallery and the reservoir were in any way associated.

[34] The Għajn Klieb spring originates from within a subterranean perched aquifer gallery with an approximate length of 286 m. See Zammit T. 1924, 9, 37.

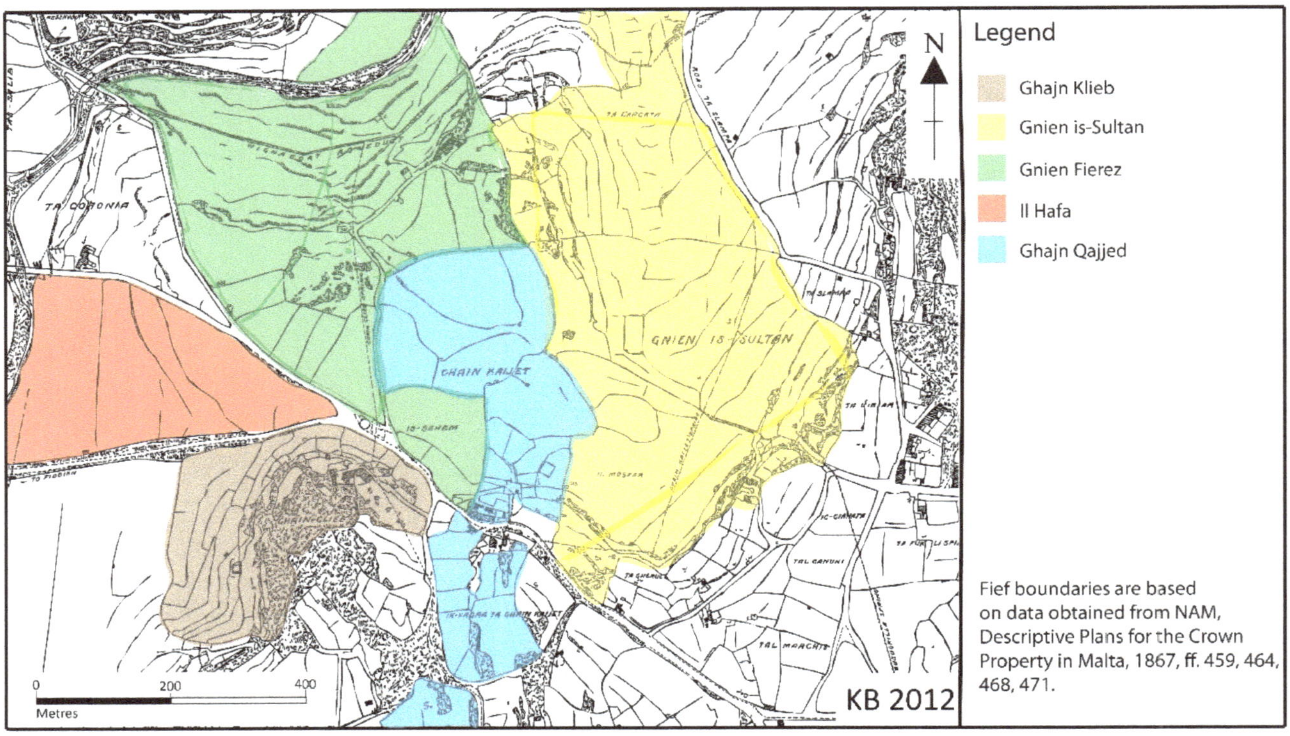

Fig. 6.22 1922 map detail of the Għajn Qajjed and Ġnien is-Sultan areas in the Rabat (Malta) territory, overlying which is *cabreo* property data showing the location of the Għajn Klieb (RA18), Ġnien is-Sultan (RA13) and the Għajn Qajjed (RA19) *giardini* based on nineteenth-century documentation. Source: Survey Sheet 72, Scale 1:2,500, School of Military Engineering, Chatham, 1922.

Nineteenth-century documentation for the Għajn Klieb estate also mentions *tre sorgenti d'acqua corrente*, *giardini* and a *caverna sotterranea* amongst other features (NAM. Descriptive Plans of the Crown Property, 1867, f. 471; Fig. 6.25). The location of the springs is not indicated on the 1867 *cabreo* plan for Għajn Klieb. Nonetheless, the approximate whereabouts of these sources can be deduced from the location of two water reservoirs as well as areas of dense reed growth which are clearly indicated on the 1867 *cabreo* plan (NAM. Descriptive Plans of the Crown Property, 1867, f. 471). Reservoir "A" (Fig. 6.25) corresponds to the location of perched aquifer gallery "A" described above (Figs. 6.5, 6.24, 6.26). Reservoir "B" (Fig. 6.25) hints at the existence of another perched aquifer gallery located at the approximate map reference of 44506 71736.

Fig. 6.23 A shaft well at Nigret in the territory of Rabat (Malta) probably giving access to the Għajn Klieb spring (RA18). Source: the author.

Fig. 6.24 The Is-Salvatur area below the west-facing escarpment at Għajn Klieb (RA18) in the territory of Rabat (Malta). Dense reed growth shows perched aquifer availability in the area. Source: the author.

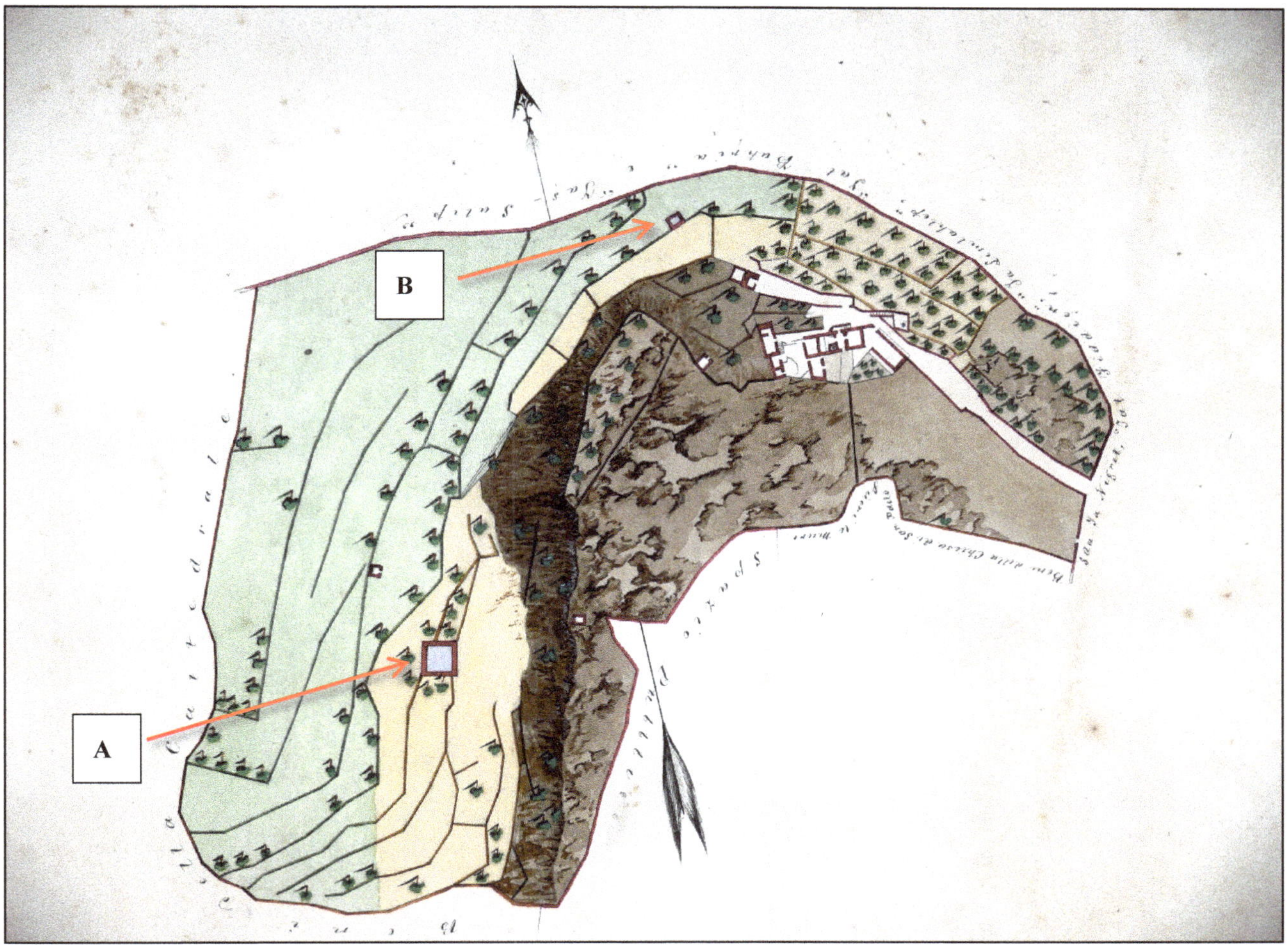

Fig. 6.25 NAM. Descriptive Plans of the Crown Property in Malta, 1867, f. 471. Plan detail of the Għajn Klieb *giardino* (RA18) in the territory of Rabat (Malta). Indicated on the plan are water reservoirs the water source of which is probably derived from perched aquifer galleries located in this tenement.

The presence of the Late Medieval church dedicated to San Mikiel is-Sinċir (RA50; Fig. 6.28) gives credence to view that the *giardini* of Għajn Qajjed, *delle due Fontane*, Ġnien Musfar, Ġnien is-Sultan and *Hayn Tausen* were intensively cultivated during the Late Medieval period. Field-research carried out by the present writer has shown the church to be strategically positioned to attend to the spiritual needs of the farming folk living and working in the various *giardini* frameworks mentioned above. The church does not feature in the Knights' period *cabreo* plans for the Għajn Qajjed territory, but a well-defined country lane flanking the *Hayn Tausen*, Ġnien is-Sultan and *Listorjali* gave the church area ease of access (NLM. Treas. B. 289, Plan 9).[35] Furthermore, it has also been recently proposed that during the late tenth and the eleventh centuries, the Għajn Klieb and the Ġnien is Sultan areas housed a cluster of rural dwellings.[36] This claim is primarily based on ceramic material retrieved from within the San Mikiel is-Sinċir site and its neighbouring *giardini*. A comparison

Fig. 6.26 A large rectangular-shaped depression at Għajn Klieb (RA18) which probably served the purpose of an open-air water reservoir. Source: the author.

[35] This single-celled structure has no apse, has a well-pronounced gabled roof and a plain west-facing voussoir doorway containing no hood-mould decoration. See Buhagiar M. 2005, 92, 94; Luttrell 1975, 177.

[36] This argument mainly centres on the discovery of glazed ceramics in an ancient tomb at Għajn Klieb and in a silo pit at Ġnien is-Sultan. The archaeological investigation of the San Mikiel is-Sinċir church held in 1981-1982 also yielded similar results. See Cutajar 2004, 60, 64.

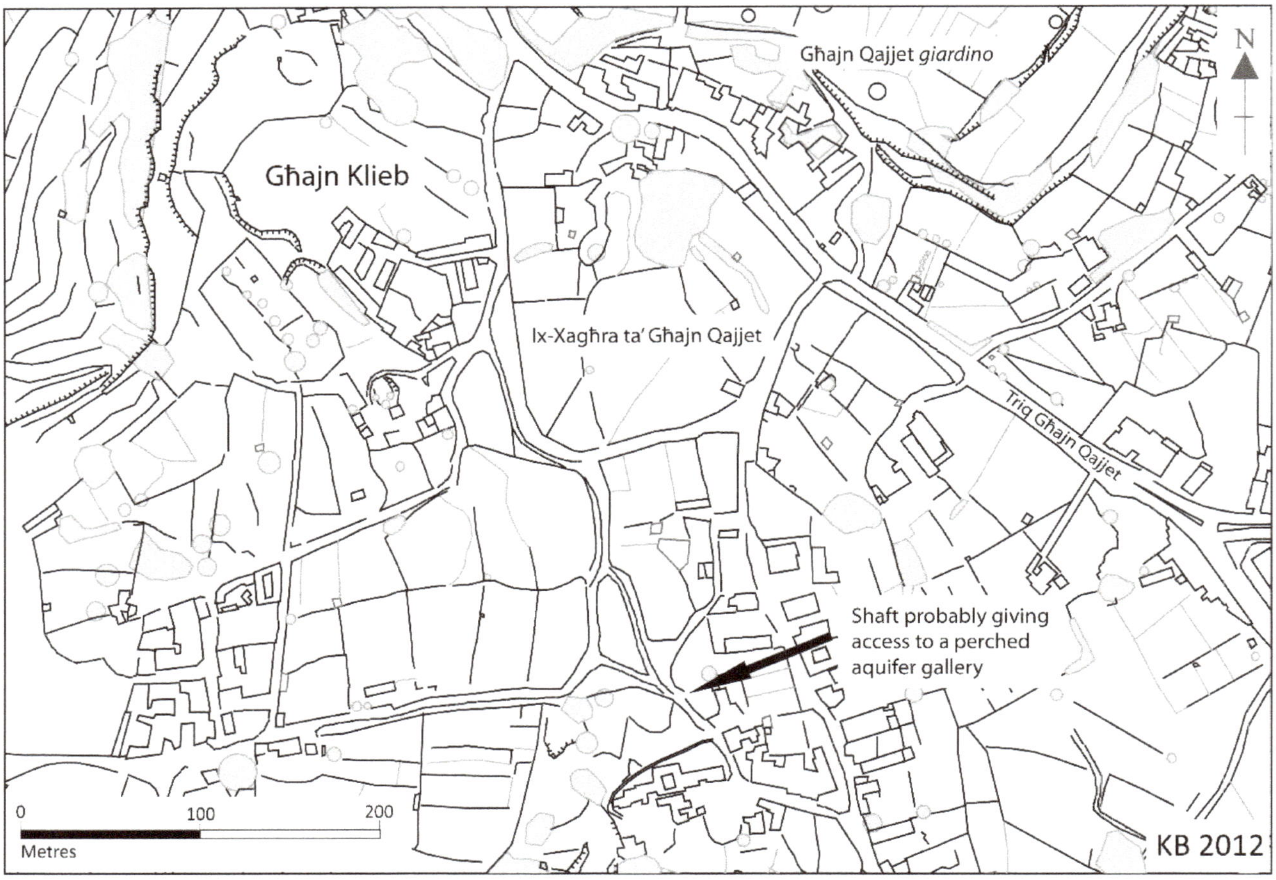

Fig. 6.27 Site plan of the Nigret area in the territory of Rabat (Malta), showing the location of a shaft well which possibly connects to the Għajn Klieb spring (RA18). Topographical map image adapted from MEPA Map Server data.

Fig. 6.28 The Late Medieval church of San Mikiel Is-Sinċir located in the *terreno S. Michaele* (RA50) in the territory of Rabat (Malta). Source: the author.

between the Knights' period and the nineteenth-century *cabreo* data shows how a group of much smaller estates had been assimilated within the territory covered by Ġnien is-Sultan and Ġnien Fierez by the latter half of the nineteenth century (Fig. 6.22).

6.1.4 The Marget Neuia *estate in the Fiddien territory of Rabat (Malta)*

Further downstream, north-west of the Għajn Klieb escarpment and bordering the Sentini territory at Fiddien is a relatively flat expanse of agricultural land known by the place-name of *Marget Neuia* (RA34; Figs. 6.5, 6.29).[37] This is a small territory located to the immediate north of the country lane known as *Triq tal-Fiskal*, which from the Fiddien Bridge leads to the small hamlet of Landrijiet. A portion of land forming part of this territory, the entry for which is in the *cabreo del Magistero*, listed as *Chibula ta turba membro di Marget meuia* (RA34),[38] makes reference to a *cisterna d'acqua sorgente*. This is possibly a reference to a shaft well piercing the ceiling of a subterranean perched aquifer water gallery. A scrutiny of the 1922 Survey Sheet 71 for this area hints at the presence of a subterranean water feature in very close proximity to *Marget Neuia* (Fig. 6.63).

[37] Marget Neuia is located at map reference 43384 71624. Also present in this territory was a *mandra* (animal pen), inside which was a cave. The *mandra* was enclosed by means of a wall. See NLM. Treas. B. 289, f. 70.

[38] The mid-seventeenth-century entry for this toponym predates the *Marġet Mewwija* place-name, first encountered in notarial archives and listed by G. Wettinger, by over a century. See Wettinger 2000, 363.

Fig. 6.29 General view of the *Marget Neuia* tenement (RA34) in the territory of Rabat (Malta). Source: the author.

Fig. 6.30 General view of the water gallery and its fronting water storage reservoir known as the Tal-Fiskal fountain (RA34). Source: the author.

A field-investigation of the *Marget Neuia* area confirmed that Survey Sheet 71 contains a reference to a perched aquifer water gallery (Figs. 6.30, 6.31, 6.32, 6.33), the ceiling of which is almost certainly pierced by at least nine shaft wells. The gallery still yields a perennial water supply and its outermost section is lined by dry-stone walls and is roofed over by means of a gabled roof-slab ceiling (Fig. 6.32). A series of masonry canals convey water from the gallery entrance to a water reservoir in the vicinity. The presence of shaft wells, which on Survey Sheet 71 appear to be spaced out at fairly regular intervals, strongly indicates in this instance the existence of a fully-fledged qanat-type water extraction system. Other similar rural water extraction systems involving a near-horizontal water gallery and a series of shaft wells piercing the ceiling have been noted in the Gomerino (RA15) and Santi (MG23) areas not far away.

Basing oneself on the data provided by Survey Sheet 71, the gallery feeding the *Chibula ta turba membro di Marget meuia* seemingly commences at map reference 43027 71918, close to the territory known as *Ix-Xeuchia* (Fig. 6.36). From this point onwards, the spring follows an almost linear course downhill, heading directly towards the *Tal Fiscal Fountain*, located at map reference 43180 71952. The presence of this water source is also hinted at by the *Il Meijeli Ta Fuk il Ghain* place-name,[39] located directly above the *Marget Neuia* territory. The gallery appears to have a total length of ca. 420 m and most of the shaft wells which probably pierce the gallery's ceiling are on average spaced ca. 25 m apart.

In what appears to be an adjacent piece of land known as *Cianta membro di Marget meuia*, there is also reference to a spring running in a ditch adjacent to the road.[40] The data given in this entry are insufficient to allow any conclusions to be drawn, but possibly make reference to a valley watercourse which was partly augmented by spring water derived from one or more perched aquifer water galleries in the area. In the Fiddien district there is mention of the agricultural property of *Tal Hueg*. Within *Tal Hueg* there passed the water canal which from Dejr Ħandul transported the spring water from the Ħofret ir-Rizz area to Fiddien and which also contained a *spiraglio*.[41] In this particular case, *spiraglio* can most likely be interpreted as an opening through which spring water from the canal was diverted into the water fountain within the *Tal Hueg* territory.

Fig. 6.31 The Tal-Fiskal gallery (RA34). Its outermost section is lined with dry-stone walls and is roofed over by means of gabled slabs. A masonry canal conveys water to a reservoir in the vicinity. Source: the author.

[39] An English translation of this place-name reads, "The undulating land above the spring or fountain".

[40] *Sotto il detto terreno in costo la' strada nel fosso del terreno l'ha trouato acqua' surgente.* See NLM. Treas. B. 289, f. 71.

[41] [...] *che porta l'acqua della fontana diar chandun alla ualletta in cann:72 quali canale é discosto d'un spiraglio di detta fontana, che é dentro il detto terreno tal hueg* [...]. See NLM. Treas. B. 289, f. 73. See also: [...] *Vi sono dentro detto terreno tal hueg tre spiragli della fontana darchandur*, NLM. Treas. B. 289, f. 74.

Fig. 6.32 Interior view of the initial masonry-built section of the Tal-Fiskal water gallery (RA34). A rock-cut water channel facilitates the extraction of the tapped affluent to the gallery exterior. Source: the author.

Fig. 6.33 General view of the Tal-Fiskal water reservoir (RA34). Source: the author.

Fig. 6.34 The Għajn Żdejda perched aquifer gallery. This tunnels below the Qasam il-Kbir (RA37) territory of Rabat (Malta), and supplies the area with a perennial water source. Source: the author.

The description for the *Fego di beberrua* (RA10) landholding, also in the Fiddien area, includes mention of *due fontane picoli* (two small springs) and is another possible indication of the presence of perched aquifer water galleries in this territory. This property also contained water troughs, animal pens and caves, the latter possibly used for the purpose of human habitation (NLM. Treas. B. 289, f. 79).

6.1.5 *Qasam il-Kbir estate in the Fiddien territory of Rabat (Malta)*

In the nearby *Territorio detto il Kasam il Kbir dello Fiddeni* (RA37) a spring supplied an animal drinking trough with water. The water was then diverted to a reservoir for later storage (NLM. Treas. B. 291, f. 18). NLM. Treas. B. 290, f. 91 illustrates the extent of this landholding, but makes no reference to the presence of water within (Fig. 6.35). A comparative analysis of the land boundaries shown in this plan, however, made it possible to pinpoint the location of this parcel of land (Figs. 6.5, 6.37). The *Territorio detto il Kasam il Kbir* is a rather extensive property known by the present-day place-name of Il-Mejjieli iż-Żgħir and is located close to Landrijiet in the territory of Rabat (Malta). It also adjoins

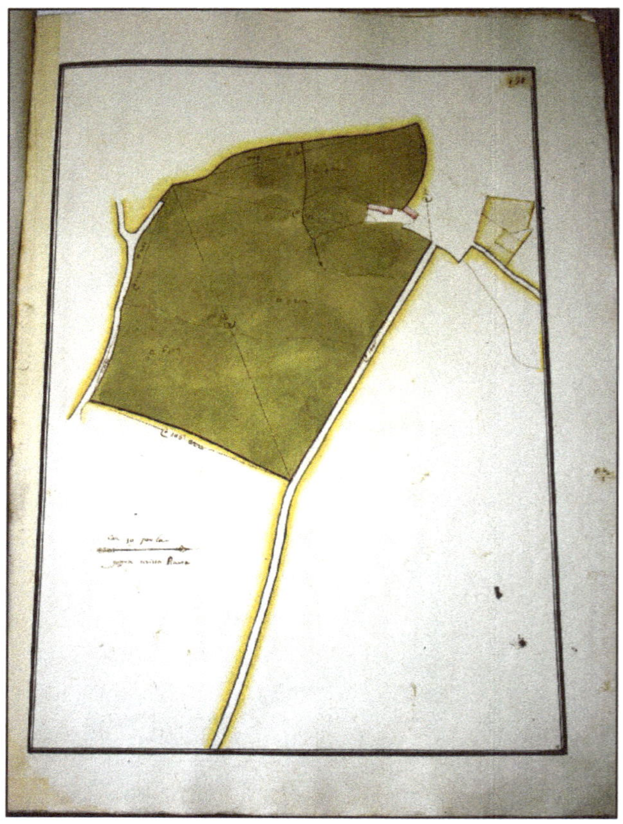

Fig. 6.35 NLM. Treas. B. 290, f. 91. Plan showing the Government-owned parcel of land at Qasam il-Kbir (RA37) in the territory of Rabat (Malta).

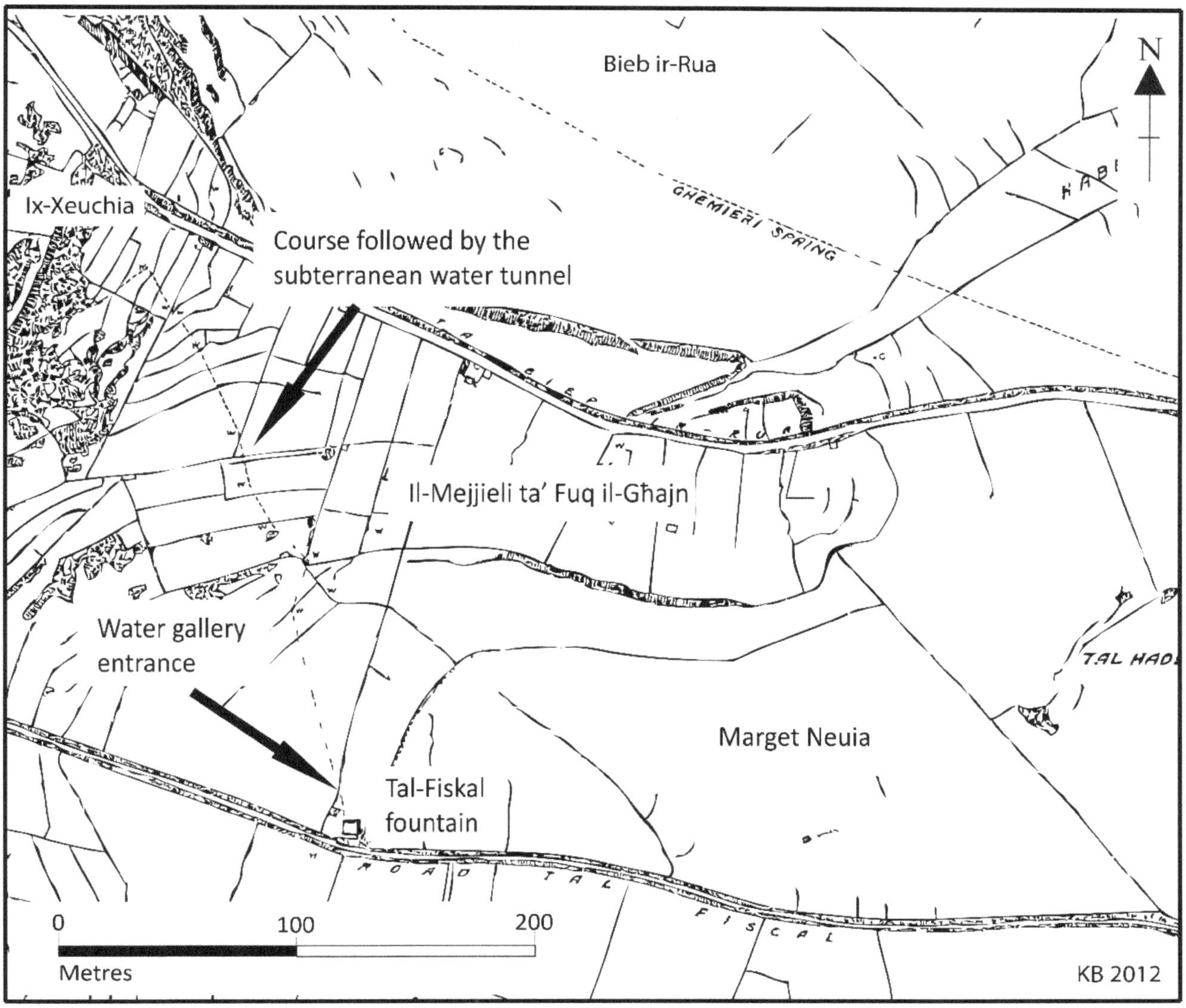

Fig. 6.36 Site plan of the *Marget Neuia* area (RA34) in the territory of Rabat (Malta). The dotted line indicated by the arrows above represents the course followed by a subterranean water tunnel, the roof of which appears to be pierced by a series of shaft wells. On the site plan these are marked as "W". The gallery is fronted by an open air water reservoir known as the Tal-Fiskal fountain. Source: Survey Sheet 71, Scale 1:2,500, School of Military Engineering, Chatham, 1922.

the Government-owned *Marget Neuia* property (RA34) in which the Tal-Fiskal perched aquifer gallery is located.[42]

Located at map reference 43005 71402, Qasam il-Kbir (RA37) is mainly composed of Upper Coralline and Blue Clay deposits which permit the accumulation of a substantial perched aquifer ground-water source in the area. An examination of Survey Sheet 71 for this area indicates a series of wells and shafts in this property, further accentuating the presence of ground-water.[43] The field survey of the Qasam il-Kbir revealed that this landholding was located well above the perched aquifer, making it unlikely that an entry point giving access into a perched aquifer water gallery was located within this parcel of land.

Access to a perched aquifer water gallery was detected to the immediate south of Qasam il-Kbir at map reference 42944 71343. Known by the name of Għajn Żdejda, this gallery tunnels right below Qasam il-Kbir and supplies the area with a perennial water source (Fig. 6.34).[44] The gallery is cut into an Mtarfa Member deposit and was provided with a canal at its base in order to facilitate water extraction from within. The retrieved water is currently being channelled into a small uncovered reservoir for temporary storage and later distribution.[45] Għajn Żdejda is of unknown antiquity but considering

[42] See section 6.1.4.
[43] See Survey Sheet 71, School of Military Engineering, Chatham, 1922.
[44] Personal communication by A. Scicluna, a part-time farmer who owns agricultural territory to the immediate west of this water source.
[45] The water storage reservoir is located at map reference 42960 71316.

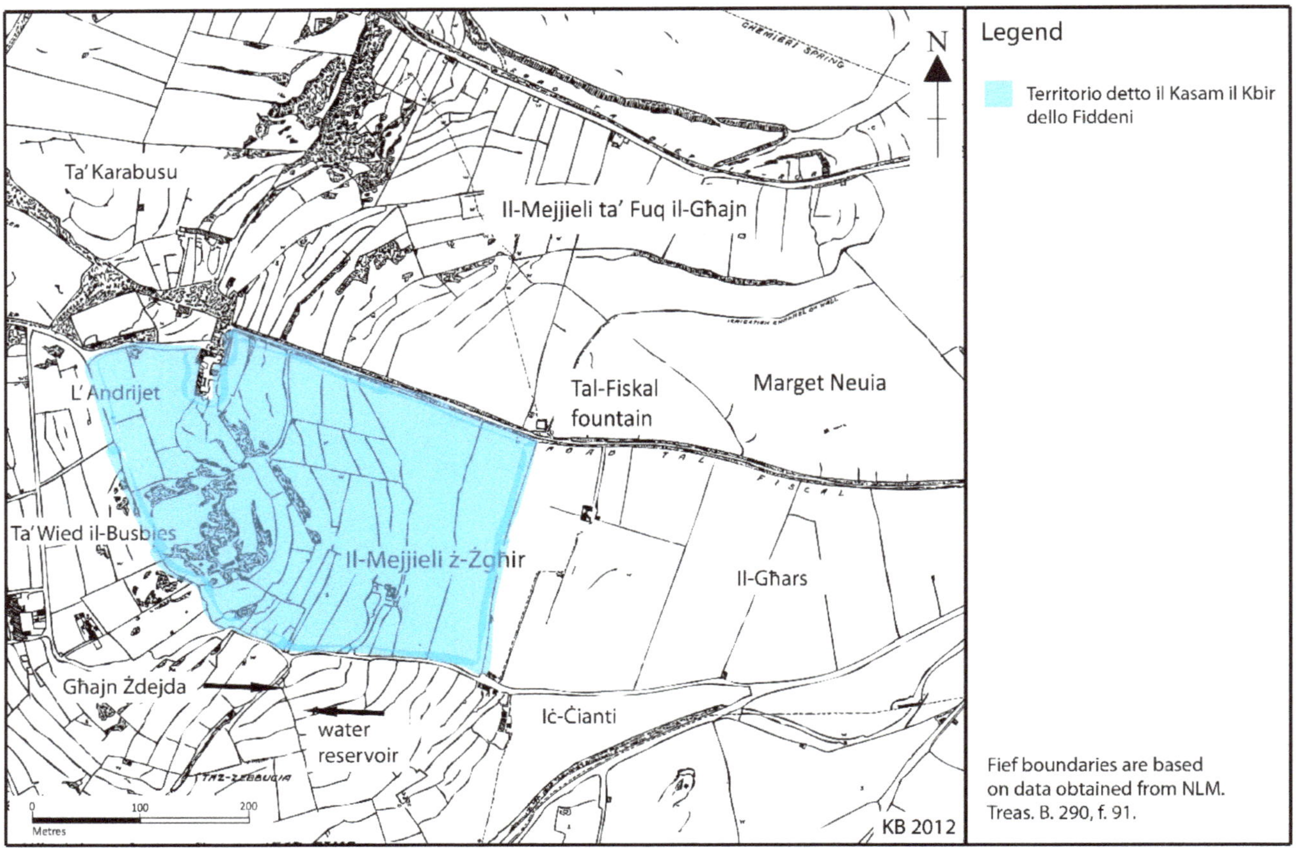

Fig. 6.37 1922 map detail overlying which is *cabreo* property data showing the location of Government landholdings at Qasam il-Kbir (RA37) in the territory of Rabat (Malta). Sources: NLM. Treas. B. 290, f. 91; Survey Sheet 71, Scale 1:2,500, School of Military Engineering, Chatham, 1922.

that the gallery's ceiling is pierced by a series of shafts,[46] it is typologically classifiable as a *qanat* system. The gallery is thought to extend up to the Tal-Fiskal road which is located ca. 300 m to the north of the gallery entrance.[47]

It is the present writer's suspicion that the water source in Qasam il-Kbir mentioned in the *cabreo* 291 text (NLM. Treas. B. 291, f. 18) was accessed by means of one or more shafts piercing the ceiling of this perched aquifer water gallery.

6.1.6 Wied ir-Rum and Wied Ħażrun

A number of *cabreo* entries dating from the early seventeenth till the nineteenth century include references to Wied ir-Rum (RA56) and its neighbouring Wied Ħażrun (RA55; Figs. 3.6, 4.3). The *Cabreo Antico della Procura delle Distribuzioni* of 1621 mentions a *cisterna rutta* located in the *misraħ* or the communal gathering point of Simblija (RA41; Fig. 4.10 "A"), a *grotta con acqua bona* (a cave containing good quality water), and a fountain located alongside the public road (NLM. Treas. A. 74, f. 136).[48] The 1621 documentation lacks a site plan for the Wied Ħażrun area, but the description included gives their approximate location. The reference to a *grotta con acqua bona* appears to correspond to the perched aquifer gallery located close to the viaduct in Wied Ħażrun (Fig. 4.10 "B").[49] Located at map reference 42970 69795, this consists of a roughly circular-shaped water collection feature. The field-investigation of the area has shown how the water collecting in this rock-cut basin originates from an adjoining perched aquifer gallery which tunnels into an exposed Mtarfa Member deposit (Fig. 4.10 "B"). Blue Clay deposits in this area give the basin excellent water retention properties. The access point to this water gallery is partly collapsed, making the investigation of its interior difficult.[50]

[46] Personal communication by A. Scicluna.
[47] Personal communication by A. Scicluna.

[48] The Simblija place-name never appeared in the seventeenth to nineteenth Government documentation for property in Wied ir-Rum and Wied Ħażrun. See Saliba *et al* 2002, 20.
[49] The current viaduct belongs to either the nineteenth or the early twentieth century. Its method of construction betrays British civil engineering works, thus making it probable that the current structure replaced a much older one. See Saliba *et al* 2002, 33.
[50] Farming folk identify this perched aquifer gallery as Id-Demus. See Saliba *et al* 2002, 35.

Fig. 6.38 A perched aquifer gallery at Wied Ħażrun (RA55; labelled "C" on Fig. 4.10), and its fronting water settling vat and drinking trough. Source: the author.

The other water source is described as: [...] *dalla parte di mezo di nella strada publica ui e una fontana la quale se bene e nel publico nulla di mezzo appartiene a detto giardino dentro li quale giardino ui e anco una gebietta che racoghe l'acqua di detta fontana per seruitio di detto giardino* (NLM. Treas. A. 74, f. 136). The water source in question can be probably associated with the perched aquifer gallery located alongside the road demarcating the southernmost extent of the Wied Ħażrun property at map reference 42970 69510. The retrieved water is channelled into two reservoirs located further downstream in the Wied Ħażrun *giardino* (Fig. 4.10 "C").[51] The name of this spring is not known.[52] The investigation of the gallery interior revealed significant quantities of calcite deposits on its floor and sides, an analysis of which can in the future give a more precise indication of its antiquity (Figs. 6.38, 6.39).

Fig. 6.39 Interior view of the perched aquifer gallery illustrated in Fig. 6.38 (RA55). Thick calcite deposits cover the gallery floor. Source: the author.

These same water sources are again indicated as a *fontana* in early eighteenth-century documentation (NLM. Treas. B. 292, ff. 27, 106, 108; Figs. 6.40, 6.41). 1867 plan data for Wied ir-Rum and Wied Ħażrun mention the presence of: (a) *due sorgenti d'acqua corrente*; (b) *una cisterna d'acqua piovana*; (c) *tre gebiette*; and (d) *una spiera d'aqua corrente* (NAM. Descriptive Plans of the Crown Property, 1867, f. 496). The water cistern in question is most likely the one located in the *misraħ* (the communal gathering area) of Simblija (Fig. 4.10 "A"). The location of the two springs is not specifically indicated on the 1867 *cabreo* site plan. Their whereabouts can be worked out from the positioning of the water reservoirs (Figs. 6.42, 6.43), which conform to the location of above described perched aquifer galleries, both located in Wied Ħażrun. The location of the *spera d'acqua corrente* remains unknown. It may possibly correspond to one of the shaft wells detected during the 2002 survey of the site.[53] This water feature may well refer to a shaft well piercing the ceiling of one of the other perched aquifer galleries known to exist in this area (Fig. 4.10 "D").

[51] See Malta Survey Sheet 99, School of Military Engineering, Chatham, 1903.

[52] J. Micallef's family have been leasing land on either side of this gallery from the Government for over a century and a half. He personally communicated that his family always referred to this spring as either the Għajn ta' Wied Ħażrun or the Mina ta' Wied Ħażrun (the Wied Ħażrun spring or gallery).

[53] Saliba *et al* 2002, 43. Refer to shaft wells "H" and "J" which in this publication have been erroneously classified as "artesian wells".

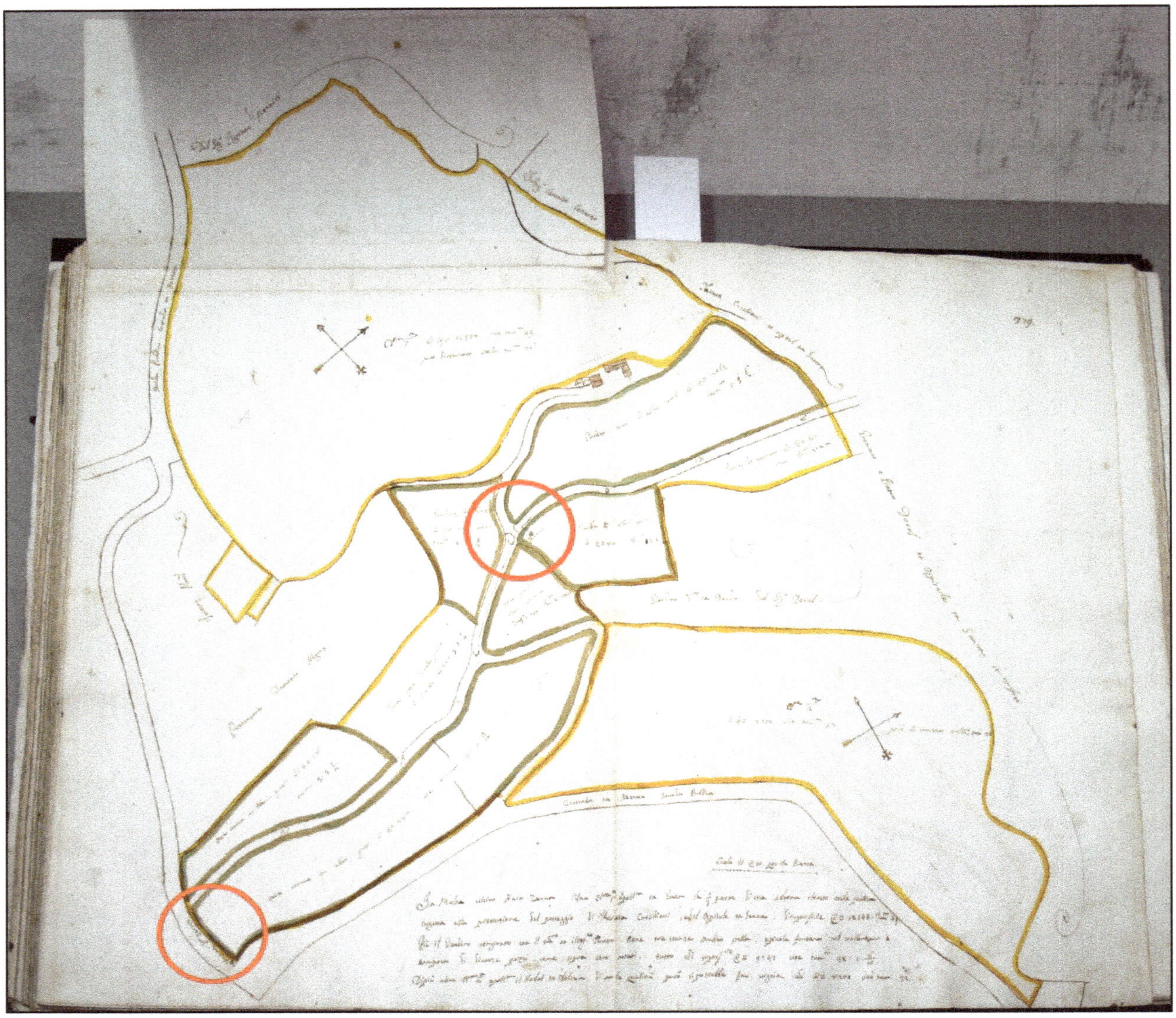

Fig. 6.40 NLM. Treas. B. 292, f. 379. The eighteenth-century *Cabreo* 290 plan for Wied Ħażrun (RA55) and Simblija (RA41) in the territory of Rabat (Malta). Clearly indicated on the plan are the two water sources which field-research has revealed to originate from within perched aquifer galleries.

6.1.7 The Tal-Callus giardino *at Wied ir-Rum*

The *Giardino di Hued irrun tal Callus* (RA39), located in Wied ir-Rum (RA56) in the territory of Rabat (Malta), also contains two water sources which the *cabreo del Magistero* 289 manuscript describes as follows: *In detto Giardino ui sono due fontane la' grande sotto il balzo dalla parte di mezzo giorno, doue la sua' gebia per dentro la' rocca, l'altra dalla parte di ponente doue ui é ancora la gebietta* (NLM. Treas. B. 289, f. 84). This is a clear reference to the presence of at least one perched aquifer water gallery, at the entrance of which was a rock-cut subterranean water reservoir.

The tenement of Tal-Callus (RA39) belonged to the physician and popular hero Josephus Callus, who was convicted of high treason and condemned to death in 1561 (Buhagiar and Fiorini 1996, 502). According to the 1575 Visitation Report compiled by Pietro Dusina, a church present in the Tal-Callus *giardino* was dedicated to St Nicholas.[54] Evidence for a church in the Callus property is further corroborated by notarial deeds.[55] Confusion concerning the identity of the church may have arisen from the fact that in one of the notarial deeds,

[54] Dusina's description of the church of St Nicholas reads as follows: Visitavit Ecclesiam ruralem Sancti Nicolai constructam intus viridarium, quod fuit quondam Josephi Callus, et modo Jllustissimi Domini Magni Magistri, sitam in dicto loco Guedrum alias in dicta Ecclesia celebrabatur officium in die festivitatis suæ, et alebant pauperes; Bene esset, etiam hoc ad memoriam eiusdem Jllustrissimi et Reverendissimi Domini Magni Magistri reducere, qui pro sua devotione potius augebit quam quicumque diminuat de huiusmodi obligatione mandabitque illi satisfieri. See Aquilina and Fiorini 2001, 169-70.

[55] A notarial deed dating to 1514 makes mention of a newly built church in the Callus property. See Fiorini 2003, 24.

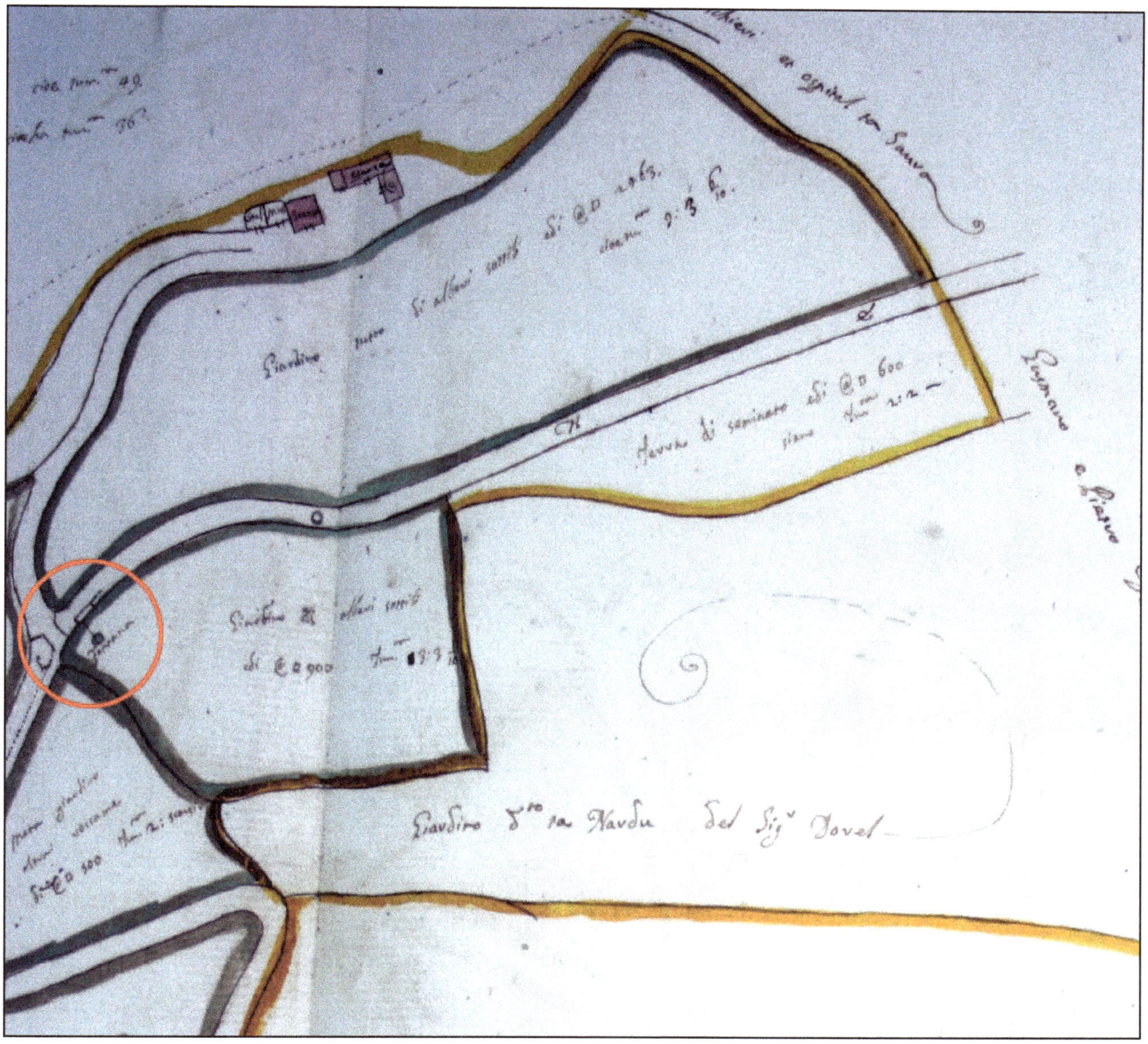

Fig. 6.41 NLM. Treas. B. 292, f. 379. Detail of Fig. 6.40. Circled in red is the *fontana* which is clearly indicated as being located close to the viaduct on the north-facing bank of Wied Ħażrun (RA55).

it is listed as Ta' Santa Marija – a title by which the church continued to be referred to until its deconsecrating in 1646 (Fiorini 2003, 24).[56]

It appears that the Tal-Callus tenement at Wied ir-Rum was already in the ownership of the Callus family in 1467 (Fiorini 2003, 24).[57] Notarial deeds for the Tal-Callus property dating to the sixteenth century succeed in giving an insight into land use practices during this period. There is documentation to show that on various occasions, Josephus Callus leased out the Tal-Callus *giardino* to third parties. In most of the examined contracts, Callus reserved for his own personal use access to the church, house, the collection of brushwood from the southernmost section of the property, and the right of access to three peach trees (Fiorini 2003, 23). In a will dating to 1548, it appears that it was customary for this property to be divided into a number of separate portions. These consisted of the already mentioned church and house, a small giardino, and another stretch of land (Fiorini 2003, 24). As evidenced by the Dusina visitation report, it appears that in 1575 the Callus property had already been confiscated and come to form part of the Grand Master's estates (Aquilina and Fiorini 2001, 169-70).

[56] Nineteenth-century documentation states that this *giardino* was located close to a church dedicated to *S. Andrea*. See NLM. Treas. B. 291, f. 15. It is unclear to which church of *S. Andrea* this *cabreo* entry is refering to. In the *Rurales Civitatis Veter* section of the Pietro Dusina 1575 report, there is no church listed as having such a dedication. See Aquilina and Fiorini 2001, 166-73.

[57] According to the 1867 *cabreo* documentation, the section of the Wied ir-Rum valley within which the Tal-Callus estate is located is called *Wiet Zaifi*. See NAM. Descriptive Plans of the Crown Property, 1867, f. 606.

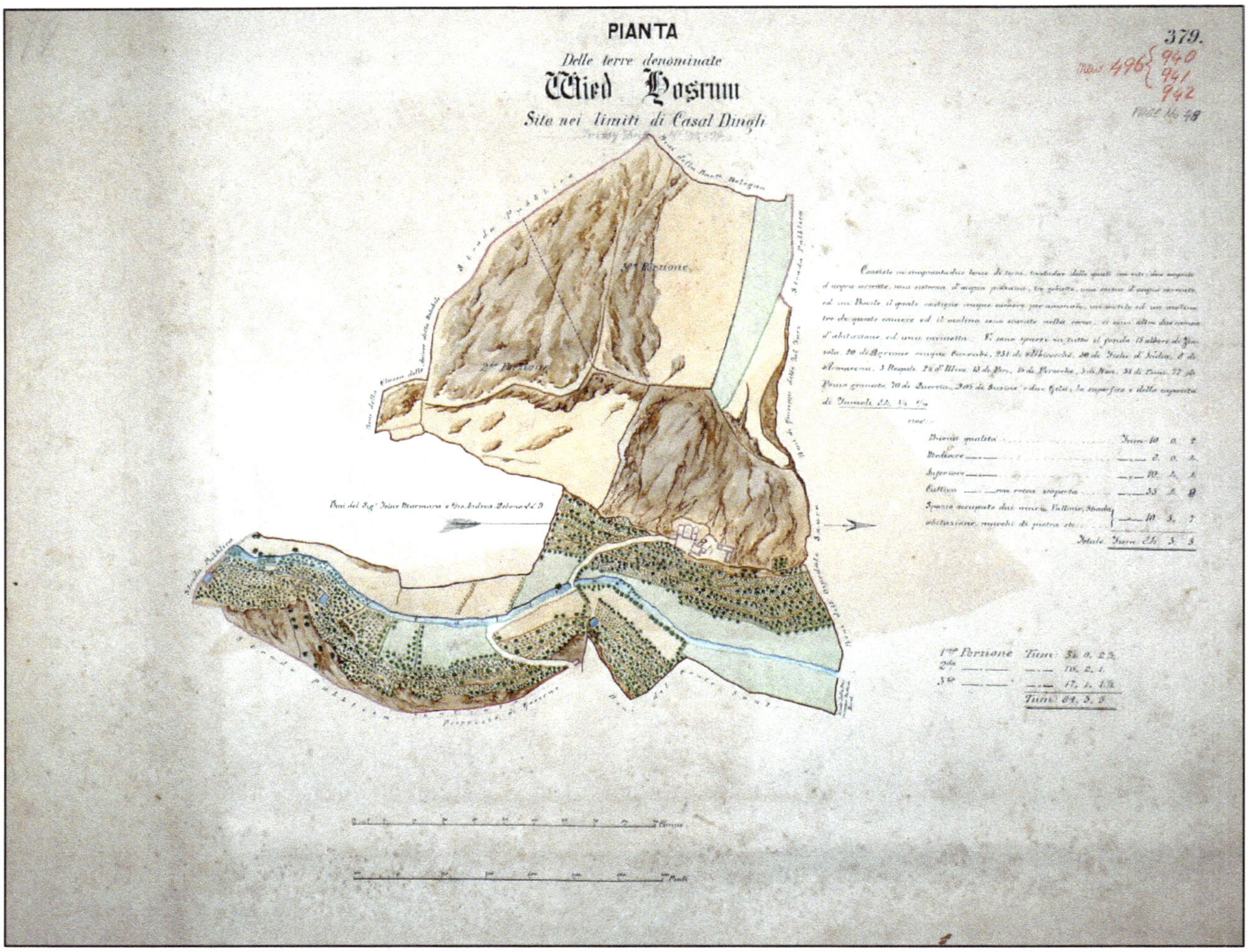

Fig. 6.42 NAM. Descriptive Plans of the Crown Property in Malta, 1867, f. 496. Plan showing the land boundaries of the Wied Ħażrun estate (RA55) in the territory of Rabat (Malta).

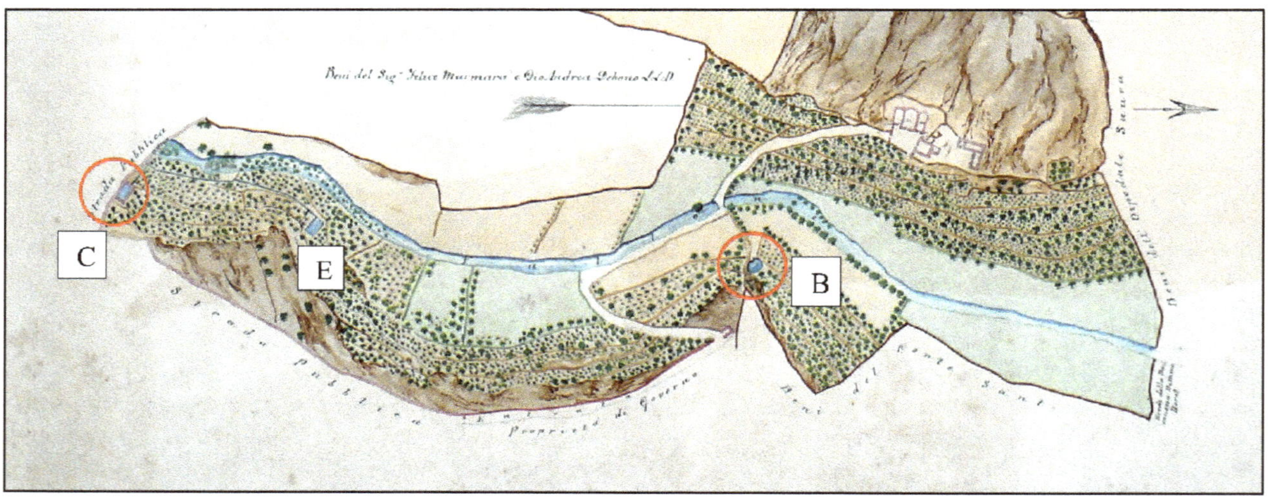

Fig. 6.43 NAM. Descriptive Plans of the Crown Property in Malta, 1867, f. 496. Detail of Fig. 6.42. The red circles indicate the location of the two perched aquifer galleries located within this *giardino* and marked "B" and "C" on Fig. 4.10. Reservoir "E" receives a perennial water source from gallery "C".

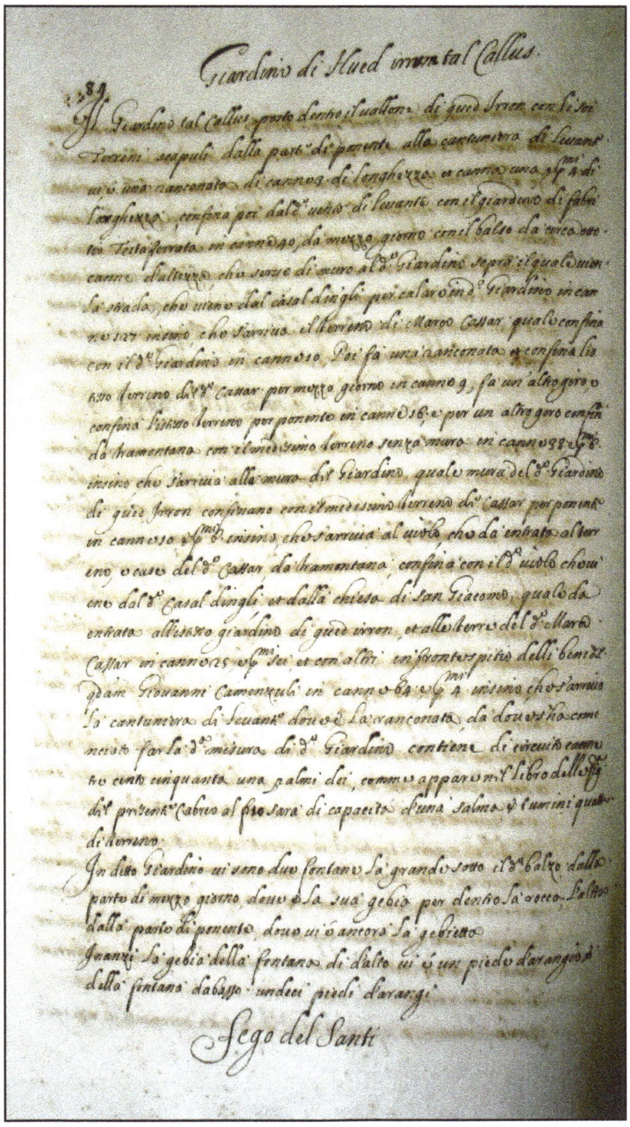

Fig. 6.44 NLM. Treas. B. 289, f. 84. Excerpt from the *Cabreo del Magistero* 289 showing the entry for the *Giardino di Hued Irrum tal Callus* (RA39).

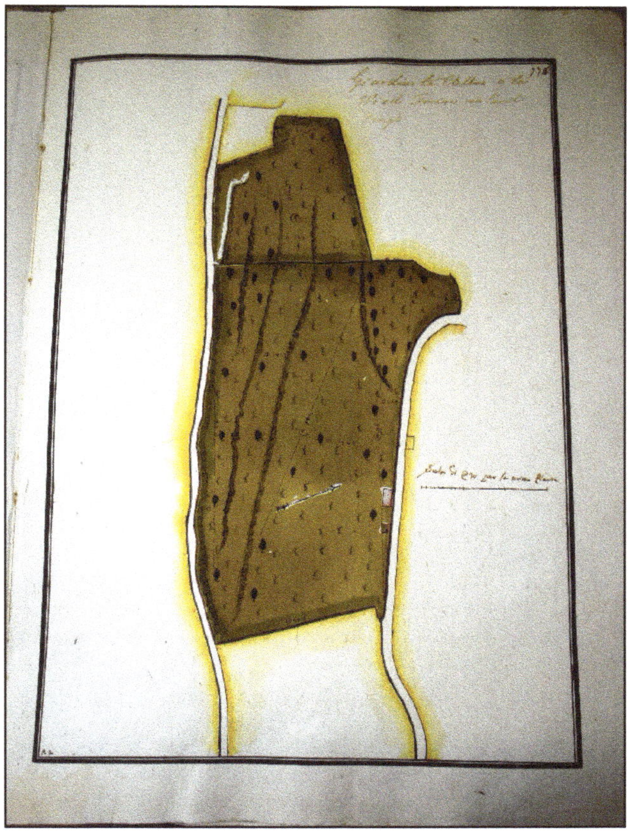

Fig. 6.45 NLM. Treas. B. 290, f. 77. Plan for the *Giardino ta Callus o ta Wied Ferin* (RA39) in Casal Dingli.

In a past publication concerning the study of the landscape and irrigation systems at Simblija (RA41) in the territory of Rabat (Saliba *et al* 2002), the *giardino* Tal-Callus was tentatively associated with the government-owned estate of Simblija (Fig. 6.42).[58] The tenement at Simblija consists of garigue and agricultural land at the point of intersection between Wied ir-Rum and Wied Ħażrun (Saliba 2008, 17-19, 66; RA55; Figs. 3.6, 4.3, 4.10). Located at map reference 42904 69895, the Simblija tenement contains a cluster of rock-excavated and masonry-built units, the most important of which were utilised as a *centimolo* (animal operated mill) and possibly a church (Buhagiar M. 2005, 51-53). The difficulty in pinpointing the location of the Tal-Callus *giardino* stemmed from the fact that both the Tal-Callus and Simblija touch the Il-Qattara *giardino* (RA32; Fig. 4.10).[59] When writing on the Callus property in 2003, S. Fiorini had already indicated that this should probably not be associated with Simblija (Fiorini 2003, 23-26). Fiorini based his argument on a notarial deed dated to 11 October 1514 which described the easternmost section of the Tal-Callus estate as bordering with the Il-Qattara territory (Fiorini 2003, 24-25; NLM Univ. 12, f. 92, 11.x.1514). The 1514 deed also stated that the southernmost boundary of the Tal-Callus *giardino* was marked by a ravine. Working on this evidence, Fiorini correctly concluded that the Tal-Callus property was located in the shelter of Wied ir-Rum, at the approximate map reference of 42500 70000 (Fiorini, 2003, 25), and not as formerly stated (Saliba *et al* 2002, 17-20), on the nearby headland of Is-Simblija.[60]

[58] This area was erroneously marked as *Is-Simblu* on the 1973 (SS 4269) and 1993 (SS 4269) editions of the Malta Survey Sheets having a scale of 1:2,500. The 1922 and 1959 Survey Sheets 85, however, clearly indicate the place-name of this area as *Is-Simblija*.

[59] The Il-Qattara territory touches the eastern flank of the Tal-Callus property and the western boundary of the Simblija estate.

[60] Mattew Falzon, a prominent late seventeenth-century judge, also owned land at Wied Ħażrun. Like Josephus Callus, Mattew Falzon was also condemned to death following accusations of heresy. His land at Wied Ħażrun was similarly confiscated and passed into the hands of the *Veneranda Assemblea* (the Venerable Congregation) of Conventual Chaplains of the major church of St John, sometime between 1565 and 1575. See Fiorini 2003, 25-26. This property also contained a church dedicated to the Nativity of the Virgin Mary, and can be tentatively associated with a large parcel of land which the 1867 *cabreo* containing information on the tenements forming part of the Crown Property lists

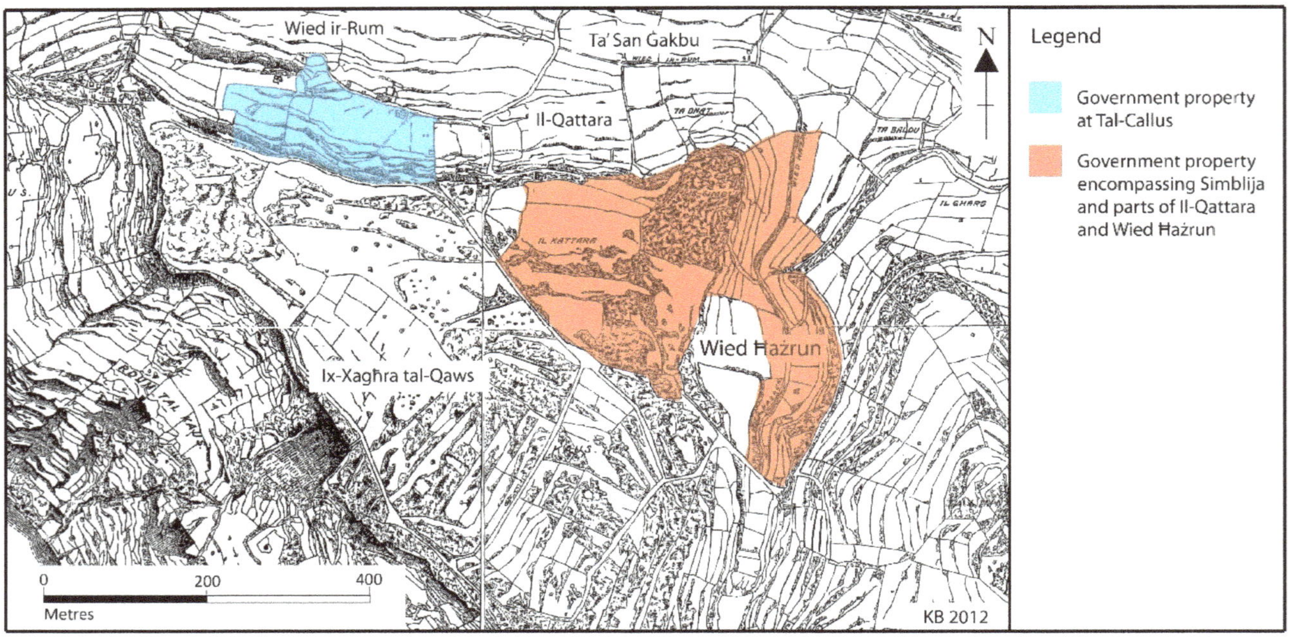

Fig. 6.46 Plan showing the Government landholdings of Tal-Callus (RA39) and Simblija (RA41) at Wied ir-Rum (RA56) and the adjoining Wied Ħażrun (RA55). Sources: NLM. Treas. B. 290, f. 77; NLM. Treas. B. 291, ff. 86-87; Survey Sheet 84, Scale 1:2,500, School of Military Engineering, Chatham, 1903; Survey Sheet 85, Scale 1:2,500, School of Military Engineering, Chatham, 1922.

Fig. 6.47 General view of the Tal-Callus estate (RA39) at Wied ir-Rum (RA56) in the territory of Rabat (Malta). The closely-knit masonry-built structures in the shaded section of the photograph mark the northernmost extent of the property. The arrow indicates the country lane which from the San Ġakbu settlement (RA49), on the north side of Wied ir-Rum, meanders down towards the Tal-Callus and the Il-Qattara (RA32) *giardini*. The shaded area indicates the area of land encompassed by this estate. Source: the author.

An entry in the *cabreo del Magistero* 289 describing the Tal-Callus *giardino* lists the properties which adjoin the Tal-Callus tenement, hence facilitating the identification of the approximate location of the portion of land in question (Fig. 6.44). The *cabreo* description for this property specifically stated that Tal-Callus could be accessed through a public road, which from the San Ġakbu area in the northern part of Wied ir-Rum leads down to the Valley bed and the Il-Qattara area on the opposite side of the valley.[61] This early seventeenth-century description of the Tal-Callus *giardino* was not accompanied by a plan of the property, so that the agricultural territory in question cannot be identified with certainty.

under the heading of *Ta' Falzun* (RA30). A comparative analysis of land boundaries made it possible to identify this with agricultural land on the eastern flank of Wied Ħażrun, at its point of intersection with Wied ir-Rum at map reference 43174 69852. On the 1922, Survey Sheet 85, the place-name for this area is indicated as Il-Għars (RA30; Fig. 21).

[61] [...] da tramontana, confina con il detto uiolo che uiene dal detto Casal dingli, et dalla chiesa di San Giacomo; quale da entrata all'istesso giardino di gued iron, et alle terre del detto Maro Cassar [...]. See NLM. Treas. B. 289, f. 84.

Fig. 6.48 The arrow marks the narrow and tortuous footpath which descends to the southern-most flank of the Tal-Callus property (RA39). The water gallery is located in the uppermost terraced field in the area underlying the footpath. Source: the author.

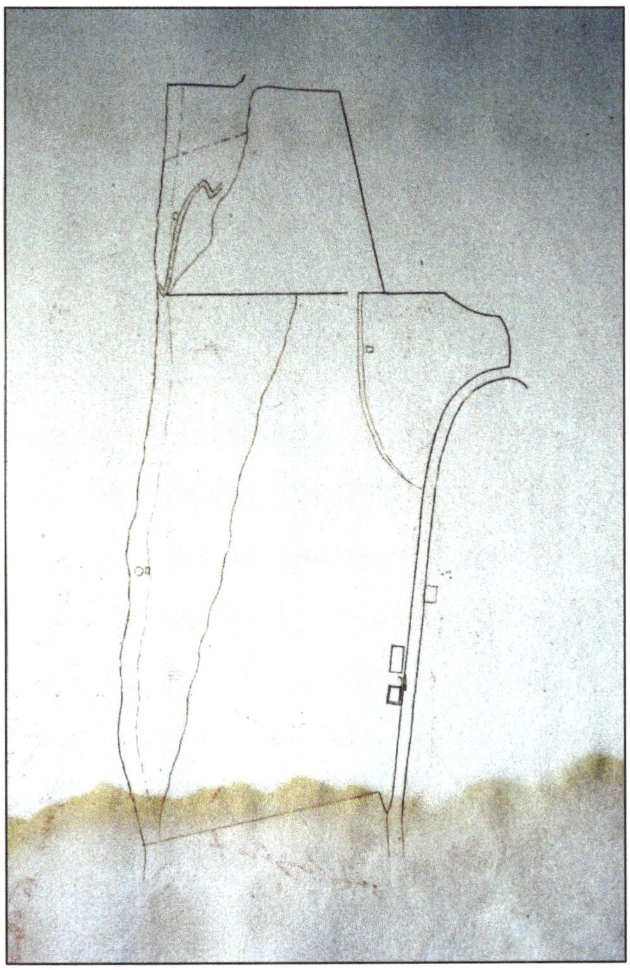

Fig. 6.49 NLM. Treas. B. 291, f. 86. This line drawing of the Tal-Callus property (RA39) is possibly the original design on which the plan included in NLM. Treas. B. 290, f. 77 was produced (Fig. 6.45).

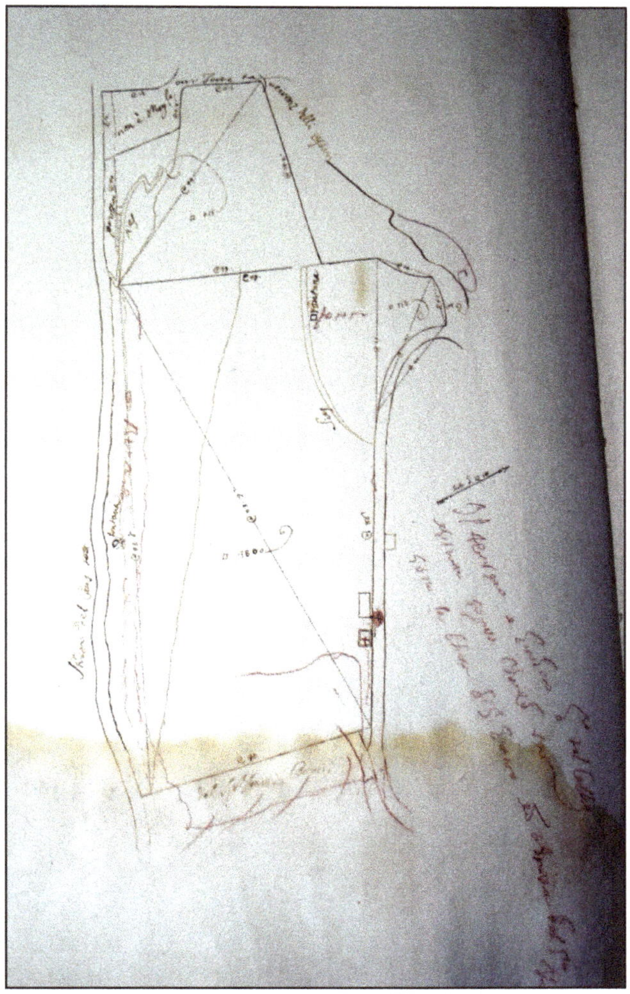

Fig. 6.50 NLM. Treas. B. 291, f. 86. This is a more detailed version of the plan of the Tal-Callus estate (RA39) on which that included in NLM. Treas. B. 290, f. 77 was produced (see Fig. 6.45). Clearly indicated is the location of two *fontane* first mentioned in NLM. Treas. B. 289, f. 84. The transverse triangulations included in the plan suggest that this was probably drawn on site.

This shortcoming was fortunately offset by the availability of three different plans for the Tal-Callus territory which the present writer managed to locate in two other *cabrei* manuscripts produced in the opening decades of the eighteenth century: the *cabreo del Magistero vol. 1* (NLM. Treas. B. 290, f. 77), and the *cabreo del Magistero "Secrezia"* (NLM. Treas. B. 291, ff. 86-87). The plan included in *cabreo* 290 is entitled *Giardino ta Callus o ta Wied Ferin in Casal Dingli* (Fig. 6.45). The inclusion of access routes and property boundaries, all clearly indicated in this plan, enabled the present writer to carry out a comparative study of the field boundaries using both the plan in question and the early twentieth-century Malta Survey Sheets for this area.[62] A tract of land on the 1903 Survey Sheet 84, located to the immediate west of the Il-Qattara tenement,

[62] Survey Sheets 84 and 85 cover most of the Wied ir-Rum territory.

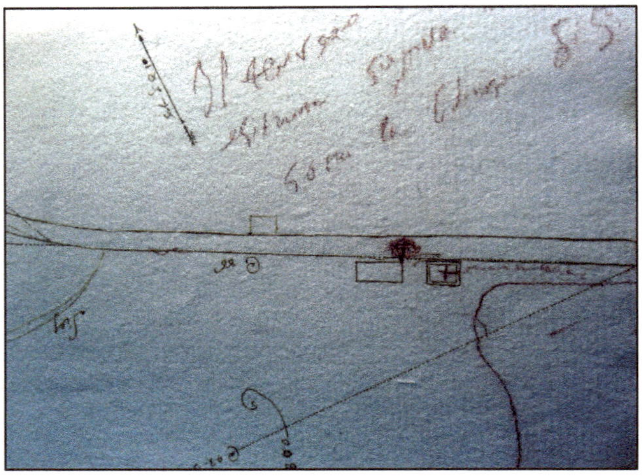

Fig. 6.51 Detail of NLM. Treas. B. 291, f. 86 showing two single-celled units flanking the northernmost extent of the Tal-Callus estate (RA39) at Wied ir-Rum (RA56) in the territory of Rabat (Malta).

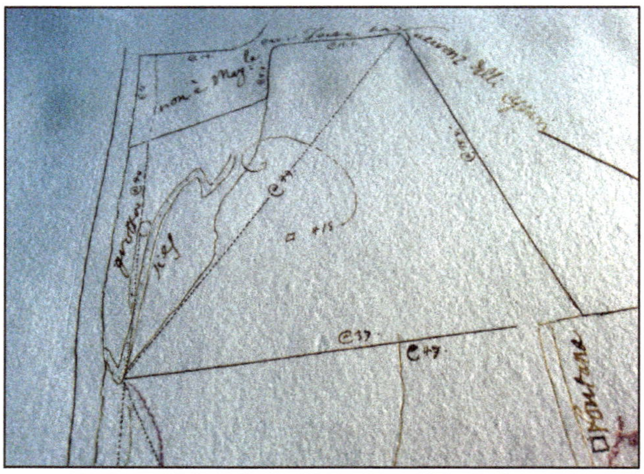

Fig. 6.52 Detail of NLM. Treas. B. 291, f. 86 showing the footpath which from the ravine flanking the south side of the Tal-Callus estate (RA39) gives access to this territory. The *cabreo* representation for this property also shows the presence of a small circular-shaped cave, labelled *grotta*.

was found to correspond perfectly to the property illustrated on f. 77 of *cabreo* 290 (Figs. 6.46, 6.47). In fact, a section of this property corresponds to the tract of land which Prof. Fiorini had, in 2003, indicated as possibly forming part of the Tal-Callus *giardino* (Fiorini 2003, 25).[63]

Delimiting the southernmost extent of the Tal-Callus *giardino* is a ravine on top of which is a public country lane giving access to the Il-Qaws and the Tax-Xieref (RA54) areas. Branching from this road (map reference 42356 69943) is a narrow and tortuous footpath which descends to the southernmost flank of the Tal-Callus property (Fig. 6.48). The other point of access is through a much wider country lane (map reference 42582 699984), which from next to the San Ġakbu settlement (RA49) on the north side of Wied ir-Rum meanders down towards the Tal-Callus and the Il-Qattara *giardini* (Fig. 6.47). The plan on f. 77 of *cabreo* 290, besides giving the dimensions of the various portions of land forming part of this *giardino*, also indicates the presence of two single-celled masonry-built structures. These flank the main access point of the Tal-Callus *giardino* which connects to both the Il-Qattara and the San Gakbu sides of the Wied ir-Rum (Fig. 6.45). Field-terracing and areas containing tree plantations in the Tal-Callus property are also depicted through the use of different greenish-brown coloured brush strokes.

The two other plans for the Tal-Callus property form part of the *cabreo* 291 manuscript (NLM. Treas. B. 291, ff. 86-87; Figs. 5.38, 5.43). It is the writer's view that these plans are contemporaneous with the one on f. 77 of *cabreo* 290, and these might in essence be the draft versions on which the more artistically elaborate *cabreo* 290 drawing was based. The plan on f. 87 (*cabreo* 291) is a simple line drawing (Fig. 6.49) which contains no

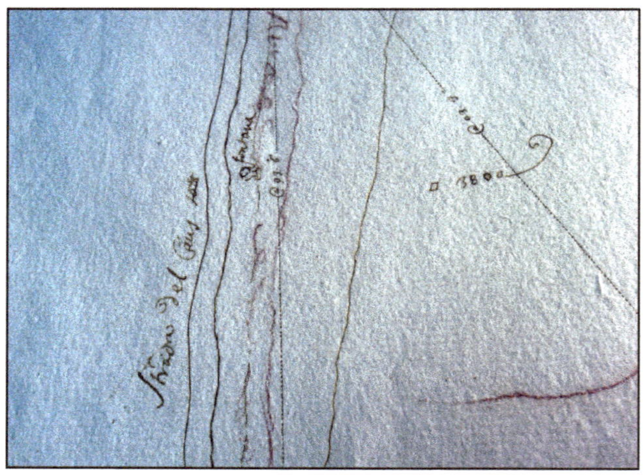

Fig. 6.53 Detail of NLM. Treas. B. 291, f. 86 showing a *fontana* (spring) originating from within a rock-cut chamber on the southern side of the Tal-Callus estate (RA39) at Wied ir-Rum (RA56) in the territory of Rabat (Malta). Field-research has shown this source to originate from inside a perched aquifer gallery.

[63] N. Ciantar has similarly concluded that the Tal-Callus property cannot be associated with Simblija but proposed no alternative location for it. See: http://www.timesofmalta.com/articles/view/20120603/life-features/Rendering-justice-to-u-eppi-Callus.422598. The Tal-Callus estate is still under Government ownership, but is currently on lease to fourteen individuals, each one of whom manages different portions of the property. The estate is no longer identified as Ta' Callus, but as Ta' Katisu or Ta' Djiegu, the latter being the name of a certain Djiegu Gauci who lived on the estate during the course of the second World War. That the tenants of this property are aware of the archaeological and historical importance of the Tal-Callus estate is intimated by the fact that they very reluctantly allow onsite access to visitors. This is due to the ongoing concern that portions of the site might be subjected to a fate similar to that experienced by the tenants of the Simblija estate. Even though adequately compensated, the tenants at Simblija still feel aggrieved that they were deprived of the continuous use of historically and archaeologically sensitive areas of the esate. Personal communication by D.M. Galea, a University of Malta archaeology graduate.

Fig. 6.54 The exterior of the *fontana* mentioned in NLM. Treas. B. 289, f. 84. Its location is indicated on a plan of the Tal-Callus estate (RA39) in NLM. Treas. B. 291, f. 86. This water source originates from a perched aquifer gallery which is fronted by a subterranean water storage reservoir. A series of steps to the right of the water sump facilitates access to the reservoir interior. Source: the author.

caption.[64] Nonetheless, the outline shape of the tenement, the passageways giving access to the *giardino*, and the two single-celled structures adjoining the roadside in the northern section of the plan, leave no doubt that the tenement in question represents the Tal-Callus property. The plan on f. 86 (*cabreo* 291) is likewise an outline scale drawing of the same property, more informative than the former (Fig. 6.50). It is accompanied by a handwritten caption, the legible sections of which read: *Il terreno e Giardino detto tal Callus et il* [...] *sotto la*

Fig. 6.55 Interior view of the *fontana* mentioned in NLM. Treas. B. 289, f. 84, the location of which is indicated on a plan of the Tal-Callus estate (RA39) in NLM. Treas. B. 291, f. 86. This water source originates from a series of galleries, a number of which were walled up at an unknown date. The extracted water is stored in a fronting subterranean reservoir. Source: the author.

[64] In the lower left-hand section of this manuscript page there appears to be an illegible vertical line of text written in red pigment.

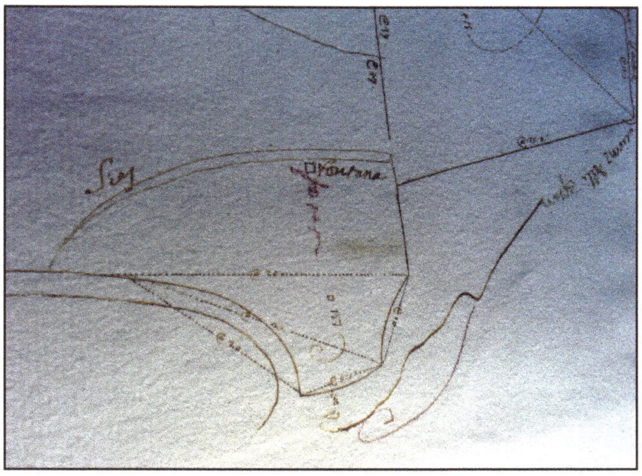

Fig. 6.56 Detail of NLM. Treas. B. 291, f. 86 showing a *fontana* located in the north-west section of the Tal-Callus estate (RA39) at Wied ir-Rum (RA56) in the territory of Rabat (Malta).

chiesa di S. Giacomo ed [...].[65] Of significance is the fact that the easternmost room contains a cross marking in a red coloured pigment, hinting that this was utilised as the church which was either dedicated to St Nicholas or Santa Marija (Fig. 6.51). Another cross depicted on one side of the same country lane, in very close proximity to the former, was apparently placed there by mistake and was haphazardly erased.

The two *fontane* or springs mentioned in the *cabreo* 289 manuscript (NLM. Treas. B. 289, f. 84) are also marked on this plan version of the Tal-Callus *giardino* (NLM. Treas. B. 291, ff. 86-87; Fig. 6.50). One of the springs is shown to be located on the northern side of a country path, close to the north-western boundary of this property. This water-related feature is marked by a square-shaped box which contains the legend *fontana*. The second *fontana* flanks the southernmost extent of the property, and an almost circular chamber which appears to be fronted by a rectangular shaped reservoir clearly indicates that in this particular instance, we are dealing with a rock-cut water-related feature.[66]

[65] A finely cut, rectangular-shaped cave in the northern side of Wied ir-Rum and which overlooks the Tal-Callus *giardino* area has been tentatively identified by the present writer as the church of St James. See Buhagiar K. 2013, 98-107.

[66] Nineteenth-century *cabreo* documentation again mentions an: [...] *altra piu piccola Fontana con vasca della capacita circa di 1900 palmi cubi, e somministra circa palmi cubi 400 al giorno munita pure di un lungo canale d'irrigazione*. The other water source also available on this property is described as an *altra Fontana con canale o vasca della capacita circa di plami 1540, e somministro palmi cubi 200 per giorno*[...]. See NAM. Descriptive Plans of the Crown Property, 1867, f. 606 (Fig. 6.62). The first water source conforms to the gallery and subterranean water reservoir excavated into the north-facing ravine of the Tal-Callus *giardino*. The rock-cut reservoir is clearly indicated on the 1867 plan as well as the network of canals which transported the water retrieved by this gallery to the easternmost extent of the Tal-Callus site. The location of the other perched aquifer gallery is indicated by the reservoir in the western part of this estate, which probably adjoins a shaft well piercing the ceiling of a gallery which passes below this section of the site. In the nineteenth century, this property was no longer identified as Tal-Callus but as *Wiet Ferin*.

Fig. 6.57 The hamlet at the Tal-Callus estate (RA39) at Wied ir-Rum (RA56) in the territory of Rabat (Malta). It is possible that structures "A" and "B" are those represented in the early eighteenth-century NLM. Treas. B. 291, f. 86 manuscript plan. Source: the author.

Fig. 6.58 Water gallery tunnelling into the north-western side of the Tal-Callus estate (RA39). It is probable that this feeds a close-by *fontana* mentioned in NLM. Treas. B. 289, f. 84 and NLM. Treas. B. 291, f. 86 with a perennial water supply. Source: the author.

Access to the Tal-Callus property is largely restricted. In recent years, entry to the country lane passing through the Wied ir-Rum valley bed at map reference 42582 699984 has been partly fenced off. No trespassing signs are also prominently displayed. Not having contacts with any of the tenants leasing this land has so far hindered my access to the central and northern parts of the Tal-Callus *giardino*. Access to the footpath adjoining the Tal-Qaws road at map reference 42356 69943 remains unhindered, enabling the present writer to investigate extensively this section of the property (Fig. 6.48). It was also an opportunity for testing the reliability and accuracy of the data provided in the *cabreo* 291 plans.

The first feature encountered on accessing Tal-Callus via the Il-Qaws footpath is a circular-shaped cave (map reference 42356 69943). This rock-cut feature was also distinctly marked on the *cabreo* 291 plans for this property (Figs. 6.52, 6.63). The cave is excavated into the brittle Mtarfa Member deposit, has rather intimate proportions and was probably either used as an animal pen or as a storage space for agricultural produce. The *fontana* was also located in the section of the Tal-Callus property indicated by the plans in *cabreo* 291 (Figs. 6.49, 6.50, 6.53, 6.63). This water spring actually originates from a subterranean gallery, the exterior section of which has been modified and remodelled into a rock-cut subterranean water reservoir (Figs. 6.54, 6.55). That this spring dates to at least the opening decades of the seventeenth century is demonstrated by the fact that its current setup perfectly corresponds with the description found in *cabreo* 289.

Due to access restrictions to the northern section of the Tal-Callus *giardino*, the second *fontana* could not be traced and investigated (Fig. 6.56). Nonetheless, the scrutiny of the northern perimeter wall of the Tal-Callus property led to the identification of another two perched aquifer galleries at map references 42456 70020 and 42524 70004 respectively (Fig. 6.58).[67] Both tunnel below the northern section of the Tal-Callus property and a shaft well piercing the ceiling of one of these subterranean water systems, possibly supplied this fontana with a perennial water source.

[67] The partial investigation of one of these two perched aquifer water capture systems was made possible through the kind assistance of D.M. Galea, whose family owns agricultural property adjacent to the northern boundary of the Tal-Callus estate.

Fig. 6.59 Exterior detail of the north-facing façade of the structure "A" illustrated in Fig. 6.57. This probably conforms to the room represented in the NLM. Treas. B. 291, f. 86 plan for the Tal-Callus estate (RA39). Source: the author.

Fig. 6.61 Exterior detail of the back façade of structure "B" illustrated in Fig. 6.57. This seems to fit in with the room represented in the NLM. Treas. B. 291, f. 81 plan for the Tal-Callus estate (RA39). Source: the author.

Fig. 6.60 Exterior detail of the back façade of structure "B" illustrated in Fig. 6.57. This seems to fit in with the room represented in the NLM. Treas. B. 291, f. 86 plan for the Tal-Callus estate (RA39), which was probably utilised as a church. Source: the author.

It was also impossible to gain access to the small hamlet located at map reference 42505 69996. The closely-knit cluster of structures which since the eighteenth century have been constructed in this part of the Tal-Callus *giardino* were nonetheless visually inspected from the overlying garigue plains of Il-Qattara and Ta' San Ġakbu located on the southern and northern flanks of Wied ir-Rum respectively. Furthermore, the masonry-built structures at Tal-Callus were also visually investigated from the nearby country lane at map reference 42536 70097 and the footpath which borders the Tal-Callus tenement at map reference 42522 70000. The close scrutiny of these structures, as well as the use of the *cabreo* 291 plans, made it possible for some preliminary conclusions to be reached. Notwithstanding, it is important to note that the data presented below will

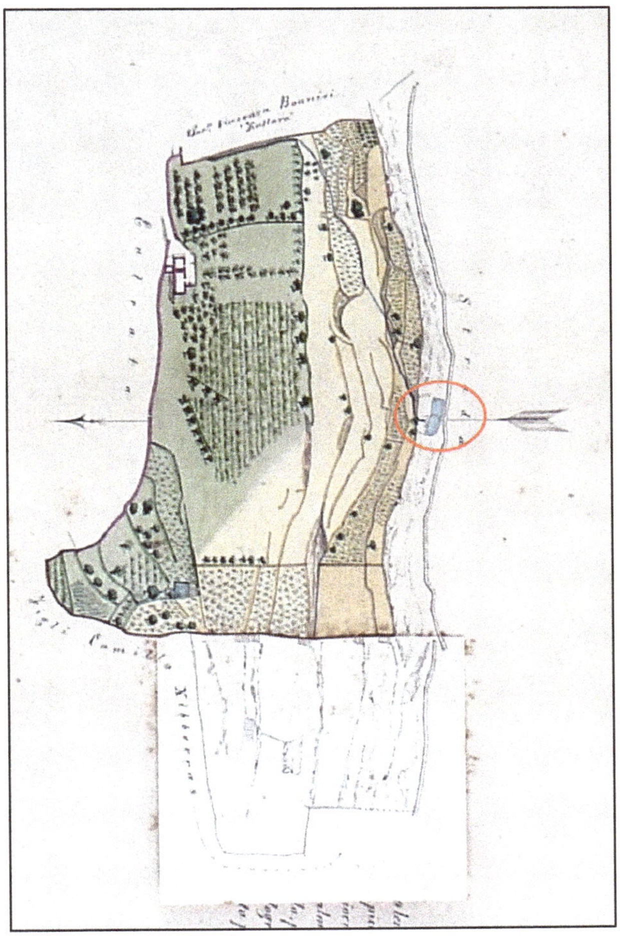

Fig. 6.62 NAM. Descriptive Plans of the Crown Property in Malta, 1867, f. 606. Plan detail of the Tal-Callus *giardino* (RA39) dating to 1867. Circled in red is the subterranean reservoir within which is collected water originating from a perched aquifer gallery.

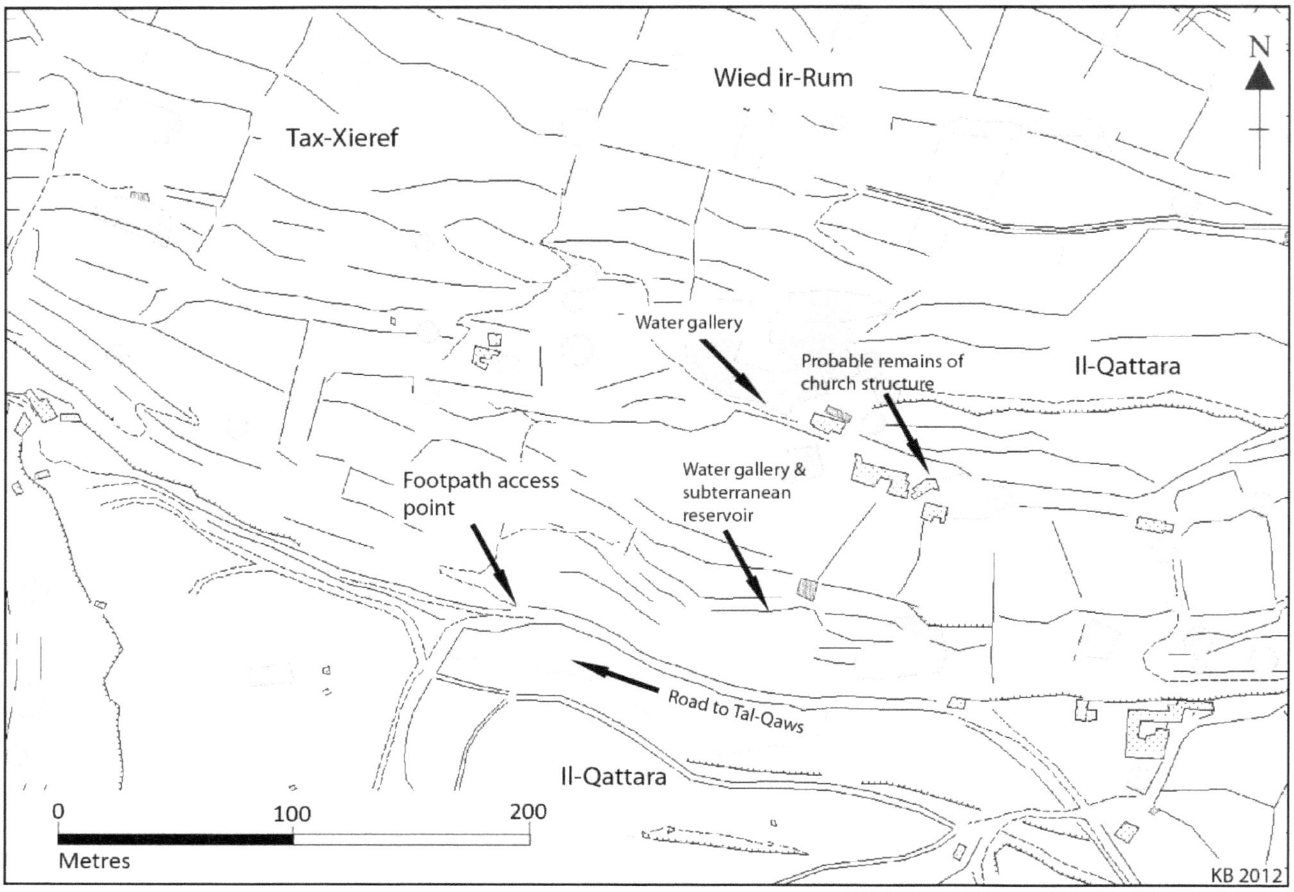

Fig. 6.63 Site plan of the Tal-Callus estate (RA39) at Wied ir-Rum (RA56) showing the location of the two perched aquifer galleries and the probable remains of the church dedicated to St Nicholas. Topographical map image adapted from MEPA Map Server data.

probably have to be revised once access is gained into this section of the Tal-Callus *giardino*.

The two single-celled structures indicated in the *cabreo* 291 plans still exist (Fig. 6.51). Both appear to survive in different degrees of structural integrity, but have been abutted by later period constructions. The larger structure appears to conform to a rectangular-shaped room, the visible exterior walls of which betray a wet-rubble type of construction (Figs. 6.57, 6.59). This room appears to be capped by a flat roof structure from which water spouts drained off any access rain water.

At a distance of ca. 12 m to its east are the surviving remains of the single-celled structure which in the *cabreo* 291 plans is marked by means of a cross. Its visible exterior walls once more betray a wet-rubble type of construction technique, but the original structure only appears to survive in part (Figs. 6.60, 6.61, 6.63). Apart from suffering a partial roof collapse, its western section was abutted by another rural construction, the walls of which were shaped out of ashlar masonry blocks. It is interesting to point out that what appears to have survived of the original fabric betrays the presence of a gabled ceiling and a narrow, slit-like window, located high up in the north façade of the structure. Both features are reminiscent of Late-Medieval church architecture (Buhagiar M. 1975, 163-80; Buhagiar M. 2005, 83-99), making it likely that these are the surviving remains of the long lost church which was either dedicated to St Nicholas or Santa Marija (Aquilina and Fiorini 2001, 169-70; Fiorini 2003, 24). It is naturally only by gaining access to the interior of this room that such conjectures can be verified.

6.1.8 The fief of Tas-Santi in the territory of Mġarr (Malta)

The *cabreo* 289 description of the *Fego del Santi* (NLM. Treas. B. 289, ff. 84-88; MG23), a government-owned estate located in the territory of Mġarr, Malta, is another source which gives a glimpse as to the hydrological potential and the water extraction strategies employed in this part of Malta. The locality of Tas-Santi was first recorded by notary Amati de Messina in 1465 (Wettinger 2000, 520). Tas-Santi is a fairly large parcel of land, the NLM. Treas. B. 289 description of which gives an account of its bordering lands, allowing a reconstruction

Fig. 6.64 NLM. Treas. B. 289, Plan 13. Plan of the Tas-Santi *feudo* (MG23) located in the territory of Mġarr (Malta).

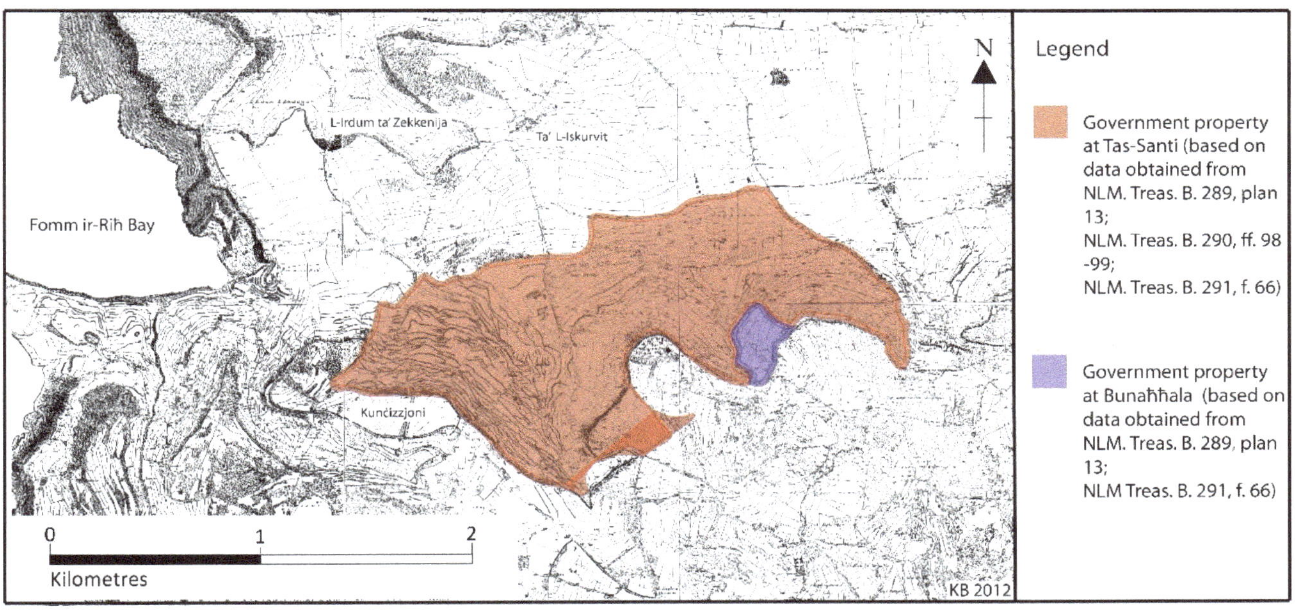

Fig. 6.65 Plan showing the approximate boundaries of Government landholdings at Tas-Santi (MG23) in the territory of Mġarr (Malta) in the seventeenth and eighteenth centuries. Sources: NLM. Treas. B. 289, plan 13; NLM. Treas. B. 291, f. 66.

of the boundaries of this estate (Figs. 6.64, 6.65). Tas-Santi is also rich in geological and ecological diversity. Mtarfa Member deposits align the southern and south-eastern flanks of the estate, at an elevation of ca. 195 m above sea level. It is roughly at this level that the perched aquifer water sources furnishing this estate with a

Fig. 6.66 NLM. Treas. B. 291, f. 66. Plan of the Tas-Santi *feudo* (MG23) located in the territory of Mġarr (Malta).

perennial water source originate. Slightly further downhill are the Blue Clay deposits and Middle and Lower Globigerina Limestone deposits. A section of the Santi estate is composed of karst terrain and was possibly used for the grazing of sheep and goats (map reference 42400 73260).

A reading of the description of this *feudo* makes it, on occasion, difficult to identify with certainty the area being mentioned in the text (NLM. Treas. B. 289, ff. 84-88). The southern flank of the Tas-Santi territory is the first to be described, albeit rather sketchily. For this portion of the estate located in the uppermost section of the valley, there is the mention of a *beueratura, e calcara del detto fego, che e sopra il vallone.*[68] *Foglio* 88 contains more information on this same valley section: *Vi sono dalla parte di mezzo giorno molte grotte e ricetti d'animali. Vi sono dentro il fego due fontane et due beverature dalla parte di ponente l'una uicino il forno della calcara l'altra un poco pui dentro il fego* (NLM. Treas. B. 289, f. 88).[69] This description conforms to land at map reference 42190 72965 in which are located a watering trough, several caves, two perched aquifer galleries and a fronting open-air water storage reservoir (Figs. 5.52, 5.54, 5.55, 5.56, 5.57, 5.58, 5.59, 5.60, 5.61). On the other hand, the location of the lime kiln remains unknown.[70]

The water trough is adjoined by an elevated rectangular enclosure in which two caves are excavated at map reference 42150 72941.[71] The eastern-most cave gives access to the first perched aquifer gallery (Figs. 5.54, 5.55). This is a very finely cut gallery which is ca. 50 m long. It is on average 0.7 m wide and in its floor is a water canal which channels the collected aquifer water to the cave's exterior. From this point onwards, a series of masonry canals channel this water into an open-air water reservoir, where it is stored for later distribution to other sections of the Tas-Santi valley (Figs. 5.56, 5.57, 5.58). The second perched aquifer gallery is located ca. 30 m to the east.[72] The initial section of this gallery is full of debris and sedimentation and could not be investigated further. It currently only yields a small water trickle. It is interesting to note that immediately overlying this gallery are at least three vertical shafts which appear to be

[68] [...] animal water troughs, and the lime kiln of this tenement, which is at the top of the valley [...]. See NLM. Treas. B. 289, f. 85.

[69] The other water trough is quite probably the one flanking the road at map reference 42094 73717, which from Baħrija and the Tax-Xini areas gives access to Tas-Santi and Mġarr.

[70] Abela, writing almost contemporaneously with the compilation of *cabreo* 289, does mention an *Ayn tal Carcara* in the area. See Abela 1647, 68.

[71] B. Mizzi, the owner of this section of Tas-Santi valley, kindly granted me permission to access this property.

[72] The existence of this water gallery was brought to my attention by M. Borda, the owner of the uppermost valley section at Tas-Santi.

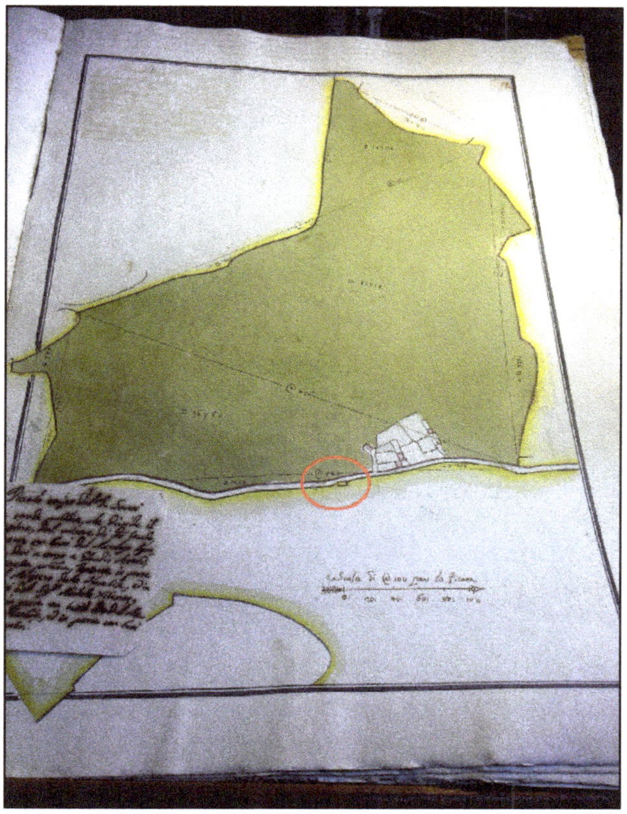

Fig. 6.67 NLM. Treas. B. 290, f. 98. Plan showing the lower portion of the Tas-Santi estate (MG23). Circled in red is the *beveratura* or a water drinking trough located on the side of the public road which divides Tas-Santi into two different portions.

piercing its ceiling (Fig. 5.61).[73] Should this be the case, it is possible that in this particular instance there is a fully-fledged *qanat*-type water extraction system. One of these two galleries is probably *Aayn tal Carcara* (MG5) mentioned by Abela,[74] a matter which is also hinted at by the mention of a *calcara* in the *cabreo* 289 text (NLM. Treas. B. 289, f. 85).

Four other cave units were observed beneath the north-west-facing ravine at map reference 42213 72975. All are fronted by finely-built dry-stone walls, but permission to access their interior was not granted. *Cabreo* 289 also makes reference to *acqua sorgente*, or spring water (NLM. Treas. B. 289, f. 85), but its exact location could unfortunately not be established. There is also mention of significant capital investment being injected into the maintenance of rubble walls and the re-construction of a water trough within the *ġnien* Tas-Santi (NLM. Treas. B. 289, f. 88).

The *cabreo* 291 description for the fief of Tas-Santi for the territory of Rabat (Malta) is short and schematic (NLM. Treas. B. 291, f. 19). Tas-Santi is in this entry divided into two separate areas: the *Santi superiore* and the *Santi inferiore*.[75] The only mentioned water-related feature for the *Santi superiore* is the *beveratura*, or water drinking trough, which adjoins the Santi public road at map reference 42098 73715 and which probably dates to the late seventeenth century. Even the plan included in *cabreo* 291 for Tas-Santi does not divulge evidence of any water features in this property (NLM. Treas. B. 291, f. 66; Fig. 6.66). The same applies to two plans of the Santi *feudo* in *cabreo* 290 (NLM. Treas. B. 290, ff. 98-99; Figs. 6.67, 6.68). Both show the location of the Santi hamlet as well as the water drinking trough flanking the public road slightly over 70 m uphill from the hamlet.

A plan for the Is-Santi estate is also included in the Descriptive Plans of the Crown Property *cabrei* (NAR.

Fig. 6.68 NLM. Treas. B. 290, f. 99. Plan showing the upper portion of the Tas-Santi estate (MG23). Circled in red is the *beveratura* or water drinking trough located on the side of the public road which bisects Tas-Santi into two different portions.

[73] Personal communication by M. Borda. It has already been specified in chapter one, section 1.6 that *qanat* "doughnuts" or "molehills" do not occur locally.

[74] See chapter five, section 5.2.1.3.

[75] Affixed to the binding spine of *cabreo* 290 is another short description of the *feudo*, which likewise states that this territory is divided into two main areas known as the Santi *maggiore* (greater) and the Santi *minore* (smaller). See NLM. Treas. B. 290, f. 98-99.

Descriptive Plans of the Crown Property, 1867, f. 599). No description of the water resources of this estate is to be found in the available documentation (Fig. 6.69). The 1867 cabreo plan clearly indicates that by the latter half of the nineteenth century the Santi estate was fragmented into a number of smaller units which were leased out to private individuals. A comparison between the Knights' period data and the 1867 *cabreo* also makes it possible to identify portions of land which by the nineteenth century became estranged from the original *feudo* setup. This is the case with a substantial portion of land to the east of Fort Binġemma. Even though this probably remained Government-owned property, it appears that this now came to form part of the *Burnohhala* territory. It is possible that property fragmentation had by the latter half of the nineteenth century also affected land in the upper reaches of the Santi valley where the two above-mentioned perched aquifer galleries perennial sources are situated. This land is currently in private ownership, but access to it is available.

6.1.9 Birhauex and the fief of Għajn Tuffieħa

For Birhauex (MG10) in the Żebbiegħ district of Mġarr (Malta), there is the scant mention of a water cistern which appears to have been located in the middle of the road leading down to the Għajn Tuffieħa *giardino* (NLM. Treas. B. 289, ff. 88-89; MG15).

The fief of Għajn Tuffieħa is located north of Mġarr (Malta) and covers a rather extensive territory (Figs. 5.68, 6.70). The Għajn Tuffieħa *giardino* has a long documented history and was mentioned in Guglielmo de Malta's will of 1299 (Wettinger 2000, 191; Dalli 2006, 152). In the mid seventeenth century, the Għajn Tuffieħa *feudo* was composed of a principal *giardino*, an extension referred to as *horto*, and a number of barren lands listed as *terreni scapuli* (NLM. Treas. B. 289, ff. 90-95). It was accessed through the road which meandered downhill from Mġarr towards Għajn Tuffieħa and then proceeded towards Mellieħa (NLM. Treas. B. 289, f. 92). Another road passing through the Għajn Tuffieħa *feudo* came from the direction of *Notabile* (Mdina), and led to the nearby area of Pwales known as the *sbarcature del Gozo* (NLM. Treas. B. 289, f. 93; Vella and Spiteri 2008, 24).

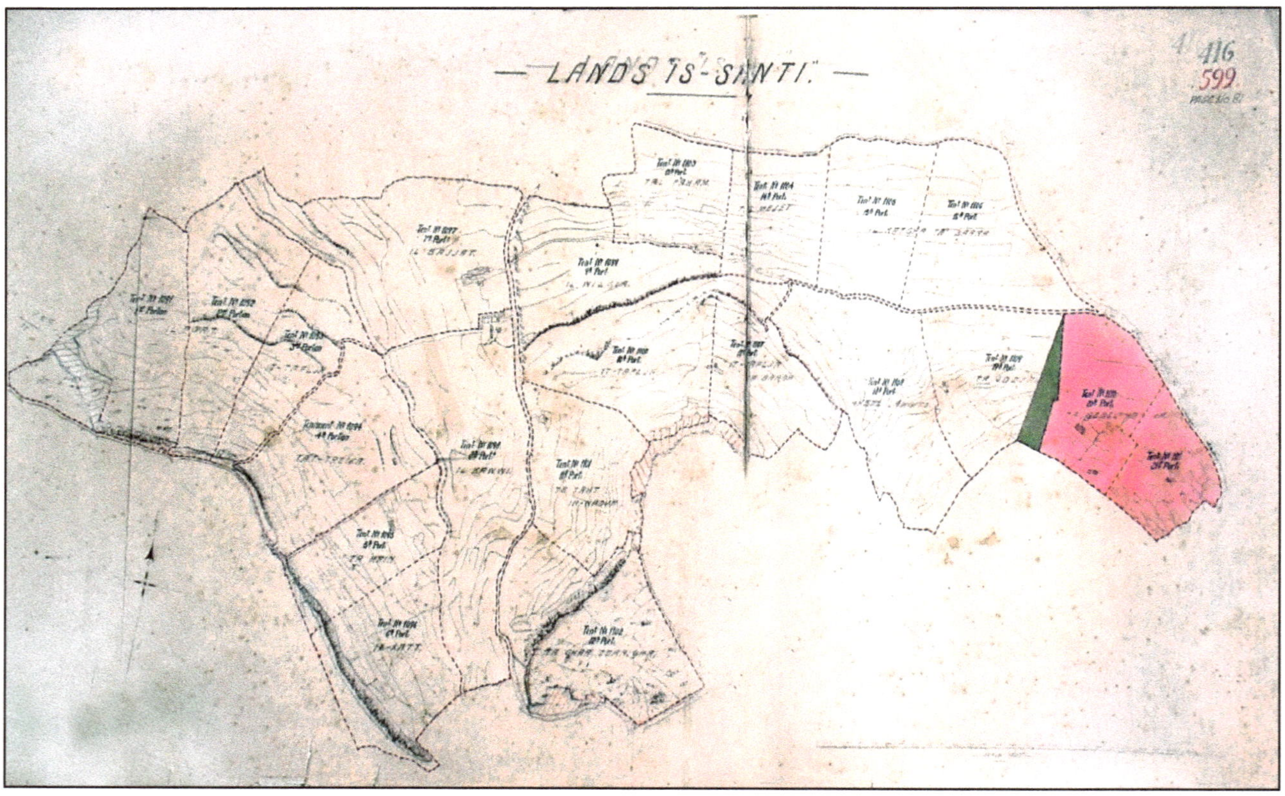

Fig. 6.69 NAM. Descriptive Plans of the Crown Property in Malta, 1867, f. 599. Plan showing the Government-owned estate of Tas-Santi (MG23).

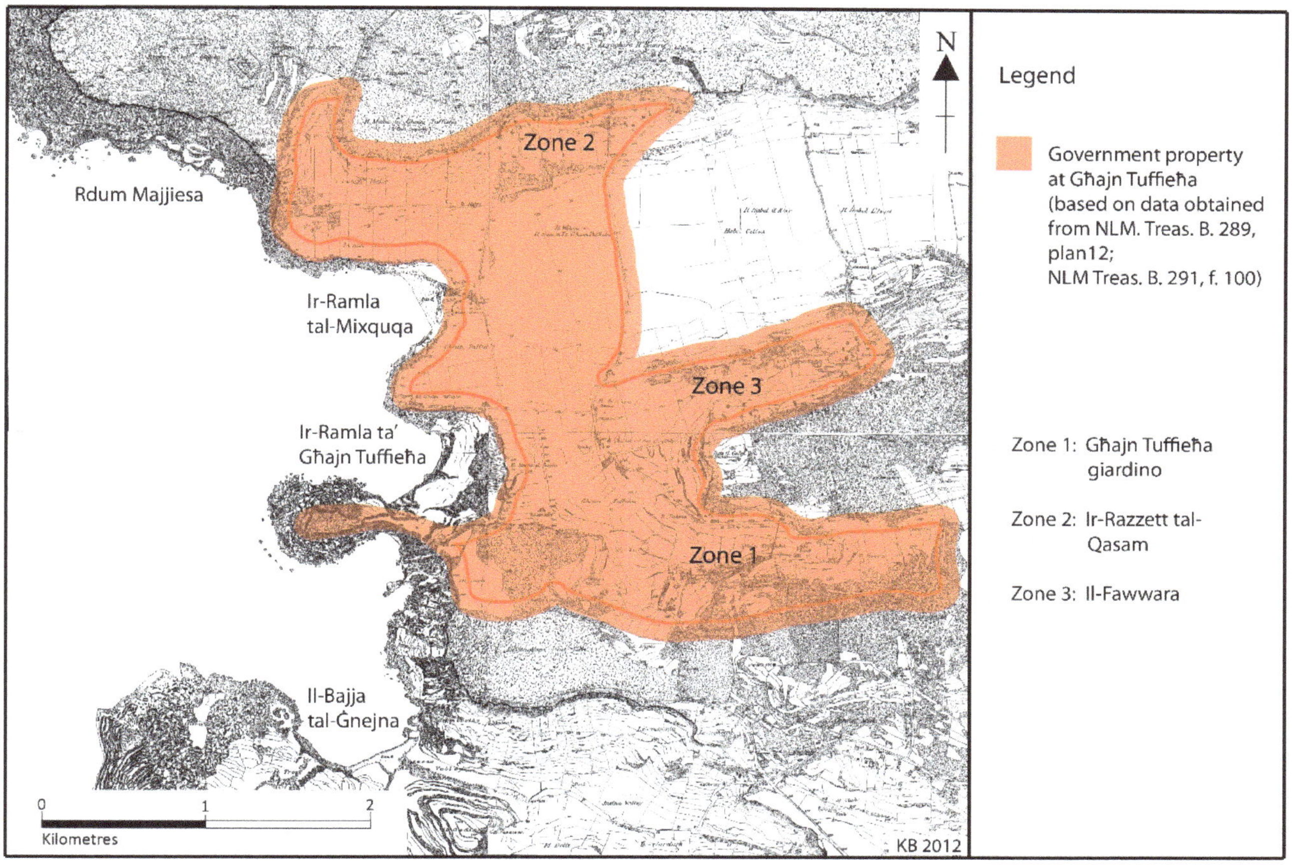

Fig. 6.70 Plan showing the boundaries of the Għajn Tuffieħa *feudo* (MG15) compiled on data obtained from NLM. Treas. B. 289, Plan 12 and NLM. Treas. B. 291.

The *cabreo* 289 description for the *Giardino e terreni scapuli di hayn toffeha* conforms to that of an agriculturally fertile territory bordering the south side of the Għajn Tuffieħa road at map reference 41861 78240 (Fig. 6.82). The *cabreo* excerpt describing this territory mentions at least two different access points, one of which was located close to the *gebia grande* – the large water reservoir (NLM. Treas. B. 289, f. 90). This can be probably associated with an old open-air reservoir located at map reference 41688 76144, the outer sides of which were buttressed for additional strength and which probably predates the early seventeenth century (Fig. 6.71). The reference to orange trees in this *giardino* also points to the presence of a perennial water source. Another feature mentioned is that of a *muraglia grande* (the great wall), which functioned as a windbreak protecting the orange tree cultivations from the *majjistral*, the north-westerly winds (NLM. Treas. B. 289, f. 291).[76] Provision was in addition taken to prevent rain water from flooding the cultivated sections of the land. Interventions in this respect appear to have consisted in the paving of the Għajn Tuffieħa valley water course in order to ease water discharge further downhill, but this no longer survives. The *muraglia grande* also appears to have been perforated by means of a series of arched passageways in order to facilitate rain

Fig. 6.71 This water storage reservoir at the Għajn Tuffieħa *giardino* (MG15) is tentatively associated with the *gebia grande* mentioned in the NLM. Treas. B. 289, f. 90 entry for this area. Source: the author.

[76] Sections of this wall still survive. Piercing this wall was an arched entrance, partly demolished in the 1960s, which gave access to the lower section of the Għajn Tuffieħa *giardino* and hamlet. The wall's building style and function closely parallel that of similar structures at Ta' Baldu in the upper valley section of Wied ir-Rum in the territory of Rabat (Malta).

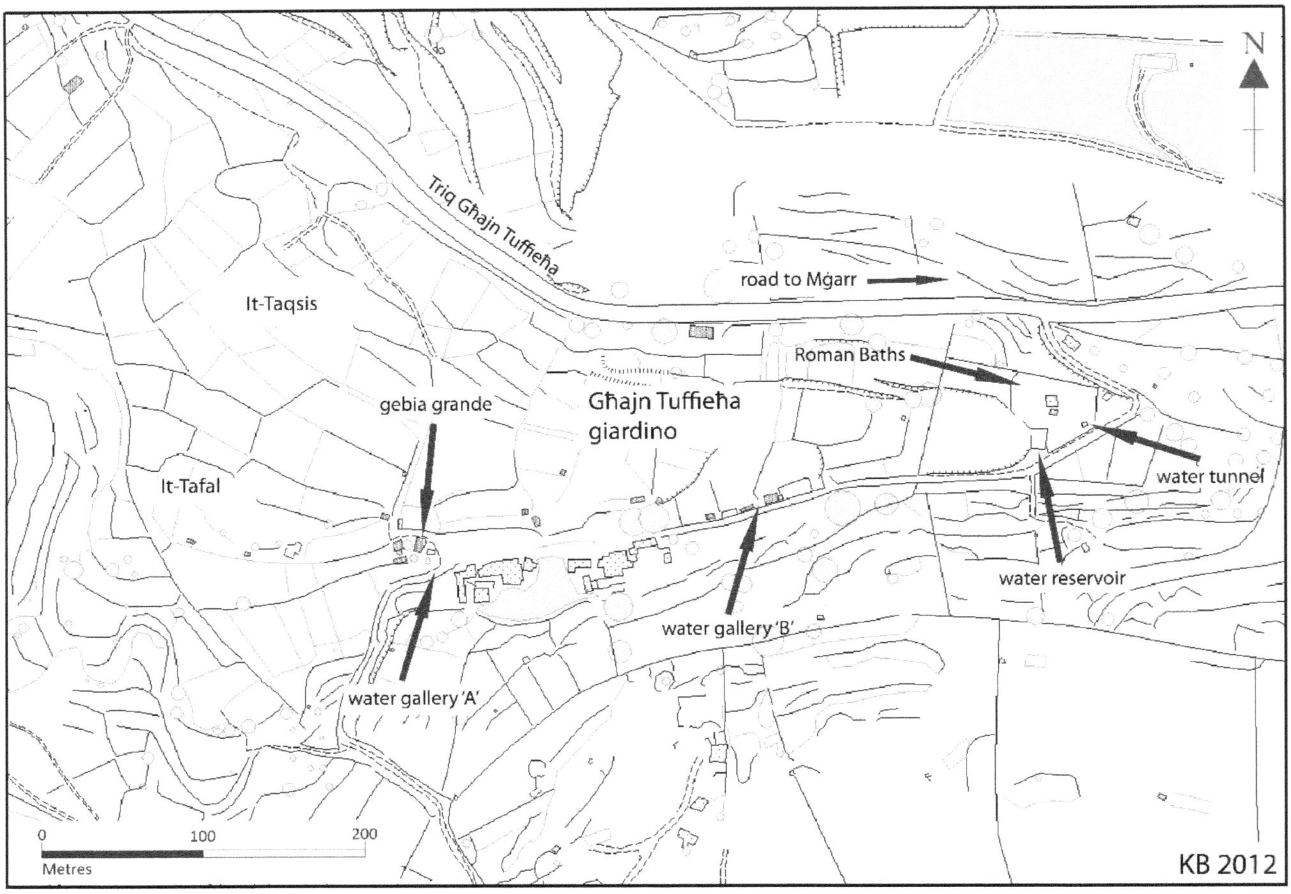

Fig. 6.72 Plan of the Għajn Tuffieħa *giardino* (MG15) indicating the water-related features observed within. Topographical map image adapted from MEPA Map Server data.

water drainage.[77]

The difficulty in interpreting any hydrological evidence related to the Għajn Tuffieħa *giardino* stems from the fact that the *cabreo* entry for this area does not make any direct reference to the presence of water within. As is attested by the neighbouring Roman period baths located slightly further upstream at map reference 42054 76211, this area appears to have always had enormous natural hydrological potential. It is probable that the Roman baths were sited in this exact location due to the presence of one or more natural water-yielding spots which provided the area with a perennial water supply.[78] Efforts by the local water company to secure an abundant subsurface perched aquifer water-yield from this area must have certainly contributed towards the lowering of the perched aquifer at Għajn Tuffieħa, transforming what was formerly a surface spring into a subterranean water supply. The former presence of spring water at the Għajn Tuffieħa *giardino* can perhaps be paralleled to the current hydrological situation in the neighbouring Wied ta' San Martin.[79] The presence of a natural uninterrupted water supply at San Martin valley is attested to by the presence of the *Potamon Fluviatile* species (the freshwater crab).[80]

Returning to the *giardino* ta' Għajn Tuffieħa area, a part-time farmer cultivating fields in this agricultural territory informed the present writer that up until about half a century ago, it was common practice amongst farmers working the land in an area known as *It-Tafal* to excavate a circular-shaped hole in the clayish soil. Water percolating above the impermeable clay deposit would accumulate and be subsequently utilised for irrigation purposes (Fig. 6.72).[81]

[77] The current layout of this section of the Għajn Tuffieħa *giardino* lacks a well-defined watercourse and excess rain runoff soaks the underlying arable land.

[78] This is based on personal observations and information obtained from M. Vassallo and elderly farmers who inhabit and till land in this *giardino*. T. Zammit writing in 1930 also infers that a natural water source provided the Għajn Tuffieħa bath complex with its water supply. See MAR, 1930, x. It is also significant to note that various water-related projects have since a century and a half ago all successfully tapped the perched aquifer deposits, which in this section of the Għajn Tuffieħa valley lie close to the surface.

[79] See map reference 44230 77420.

[80] In the case of the Wied ta' San Martin, the natural water source flowing through the valley bed is partly augmented by a series of at least three perched aquifer water galleries of unknown antiquity located in different key areas of this valley. The *Potamon Fluviatile* is an endemic subspecies which up to a few centuries ago had a much wider distribution base in Malta. See Sultana and Falzon 2002, 256.

[81] Personal communication by J. Camilleri, a part-time farmer and Head of Department for Mathematics with the Education Division at Floriana. The approximate map reference for the *It-Tafal* area is 41577 76212. The *It-Tafal* place-name can be loosely translated into English as meaning "The Clay".

Fig. 6.73 Detail of the entrance to the water tunnel located in the Roman Baths site at the Għajn Tuffieħa *giardino* (MG15) at Mġarr (Malta). Source: the author.

Fig. 6.74 Interior view of the initial section of the water tunnel accessed from within the Roman Baths site at Għajn Tuffieħa (MG15). The passage side walls are masonry-built in a construction method and style which emulates wet-rubble. The tunnel ceiling and floor are capped by means of ashlar slabs. Source: the author.

Fig. 6.75 Interior view of the innermost section of the water tunnel accessed from within the Roman Baths site at Ghajn Tuffieħa (MG15). Source: the author.

6.1.9.1 Water tunnel at the Roman Baths site

Personal field-research at the *giardino* ta' Għajn Tuffieħa (MG15) revealed the existence of three man-made water-related interventions in the landscape, all aimed at increasing the hydrological potential of the area. The most recent appears to date to the opening decades of the twentieth century and is accessed from within a small rectangular-shaped ashlar masonry structure positioned close to the eastern perimeter wall of the Roman baths site (Fig. 6.72).[82] This gives access to a masonry-built subterranean tunnel above the entry of which is a marble plaque bearing the date "1910" (Fig. 6.73).[83]

The tunnel interior is composed of three largely distinct segments and gives direct access to a subterranean perched aquiferous source. The outermost sector of the tunnel is ca. 37.51 m long and it has a height of ca. 1.80 m. Its width varies between 0.89 and 0.77 m. The passage side walls are masonry-built in a construction method and style which emulates wet-rubble. The initial tract of the tunnel's floor and ceiling is paved and capped with ashlar slabs (Fig. 6.74). The innermost section of the first segment of the tunnel is capped by a barrel-vaulted-type ceiling constructed out of irregularly shaped masonry (Fig. 6.75). The construction method employed in this stretch of tunnel was probably conditioned by the unstable deposits in which it was excavated. Located right in the middle of the valley floor, this section of the subterranean passageway probably tunnels through Terra Rossa soil deposits. The tunnel follows an almost linear course and gives access to a wider intermediate chamber.

[82] The water tunnel access through the Roman baths site is located at map reference 42000 78000.

[83] It appears that this water system has in recent years not been studied in detail. The gallery is driven into an Upper Coralline deposit parallel to the Għajn Tuffieħa – Żebbiegħ road with its location being determined by a surface spring which formerly originated at this point and which formerly supplied the adjoining Roman baths site with a perennial water supply. The purpose of this water tunnel was that of augmenting the St Paul's Bay and Mellieħa water supply resources. Morris, 1952, 92-93.

Fig. 6.76 The water tunnel accessed from within the Roman Baths site at Għajn Tuffieħa (MG15). Interior detail of the middle chamber, the rock walls of which are covered with calcite deposits. In the background is the starting point of the third section of this water tunnel. Source: the author.

Fig. 6.77 Water tunnel accessed from within the Roman Baths site at Għajn Tuffieħa (MG15). Detail of an almost square-shaped shaft piercing the ceiling of the middle chamber. Source: the author.

The intermediate chamber measures 2.89 m by 1.67 m and is almost completely excavated into a brittle Mtarfa Member deposit. Pick marks are visible on the exposed rock-face which nonetheless has a coloration that visually distinguishes it from the third and final section of the water tunnel (Fig. 6.67). A detailed visual examination revealed the presence of thin calcite deposits covering most wall sections within this chamber. There is a walled-up doorway in its north-eastern end, on which an inscribed graffito reads: "Angelo Bartolo 1929".[84] It is likely that Bartolo was one of the workmen entrusted with the water tunnel works and the date possibly recalls the year in which this passageway was blocked. A square-shaped vertical shaft measuring 0.56 by 0.56 m connects this section of the tunnel to the surface (Figs. 6.77, 6.78). The uppermost portion of the shaft is currently blocked, but it extends upwards for ca. 1.17 m. The rock coloration and pick markings hint that the shaft is coeval in date to the intermediate chamber in which it is located.

It is also significant to note that applied to the north rock-wall of this chamber there are traces of cream-coloured mortar, over which a dark red coloured pigmentation has been applied.[85] The best extant evidence is found at the point where the wall containing the "Angelo Bartolo" graffito touches the chamber's ceiling (Figs. 6.79, 6.80). From the little which survives, it is possible to cautiously hypothesise that this section of the tunnel was extensively decorated by means of a series of linear decorative motives, the coloration and style of which are similar to what survives in various Palaeochristian period hypogea.

This intermediate chamber in turn gives access to a neatly excavated tunnel, similarly cut into an Mtarfa Member deposit. The tunnel is ca. 0.60 m wide and 1.77 m high and extends in an eastern direction for approximately another 30 m, at the end of which are three walled-up passageways (Figs. 6.81). Modern concrete bricks were used to conceal access to these extensions, giving weight to the probability that this was done during the last 40 years or so. A steady trickle of water could be heard from behind the northern walled-up tunnel. Another graffito in this section of the tunnel reads, "AB 1928", and it probably recalls the aforementioned Bartolo.

Fig. 6.78 The water tunnel accessed from within the Roman Baths site at Għajn Tuffieħa (MG15). Detail of the almost square-shaped shaft piercing the ceiling of the middle chamber. Source: the author.

[84] It appears that between 1927-1928 work was still being carried out on this water tunnel. Morris, 1952, 93. This is also hinted at by the "Angelo Bartolo" graffito next to which is inscribed an almost coeval date.

[85] The traces of mortar and red pigmentation had until my investigation of this subterranean space gone unnoticed. D. Cardona, the site's curator, was informed of my observations.

Fig. 6.79 The water tunnel accessed from within the Roman Baths site at Għajn Tuffieħa (MG15). A close-up view of the mortar and red pigment which were observed by the author to cover sections of the rock-walls of the middle chamber. Source: the author.

Fig. 6.81 General view of the innermost section of the water tunnel accessed from within the Roman Baths site at Għajn Tuffieħa (MG15). It is possible that concealed behind the walled-up areas of this tunnel are the water collection chambers. Source: the author.

Fig. 6.80 The water tunnel accessed from within the Roman Baths site at Għajn Tuffieħa (MG15). Detail of the mortar and red pigment which were observed by the author to cover sections of the rock-walls of the middle chamber. Source: the author.

6.1.9.1.1 Interpreting the evidence

It is hypothesised that both the initial and the final tracts of this water tunnel date to the first three decades of the twentieth centuries. This subterranean tunnel cannot be technically classified as a water gallery or subterranean aqueduct as it does not channel and transport an aquiferous water source to its entrance area. By contrast, as evidenced by the three walled-up chambers in the innermost section of this tunnel, it appears that it does intercept a perched aquifer water source which was possibly pumped directly to the surface by means of one or more diesel-operated pumps.[86] It would seem that the primary function of the water tunnel was to provide ease of access to the subterranean area where the ground-water source was intercepted.

The intermediate chamber probably belongs to an earlier period and possibly formed part of a miniature catacomb which was accidentally cut through and mutilated almost beyond recognition while the water tunnel excavation works were underway. Rock-cutting in the north-east section of this chamber, mainly preserved in the ceiling area, bears a close resemblance to passageways found in other Maltese Palaeochristian hypogea. Gaining access to the walled up area behind which the Bartolo graffito is inscribed, and carrying out the analysis of a mortar sample retrieved from this section, might be more revealing in this respect. Funerary-related subterranean chambers, which on stylistic grounds were dated to Late Antiquity, are indeed present in the surrounding countryside. Located at a linear distance of 3.18 km away, Binġemma (MG19) is a case in point. At an even closer proximity to Għajn Tuffieħa is the hypogeum of Tar-Ragħad at Mġarr (Malta) as well as a number of single-celled burial chambers excavated within caves present in the Manikata and Pwales areas (Buhagiar M. 2000, 23-37; Cardona 2002, 156-58).

[86] The Mġarr (Malta) territory was connected to the power grid only in 1955.

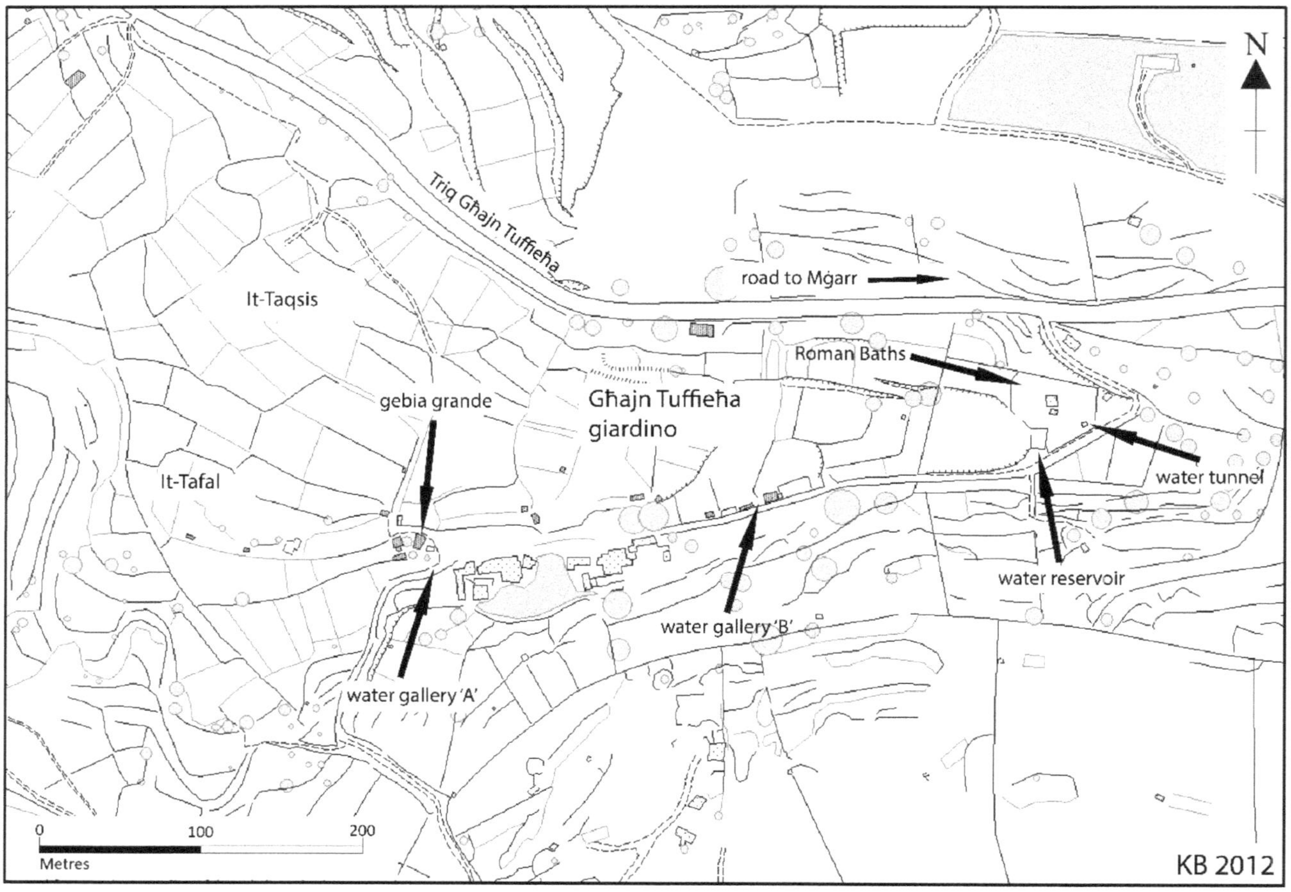

Fig. 6.82 Plan of the Ghajn Tuffieħa *giardino* (MG15) indicating the water-related features observed within. Topographical map image adapted from MEPA Map Server data.

6.1.9.2 *The perched aquifer galleries at the Għajn Tuffieħa* giardino

The other water extraction features noticed in the Għajn Tuffieħa *giardino* (MG15) consist of two perched aquifer galleries situated further downstream (Fig. 6.82, water galleries "A" and "B"). The first gallery (Fig. 6.82 "B") is located at map reference 41911 76170 and tunnels into the southern side of the *giardino* di Għajn Tuffieħa valley at a distance of only a hundred metres from the western boundary wall of the Roman Baths site. The second (Fig. 6.82, "A") is found at map reference 41704 76147 and situated only a few metres away from the *gebia grande* and appears to be closely associated with a series of Knight-period farmsteads. This water spring originates from a perched aquifer water gallery excavated into the south side of the Għajn Tuffieħa *giardino* and supplies the *gebia grande* with a perennial spring.

Farmers tilling land in this *giardino* provided me with conflicting evidence as regards the approximate dating of these water features. Two elderly farmers, who wished to remain anonymous, claimed that these two perched aquifer galleries are early twentieth-century interventions within the landscape.[87] Their version of events recounted how the excavation of these water springs was motivated by the reduction of the surface perennial water flow which originated from further upstream. Allegedly, this reduction of water supply stemmed from the excavation of the water tunnel located in the Roman baths site discussed in section 6.1.9.1.[88] A discussion on these two perched aquifer systems with J. Camilleri disclosed his belief that these must at least date to the Knights' period.[89] These are claims which necessitate further investigation, but a scrutiny of the available evidence allows some preliminary conclusions to be reached.

The water stored in the *gebia grande* (Fig. 6.71) must have been derived from either a natural water-yielding spot or from a perched aquifer water gallery. The relief and gradient of the Għajn Tuffieħa landscape is a matter which has to be given serious consideration. The channelling of a natural water source flowing through the valley bed into the *gebia grande* without the employment of water lifting machinery would inevitably involve the use of open masonry canals tapping spring water from

[87] This version of events was also corroborated by M. Vassallo, who was basing himself on evidence given to him by farmers tilling land in the Għajn Tuffieħa *giardino*.
[88] Personal communication by M. Vassallo. See also section 6.1.9.1.
[89] Personal communication by J. Camilleri.

Fig. 6.83 The entrance of perched aquifer gallery "A" (Fig. 6.82), located in the Għajn Tuffieħa *giardino* (MG15). This yields a perennial water supply that is fed directly to an open water reservoir close by which this study tentatively identifies with the *gebia grande* mentioned in NLM. Treas. B. 289, f. 90. Source: the author.

Fig. 6.84 Interior view of perched aquifer gallery "A" (Fig. 6.82), located in the Għajn Tuffieħa *giardino* (MG15). Water is currently being channelled out of this water system by means of a cast iron pipe, possibly installed at the turn of the twentieth century. Source: the author.

further upstream. The point of origin of this source would approximately coincide with the location of the Roman baths site. Should this have been the case, it would signify that the *gebia grande* was located ca. 400 m away from the water source which a number of farmers claim to have originally irrigated the Għajn Tuffieħa *giardino*. Personal experience supports the view that it is unlikely that the principal water storage and distribution tank, the *gebia grande*, would be located almost half a kilometre away from the source giving life to this valley system.

Farmsteads bearing the coat-of-arms, probably of Grand Masters Alof de Wignacourt (1601 – 1622)[90] and Ramon Perellos Y Roccaful (1697 – 1720), are pointedly located at about 20 m away from a perennial source originating from within a perched aquifer gallery (map reference 41704 76147; Fig. 6.82 "A"). The farmstead location strongly favours the argument that, in the seventeenth century, this was the principal source feeding this section of the Għajn Tuffieħa *giardino* with a perennial water supply. Considering that water is one of the primary variables for any successful settlement, it is possible that it was this water presence in this particular point which actually conditioned the location and construction of these farmsteads. It is also significant that the settlement of the Għajn Tuffieħa *giardino* probably predates the seventeenth century. A farmer still inhabiting this *giardino* confirmed the present writer's suspicion that a number of seventeenth-century farmsteads were actually built in front of a number of caves, some of which could have been utilised for the purpose of human habitation.[91] If this proves to be the case, settlement development at the Għajn Tuffieħa *giardino* closely parallels that of the Razzett tal-Qasam (MG18) close by, located within the northern boundary of the Għajn Tuffieħa *feudo*.

Water gallery "A" (Fig. 6.82) is located at a distance of only 25 m from the *gebia grande* (Fig. 6.83). Dry-stone walls line the initial sections of this gallery, which is roofed over by means of a low vaulted ceiling (Fig. 6.84). The investigation of the outermost section of this water gallery revealed the presence of a canal embedded within its floor, the purpose of which was to facilitate and enhance water transportation to the gallery's exterior. This canal is no longer in use and is at present concealed by sediment accumulation. The current setup involves the transportation of water to the gallery's exterior through the employment of a cast iron pipe (Fig. 6.84) probably installed in this gallery at the turn of the twentieth century.[92] Furthermore, on stylistic grounds, this gallery typologically conforms to others which, on the basis of this study, can be dated to at least the seventeenth century. The masonry canal system connecting the gallery to the *gebia grande* and other areas of the Għajn Tuffieħa *giardino* appears to date to the same period as the open water reservoir itself.

[90] An accurate identification of one armorial shield was not possible. Being almost identical to the insignia of Grand Master Adrien de Wignacourt, who was Grand Master in between 1690-1697, the armorial shield at the Għajn Tuffieħa *giardino* lacks the necessary detailing to help distinguish between the two Grand Masters.

[91] The presence of caves in this area can be elucidated from the toponomastic evidence. The place-name for the area immediately behind the Knights' period farmsteads is known as *Ix-xagħara Ta' Għar is-Sinia*, which incidentally also indicates a water presence. See Survey Sheet 45, School of Military Engineering Chatham, 1903.

[92] Similar cast iron pipes were installed elsewhere in Malta towards the end of the nineteenth century or in the first two decades of the twentieth century. In the case of perched aquifer gallery "A" at Għajn Tuffieħa, the cast iron pipe installation within indicates that this was one of the spring water sources being channelled towards the Għajn Tuffieħa military camps and the St Paul's Bay and Mellieħa settlements. See Morris 1952, 92-93.

Fig. 6.85 Exterior view of perched aquifer gallery "B" (Fig. 6.82), located in the Għajn Tuffieħa *giardino* (MG15). Source: the author.

Fig. 6.86 One of the uncovered reservoirs adjoining perched aquifer gallery "B" (Fig. 6.82), at the Għajn Tuffieħa *giardino* (MG15). Source: the author.

The other perched aquifer gallery (Fig. 6.82 "B") is located at map reference 41911 76170. Even though this consists of a rock-cut tunnel the purpose of which was to extract a perennial water supply from the perched aquifer, it is typologically distinct from any of the other perched aquifer galleries investigated so far. Large ashlar blocks of masonry were utilised as door jambs. These in turn support another large masonry slab which functions as a lintel (Fig. 6.85). The entrance to the gallery is also wider and has a more regular nature than the other galleries included in this study. The same applies to the gallery's interior, which is lined with neatly quarried Upper Coralline Limestone ashlar masonry slabs (Fig. 6.87).

Considering the size of the ashlar masonry utilised in its entrance section and in parts of the interior, it is tempting to propose that this was recycled stone material lifted from the Roman baths site in the neighbourhood. It is indeed likely that this gallery might date to the opening decades of the twentieth century. This is attested by the use of concrete on its exterior which seemingly integrates with the structural integrity of the outer section of this water extraction system. The use of ashlar masonry slabs lining both the exterior and the interior of this gallery likewise hints at an early twentieth-century date.[93] The canals transporting water from this gallery to fields in the upper section of the *giardino* Ta' Għajn Tuffieħa also appear to be of a more recent date than those related to the previously described gallery located further downstream (Fig. 6.82 "A"). The same applies to a series of open-air water reservoirs flanking either side of this water gallery (Fig. 6.86). These are certainly of a much more recent date than the reservoir which this study tentatively identifies with the *gebia grande*.

The entry for the Għajn Tuffieħa territory describes the existence of the *fontana faura*, which conforms with that of an area still known by the place-name of *Fawwara* (NLM. Treas. B. 289, f. 91; MG11; Fig. 6.70, Zone 3; Fig. 6.88). A scrutiny of Malta Survey Sheet 26 for this area, dating to 1903, disclosed the presence of a fountain located at the approximate map reference of 41492

Fig. 6.87 Interior view of perched aquifer gallery "B" (Fig. 6.82), located in the Għajn Tuffieħa *giardino* (MG15). Source: the author.

[93] Water is currently being channeled out of this gallery by means of a series of cast iron pipes identical to those installed inside the nearby water gallery "A" (Fig. 6.82).

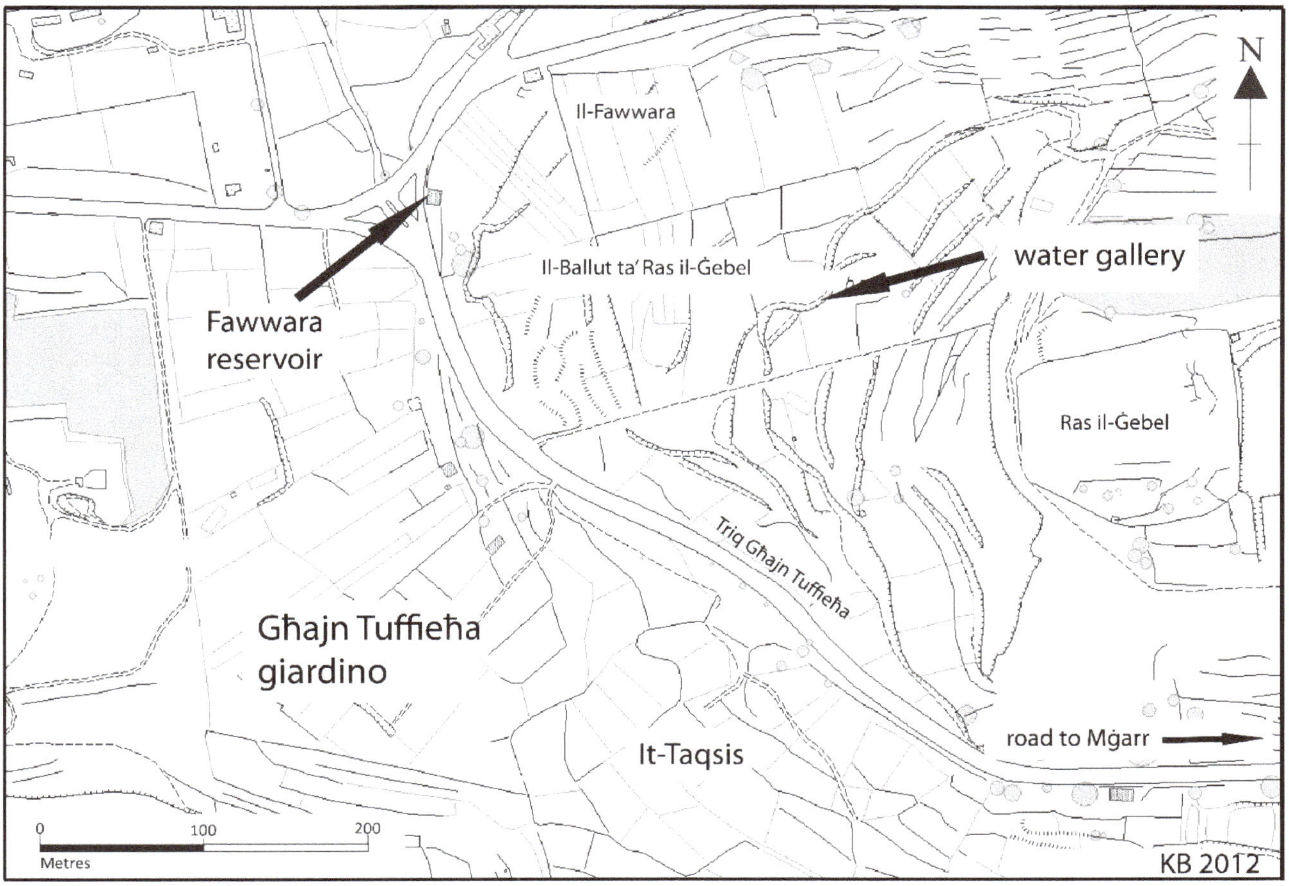

Fig. 6.88 Plan of the Fawwara (MG11) and the Il-Ballut ta' Ras il-Ġebel areas which adjoin the north-east side of the Għajn Tuffieħa *giardino*. Topographical map image adapted from MEPA Map Server data.

76668.[94] Farmers cultivating land in the area maintain that the Fawwara fountain used to receive surplus water from the Għajn Tuffieħa *giardino* (Fig. 6.92).[95] Personal field-research, however, revealed the presence of another perched aquiferous source at map reference 41655 76583, at an altitude of ca. 74 m above sea level (Fig. 6.89). This could have likewise supplied the Fawwara fountain with water.

The nineteenth-century Descriptive Plans of the Crown Property *cabrei* for the Għajn Tuffieħa *giardino* mention the presence of: (a) irrigation canals and other *condotti* for the conveyance of water; (b) *vasca*; (c) *tre fonti d'acqua sorgente*; (d) *tre gebiette*; and (e) *scavo in ricerca d'acqua* (NAM. Descriptive Plans of the Crown Property in Malta, 1866, ff. 209-13; Fig. 6.91). The location of the three springs is not indicated on the accompanying plan of the Għajn Tuffieħa *giardino*, but their approximate location can be probably ascertained by the positioning of the three reservoirs, all of which are indicated in the 1866 plan. The canalisation system which conveyed water from the reservoirs to the various

Fig. 6.89 The modern brick structure in the background probably conceals access to a perched aquifer gallery in the Ballut ta' Ras il-Ġebel area. Source: the author.

[94] See Survey Sheet 26, School of Military Engineering, Chatham, 1903. This water source is captioned "Fauara". The location of *fontana faura* was also indicated to me by M. Vassallo. Masonry canals which fed this tank with water were almost completely destroyed with the widening of the adjoining Għajn Tuffieħa road.
[95] Personal communication by M. Vassallo.

cultivable areas of this *giardino* is likewise indicated.[96] The positioning of the nineteenth-century water springs appears to correspond to those detected during the field survey of the Għajn Tuffieħa *giardino* area carried out by the present writer. These include water galleries "A" and "B" (Fig. 6.82) as well as the natural spring flowing through the upper reaches of the Għajn Tuffieħa *giardino* where the Roman baths site is located. The setup of this spring was modified in 1910 with the excavation of the water tunnel described in section 6.1.9.1 (NAM. Descriptive Plans of the Crown Property, 1866, f. 209), and might be tentatively associated with the *scavo in ricerca d'acqua* mentioned in the late nineteenth-century Crown Property documentation for this *feudo*.

6.1.9.3 The Ir-Razzett tal-Qasam area at Għajn Tuffieħa

The description of the northern part of the Għajn Tuffieħa territory (MG15) mentions the presence of a cluster of caves excavated into an Mtarfa Member deposit at map reference 41199 77597.[97] This is an area which falls immediately outside the boundaries of the recently declared *Il-Majjistral* nature park and is referred to by the inhabitants of the neighbouring Manikata settlement as *Ir-Razzett tal-Qasam* (MG18; Figs. 5.68, 6.70 "Zone 2", 6.94).[98] Fronting the caves is a large open space which is reminiscent of a *misraħ* or gathering point, and indeed appears to have functioned as the focal node of this settlement. The area fronting each of the various cave units was, probably during the Early Modern period, partitioned into a series of rectangular-shaped enclosures by means of dry-stone wall structures (Fig. 6.93).

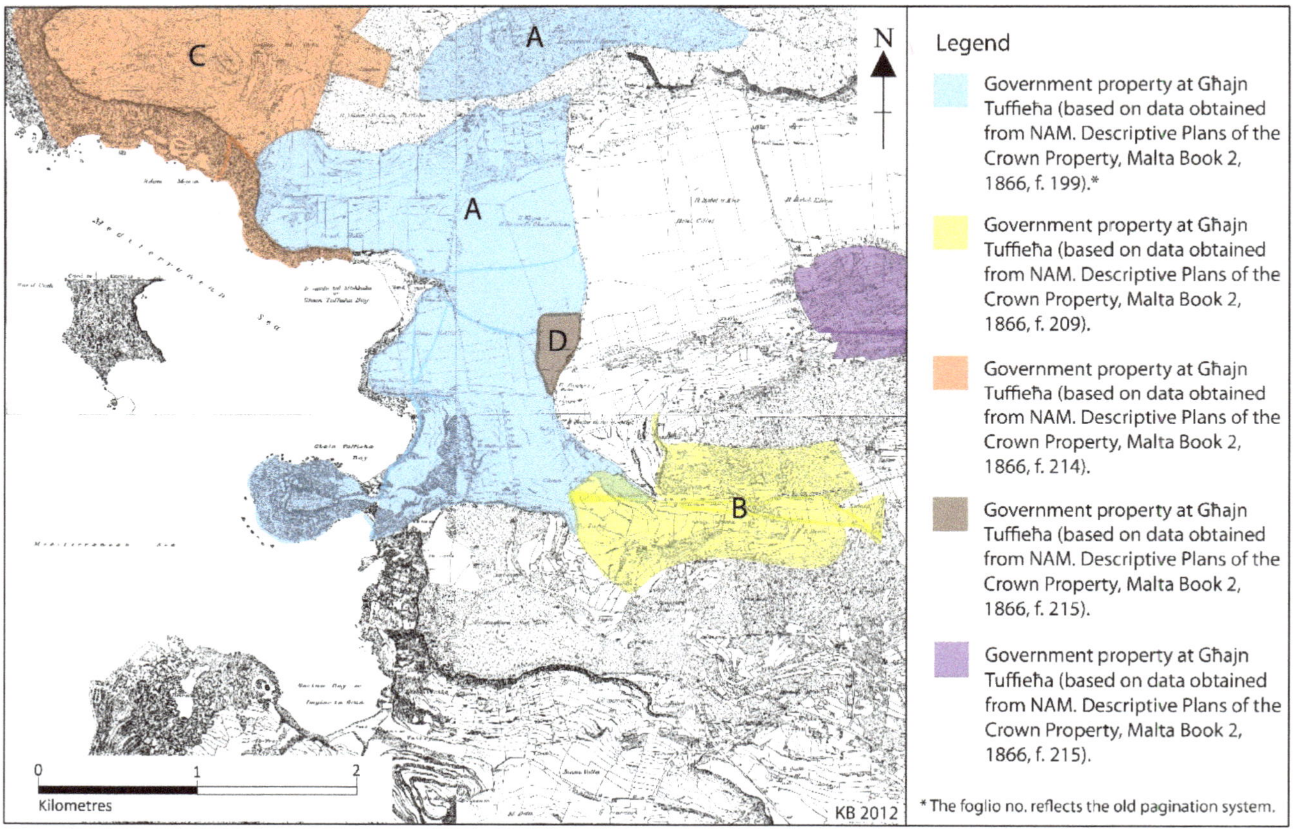

Fig. 6.90 Plan showing the extent of the Għajn Tuffieħa tenement (MG15) in the nineteenth century. Source: NAM. Descriptive Plans of the Crown Property in Malta, ff. 119, 209, 214, 215.

[96] As specified in chapter one, sections 1 and 1.1, water distribution and allocation for agricultural usage falls outside the scope of this publication. As a consequence, the mapping of the water distribution systems in this "hydraulic space" will only be taken up during future research work in the Għajn Tuffieħa area.

[97] [...] *Di la il detto fego confina da tramontana con il spasio publico in ui e* [?], *et insino che s'arriua al valone sopra le grotte del detto fego in canne 420* [...]. See NLM. Treas. B. 289, f. 93.

[98] The Razzett tal-Qasam has been managed by the Koperattiva Rurali Manikata since 2008. See http://www.manikatafarmers.com/historic_farmstead.html.

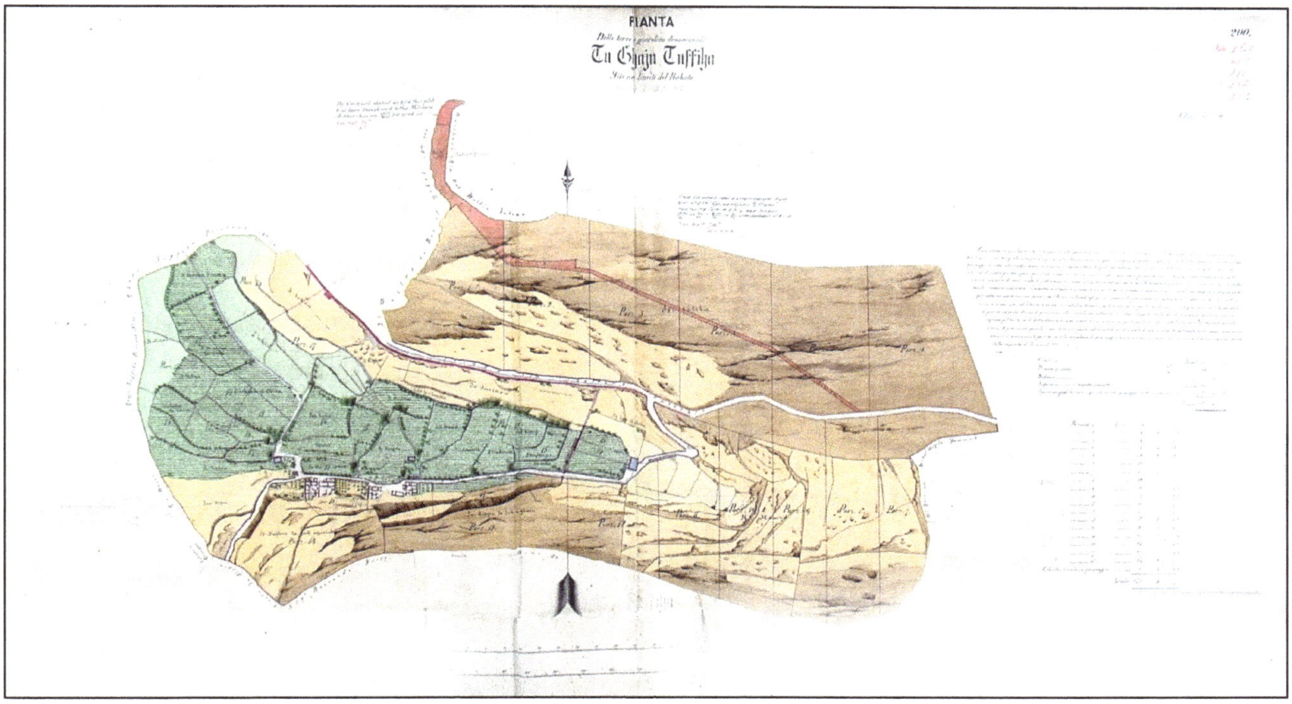

Fig. 6.91 NAM. Descriptive Plans of the Crown Property in Malta, 1866, ff. 209-13. Plan for the Għajn Tuffieħa *giardino* (MG15) in the territory of Mġarr (Malta).

Fig. 6.92 An uncovered water reservoir in the Fawwara (MG11) area of the Għajn Tuffieħa *feudo* (MG15). This water storage tank is no longer in use and lies in a state of disrepair. Source: the author.

The caves are today fronted by a cluster of rural farmsteads, which the Maltese language identifies as *razzett* (singular) or *rziezet* (plural). These vernacular structures are difficult to date, but in the case of the *Razzett tal-Qasam* setting, they possibly originated during the course of the seventeenth century or even later. Their construction was perhaps motivated by the abandonment of the cave units as the place of habitation. It might be surmised that other Maltese troglodytic sites were during this period experiencing a similar desertion process with the vacated caves being reutilised as animal pens and/or storage recess for agricultural produce.[99] It is equally possible that some of the smaller caves in the north and north-west sections of Malta were, during this period, also converted into apiaries (Buhagiar K. 2008B, 14).

Fig. 6.93 The Razzett tal-Qasam (MG18) settlement at Għajn Tuffieħa (MG15), Malta. General view of the dry-stone wall enclosures fronting several of the caves present in this area. Source: the author.

[99]See also Luttrell 1979, 464, who likewise supports this sequence of events.

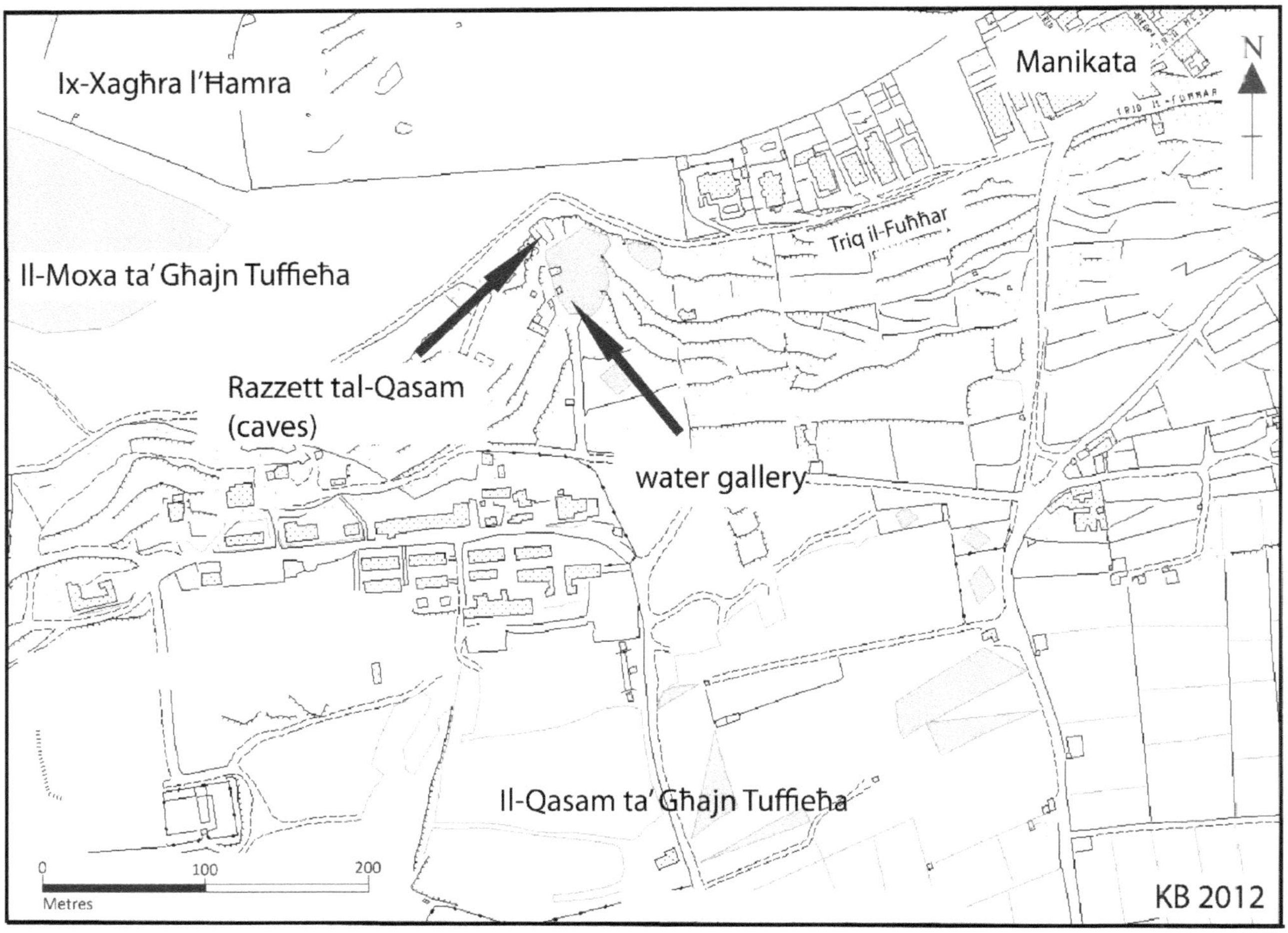

Fig. 6.94 Plan showing the location of the Razzett tal-Qasam (MG18) area in the territory of Għajn Tuffieħa (MG15), Malta. Topographical map image adapted from MEPA Map Server data.

Several of the caves at the *Razzett tal-Qasam* remained in use until the pre-World War II period (Cardona 2002, 157), but many of these are actually multi-period sites. Two bi-partite window tombs located in one of the caves are indicative of the site's usage as a place of burial during the Palaeochristian period (Cardona 2002, 157-58), and possibly contained several small catacombs which were destroyed during a later occupation of the area.

The *cabreo* 289 description of this site also refers to a *reparatione particolarmente della mina e beueratura sotto li grotte del detto fego* (NLM. Treas. B. 289, f. 94). This mention of a water source at the *Razzett tal-Qasam* is an extremely significant entry. It describes a still extant water gallery which tunnels directly into the area below the settlement's *misraħ* and caves (Fig. 6.95).[100] The gallery no longer yields a water supply and access to it was restricted by means of a masonry wall blocking access to its interior. It was also fronted by an open-water storage reservoir from which water could be channelled to agriculturally productive land which needed to be irrigated by means of this gravity-fed water system. The fact that maintenance works were carried out on the water gallery system would indicate that this belonged to an earlier period. The discovery of a perched aquifer water source at this point was probably a determining factor in conditioning the location of this troglodytic settlement during the Early Modern and earlier periods.

6.1.10 *The Buskett and Ġnien il-Haggel areas in the territory of Rabat (Malta)*

The *cabreo del Magistero* 289 documentation of the *Bosco* (Buskett – RA6) property makes reference to the existence of a *gebia noua* (the new water reservoir), and *la fontana grande* (the large water spring or fountain). The location of

[100] Further information on the Razzett tal-Qasam area of the Għajn Tuffieħa *feudo* (MG15) is included in the Descriptive Plans of the Crown Property *cabrei* plan entitled *Pianta del Terittorio denominato Ghain Tuffieha* and which date to the nineteenth century (Fig. 6.90 "A"; Fig. 6.96). The legend describing the hydrological assets of this territory mentions: (a) *linea di canali*; (b) *un altro canale*; (c) *tre cisterne d'acqua piovana*; *tre gebiette*; *due fontane d'acqua sorgente di una delle quali ne ha il diritto di avere la congiunta Clausura detta 'Il Fauara' un giorno ogni quindici giorni*; and (f) *una cisterna d'acqua piovana*. See NAM. Descriptive Plans of the Crown Property, 1866, f. 199. Even though unspecified, the springs in question probably refer to the one tunnelling below the Razzett tal-Qasam settlement and the water source feeding the Il-Fawwara area (MG11).

Fig. 6.95 The Razzett tal-Qasam (MG18) settlement at Għajn Tuffieħa (MG15), Malta. The arrow indicates the entrance to the water gallery mentioned in the NLM. Treas. B. 289, f. 94 documentation. Source: the author.

Fig. 6.97 The nymphaeum at Buskett (RA6). The small opening indicated above is one of the access points to the perched aquifer gallery which originally supplied this fountain with a water supply. Source: the author.

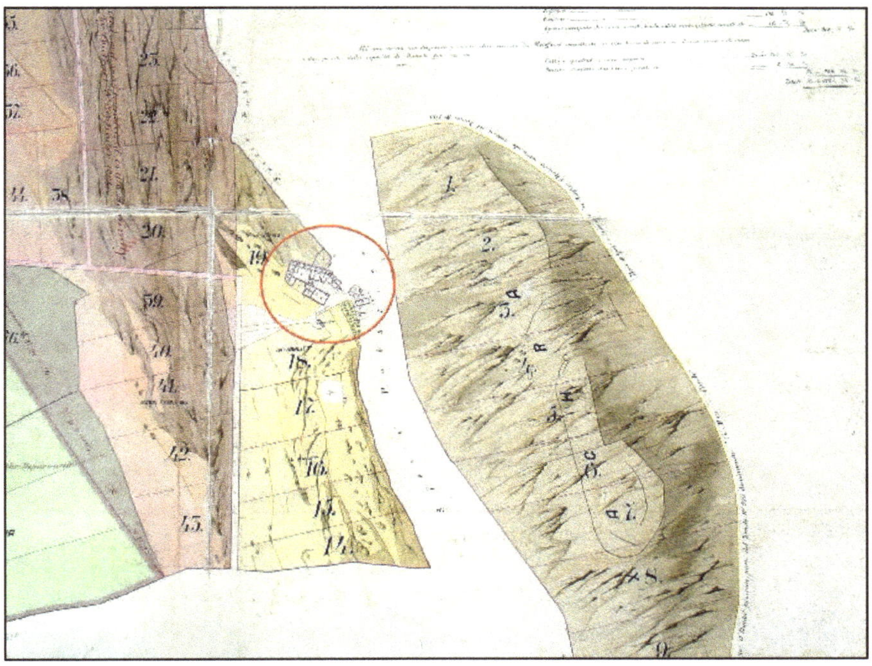

Fig. 6.96 NAM. Descriptive Plans of the Crown Property in Malta, 1866, f. 204. Circled in red is the Razzett tal-Qasam area (MG18) at Għajn Tuffieħa (MG15), Malta.

the *gebia noua* remains unknown (NLM. Treas. B. 289, ff. 101-3). *La fontana grande* can be tentatively associated with the current monumental fountain, located at map reference 45610 68348, right next to the principal access point to Buskett (Figs. 3.6, 6.97, 6.98). This originates from a perched aquifer water gallery adjoining one of the sides of the fountain.[101] Another reference to this spring can be found in the section of *cabreo* 289 describing the *Clausura tal Vueta.*[102] This source appears to be currently yielding only a seasonal flow of water. The document also refers to a *gebia grande* (a large water reservoir), known as *il mare*, perhaps due to its large size, the location of this water reservoir is also unknown (NLM. Treas. B. 289, f. 108).

Understanding the hydrological landscape evolution at Buskett is a complicated task. During the nineteenth and

[101] Personal communication by an agricultural division employee who works at Il-Buskett and who in the past was involved in the carrying out of maintenance works in the initial tracts of the gallery passageway.

[102] This reads: [...] *et insino che s'arriva alispatio della gran porta del bosco e della casa sopra la fontana ui sono canni 97*. See NLM. Treas. B. 289, ff. 112-13.

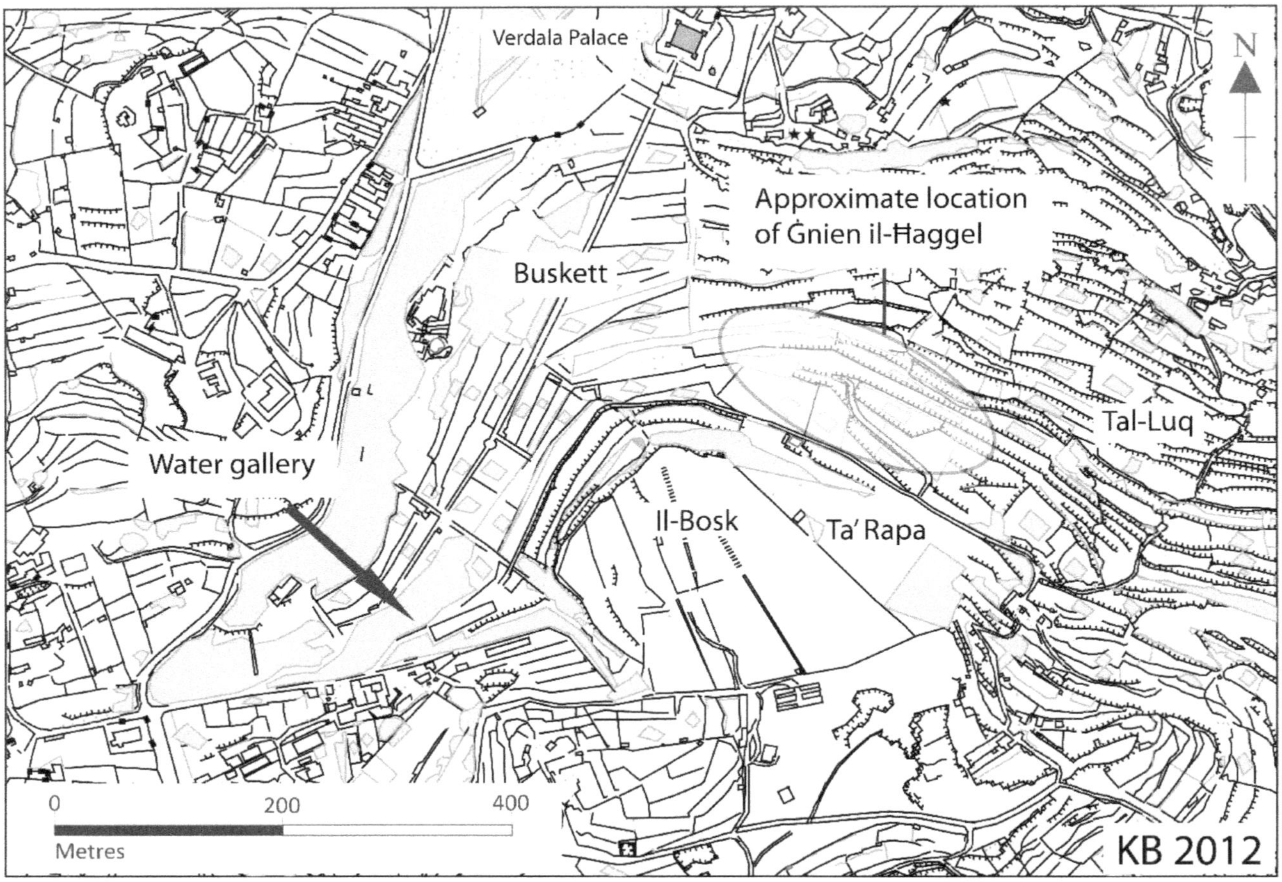

Fig. 6.98 Plan of the Buskett area (RA6) in the territory of Rabat (Malta). The water gallery indicated above has in this study been tentatively associated with the *Fontana Grande* mentioned in NLM. Treas. B. 289, f. 101-3. Topographical map image adapted from MEPA Map Server data.

early twentieth centuries, the Buskett and the adjoining Wied il-Luq areas were intensively exploited. The hydrological landscape of these areas was largely modified in an attempt to harness the potential of the perched aquifer ground-water source within. Various water tunnels were excavated, with many of these being marked on the 1919 Malta Survey Sheets for this area.[103] Also represented is a British period aqueduct which channels the extracted water towards the Girgenti area in the territory of Siġġiewi.[104] British period hydrology-related interventions in the Buskett area are discussed separately in chapter four, sections 4.2 and 4.3.

The exact location of *Ġnien il-Haggel* (NLM. Treas. B. 289, ff. 106-12; RA12) is not known, but it is possible that this was located on the north-facing flank of Buskett and touched agricultural land known as Ta' Rapa (RA48; Fig. 6.98).[105] *Ġnien il-Haggel* was divided into a *Parte Palombara*, *Parte ta Nofs* and a *Parte Soprana*. Within the *Ġnien il-Haggel* territory was a water reservoir known as *gebia negra*, and there was also a *fontanella* – a possible reference to a small spring (NLM. Treas. B. 289, ff. 108-12). The mention of a nearby *gebia fonda* also highlights water availability in this area (NLM. Treas. B. 289, f. 109). It appears that *Ġnien il Haggel* was at least in part used for fruit tree cultivation, further stressing the availability of a perennial water-yield in this estate.

6.1.11 Water sources on the Island of Gozo

The *cabreo 289* manuscript section for Gozo portrays a hydrological situation similar to the one in Malta. *Cabrei* entries for *fontana* or *fontane* amount to a little over a handful and show how these are similarly located within *giardino* frameworks. In the case of Gozo, it appears that springs are mostly mentioned as a principal landmark within the territory, especially when demarcating

[103] See Survey Sheet 101, School of Military Engineering, Chatham, 1919.

[104] The extensive hydrology-related exploitation that the Buskett area was subjected to since the British period might partly account for its reduced ground-water potential.

[105] Ta' Rapa is the place-name of an area above the southern flank of Buskett valley located at the approximate map reference of 46195 68481.

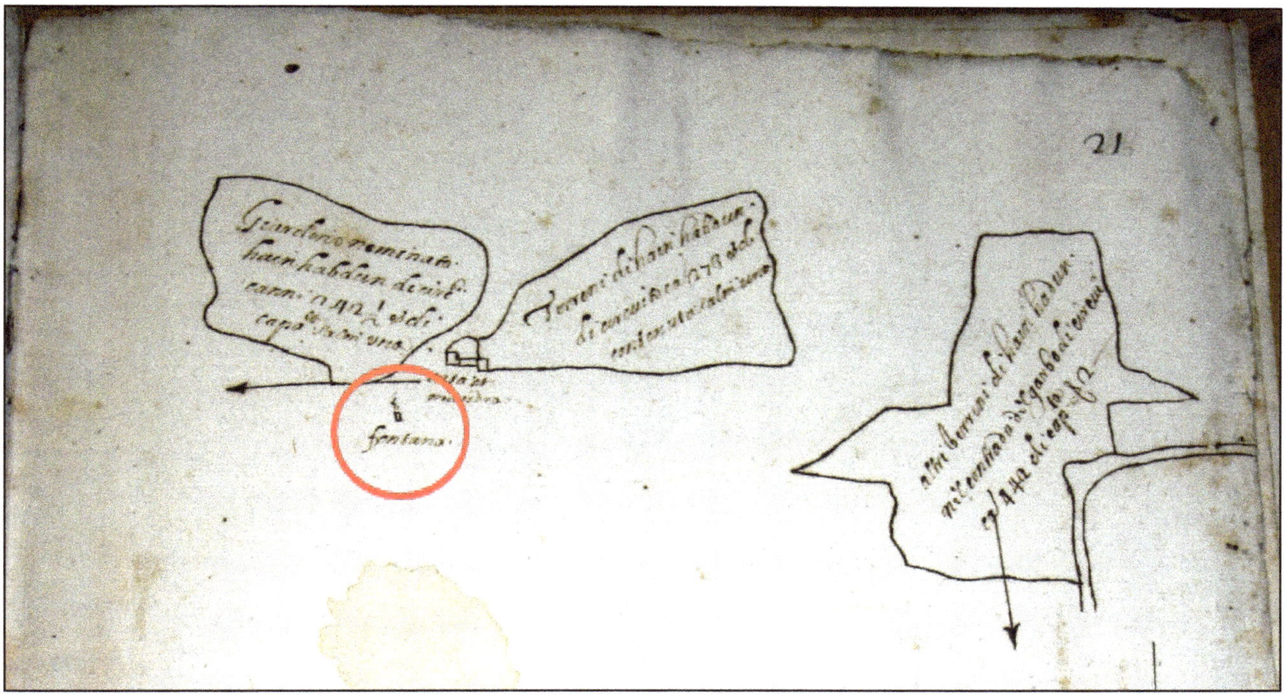

Fig. 6.99 NLM. Treas. B. 289, Plan 21. Plan showing Government-owned territory at Għajn Għabdun (GO8), Gozo. The circle indicates a rectangular feature which on the plan is marked as *fontana.*

properties. As is the case with Malta, barely any information is given on the physiognomy of the water springs in question. The relevant water-related entries for Gozo make mention of the *hain Catet* (GO15) spring, present in the *Fego della taflia* (NLM. Treas. B. 289, ff. 121-23; GO15),[106] and the *Giardino delle fontane* in the entry for the *Fego ta Ghamet,* in which there is a reference to a cistern.[107] Another entry makes a passing reference to *Ghajn Melech* (NLM. Treas. B. 289, ff. 142-43; NLM. Treas. B. 289, Plan 21.), which probably conforms with the territory of Għajn Mielaħ in the locality of Marsalforn, the first known recorded instance of which dates back to 1574 (Wettinger 2000, 187).[108] There is only one instance where direct reference is made to the presence of a water cistern.[109] In the description of the *Fego ta Meimun* (GO3), passing reference is made to the *Fontana Grande,*[110] which can possibly be associated with land to the south-south-east of Rabat (NLM. Treas. B. 289, ff. 138-39).

The *cabrei* 290 and 291 manuscript entries for Gozo are more informative, and in general corroborate further on the data given in *cabreo* 289 which are discussed in section 6.1.11. Property plans and their respective descriptions offer evidence for the presence of cisterns, wells and springs. With the exception of data for water-yields in the case of the Għajn Qasab (GO12) and Għajn Tuta (GO21) springs (NLM. Treas. B. 291, f. 22; NLM. Treas. B. 291, f. 37), almost no additional information is given on the physiognomy of the mentioned well and spring sources other than what was recorded in the *cabreo* 289 documentation.

In *cabrei* 290 and 291 documentation, cisterns in Gozo are recorded within the *Territorio ta Kamet* (GO18), a locality at Xewkija and a couple of other parcels of land in its neighbourhood (NLM. Treas. B. 290, f. 114; NLM. Treas. B. 291. f. 35.).[111] Other cisterns were to be found in the *Clausura ta Porterial* (NLM. Treas. B. 290, f. 126), the *Feudo tal Hamria* (NLM. Treas. B. 290, f. 120), *Feudo Gnien Issultan* (NLM. Treas. B. 290, f. 121), *Feudo San Georgio* (NLM. Treas. B. 290, f. 128), *Territorio di Hain Mihelhel* (NLM. Treas. B. 290, f. 132) and *Ras il Wilgia* (NLM. Treas. B. 290, f. 133; NLM. Treas. B. 291, f. 38). All these entries succeed in showing the importance of water storage tanks as an aid to agriculture in areas where perhaps springs and perched aquifer water galleries were not available.

[106] Ghajn Qatet is a district in Gozo located to the south-east of Rabat. See Wettinger 2000, 188.

[107] This reads: [...] *confina con la strada che uiene dal Giardino delle fontane, et ua à ta xauchia infontespicio delle clausure d'Antonio Apap e Lucio Apap, e d'un spatio doue é la gisterna del fego.* See NLM. Treas. B. 289, ff. 126-27.

[108] The whereabouts of this area is unknown, but possibly conforms to the territory of *Ghajn Melel* (GO10) at Żebbuġ (Gozo), located at map reference 31520 92936.

[109] The NLM. Treas. B. 289 volume in which this plan is located is not numbered. The plan contains information of the *Fego della Taflia* and its adjoining areas. The cistern is located in close proximity to the boundary wall of the *Clausura* owned by Andrea Rapa.

[110] See map reference 32285 87913.

[111] See also NLM. Treas. B. 290, ff. 115-16, which indicate the location of these same cisterns. Reference to a cistern at *Fego ta Ghamet* was already made in the *cabreo* 289 for this fief. See NLM. Treas. B. 289, ff. 126-27.

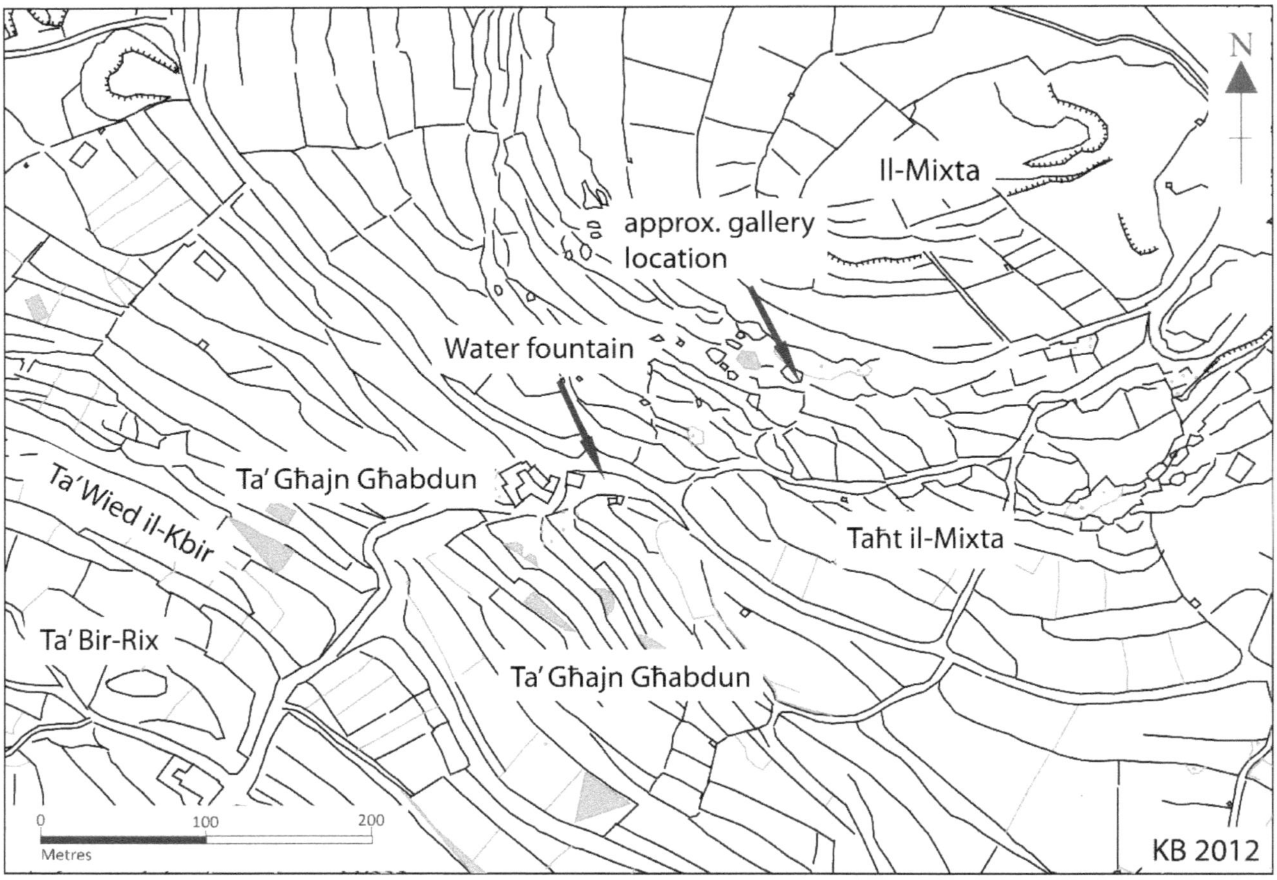

Fig. 6.100 Plan of the Għajn Għabdun (GO8) area in Gozo, showing the roadside fountain and the approximate entry point for the water gallery which feeds this with a perennial water supply. Cliff face detachment has made the Għajn Għabdun gallery inaccessible. Topographical map image adapted from MEPA Map Server data.

There is only one instance where specific mention is made of a *mina*, or water gallery. This can be found in the *cabreo* 291 entry for the *Feudo della Madia* (GO16) at Mġarr in Gozo (NLM. Treas. B. 291, f. 21). In other instances, even if the name of a listed property is water-related, there is no mention of springs or other associated hydrological assets in the respective *cabreo* plans and text entries concerning that particular landholding. A case in point is the *cabreo* entry for *Hain Hosna* (NLM. Treas. B. 290, f. 111; NLM. Treas. B. 291, f. 35),[112] located in the eastern section of the Xagħra settlement and which adjoins Ġnien Xibla.

6.1.11.1 The Għajn Għabdun spring

In the description of the *Fego di San Georgio* estate, there is a passing reference to the *Hain Habdun* (Għajn Għabdun – GO8) spring (NLM. Treas. B. 289, f. 134; NLM. Treas. B. 289, ff. 148-49), which at present originates from a perched aquifer water gallery. A plan for the *Hain Habdun* property in this area shows a tract of land bisected by a road, along the southern flank of which are farmsteads, marked as *casa et mandra* (NLM. Treas. B. 289, Plan 21; Figs. 2.17, 6.99).[113] A rectangular feature to the north-west is marked as *fontana*. This setup possibly conforms to the location of a group of rural structures and a water basin located at the approximate map reference of 28597 89515 alongside *Triq ta' Għajn Abdul* (Fig. 6.100). The water source feeding this fountain originates from a perched aquifer gallery located further upstream at the approximate map reference of 28709 89565. Due to the natural process of cliff-face detachment, the gallery is today inaccessible, but it still yields a perennial water supply.[114] The antiquity of this water source is not known, but it was possibly a principal factor conditioning the occupation of a series of caves excavated into an Mtarfa Member deposit into the side of a ravine at Il-Mixta. In the *Real Cancelleria* manuscript dating to 1373, there is already the mention of a *fons*

[112] The first known recorded instance for Għajn Ħosna dates back to 1373. See Wettinger 2000, 184.

[113] The land boundaries for these properties could not be located with certainty on a modern 1:2,500 scale map of Malta.

[114] Personal communication by a farmer cultivating land in the Għajn Għabdun area.

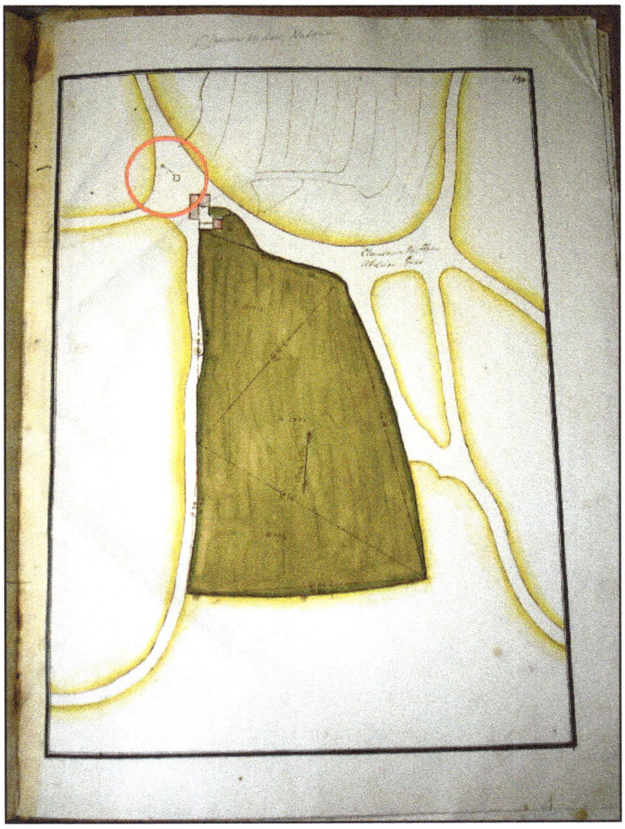

Fig. 6.101 NLM. Treas. B. 290, f. 130. The *Clausura ta Hain Abdun* (GO8) in Gozo. The red circle indicates a water-related feature, possibly representing a subterranean canal feeding an uncovered reservoir with water. This probably originates from within the Għajn Għabdun perched aquifer gallery located further uphill.

abiduni (Wettinger 2000, 183). The *cabreo* 289 entry for the Għajn Għabdun *giardino* also makes reference to a water reservoir, houses, canals transporting water into the reservoir, caves, bakery and animal pens (NLM. Treas. B. 289, ff. 148-50).[115]

Cabreo 290 and 291 documentation describes the water assets of Għajn Għabdun as follows: [...] *nella quale ci ua l'aqua della Fontana Hajn Habdun, che nel mese di Novembre 1720 [?] corse in una [?] palmi cubi 50* (NLM. Treas. B. 291, f. 38). Despite the inability to decipher the section containing data concerning the water-yield of this spring, it is pertinent to note that water was probably brought over to the *Orto Hain Abdun* by means of a subterranean canal, connected to the above-mentioned perched aquifer gallery (Fig. 6.101).[116]

[115] Only few surviving remnants account for the original early eighteenth-century water distribution system. As water distribution and allocation falls outside the scope of the present study (see chapter one, sections 1 and 1.1), this will be only taken up during the course of future research work in the Għajn Għabdun area.

[116] The hydrological setup described in *cabreo* 290 is also confirmed by nineteenth-century documentation. See NAM. Descriptive Plans of the Crown Property, 1861, ff. 113-14.

Fig. 6.102 View of the Għajn Qasab territory (GO12) in Nadur, Gozo. Dense reed growth indicates the presence of a perennial water source. Slightly further downhill is Wied Għajn il-Qasab, one of the tributaries of Wied ir-Ramla. Source: the author.

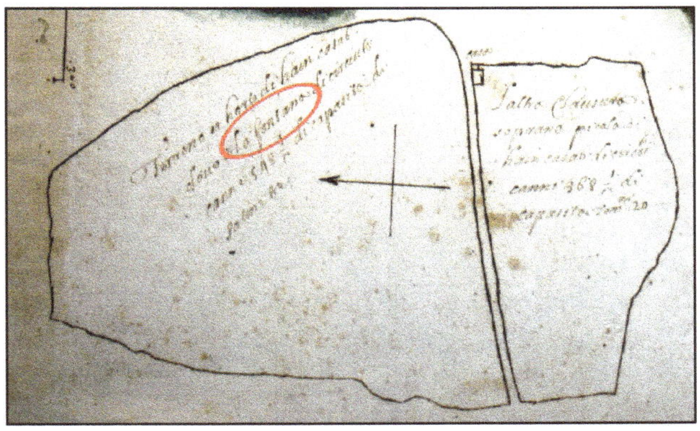

Fig. 6.103 Plan showing the approximate land boundaries of the Government-owned estate of Għajn Qasab (GO12) in the territory of Nadur, Gozo. Topographical map image adapted from MEPA Map Server data.

6.1.11.2 The Għajn Qasab territory

The *Terreno di hain Casab* is a parcel of land on the Nadur side of the fertile Ramla Valley in Gozo (GO12; Figs. 2.17, 6.102; 6.104). The *cabreo* description for the *hain Casab* territory does not contain any reference to water sources present within (NLM. Treas. B. 289, f. 141). The accompanying plan for this territory, however, does give an indication as to the presence of a water source (NLM. Treas. B. 289, Plan 15; Fig. 6.103). It consists of two parcels of land bisected by a country lane, with a *fontana*, or spring, present in the northern most field.

The *cabreo* 290 for the Għajn Qasab territory is more informative. Apart from confirming that the *fontana* was

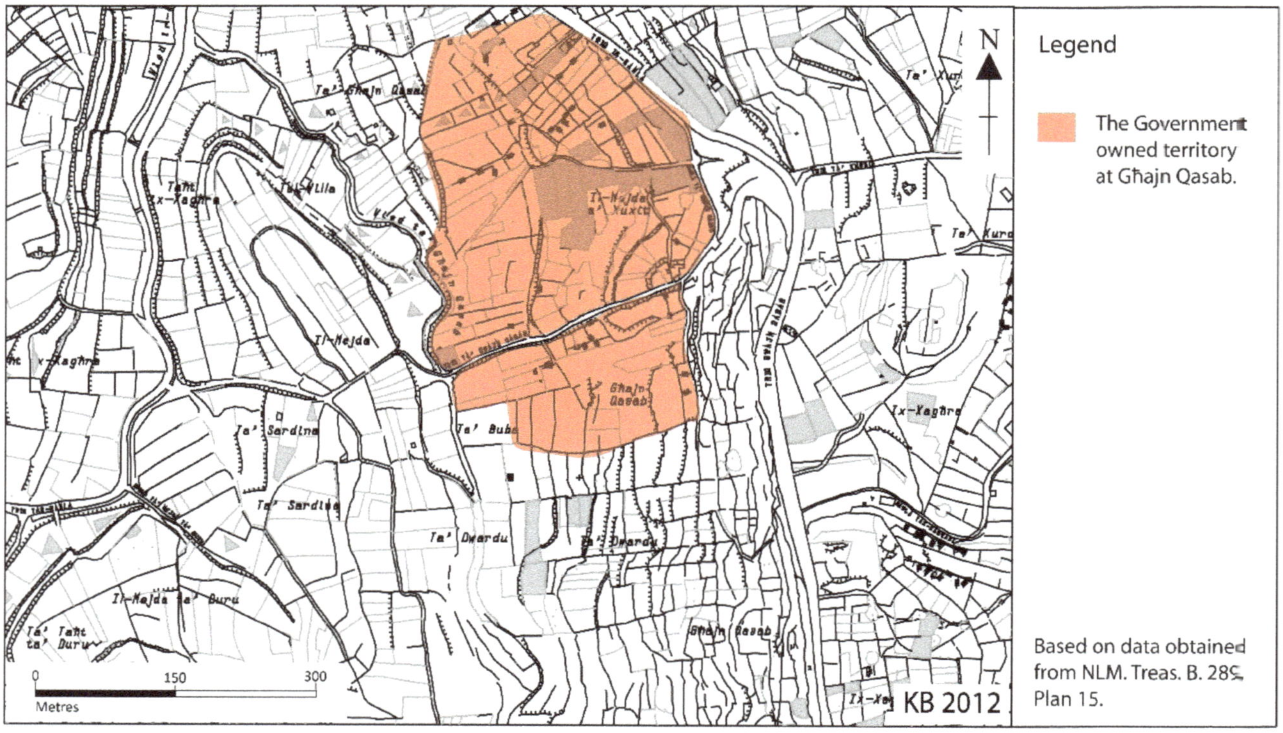

Fig. 6.104 Plan showing the approximate land boundaries of the Government-owned estate of Għajn Qasab (GO12) in the territory of Nadur, Gozo. Topographical map image adapted from MEPA Map Server data.

located within the northernmost field at Għajn Qasab, it also gives a glimpse into the water-harvesting strategies being practised in the area (Figs. 6.105, 6.106). A series of water canals is shown to connect the *fontana* to two open water reservoirs located slightly further downstream. The relevant text entry in *cabreo* 291 describing Għajn Qasab reads: *Il territorio ò sia finsario(?) di Hajn Casap, e diuiso in due [?] da una strada pubblica nella maggior parte che uien dirotto uie una sorgente d'acque detto Hain Casab. [...] la quantità dell' aqua che scorre trouai nel mese di nouembre 1720 che scorre @ cubi 54 et era annata scarsa di piogge* (NLM. Treas. B. 291, f. 22). Even though the unit of measurement and the time-frame for the water-yield are not specified, it is likely that *cubi 54* is a reference to 54 cubic *palmi* of water, possibly collected in a day. A large modern water reservoir located in the Għajn Qasab territory at map reference 36074 90301 still collects water from this same source.[117]

The personal field-investigation of the Għajn Qasab territory has successfully associated the parcel of land illustrated in *cabreo* 289 with fields at map reference 36081 90314, the western boundary of which lies in close proximity to *Wied ta' Għajn Qasab*. A perched aquifer gallery, the outer section of which is masonry-built,

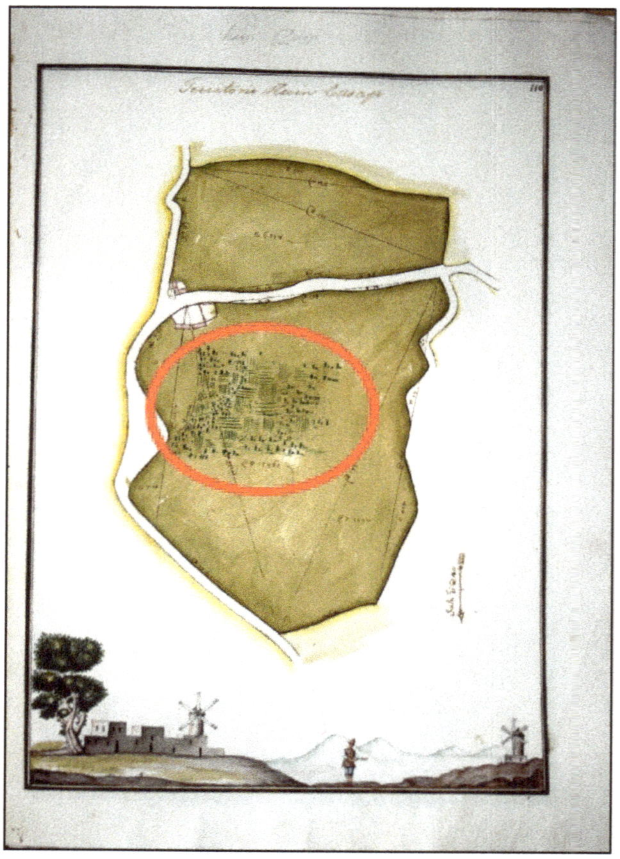

Fig. 6.105 NLM. Treas. B. 290, f. 110. Seventeenth-century plan for the Għajn Qasab territory (GO12). The circle indicates the presence of a gravity-fed water system which originates from a *fontana*.

[117] Nineteenth-century documentation confirms the eighteenth-century hydrological setup and clearly indicates the water canalisation system which from the perched aquifer gallery conveys water to two reservoirs which are likewise indicated in the *cabreo* 290 entry for Għajn Qasab. NAM. Descriptive Plans of the Crown Property, 1861, f. 63.

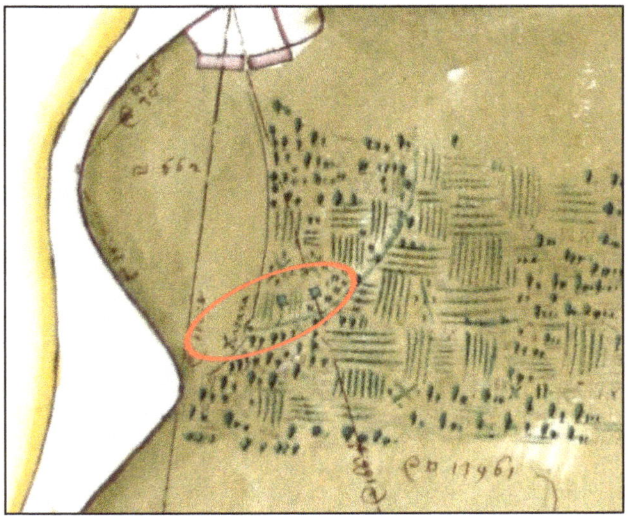

Fig. 6.106 Detail of NLM. Treas. B. 290, f. 110 (Fig. 6.105). The *fontana* of Għajn Qasab (GO12) at Nadur, Gozo, is clearly shown to feed two reservoirs with a water supply.

Fig. 6.107 General view of the Għajn Qasab territory (GO12). Dense reed growth indicates the presence of a perennial water source. Source: the author.

diverts perched aquifer water intercepted further upstream into this section of the valley (Fig. 6.110). One gallery shaft was intercepted, and this was also observed to be a dry-rubble construction.[118] Water retrieved from the Għajn Qasab gallery is presently being directed into an open-air reservoir in which it is stored for later distribution amongst the fields forming part of this landholding (Figs. 6.107, 6.108, 6.109, 6.111). It is indeed significant that the Għajn Qasab place-name was first recorded in 1374 (Wettinger 2000, 188).[119]

6.1.11.3 The Għajn Naħrin territory

The same *foglio*, on which the plan for the *hain Casab* territory is included, also contains the land boundaries for land at *hain nacharin* (Għajn Naħrin – GO11; NLM. Treas. B. 289, Plan 15), first mentioned in 1374 documentation (Wettinger 2000, 187). In 1583 this was described as *Hayn nahrin, [funtana] ab oriente di dahlet corrot* (Wettinger 2000, 187; Fig. 6.1).

Cabreo 290 plan data for the two parcels of land in this territory are much more informative and mark the location of the two water sources by means of circles coloured in green (NLM. Treas. B. 290, f. 106; Fig. 6.113). Each source appears to originate from within a rock outcrop. The *cabreo* 291 entry for the *Territorio detto Naerin* states that: *Questo territorio piglia il nome da due sorgente di poca quantita d'acque, li qualli*

Fig. 6.108 The Għajn Qasab (GO12) perched aquifer water at Nadur, Gozo, is channelled into this open-air reservoir for storage and later distribution. Source: Source: the author.

Fig. 6.109 Exterior view of one of the shafts giving access into a masonry-built subterranean section of the Għajn Qasab water gallery (GO12) at Nadur, Gozo. Source: the author.

[118] The location of the gallery and shaft entrance was indicated to the author of this text by the farming family who for the past century has been leasing this property from the Government.
[119] The implications of this assertion will be further explored in chapter seven, section 7.6.1.1.

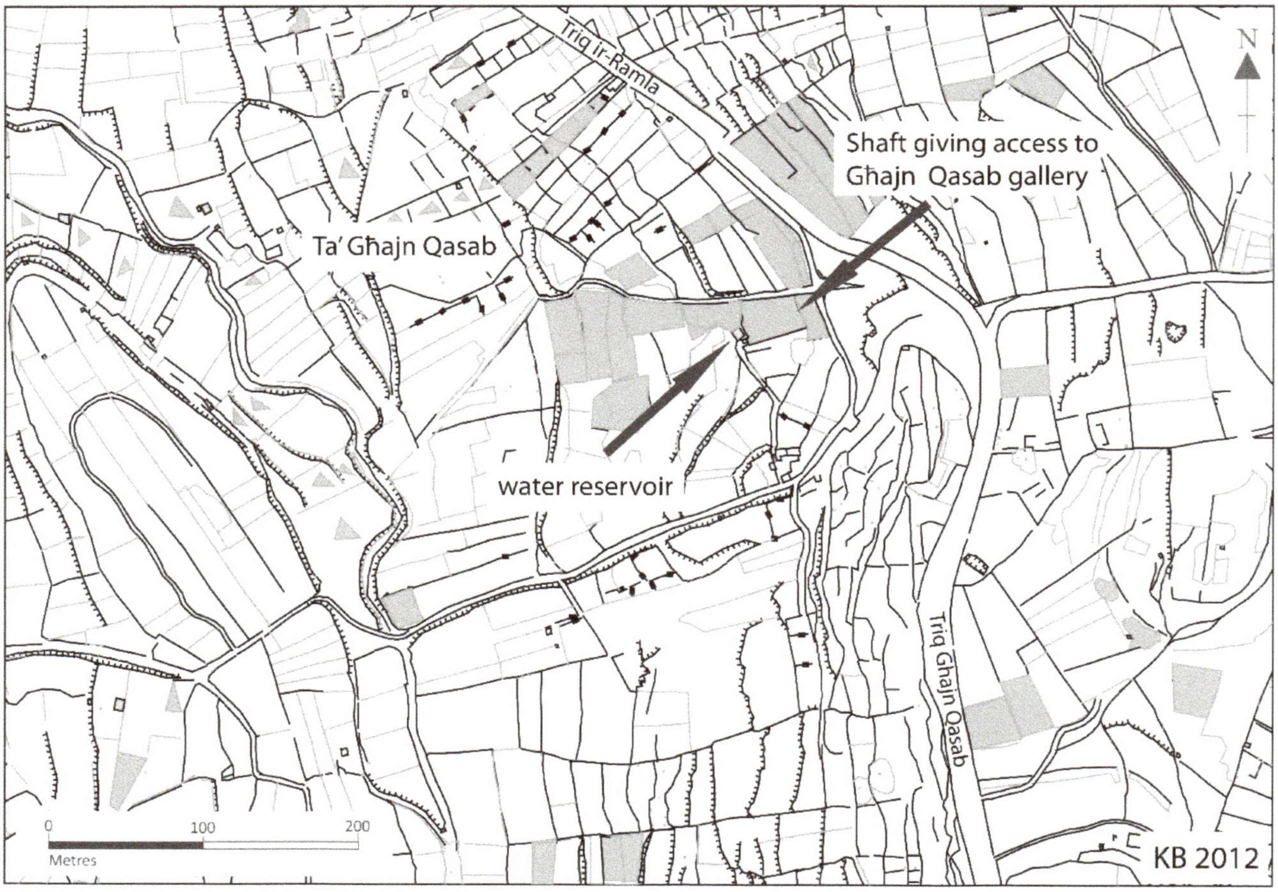

Fig. 6.110 Plan of the Għajn Qasab estate (GO12) showing the location of the shaft well entrance and its water collection reservoir. Topographical map image adapted from MEPA Map Server data.

Fig. 6.111 Interior view of the masonry-built shaft and perched aquifer gallery section at Għajn Qasab (GO12) at Nadur, Gozo. Source: the author.

nascono da sotto due vasche regavati ma nel med[esimo] terreno supraviene essendo diviso in due da una strada publica. La detta suprana dove sono le due sorgenti contiene superfice [...] (NLM. Treas. B. 291, f. 22). The land in the uppermost portion, in which the water sources are located, is described as *di buonissima qualità* (of a very good quality), whilst the lower portion is described as *malissima qualità* (of a very poor quality; NLM. Treas. B. 291, f. 22).

The identification of the whereabouts of the Għajn Naħrin site was only made possible through the use of nineteenth-century *cabreo* documentation (NAM. Descriptive Plans of the Crown Property, 1861, f. 82), on which are also indicated the approximate whereabouts of the two springs indicated in the *cabreo* 290 plan. The names of these springs are given as Għajn Naħrin and Għajn Ħaġar (GO11; Figs. 2.17, 6.112, 6.114). Both springs are on plan represented by a thin blue-coloured line leading towards a water collection reservoir. Access to the upper portion of land in which the springs are located was not permitted.

6.1.11.4 The Madia territory

The *foglio* on which the Għajn Qasab (GO12) plan is included also illustrates the land boundaries for a parcel of land called *Madia*, which due to the mention of a *magaseno nella' marina*, can be safely associated with

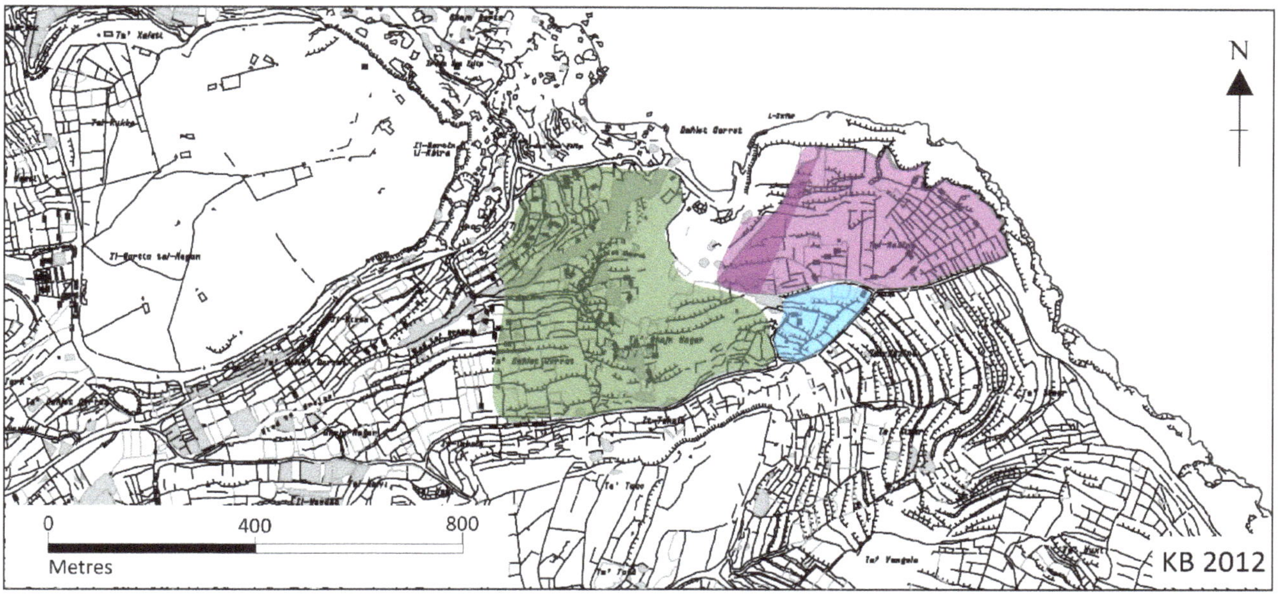

Fig. 6.112 Plan of the Daħlet Qorrot area of Gozo. The parts shaded on green, purple and blue show the approximate land-area covered by the Għajn Naħrin estate (GO11) based on data obtained from the Descriptive Plans of the Crown Property in Malta, 1861, f. 82. Topographical map image adapted from MEPA Map Server data.

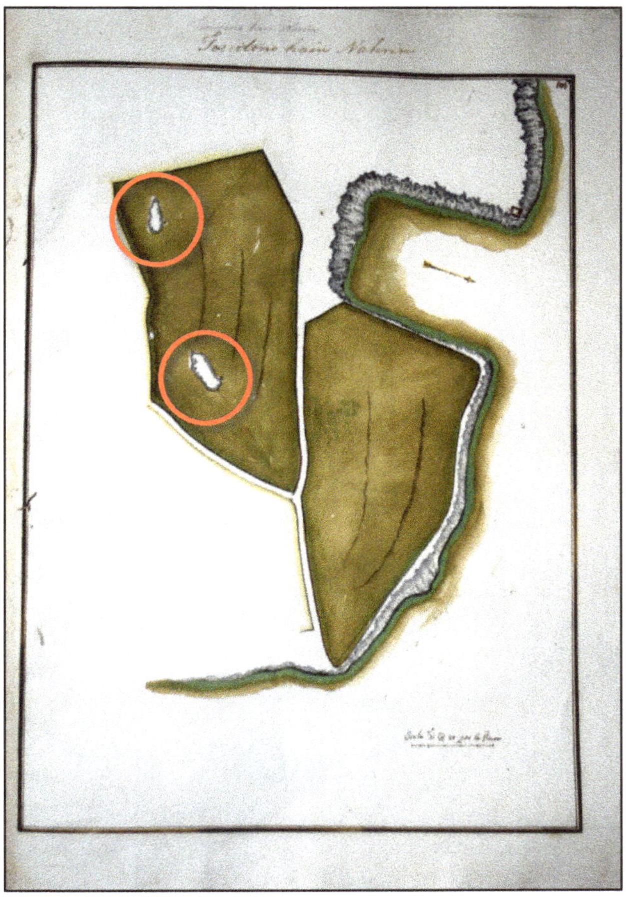

Fig. 6.113 NLM. Treas. B. 290, f. 106. Plan of the Għajn Naħrin estate (GO11) in the territory of Daħlet Qorrot, Gozo. Circled in red are two water-related sources, tentatively associated with water collection points within Blue Clay deposits.

physiognomy of a fully-fledged *giardino*, with perched aquifer water galleries known to be present in the northern part of the Mġarr harbour at map references 36667 87303 and 36704 87289. These are excavated within Mtarfa Member deposits located below the headland shaped by the Wied ta' Martin and Wied Biljun valley formations. Another water gallery of unknown antiquity which still supplies Wied Imġarr with a perennial water supply was rediscovered in late 2011.[120]

The two water sources portrayed in the *cabreo* 289 plan (Fig. 6.1) appear to represent valley water courses, possibly originating from either the Wied Imġarr, Wied ta' Martin or Ta' Wied Biljun. The *cabreo* 289 entry for the *Terreno della Madia* makes reference to *hain Raies* and *Vallone di hain Raies* without giving further information (NLM. Treas. B. 289, ff. 151-53).[121]

The cabreo 290 and 291 entries for this estate are more informative and show a density of water-related features (NLM. Treas. B. 290, f. 105; and NLM. Treas. B. 291, f. 21). The relevant sections of the *Feudo della Madia* (GO16) entry reads as follows: *[...] da cinque fontane ed ancora con la spandense di Hain Selma posta nella strada publica in servizio del publico. Le fontane una e nominata Tal Liesuru, et altra ta Rajes quali sono in mezzo del vallone, altra Fontana deriva da una o due sorgenti che sitrovano nella strada publica della quale si salisce dalla spiaggia. Di piu in mezzo del vallone una mina quale ui e altra sorgente della quale si aberreva(?) parte del vallone [...]*(NLM. Treas. B. 291, f. 21).

[120] This perched aquifer gallery is located at map reference 36179 87435 and was brought to the attention of the present writer by M. Attard, an archaeology graduate from the University of Malta.
[121] The approximate location of *hain Raies* remains otherwise unknown.

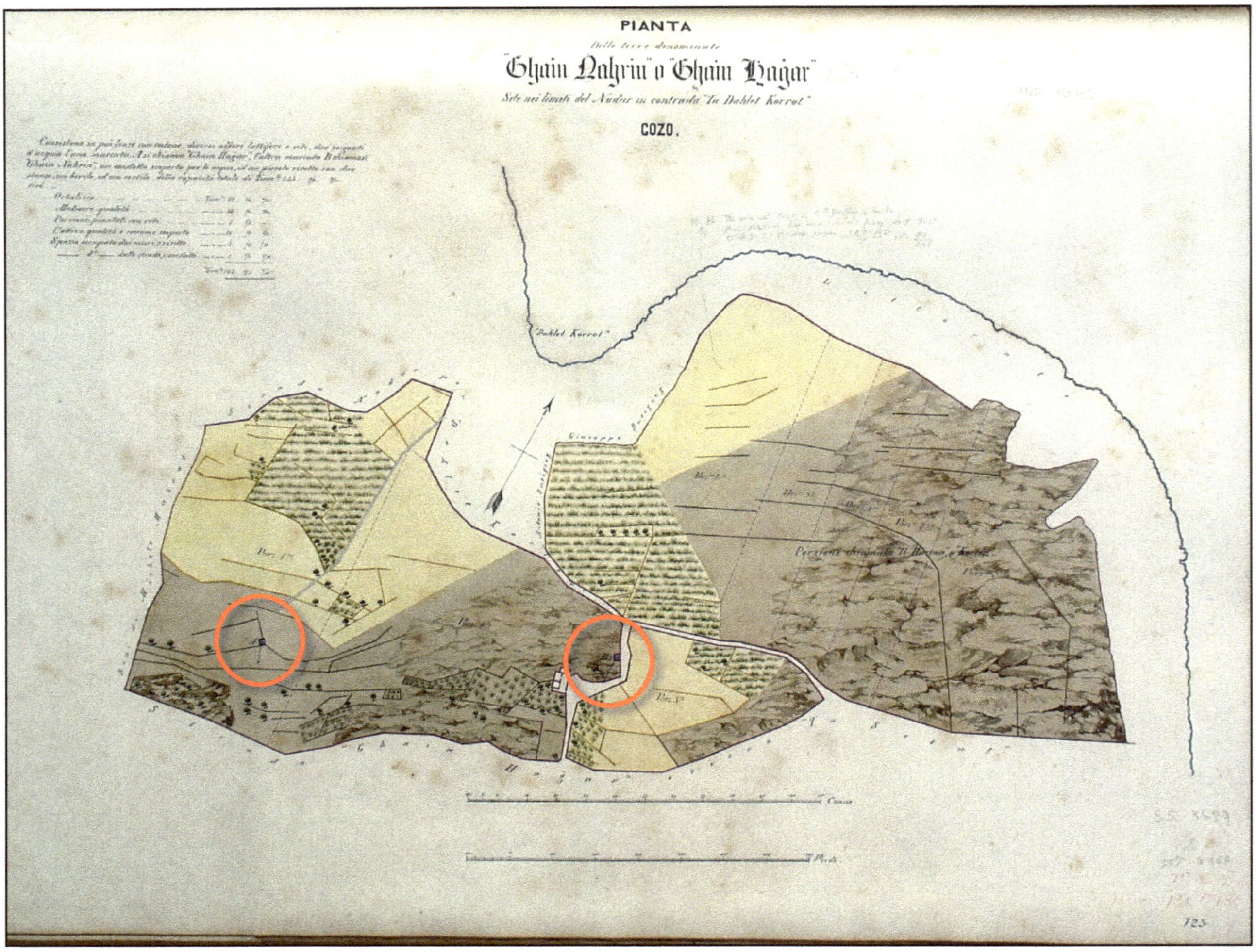

Fig. 6.114 NAM. Descriptive Plans of the Crown Property in Gozo, 1861, f. 82. Plan of the Għajn Naħrin and Għajn Ħaġar estate in the Dahlet Qorrot area of Gozo (GO11). The location of the Għajn Naħrin and Għajn Ħaġar springs is approximately indicated by their fronting water collection reservoirs.

The location of the five springs mentioned above are probably indicated in the corresponding *cabreo* 290 plan through a series of five rectangular enclosures. These are coloured in greenish-blue pigment and probably conform to open-air water reservoirs in which the harvested spring water was stored for later distribution (Fig. 6.116 "A" to "E"). Additionally, the text entry in *cabreo* 291 for the *Feudo della Madia* asserts that two of the fountains were located on public roads – a probable reference to reservoirs "A" and "E" (Fig. 6.116). Only the names of the *Tal Liesuru* and the *ta Rajes* water sources are given, but nineteenth-century documentation on the water resources of Mġarr (Gozo) lists the availability of seven springs: (a) Għajn Barrani (GO5; GO16); (b) Għajn ta' fejn il-Barri (GO16); (c) Għajn Ħluk (GO16); (d) Għajn Sielem (GO16); (e) Għajn ta' Franġisk Spiteri (GO16); (f) Għajn Rajes (GO16); and (g) Għajn Fawwara (GO16).[122]

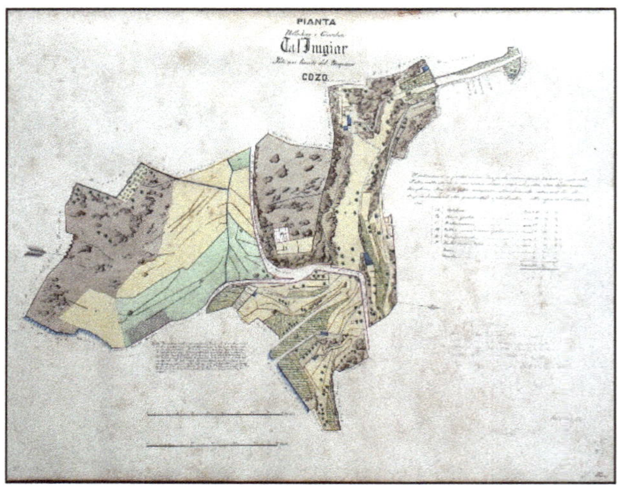

Fig. 6.115 NAM. Descriptive Plans of the Crown Property in Gozo, 1861, f. 96. Plan of the area covered by the Mġarr (Gozo) *giardino* (GO16). Indicated in blue on the plan are the reservoirs which collected water from the various springs in the area, a number of which probably originated from within perched aquifer galleries.

[122] See Schinas 1864; Zammit T. 1924, 31. It is likely that most of these originated from within perched aquifer galleries. See also Fig. 6.115.

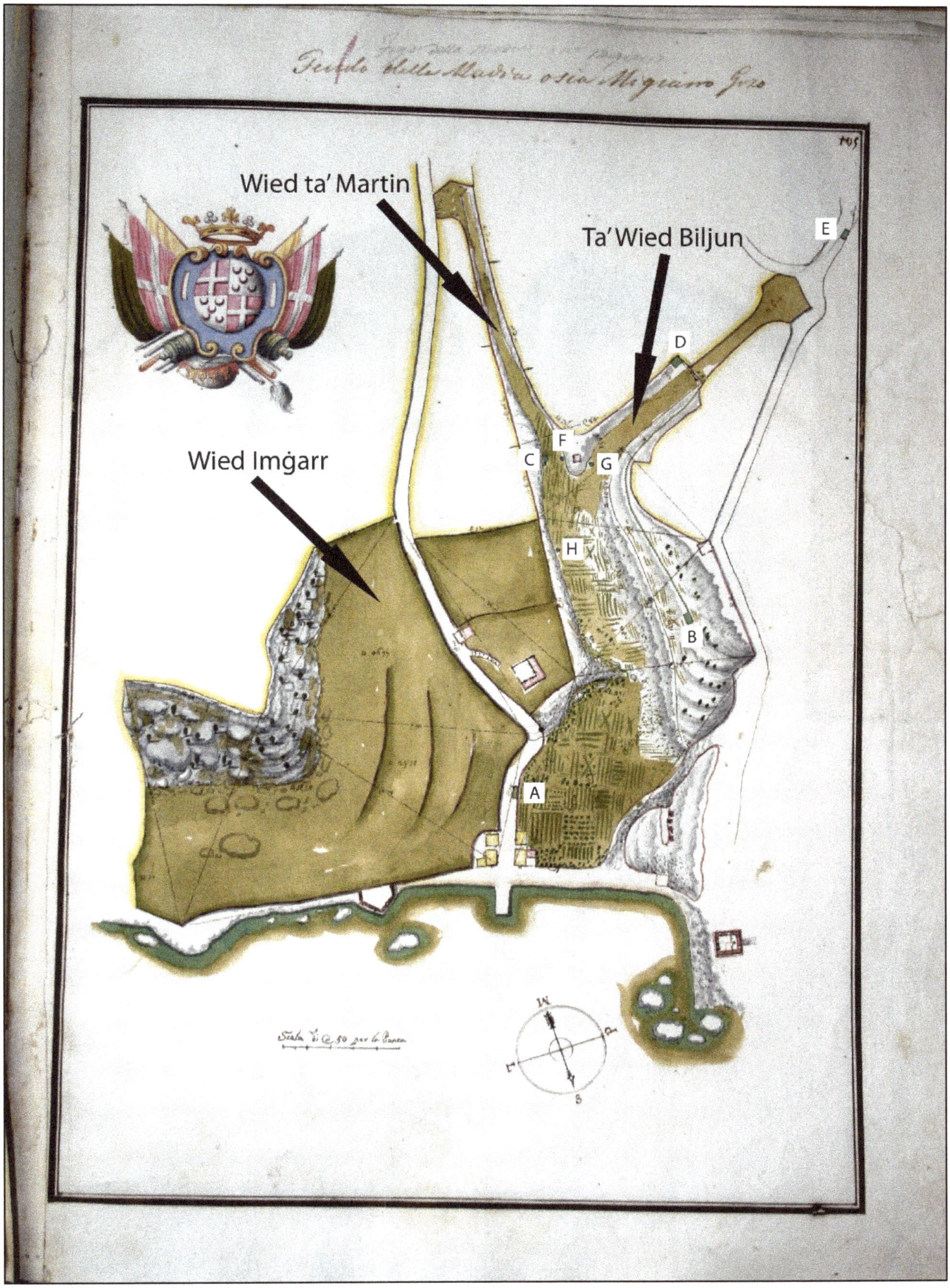

Fig. 6.116 NLM. Treas. B. 290, f. 105. The seventeenth-century plan of the *Feudo della Madia* (GO16) at Mġarr (Gozo). Marked "A" to "E" are reservoirs possibly obtaining their water supply through a series of perched aquifer galleries. "F" to "H" possibly indicate the location of wells which tap a perched aquiferous source.

Of significant importance is the inclusion of a *mina* – a probable reference to a perched aquifer water gallery. The setting and location of reservoirs "C" and "D" (Fig. 6.116) at the foot of a ravine make it equally likely that these were receiving a water source originating from water galleries tunnelling into the ravine at this point. The plan in *cabreo* 290 also includes a series of three circles, the greenish-blue colouring of which probably correlates to water-related features, possibly wells (Fig. 6.116 "F" to "G"). Considering the abundance of Blue Clay deposits in this area, it is also possible that these represent a series of circular-shaped holes excavated in the clayish soil and inside which runoff perched aquifer water used to accumulate.[123]

The area portrayed in *cabreo* 290 appears to correspond to that of the natural shoreline of the Mġarr (Gozo) harbour area at the approximate map reference of 36702 87204. The plan shows three valley systems, which can probably be associated with Wied Imġarr, Wied ta' Martin and Ta' Wied Biljun (Fig. 6.116). The largest and most developed of these valley systems is Wied Biljun which is ca. 1.3 km long. With a reasonably extensive

Fig. 6.117 The Għajn Tuta perched aquifer gallery and its fronting Għajn tal-Ħasselin (GO21). This is one of the galleries which at present feeds the Wied il-Lunzjata valley system with a perennial water supply. Source: the author.

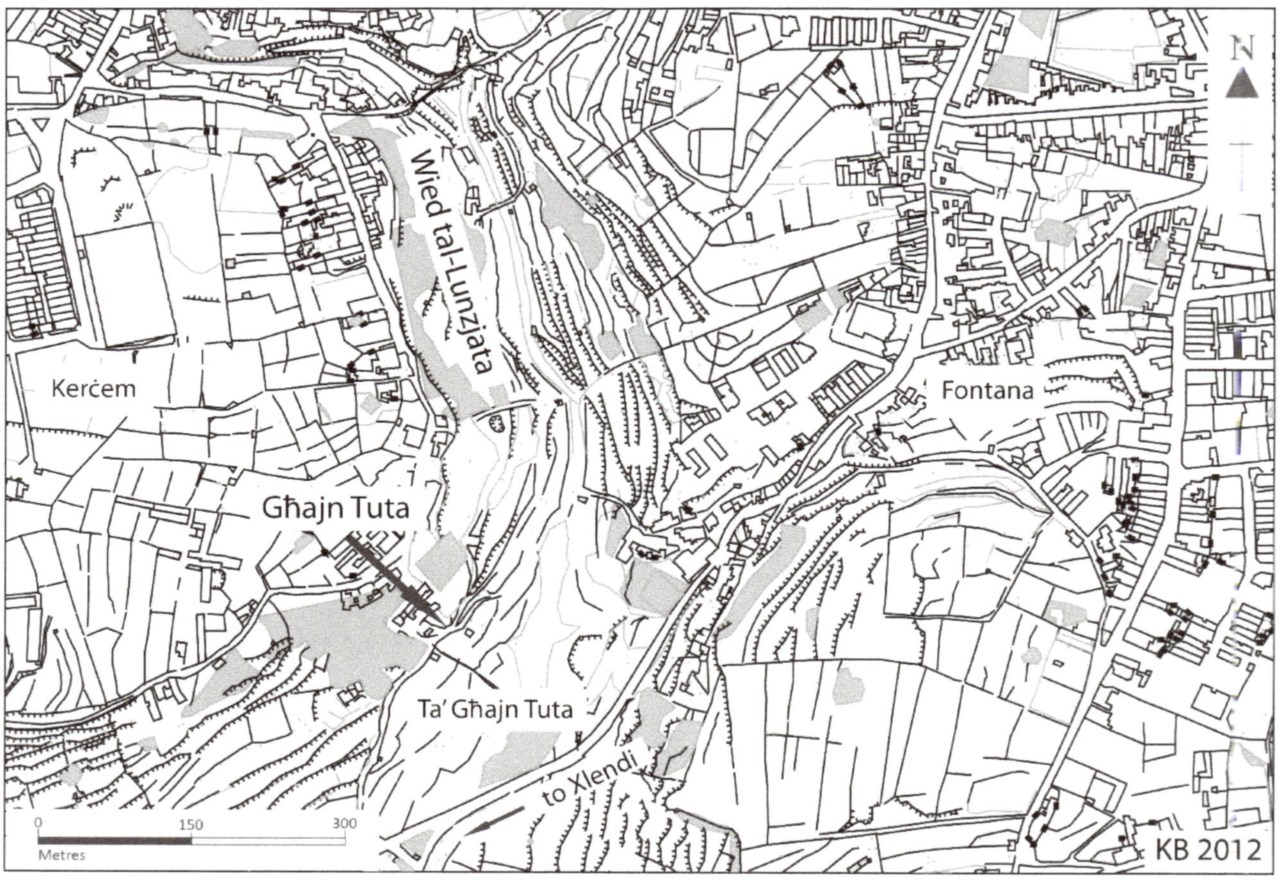

Fig. 6.118 Plan of the Wied tal-Lunzjata (GO21) area at Kerċem, Gozo, showing the location of the Għajn Tuta perched aquifer gallery. Topographical map image adapted from MEPA Map Server data.

[123] Such a mode of water collection was already documented for the Għajn Tuffieħa *giardino*. See section 6.1.9.

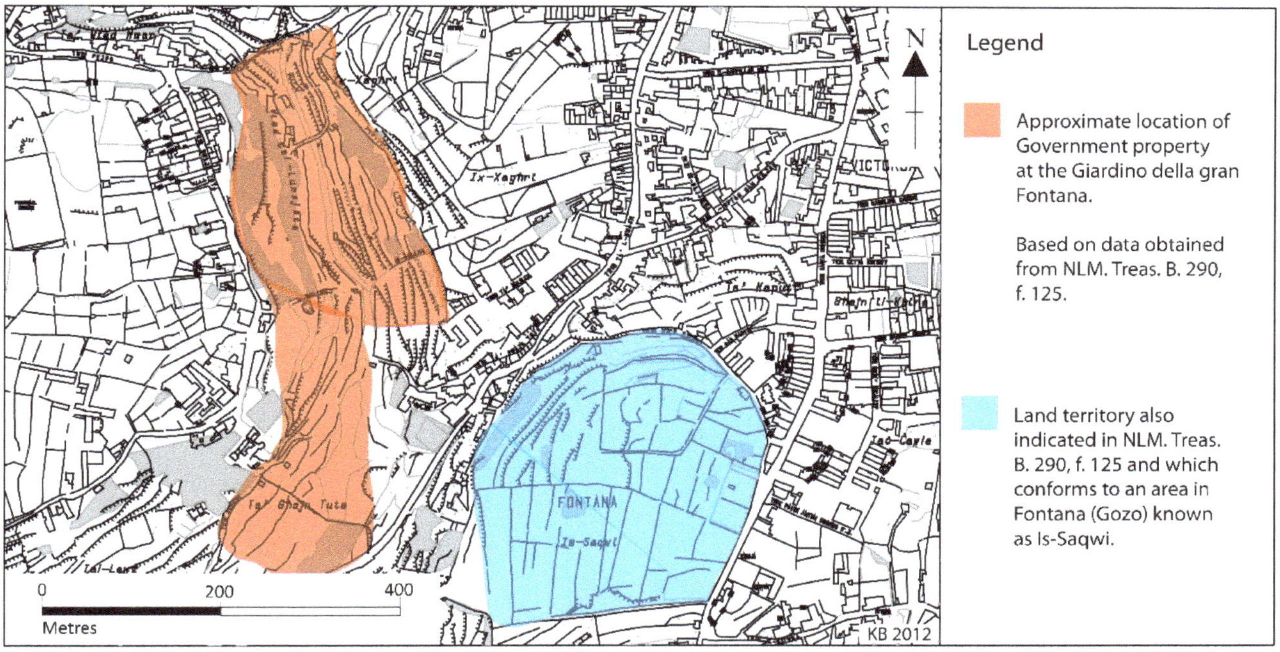

Fig. 6.119 Plan showing the approximate location of Government property at the *Giardino della Gran Fontana* (GO21) located close to Rabat (Gozo). Topographical map image adapted from MEPA Map Server data.

watershed area, this provides the largest surface catchment for the parcel of land in question. The *Feudo della Madia* area, together with its surrounding area, has been subjected to land faulting and micro-plate movement.[124] This means that naturally yielding water-bearing Mtarfa Member and Blue Clay strata can in this section of Gozo be distributed over a wide-ranging area and at a varying height above sea level.[125]

6.1.11.5 The Wied tal-Lunzjata territory

The *Giardino della gran Fontana*, also known as *Wied tal-Lunzjata* (GO21; Figs. 2.17, 6.118, 6.119), was during the Late Medieval and the Early Modern period, a principal agricultural site located to the immediate south-east of Rabat (Gozo). The presence of the *Potamon Fluviatile* (fresh water crab species) at Lunzjata demonstrates that the perennial water course flowing through the valley bed has done so uninterruptedly for centuries. The hydrological framework of the valley has also been drastically altered during the past centuries through the excavation of a series of perched aquifer galleries which allow the existence of a perennial gravity-fed water distribution system throughout most of the valley. That this hydrological framework was already in place during the early seventeenth century can be evidenced by *una fontanella nominate hain tuta della quale l'acqua cala dentro una gebietta del detto Giardino* (NLM. Treas. B. 289, f. 144). Għajn Tuta (GO21) is a spring located at map reference 30909 88475 originating from a perched aquifer water gallery (Figs. 6.117, 6.118). There is further reference to *fontane* and at least one reservoir in this area (NLM. Treas. B. 289, f. 144). The *Giardino della Gran Fontana* (GO21) bordered the *Giardino delle due fontane* (NLM. Treas. B. 289, f. 144). To the east it also bordered territory, in which three other fountains appear to have been located, as well as a drinking trough and two reservoirs (NLM. Treas. B. 289, f. 143).

According to data given in *cabreo* 290 and 291 for this landholding, it appears that only the spring of Għajn Tuta was officially situated within the boundaries of the *Giardino della gran Fontana* (NLM. Treas. B. 290, f. 125; NLM. Treas. B. 291, f. 37.). Specific mention is made of four *fontane*, three of which were likely located immediately outside the northernmost border of this tenement (NLM. Treas. B. 291, f. 37). These were nonetheless instrumental in supplying agricultural territory on the eastern bank of the *Giardino della Gran Fontana* with a perennial water supply. The plan clearly shows the springs originating from within a rock face, thus strongly suggesting perched aquifer galleries as their source.[126]

[124] Debono and Xerri 1993, Geological Map of Malta, Sheet 2.

[125] It is within such deposits that perched aquifer water galleries would be excavated.

[126] The illustration of *fontane* originating from a rock face is in this case similar to that presented in the *cabreo* 291 plan for the *Tal Callus giardino* at Wied ir-Rum in Rabat (Malta). NLM. Treas. B. 291, f. 86.

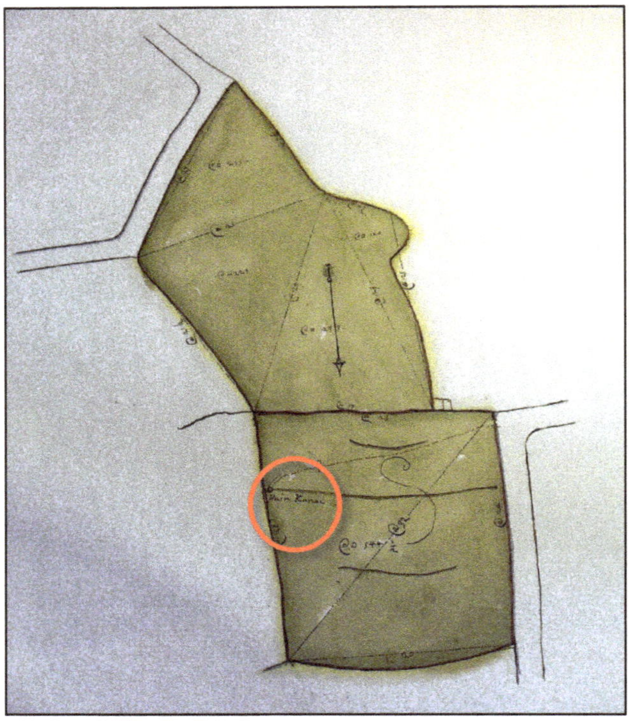

Fig. 6.120 NLM. Treas. B. 290, f. 108. Eighteenth-century plan for two parcels of land at Ta' Ċimblu Rasu in Gozo (GO19). Circled in red is a water source which on the plan is marked as *Hain Kanac.*

Fig. 6.121 NLM. Treas. B. 290, f. 109. A parcel of land at Ta' Ċimblu Rasu in Gozo (GO19). The circle in red indicates a water-related entry which on the plan is labelled as a *cisterna con aqua sorgente* (a cistern of spring water).

6.1.11.6 Government-owned properties at Ta' Ċimblu Rasu

The location for the Ta' Ċimblu Rasu (GO19) area has still to be established, but it appears to have been a sizeable territory. In both *cabreo* 290 and 291 there are two entries for two sets of different *clausure* as being located in this area. Both contain evidence for the availability of water sources.

Hain Kanac (GO19) is in *cabreo* 290 f. 108 marked as a circle outlined in bold black ink and with a dotted line emerging from it (NLM. Treas. B. 290, f. 108; Fig. 6.120).[127] It is again mentioned in the accompanying text entry in *cabreo* 291, where it is referred to as a *sorgente* (NLM. Treas. B. 291, f. 22). The available evidence is insufficient to allow any reliable conclusions to be drawn. The water-related data concerning another set of properties at Ta' Ċimblu Rasu present a more revealing report.

The text entry for these two parcels of land speaks of the presence of a *cucina e cisterna* (kitchen and cistern; NLM. Treas. B. 291, f. 22). On the equivalent *cabreo* 290 plan, this cistern is marked by means of a blue-coloured

Fig. 6.122 Detail of NLM. Treas. B. 290, f. 109. The red circle indicates a shaft well giving access to a probable subterranean perched aquifer gallery.

dot and labelled as *una cisterna con aqua sorgente* (NLM. Treas. B. 290, f. 109; Figs. 6.121, 6.122). Significance can be attached to this insofar as it makes a case for the likely presence of a perched aquifer gallery containing shafts excavated below the property in question. The cistern is, most probably, a shaft well piercing the roof of a perched aquifer water gallery

[127] *Kanac* is phonetically similar to *qana* which in Maltese means conduit and might represent a subterranean perched aquifer gallery.

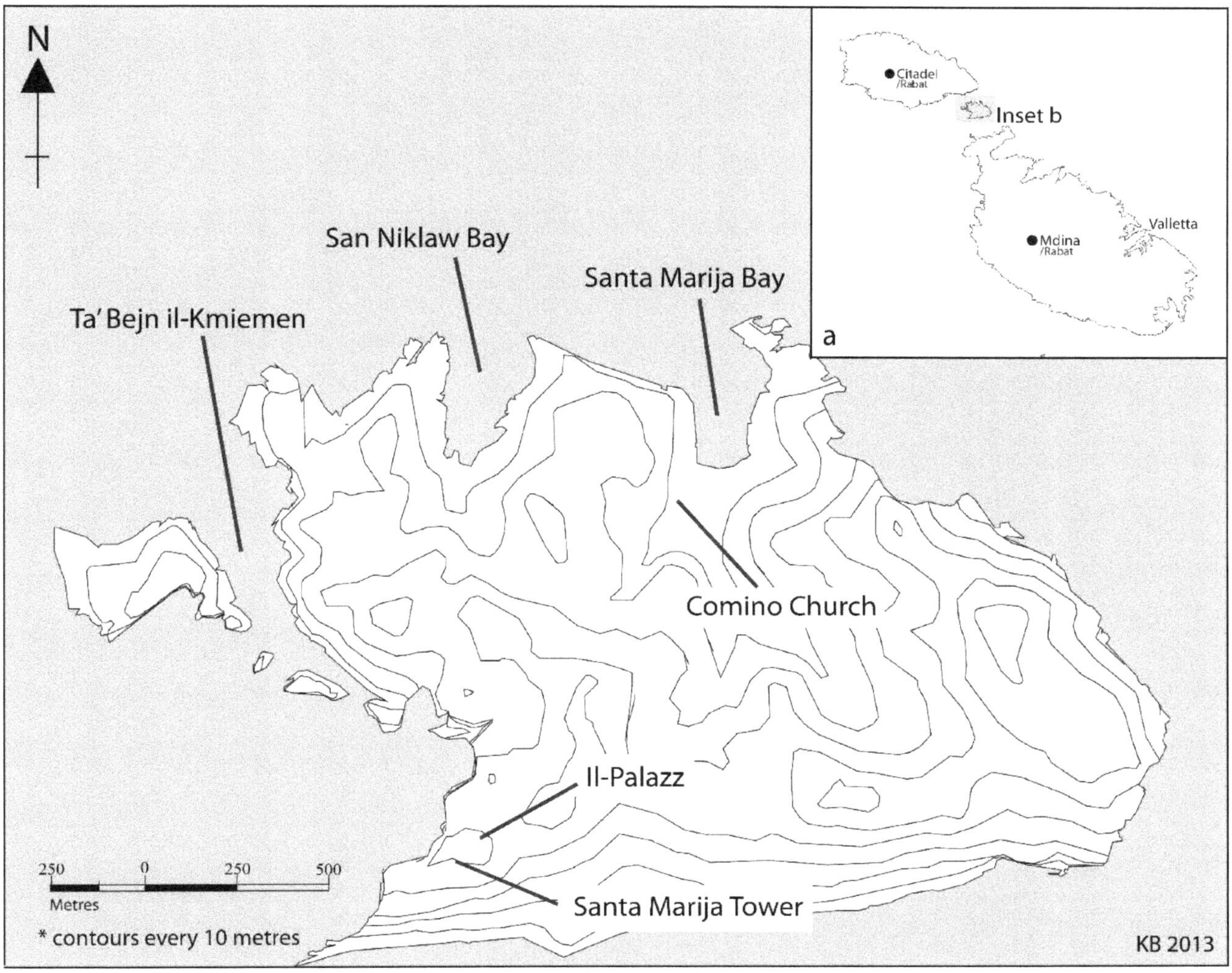

Fig. 6.123 Map of the Maltese islands (a), with inset (b) showing the location of place-names on Comino Island (GO2) mentioned in the text. Base map data: MEPA.

6.1.12 Water sources on Comino

From a hydrological perspective, the data for Comino (GO2) provided in *cabrei* 290 and 291 are more informative than the earlier *cabreo* 289 documentation. The latter seventeenth-century manuscript contains a somewhat detailed account on the main topological features of both *Cumino* (Comino) and *Cominetto* (Cominotto), but omits references to any water management practices (NLM. Treas. B. 289, ff. 156-59).[128]

The *cabreo* 290 plan for Comino indicates a substantially large open-air water storage reservoir located on the side of the path which connects Santa Marija Tower with the adjoining quarters known as Il-Palazz (NLM. Treas. B. 290, f. 104; Figs. 6.123, 6.124, 6.125).[129] *Cabreo* 291 refers to the reservoir as *una gebia d'acqua per comodita del gabellotto e nutrimento delli sui animali* (NLM. Treas. B. 291, f. 21). Since both the tower and Il-Palazz were built in the early seventeenth century, it is safe to assume that this water storage feature also dates to this period and was one of the water-supporting systems for both the tower and the Il-Palazz complex. No surviving remains pertaining to the reservoir have been noted by the present writer, but it is safe to assume that it harvested rain water runoff from the more elevated Santa Marija Tower area. Santa Marija Tower was also provided with a large subterranean water cistern, of which there is no

[128]Likewise, the attached plan of Comino and Cominotto contains no water-related evidence. See NLM. Treas. B. 289, Plan 22. For geological data on Comino see chapter two, section 2.6.1.

[129] *Il-Palazz* was originally constructed by Grand Master Alof de Wignacourt. It was extended and converted into an isolation hospital for the treatment of patients suffering from plague and cholera during the early twentieth century. See Farrugia Randon and Farrugia Randon 1993, 6-7.

Fig. 6.124 NLM. Treas. B. 290, f. 104. Plan of Comino Island (GO2). The only two indicated water-related features are a reservoir situated close to St Mary Tower and a well at Santa Marija Bay.

Fig. 6.125 Plan detail of NLM. Treas. B. 290, f. 104 showing the reservoir located in close proximity to St Mary's Tower at Comino (GO2).

mention in the *cabrei* 289, 290 and 291 entries for Comino island.[130]

Note should be taken of a tiny bluish coloured dot on the inner west side of Santa Marija Bay.[131] It marks the presence of what appears to be a bell-shaped subterranean water storage tank (Fig. 6.126). Its location tallies with that of an old cistern located in the immediate proximity of the Late Medieval church dedicated to the Holy Family on the Flight to Egypt (Buhagiar K. 2008A, 11; Fig. 6.127). This water-storage feature, located at map reference 40216 86027, is described in *cabreo* 291 as: *Vicino ad essa Chiesa si trova un pozzo d'acqua inseruizio delli pastori* (NLM. Treas. B. 291, f. 21). The use of the term *pozzo* is a clear indication of its usage as a well. On the other hand, a partial investigation of this water feature through the employment of an infrared camera which was lowered inside, shows a rock-cut bell-shaped cistern which was originally clearly intended for the harvesting of rain water.

It is indeed possible that this water feature was originally conceived as a cistern but due to the proximity of the mean-sea-level aquifer to the surface at this particular point, this started to accidentally tap a ground-water source, and so took on the function of a well. This fresh water retrieval point is closely associated with the adjoining church of the Holy Family and might likewise date to the Late Medieval period.

6.2 Concluding Remarks

The methodology followed in the carrying out of this research, together with its limitations, is discussed in chapter one, sections 1.6 and 1.7. The analysis of *cabreo* data presented in this section was aided by the fact that, in most instances, there was sufficient historical coverage to enable the exact identification of the majority of the estates discussed in the case studies presented above. Apart from permitting the implementation of a retrogressive analytical approach, *cabreo* data enabled me to carry out detailed field investigation sessions in which historical and cartographic data were brought into check with the present-day landscape and its hydrological features. It is through the adoption of such an investigative approach that the Tal-Fiskal spring at

[130] The existence of the water cistern below Santa Marija tower was brought to my attention by E. Depasquale, who also accompanied me on site.

[131] It is interesting to note that *cabreo* 291 refers to Santa Marija Bay as *la cala maggiore detta di egitto*, due to the presence of the Late Medieval church on the western flank of the valley dedicated to the *Beatissima Vergine del egitto*. See NLM. Treas. B. 291, f. 21.

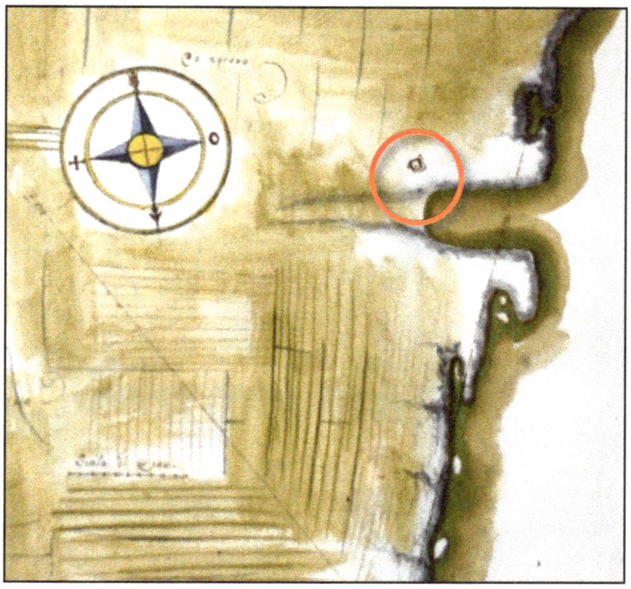

Fig. 6.126 Plan detail of NLM. Treas. B. 290, f. 104 showing the location of well probably tapping a mean-sea-level aquiferous source at Santa Marija Bay, Comino (GO2).

Fig. 6.127 General view of the well indicated in NLM. Treas. B. 290, f. 104 at Santa Marija Bay, Comino (GO2). In all probability, this was originally intended as a cistern but accidentally tapped the mean-sea-level aquifer, which at this point lies only a couple of metres below surface level. Source: the author.

Marget Neuia (RA34) and Għajn Żdejda at Qasam il-Kbir (RA37) were identified. Both springs were found to originate from *qanat*-type Upper Coralline Limestone galleries, the ceiling of which is pierced by a series of shaft-wells.[132]

This method of research also permitted the identification of the Tal-Callus estate (RA39), where it was established that two *fontane* located within this property originated from Upper Coralline Limestone galleries.[133] *Cabrei* sources only rarely make mention of the availability of a *mina* or subterranean gallery. Nonetheless, an analysis of *cabreo* documentation makes it possible to correlate several *fontana* sources with *għajn*-type springs originating from perched aquifer galleries within Upper Coralline Limestone. There is seventeenth-century *cabreo* documentation which specifically refers to *Hayn Tausen* as a *mina*, for which the length is also given. Quite significantly, however, in another section of the same *cabreo*, *Hayn Tausen* is only referred to as a *fontana*.[134] Similarly, whereas the seventeenth-century *cabreo* 289 refers to Għajn Qajjed (RA19) as a *fontana*, in an 1867 description of the Għajn Qajjed *giardino* there is the mention of a *mina*. Likewise, seventeenth-century documentation on the Razzett tal-Qasam area (MG18) at Għajn Tuffieħa refers to maintenance works being carried out on a *mina* and water drinking trough located within that property.[135]

Amongst the most noteworthy land-holdings discussed in this chapter are the *giardini* of Għajn Qajjed, Għajn Klieb (RA18), Ġnien is-Sultan (RA13), Ġnien Musfar (RA31) and *Hayn Tausen*. All are Government landholdings and the available seventeenth and eighteenth-century *cabreo* documentation seemingly intimates that the administrative entity in charge of the Valletta aqueduct project specifically directed its attention towards these *giardini* because perennial water sources were already to be found there.[136] That during the early seventeenth century, a number of these springs originated from within perched aquifer galleries is a matter that has already been confirmed and implicates that the *giardino* framework within which these springs are located, quite likely predates the Knights period. The strategic placement of the Late Medieval church of San Mikiel is-Sinċir (RA50) close to the boundaries of most of the above-mentioned *giardini* gives further credence that these were intensely cultivated during the Late Medieval period.

On the other hand, a major limitation concerning some of the rural estates discussed in this chapter was the difficulty encountered when trying to distinguish between "well" and "cistern" hydrological features. This was especially made difficult in instances where *cabrei* plan data were not corroborated further by any accompanying text or a description of the properties in question. Similarly, *cabreo* documentation made no technical distinction between perched aquifer and mean-sea-level water sources. The fact that numerous properties listed in the earliest *cabreo* 289 manuscript were not accompanied by plan drawings was another setback. When available, these were of a much poorer quality than the plans presented in the eighteenth-century *cabrei* manuscripts, often making their interpretation difficult. The lack of a key, absence of standardised norms and colour-coding in

[132] See sections 6.1.4 and 6.1.5.
[133] See section 6.1.7.
[134] See section 6.1.3.
[135] See section 6.1.9.3.
[136] See also chapter five, section 5.2.4.

some eighteenth-century *cabreo* plans proved equally problematic. In a number of *cabreo* entries, even though the name of a listed property is water-related, there is no mention of springs, nor other associated hydrological assets associated with that property. In conclusion, all objectives declared at the outset in chapter one, section 1.1 are touched upon in this chapter.[137]

[137] For a reflection on the main objectives of this publication and an evaluation on whether these were successfully addressed see chapter eight, section 8.2.

Chapter Seven

Water Management in Late Medieval Malta

7.0 Introduction

The economy of the Maltese islands in the Late Medieval period was directly dependent on agriculture (Bresc 1975, 131-32; Wettinger 1982, 2). Considering that an adequate hydrological supply forms the basis of a sound agricultural infrastructure, such a circumstance suggests that prime production sites in Malta had access to various reliable water sources. Even though the islands did not produce enough wheat to feed the resident population,[1] Maltese agriculture was sufficiently productive to produce a surplus of cotton and cumin for exportation purposes. Revenue from the sale of the cotton and cumin crops in foreign markets generated the profits necessary to offset the expenses incurred by the importation of wheat, other foodstuffs and textiles (Wettinger 1982, 3; Dalli 2006, 158).

The known historical documentation for water sources and management strategies employed in Malta during the High and Late Medieval periods between the eleventh and the first half of the sixteenth century is limited and fragmentary. The semi-arid climate of the Maltese archipelago made agriculture almost exclusively dependent on the annual rainfall.[2] Without any direct human hydrological-related intervention on the landscape, the cultivation of crops would be mainly restricted to the winter/spring growing season. A detailed analysis of the Maltese hydro-geological situation already presented in chapter two has outlined the dynamics as to why perennial water sources are only present in specifically designated areas to the north and north-west of Malta. The Maltese words *saqwi* and *bagħli* are terms which distinguish the irrigated from the non-irrigated landscape and suggest a close affinity to the medieval Muslim practice of land classification into *saqi* and *ba'l* in order to be able to tax the land accordingly (Glick 1979, 83-84; Wettinger 1982, 3).

Agricultural territory located in the central, southern, south-eastern and south-western plains of Malta was during the Late Medieval period dependent exclusively on seasonal rainfall. Dry farming was in these areas aided by the adoption of contour ploughing and the harvesting of runoff surface water in either open-air water reservoirs or subterranean cisterns.[3] Peasants suffered widespread financial hardship when rains failed, as was the case for three consecutive years in the latter half of the 1460s, when landlords had to grant their tenants a remission of rent (Wettinger 1982, 3).

The fiefs or royal estates were the largest landholdings in fourteenth and fifteenth-century Malta. It is significant to note that a particular concentration of royal estates[4] including those of Burnoħħala (MG7),[5] Binġemma (MG19), Għajn Tuffieħa (MG15), Ħemsija (RA58), Buqana, Dejr il-Bniet (DI1), Fiddien (RA11), Tabrija (RA44), Għajn Riħana (MO4) and Budaqq were all located in perched aquifer areas, the ground-water deposits of which were, in relative terms, easier to harness than those located elsewhere. From an agricultural perspective, the importance of such landholdings can be detected from annual rent values. Whereas the rent of a typical peasant's field was in the one *uncia* bracket, revenues from rents and other incomes related to Crown Property were much higher.[6] The hefty rental price for fiefs implies that these were major agricultural production sites, the successful cultivation of which could only be ensured through the availability of a perennial water source. Many *Cancelleria Regia* and notarial documents clearly document a water presence in these landholdings. It will be argued in this chapter that since the Late Medieval period, a number of these royal estates obtained their hydrological supply through perched aquifer galleries. Many of the fiefs are in the *Cancelleria Regia* and the notarial documents classified as *viridaria* (sing. *viridarium*) – the Latin equivalent of the Italian *giardino*.[7]

Hydrology-related information for the Late Medieval Maltese period can be obtained from a number of varied sources, but predominantly through the scrutiny of Maltese notarial documentation,[8] the *Cancelleria Regia* archives in Palermo (Fiorini 1999B; Fiorini 2004; and Fiorini 2007), and the local place-name evidence.[9] Most of the material presented in this chapter dates to the fifteenth

[1] For a demographic overview on the estimated Maltese population during the Late Medieval period see chapter two, section 2.2.

[2] For a discussion on climatic fluctuations since antiquity see chapter two, section 2.1.

[3] Various notarial deeds discussed in section 7.5 of this chapter give evidence on the utilisation of cisterns as an aid to agriculture.

[4] During the fifteenth century, the majority of the fief-holders were residents of the *Civitas* (Mdina). Precise data on the size of fiefs during this period are lacking. See Bresc 1975, 152-55; Wettinger 1982, 4-5.

[5] The perched aquifer gallery at Burnoħħala is documented as being ca. 70 metres long. Zammit T. 1924, 39.

[6] For a list of fief rents and incomes for the late fifteenth and early sixteenth centuries see Wettinger 1982, 5-7.

[7] Another documented Latin alternative for the term *viridaria* is *pomeria*. See Wettinger 1982, 27.

[8] The more relevant published notarial texts include the contracts of Notaries Paulo Bonello and Giacomo Zabbara which cover the second half of the fifteenth century. See Fiorini 1996; Fiorini 1999A; and Fiorini 2005.

[9] On the place-name evidence of the Maltese Islands see Wettinger 2000; and Wettinger 2006.

century, but various relevant *Cancelleria Regia* entries go back to the mid-thirteenth century. The oldest surviving Maltese register, that compiled by Notary Paulo Bonello, does not go before 1467. There are five other registers which date to the first decade of the 1500s and another twenty-six which date to 1530. Of these registers, a number are in such a state of deterioration that they cannot be examined without causing serious damage to the manuscripts.[10]

The point of departure for this study on water management in Late Medieval Malta is the Knight Commissioners' Report of 1524 (Bosio 1594-1602, 28-31), which paradoxically fails to give an appropriate insight into the Maltese hydrological scenario for the closing decades of the Later Medieval period. A discussion of the water-related content presented in Jean Quintin's description of Malta (Quintinus Haeduus 1536; Vella H.C.R. 1980), published in 1536, is also undertaken. While there is general acceptance that the year 1530 demarcates the end of the Late Medieval period in Malta, it is to be understood that Quintin's account of Malta portrays a situation synonymous with the pre-Knights' scenario. An examination of the water-related place-name evidence till 1560 is also attempted in the last section of this chapter.

In conclusion, it should once again be brought to the attention of the reader that water distribution and allocation for agricultural usage falls outside the scope of this publication.[11] For this reason, these modes of water management are not represented in the case studies presented hereunder. Moreover, in order to facilitate place-name readability, which due to Late Medieval and Early Modern unstandardised norms are in the original documentation frequently transcribed in differing ways, reference to the comparative table of toponyms (Appendix 2) is recommended. Appendix 5 is a philological appreciation of water-related Latin terminology included in sections 7.3, 7.4 and 7.5 of this chapter. Reference to it will hopefully facilitate a better comprehension and help contextualise the usage of these terms.

7.1 The Knight Commissioners' Report of 1524

Charles V's offer of the Maltese archipelago and the fortress of Tripoli to the Knights of St John in 1524 prompted the setting up of a fact-finding mission to report on the state of the tendered territory (Bosio 1594-1602, 28). This was the first in a series of events which would see the archipelago emerging from the backwater framework which characterised its existence throughout the Late Medieval period. The eight commissioners entrusted with this task set sail from Palermo to Malta and Tripoli on 13 July 1524 and were back in Palermo by 15 August of the same year (Vella H.C.R. 1980, 5). They reported negatively on the general Maltese situation which was described as [...] *aridissima, e d'ogni sorte di selve, di boschi, e di legna spogliata, e priva* (Bosio 1594-1602, 30). The centres of habitation were the *Civitas* (Mdina) and Birgu, together with a number of sparse and open *casali*. The islanders were periodically taken as slaves by corsairs and the landscape, which contained no forested areas, was described as barren. Thistles and brushwood were collected from the countryside areas for use as fire fuel (Bosio 1594-1602, 28-31).[12]

The commissioners' report gives the impression that agricultural land in Malta was scarce. Agricultural productivity was unfavourable, with the area of land west of the *Civitas* being sterile. The territory to the east of Mdina was good for cultivation, even though abundant in loose stones. Malta's sister island of Gozo was described more favourably. Apart from being more pleasing to the eye than Malta, Gozo was recognised to be more fertile. It was pointed out that the islands forming part of the archipelago could only be self-sufficient for four months a year. During the remaining eight months of the year food supplies had to be shipped over from Sicily. Although acknowledging that water resources in Malta were limited, the report noted that the inhabitants of the *Civitas* harvested water in cisterns and collected water from two springs located outside the city walls (Bosio 1594-1602, 30).

Whilst possibly referring to the Għajn Ħamiem (RA16) spring located just beyond the north-west-facing fortification walls and ditch at Mdina,[13] and overlooking the railway bridge at Mtarfa, it is equally possible that the Knight Commissioners were here describing at least two perched aquifer galleries tunnelling below the north-east and north-facing bastions at Mdina.[14] Other springs were located to the west of Mdina and irrigated some of the orchards. The Birgu (BI1) settlement contained a good number of cisterns and some brackish wells (Bosio 1594-1602, 28-31). The mention of brackish water sources at Birgu is insightful for it implicitly signifies that by the opening decades of the sixteenth century, the mean-sea-level aquifer had already been intercepted and exploited as an alternative water source.

From the above, it can be deduced that the Commissioners were grossly misinformed about the hydrological potential

---

[10] Wettinger points out that these do not give any holistic statistical information on the size of the Maltese peasant population, fiefs, fields and crop production figures. See Wettinger 1982, 1-3.

[11] See chapter one, sections 1 and 1.1.

[12] This is indicative of the karst areas located to the north and north-west, characterised by garigue-type vegetation. During the Late Medieval period these areas were used for rough grazing and in the majority of cases were considered to be common property. See Wettinger 1982, 31-32. Brushwood was similarly collected from these areas. Fiorini 1999A, 63.

[13] The existence of Għajn Ħamiem was first mentioned in late medieval documentation dating to 1436. See Wettinger 2000, 184. All the available evidence indicates that Għajn Ħamiem originates from a perched aquifer gallery. See chapter seven, section 7.6.1.2.

[14] Some of these perched aquifer systems were mentioned in early eighteenth-century documentation. See chapter five, section 5.3.

of areas in west and north-west Malta.[15] No direct reference is made to the perched aquifer perennial water springs flowing through valleys. Cases in point are perennial water sources flowing through the Wied tal-Marċa at Baħrija, various areas of Mtaħleb and through San Martin Valley in the territory of Mġarr (Malta). The continued survival of the rare and endangered *Potamon Fluviatile* species in these valleys is proof of an unfailing water presence in these areas over the centuries.

It is likely that the negative hydrological situation as portrayed by the Knight Commissioners stems from an extremely short sojourn in Malta. The Commissioners' thirty-three-day undertaking included a circular journey from Palermo to Malta, Tripoli and then back to Palermo. It has been recently pointed out that after deducting the approximate travelling time from the total duration of the journey, it emerges that the commissioners spent roughly a week in Malta (Saliba 2008, 49). Should this be the case, it is no wonder that the Knight Commissioners acquired such a distorted picture of the hydrological potential of the territories situated to the west and north-west sectors of Malta. This apparent poor knowledge of the Maltese rural landscape was possibly hindered by the fact that much of their efforts were channelled into investigating the harbour areas on the south-east coast of Malta (Saliba 2008, 49). In all probability, the Commissioners remained unfamiliar with the hydraulic and agricultural potential that the rural territory in west and north-west Malta had to offer.[16] Another matter worthy of consideration is the fact that the Commissioners' visit to Malta coincided with the height of the summer season. Meteorological data for the nineteenth and twentieth-century Malta single out the month of July as being on average not only deprived of any rainfall, but also one of the hottest (Fig. 2.1).[17]

7.2 Jean Quintin's 1536 Description of Malta

Nearly contemporaneous with the Knights Commissioners' report is Jean Quintin's description of Malta compiled in the early 1530s (Quintinus Haeduus 1536; Vella H.C.R. 1980, 1-2.).[18] From a hydrological perspective, Quintin's brief report on Malta conflicts significantly with the impression of aridity given in the Commissioners' report. This could be partly the result of better first-hand knowledge of the archipelago gained from a much longer stay. Quintin had already been in Malta a few months before the arrival of Fra Philip Villiers de l'Isle Adam and his retinue on 26 October 1530 and he stayed on until at least August 1535 (Vella H.C.R. 1980, 1). The years spent in Malta would have presumably enabled Quintin to venture into districts not visited by the Knight Commissioners during their very brief stay on the island.

It is extremely unlikely that any major agriculture-related investment would have been undertaken in the six-year interlude between 1524 and 1530, especially in the north and north-west sectors of Malta. In this respect, Quintin's assertion that, "The island is irrigated with springs, and it has garden [*giardini*] plantations which produce palm-trees [...],"[19] is probably based on direct personal knowledge and experience of what the archipelago had to offer.[20] The garden plantations can probably be associated with the *giardini*-type agricultural sites characteristic of north and north-west Malta, the water sources of which originate from a surprisingly large number of perched aquifer galleries.[21] Quintin's assertion that in *giardini* grew roses of a very sweet smell (Quintinus Haeduus 1536; Vella H.C.R. 1980, 33), was recently interpreted as being a reference to the *Rosa Sempervivens* species, more commonly known as the wild fragrant Evergreen Rose (Saliba 2008, 53). The natural habitat of this species is restricted to areas of maquis-type vegetation and the sides of sheltered valleys. Should this be the case, the evident implication is that the geographic distribution of a number of *giardini* in the 1530s was indeed located within the context of valleys having an easy access to perched aquifer sources – thus pointing to the existence of an already well-established Late Medieval *giardino*-type framework (Blouet 1963, 75-76).

A number of crops described by Quintin as being cultivated in a *giardino* framework include fig and olive trees, together with vines (Vella H.C.R. 1980, 31-33). Whilst fig and olive trees can indeed be cultivated within a *giardino* context, the presence of these two tree species cannot be taken to equate with water availability. Both can withstand drought conditions and in a Maltese setting were often planted in non-irrigated tracts of cultivable land.[22] Vineyards, on the other hand, can in the fifteenth and sixteenth centuries be commonly associated with irrigated land (Wettinger 1982, 24). Cotton cultivation and the handsome profits generated by the sale in foreign markets of this local produce is also mentioned. Quintin gives some detail on cotton cultivation: "It grows better in shallow soil

[15] Similar conclusions were reached by Morris 1952, 2.

[16] Writing in 1952, T. O. Morris similarly stated that Malta has no perennial surface streams, suggesting an unfamiliarity with areas in north-west Malta such as Il-Wied tal-Marġa at Baħrija and the Wied ta' San Martin in the territory of Mġarr (Malta) amongst a few others. Morris 1952, 48.

[17] Rainfall readings for Valletta from 1841 to 1957 only registered an average of 1.016 millimetres of rainfall for the month of July. Bowen-Jones *et al* 1962, 48-72. See also chapter two, section 2.1.

[18] The original text by Quintinus Haeduus lacks pagination numbers.

[19] *Fontibus rigua est, hortis consita, palmas* [...] *ferentibus*. Quintinus Haeduus, 1536; Vella H.C.R. 1980, 31.

[20] Similar conclusions were reached by Saliba 2008, 52-53.

[21] Other *giardini* located to the south of Malta had an essentially different setup. The Marsa *giardino* (MA9) for instance was depicted in mid-sixteenth-century maps, including Quintin's map of 1536, as located on a plain and surrounded by high walls in order to shelter the crops cultivated within from strong winds. See Agius-Vadalà and Ganado 1986, 14.

[22] Olive trees were widely grown in the Mediterranean region since antiquity due to the ease by which they can be cultivated on land unsuitable for the growth of arable crops. Moreover, olive trees have the ability to withstand moderate frosts, and their wood was an excellent source of fire fuel. Apart from being long lived, olive trees are also drought resistant and through the ages were frequently planted as part of terraced cultivation with a cereal crop beneath the canopy. Allen 2001, 187. See also chapter five, section 5.2.

than in a deep one, in a climate which is decidedly dry, and in the field of Malta, which one hardly finds elsewhere" (Vella H.C.R. 1980, 33-35). Even though cotton cultivation requires a reliable water supply, this variable is not reflected in Quintin's observations of the local cotton industry. That local fields often yielded two harvests a year was indicative that in such areas cultivation was completely dependent on rainfall (Vella H.C.R. 1980, 37).

Quintin also differentiates between salty water and the springs of sweeter water found in Malta (Vella H.C.R. 1980, 38-39).[23] Quintin appears to have been familiar with both mean-sea-level aquifer points of extraction and the perched aquifer sources. The former must have been located in the Marsa-Qormi area (MA10) and the inner shores of the Grand Harbour region. The spatial distribution of the perched aquifer sources was determined by the presence of Upper Coralline Limestone and Blue Clay deposits. In all probability, the mention of sweeter springs refers directly to the perched water sources located in the Upper Coralline regions of Malta. Admittedly, no mention is made as to whether any of the sweeter springs originated from perched aquifer galleries, yet Quintin did point out that these water sources were recharged through precipitation (Vella H.C.R. 1980, 39). He also commented on the fact that, "Their source is not deep, for in summer they frequently dry up and always diminish."[24] These observations on perched aquifer spring systems imply that Quintin was knowledgeable about the basic dynamics of ground-water discharge and recharge.

Drinking water was derived from rain and was stored in cisterns and pits (Vella H.C.R. 1980, 39).[25] The meaning of the latter is unclear, but might also refer to the storage of rainwater in subterranean cisterns. Quintin's observation that sowing during spring, "[...] is not done unless the field has been watered the day before and tilled at once",[26] might be a possible reference to irrigation, in which case water would either be obtained from cisterns, reservoirs and/or possibly perched aquifer water galleries. The tilling of the soil could alternatively also take place following a short episode of rainfall.

That spring water and harvested rain water were used to facilitate agriculture can be detected from Quintin's assertion that the soil could "[...] be sown throughout the whole year, and so there is always some harvest."[27] Quintin attributes local soil fertility to the good practice of crop rotation, but in stating that "[...] fields of the whole island produce a return of ten or twelve times of what is sown",[28] he also implies that good water-harvesting and irrigation practices were adhered to. Quintin does not give an account of field irrigation being aided by means of open masonry canals which fed water by gravity to *giardino* areas or to fields requiring irrigation.

Commenting on Malta's climate, Quintin states that the island is directly affected by the, "[...] Libyan heat",[29] presumably a reference to the summer heat. Rainfall patterns are described as erratic and farmers as being afraid of rainfall during the harvesting period (Vella H.C.R. 1980, 35, 39). Northerly winds frequently brought showers during the winter months. The role played by night dew in sustaining vegetation during the dry summer season was also acknowledged (Vella H.C.R. 1980, 41).

Reference to Malta's rocky and rough terrain (Vella H.C.R. 1980, 37), is possibly an allusion to the garigue plains of the islands. Quintin describes fields with soil deposits that are "[...] scarcely more than two or three feet in depth",[30] but also states that, "[...] most of the other places are uncultivated and unutilised."[31] As the vegetation cover mentioned in the latter instance is "[...] wild thyme, broom and thyme" (Vella H.C.R. 1980, 39),[32] it is likely that the regions of Malta under discussion were those located to the north and north-west of the island, where such shrubs are widespread. Furthermore, the mention of troglodytic settlements in this section is probably no coincidence, as it is in these areas of Malta that the majority of cave settlements are located.

When commenting on the settlements and the inhabitants of Malta, Quintin observed that both were greater than expected especially when "[...] considering the quality of the land" (Vella H.C.R. 1980, 29). This assertion was possibly motivated by disproportionate distribution of the surface hydrological resources of the island in respect of the built settlements. Besides the settlement of Mdina, Malta was in the 1530s divided into eight different parishes and described as being sparsely populated (Vella H.C.R. 1980, 29).

Quintin's description of Malta gives the impression that, until the early 1530s, many areas of the island with garigue-type vegetation had not yet been transformed into fields. It is likely that the majority of the fruit-orchards or

[23] The original Latin text reads: *Salsae et feculentae aquae; et qui sunt in insula dulcioribus aquis fontes, credo hibernis plerosque constare imbribus* [...]. See Quintinus Haeduus 1536.

[24] [...] *quorum non in altum origo est, frequenter enim aestate deficiunt, semper subsidunt.* See Vella H.C.R. 1980, 39.

[25] H.C.R. Vella translates *scrobibus* as "ditches", but this is unlikely. Personal communication by A. Bonanno.

[26] *Vere seritur, nec nisi pridie madefactum et in aruo statim restibili; illo enim agrum pinguescere ferunt: mox aratum integitur crate, uel rastro quod supertrahunt.* See Vella H.C.R. 1980, 35; Quintinus Haeduus 1536.

[27] *Super omnia est, hanc soli exilitatem, nec ipsam agricolis operosam, toto anno seri, et inde aliquid meti.* Vella H.C.R. 1980, 37; Quintinus Haeduus 1536.

[28] *Ordinarium autem ut cum decimo ac duodecimo faenore totius insulae campi fundant: ob id (credo) ab Ouidio fertilitatis titulo decoratur Melite.* See Vella H.C.R. 1980, 37; Quintinus Haeduus 1536.

[29] *Aestu uere Libyco infesta* [...]. See Vella H.C.R. 1980, 39; Quintinus Haeduus 1536.

[30] *Petrosa pene tota est et salebrosa; etiam (quod non crederes) ubi aliquid gignit, terra uix binum aut trinum cubitorum altitudine, ut saepe mirari succurrat* [...]. See Vella H.C.R. 1980, 37; Quintinus Haeduus 1536.

[31] *Pleraque eius alia inculta, et deserta omnia. See* Vella H.C.R. 1980, 39; Quintinus Haeduus 1536.

[32] The original Latin text reads: *Atque hae rupes magna ex parte serpillo, cytiso, thymo scatent.* See Quintinus Haeduus 1536.

giardini mentioned in Quintin's account would have been located in valleys (Blouet 1963, 75-76). The overlying garigue plains were in all probability still not reclaimed for agricultural usage up to the early Knights' period.

Quintin's brief report of Malta also contains a cartographic representation of the archipelago (Fig. 7.1). From a hydrological point of view, with the exception of the *Marsa Hortvs* and what appears to be a nearby water canal draining water into the inner reaches of the Grand Harbour (Fig. 7.2), the map does not list any other relevant water-related features. Malta is represented as a roughly circular-shaped island with the only indicated place names being the *Civitas*, the *Marsa Hortvs*, Fort St Angelo and a church in St Paul's Bay. Churches were depicted in areas identified as Ponta S. Maria in Sliema, the St Elmo area on the Sceberras peninsula and the Bir Miftuħ (GU1) and the old St Catherine of Alexandria churches in the Gudja and Żejtun territories respectively. Other settlements

Fig. 7.1 Map of Malta in Jean Quintin d'Autun: The first known printed description of the Maltese archipelago. Source: Quintinus Haeduus 1536.

Fig. 7.2 Jean Quintin d'Autun: map detail showing the *Marsa Hortvs* area and what appears to be a water drainage channel. Source: Quintinus Haeduus 1536.

represented in Malta can be tentatively associated with those of Birgu, Żejtun, Żabbar, Siġġiewi and Żebbuġ.

7.3 Documents of the Cancelleria Regia *at the State Archives of Palermo: The fourteenth century*

The historic evidence pertaining to the first half of the sixteenth century presented so far provides inconclusive evidence as to the hydrological strategies practised locally during this period. While Quintin's account of Malta in the 1530s already indicates a number of well-established *giardini*, the existence of which depended on a series of perennial water sources, it is only the consideration of earlier documentary source material which can yield more conclusive evidence in this respect. Prior to 1530 the Maltese archipelago was an integral part of the Kingdom of Sicily and all decisions concerning Malta and Gozo were taken by the Sicilian authorities. The *Cancelleria Regia* (Royal Chancery) documents in Palermo of direct relevance to Malta cover the 1259 to 1485 timeframe.[33] Their study can indeed provide the Late Medieval archaeologist and historian with a better understanding of the socio-political and cultural cross-currents influencing Malta during the Later Medieval period. Various documents of the *Cancelleria Regia* discussed in this section predate the formation of the Maltese and Gozitan *Universitates*, or municipal governments, the framework for which first appears in the 1360s and 1370s.

The major drawback with the *Cancelleria Regia* documents, as with the majority of other sources dating to this period, is the lack of direct reference to the nature of the water sources mentioned in this section. The apparent slackness of the Sicilian chancery practice is another factor that has to be given some consideration. Years could sometimes pass before drafts of royal mandates marked *registrandum* or *registranda* were copied into a register, hence increasing the possibility of scribe-related inaccuracies (Luttrell 1977, 33). On the positive side, the emerging picture points decisively towards the existence of a series of well-established fiefs, commonly described as *giardini*, the lifeline of which was determined by the presence of a perennial water source. The fiefs of Dejr il-Bniet (DI1), Gomerino (RA15), Għajn Tuffieħa (MG15) and Ġnien is-Sultan (RA13) are cases in point. *Giardini* were frequently documented to contain one or more *fontane*, without qualifying the use of this term. It has already been convincingly argued in chapters five and six that within the context of a *giardino* framework during the Knights period, *fontane* can in most cases be safely associated with perched aquifer galleries.

That the availability of a water source in Malta was held in high esteem in the latter half of the fourteenth century is attested by a 1360 document commemorating the bestowal

[33] The *Cancelleria Regia* documents are today housed in the *Sede 'La Catena'* of the *Archivio di Stato* in Corso Vittorio Emmanuele in Palermo. The roots of the Royal Chancery lie in the period of the Sicilian Norman conquest, during which the political and administrative organs of state were established. See Fiorini 1999B, xvii-xv.

by King Frederick of the title of the Count of Malta to Guido de Ventimiglia (Fiorini 1999B, 8-10).[34] This submission contains a detailed list of all the rights and prerogatives of the County of Malta, and it includes mention of pastures, hunting, water and watercourses.[35] Of particular significance is the differentiation made between: (a) water; and (b) watercourses. The same applies to another *Cancelleria Regia* entry which appears to document the presence of a watercourse in Gozo.[36] Perennial watercourses are typically located in the Upper Coralline valleys of north and north-western Malta. Watercourses located in the central plains and the southern sections of Malta are seasonal, a case in point being the Wied il-Kbir (MA12) valley system at Qormi. In the case of the other water sources mentioned in (a) above, these might possibly encompass a variety of hydrological sources, including *fontane*, wells and harvested rainwater stored in cisterns.

7.3.1 The fiefs of Dejr Ħandun and Għajn Tuffieħa

Other *Cancelleria Regia* entries list important Crown properties, including fiefs. The mention of the fiefs of Dejr Ħandun (Dejr Ħandul) and Għajn Tuffieħa (MG15) in an entry dating to 17 June 1361 concerning royal concessions in favour of Johannes de Sancta Sophia and his heirs is particularly significant.[37] The implication is that both the Dejr Ħandun and Għajn Tuffieħa estates were already prime agricultural sites during this period.[38] One can also note that there is the specific mention of a water source or *fonte* at Dejr Ħandun and two at Għajn Tuffieħa. Although the nature of these water-yielding spots was not specified, in the case of the Għajn Tuffieħa sources water is likely to have originated from the two perched aquifer galleries, a detailed account and description of which is given in chapter six, section 6.1.8 (Figs. 5.68, 6.71, 6.82, 6.83, 6.92, 6.94, 6.95).

A modified version of the same entry, dated 19 June 1361, once again confirms the presence of water sources at Dejr Ħandun and a section of the Għajn Tuffieħa *giardino* known as Msieraħ.[39] In *Cancelleria Regia* documentation dating to 10 March 1398, Għajn Tuffieħa is classified as a *viridarium* or *giardino*.[40] The granting of a parcel of land at Għajn Tuffieħa with the place-name of Il-Għalqa ta' Għajn Tuffieħa is mentioned in an entry for 16 May 1399. This makes specific reference to the water resources of the Għajn Tuffieħa area described as *aquarum decursibus*.[41] This can be taken to mean either "flowing water" or possibly a "watercourse". An entry from July 1399 documenting the grant of some thirteen *salme* of land confirms a water presence at Dejr Ħandun. Terms like *margijs*, *aquis*, *aqueductibus* and *aquarum decursibus*, used in relation to the Dejr Ħandun property, are all indicative of fresh water in the area (Fiorini 1999B, 320).

7.3.2 The fiefs of Ġnien is-Sultan, Tal-Qlejgħa and Benwarrad

The *Cancelleria Regia* entry dating to 18 June 1361 chronicles the concession of various lands in the archipelago in favour of the knight Jacobus de Peregrino, Captain and Castellan of Malta. These lands include the enclosed *viridarium* or *giardino* of *Di Lu Re* (Ġnien is-Sultan – RA13) and its two fountains of Għajn Qajjed (RA19) and Għajn Tewżien (RA21).[42] The fief in question is located in the territory of Rabat (Malta).[43] The final version of this concession once more includes mention of the Għajn Qajjed and Għajn Tewżien fountains as well as fields called *La Chafe* and *Fonte*, both of which appear to have been located in the Ġnien is-Sultan area (Fiorini 1999B, 17-18). No data on the physiognomy of Għajn Qajjed and Għajn Tewżien springs or fountains are given. Keeping in mind that the description evokes a fully-fledged *giardino* framework, it is likely that perched aquifer galleries were in the early 1360s already supplying the Ġnien is-Sultan area and its neighbouring *giardini* with a perennial water source.

[34] This right was revoked soon after Guido de Ventimiglia's death in 1362. See Luttrell 1975A, 45.

[35] [...] *prefato nobilj Guidonj et suis heredibus de suo corpore legitime descendentibus Jmperpetuum Comitatum Jnsularum nostrarum Melivetj et Gaudisij cun castris, civitate, villis, casalibus, vassallis, mero et mixto Jmperio, territorijs, tenimenti, pertinencijs, divisis, limitibus, pascuis, venacionibus, aquis, aquarum cursibus, molendinis et molendinorum* [...]. See Fiorini 1999B, 8-10.

[36] *La Grua prope flumen dictum de Munsunj* [...]. See Fiorini 1999B, 59.

[37] [...] *salmarum duodecim laboratoriarum ad dictam generalem mensuram sicilie cum fontibus tribus existentibus in dictis tenimentis terrarum uno scilicet fonte in dicto primo tenimento et duobus in altero secundo* [...]. Fiorini 1999B, 10-11. Another *Cancelleria Regia* entry which probably refers to the Għajn Tuffieħa *feudo* dates to 21 May 1399. This makes no reference to the availability of water sources. See Fiorini 1999B, 310-11.

[38] The excat location of Dejr Ħandul was formerly unknown, but was identified by the author of this book with agricultural territory close to Ħofret ir-Rizz (RA27). See chapter five, section 5.2.4 and Fig. 3.6.

[39] [...] *Jn territorio dicte Jnsule Melivetj certis finibus limitata videlicet tenimentum unum terrarium laboratoriarum vocatum Deichandum cum fonte uno, margis, spacijs alijsque Juribus et tenimentum aliud terrarium laboratoriarum vocatum Ayn Tifacha cum duobus fontibus nec non terries non laboratorijs vocatis Li Spacij seu Misira* [...]. See Fiorini 1999B, 14.

[40] [..] *eidem Lancee fructus, reditus et proventus viridarij unius sive clausure dicte de Antufac site et posite in dicta insola Melivetj que est reditus unciarum septem pro quolibet anno* [...]. See Fiorini 1999B 260.

[41] [...] *La Chalca di Ayn Tuffuca, situm et positum Jn \in/sula Melivetj undique via puplica circumdatum, quodquidem tenimentum terrarium ut nobis plene aquarum decursibus, spacijs, terries, cultis et Jncultis ac arboribus tam domesticis quam silvestribus, Juribus, tenimentis, proprietatibus et pertinencijs universis sub annuo censu et honere* [...]. See Fiorini 1999B, 309.

[42] [...] *corpore legitime descendentibus Jmperpetuum subscripta bona curie nostre posita Jn predicta Jnsula \Melivetj/ Jnfrascriptis finibus limitata videlicet viridarium unum <situm et positum in dicta Jnsula Melivetj> vocatum Di Lu Re muris undique circumdatum cum fontibus duobus Jpsius viridarij, quorum unus vocatur Ayn l Cayd et alter Ayn Tosen, nec non terries laboratorijs vacuis existentibus Juxta dictum viridarium* [...]. Fiorini 1999B, 12-13. For a slightly modified version of this same entry dating to 18/19 June 1361 see Fiorini 1999B, 14-15.

[43] See chapter six, section 6.1.3 and Fig. 3.6.

The adjacent estate of Tal-Qlejgħa (RA57)[44] is similarly documented to contain various types of water sources.[45] Once again, a distinction is made between "waters", "aqueducts" and "water courses". In the case of the Tal-Qlejgħa estate, it is rather safe to deduce that *aquarum decursibus* most probably refers to the natural seasonal spring water discharge flowing through the Il-Qlejgħa valley basin. The term *aquis* probably refers to different types of water sources and possibly also encompasses water obtained from the various perched aquifer galleries known to exist in the Qlejgħa area. The use of the term *aqueductibus* is open to interpretation, but possibly refers either to perched aquifer galleries or to open channels conveying the water retrieved from the perched aquifer to fields needing irrigation water. The names of other parcels of land neighbouring Il-Qlejgħa and Ġnien is-Sultan also feature in other *Cancelleria Regia* entries, such as Ħabel Sula, tal-Ħaddedin, Sintini, Il-Bajjada and Il-Ħofra (Fiorini 1999B, 60), and the fief of Għafrid Ġinwi in the district of Fiddien (Fiorini 1999B, 115-16). This would suggest that the Sicilian Crown held land in this area in high esteem. These entries, however, fail to provide any detail of water-related characteristics within such estates.

A hydrological situation similar to the one portrayed in the description of Tal-Qlejgħa is also given for the fief of Benwarrad in *Cancelleria Regia* documentation for 24 February 1398. The Għajn Riħana fountain (MO4) as well as the other water sources feeding this fief are here listed as *aquis*, *aqueductibus* and *aquarum* (Fiorini 1999B, 256).

7.3.3 The Gomerino estate

Particularly significant is the 1398 entry describing the concession of the Gomerino *giardino* (RA15) located in the limits of Fiddien in the territory of Rabat (Malta) to Antonellus de Sancta Sophia. This lengthy document contains transcripts for earlier concessions of the Gomerino fief dating to 1317 and 1318 (Fiorini 1999B, 212-18),[46] strongly hinting that a *giardino* framework was already in place during the opening decades of the fourteenth century.[47] The text of the 1318 document recording the sale of Gomerino was drawn up and validated by the royal public notary of Malta and Gozo, Rogerio de Alamanno, and is dated to 8 May of the same year (Luttrell 1978, 172-77). From this contract it emerges that Gumerin, or Gomerino, had been previously granted to Skius de Solimella, but had devolved to the Crown since he had died without any legitimate heirs. This 1318 contract clearly describes Gomerino as a *giardino*, highlighting the water sources within: [...] *Iardinum de gumerin cum terris mandris criptis spacijs misiratis aquis Ingressibus et egressibus accessibus Iusticijs et pertinencijs suis et omnibus alijs Iuribus ipsi Iardino* [...] (Luttrell 1978, 173). The use of the word *criptis* is intriguing for in this circumstance may allude to a rock-cut burial site or another subterranean space. Two other *Cancelleria Regia* entries on the Gomerino fief, both dated to 9 February 1398, again make reference to the water sources in this territory.[48]

The present-day hydrological setup up of the Gomerino estate involves the presence of two different water sources (Figs. 5.52, 5.62, 5.65). The first consists of a seasonal spring flowing through the valley bed which dries up completely during the warmer, rainless summer season. The second water source is perennial in nature and originates from within a perched aquifer gallery, strategically located in the upper reaches of the valley (Fig. 5.63). Water retrieved from the gallery system appears to have been stored in the past within a large, open-air reservoir of unknown antiquity fronting the gallery. A series of masonry canals gravity-fed the gallery water to adjoining and underlying fields. Of significance is the fact that the ceiling of the Gomerino gallery is pierced by a series of rock-excavated shafts. Three of these are still visible from the surface.[49] Two of the shafts are circular in shape, the other is rectangular. Having larger dimensions, the rectangular shaft could easily allow access to the gallery interior – a matter attested to by a series of rock-hewn handholds and footholds leading down to the bottom of the shaft (Figs. 5.64, 5.66).

Ongoing contentions on matters related to water usage rights between the current Gomerino estate landowners and long-lease farmers holding agricultural property on the estate led to the installation in 2009 of a metal gate blocking access to the gallery interior. This prevented the present writer from personally investigating the Gomerino gallery system. However, documentation made available by V. Tabone, as well as reliable claims made by a group of amateur speleologists who had partially investigated the Gomerino gallery in the past, and the present writer's personal observations, lend substance to the view that the subterranean water-retrieval system is a fully-fledged *qanat* system.[50] A copy of a map based on the 1897 survey

[44] Tal-Qlejgħa (RA57) is nowadays more commonly known as Chadwick Lakes, which covers a substantial portion of the surface area of this fief.

[45] [...] *hedificijs, aquis, aqueductibus, aquarum decursibus, piscacionibus* [...]. See Fiorini 1999B, 252.

[46] A 1465 *Cancelleria Regia* document once again confirms the *giardino* framework present within the Gomerino landholding. See Fiorini 2007, 74.

[47] The fief of Eghmieri or Gomerino was once again listed as a *giardino* in a much later entry dating to January 1480. See Fiorini 2007, 400-1.

[48] [...] *in perpetuum predictum Jardinum de Gumerin cum terris, mandris, criptis, spacijs, misiratis, aquis* [...]. See Fiorini 1999B, 214; and [...] *de Gumerin situm et positum in Jnsola nostra Meliveti cum terries, spacijs, criptis, mandris, misiratis, aquis, Ingressibus et regressibus* [...]. See Fiorini 1999B, 219.

[49] V. Tabone, one of the co-owners of the Gomerino estate, had in December of 2008 indicated the location of the shafts to the present writer. The family's eighteenth-century *cabreo* for the Gomerino estate was also made available. The perched aquifer gallery entrance is not marked on the plan of the estate, but the fronting rectangular structure is coloured in blue ink, thus indicating that this was a water-related feature. On the other hand, *foglio* 94 of the Gomerino *cabreo* does make direct reference to the *fonte d'aqua sorgente*.

[50] Geotechnical Engineer A. Mifsud, A. Camilleri and A. Falzon had partially investigated this perched aquifer gallery about fifteen years ago

sheet of the Gomerino area in the possession of V. Tabone is very informative (Fig. 4.5). The map probably dates to the post World War II years and has ink markings indicating the west and north-facing boundary walls of the Gomerino estate.[51] Similarly included are an intermittent line and a series of circles which represent the course followed by the subterranean gallery system at Gomerino. Marked on the map as *Ghemieri Spring*, this perched aquifer gallery is shown to be made up of at least thirteen shaft-wells. The outermost section of the gallery first follows a west-north-west direction but then turns sharply south and continues heading in a south-eastern direction towards the Bieb ir-Ruwa area of Fiddien. According to the data presented in the map, the Gomerino gallery is over 424 m long and seemingly extends further south-eastwards in an area not covered by this plan. The shafts are on average slightly more than 30 m apart, suggesting the existence of a *qanat*-type setup.[52]

7.3.4 *The fief of Ħemsija and other* viridaria *in north-west Malta*

Other entries dating to 1375 make direct references to *viridaria* or *giardini*. A case in point is the mention of land at Buqana below Mdina and Għeriexem (RA26) at Rabat (Malta) (Fiorini 1999B, 132-33). Even though the available historical documentation does not refer to the availability of water sources in the Il-Qattara area of Buqana (MO1), these are indeed available in the form of localised perched aquifers within exposed Upper and Middle Globigerina deposits in the Il-Qattara area of Buqana. These aquifers are undoubtedly an integral part of crop cultivation in the area.[53]

The entry dated 14 February 1398, eliciting the granting of the fief of Ħemsija (RA58) to Franciscus Gatto and his heirs in perpetuity, describes the availability of water sources (Fiorini 1999B, 225-28). Particularly noteworthy is the distinction made between waters, water channels and water outlets.[54] The impression given is that of a well-established *giardino* framework. The fief of Ħemsija can probably be associated with the il-Wilġa tal-Ħemsija area, located to the immediate east of Mdina (Figs. 5.115, 5.116). A perched aquifer gallery tunnels into an exposed Mtarfa Member deposit located below the De Redin bastion walls in the Bishop's palace (MD1) area of Mdina at map reference 46307 71681, and could have supplied the Ħemsija estate with a perennial water source (Fig. 5.117).[55] The fief of Ta' Brija (RA44), located between Buskett and Siġġiewi, is likewise described as having water, water channels and water outlets.[56] The same description applies to the fief of Budaqq located in the whereabouts of Burmarrad, for which there is mention of water, water channels and water outlets.[57]

Is-Saqqajja, Il-Mislub, Ġnien il-Far and Ġnien il-Firien are all place-names listed in *Cancelleria Regia* entries for 18 and 22 June of 1361 (Fiorini 1999B, 13-14, 18-19). Most of these toponyms appear to belong to areas in the whereabouts of Rabat (Malta) and Buskett.[58] The place-name *Saqqajja* is indicative of water availability and its use for the purpose of irrigation (Wettinger 2000, 522). The *Cancelleria* entry for 22 June actually does mention water availability here,[59] but does not specify whether this originates from within a perched aquifer gallery, as is the case with the current setup at Saqqajja (RA40) located to the north-eastern side of Rabat (Malta).[60] It is only in documentation starting from 1530 onwards that the existence of a perched aquifer gallery at Saqqajja in Rabat is referred to.[61] The Ġnien il-Far and Ġnien il-Firien place-names are also indicative of a *giardino* framework already in these areas in the mid-fourteenth century.

The 8 July 1376 *Cancelleria Regia* entry for the fief of Dejr il-Bniet (DI1) likewise classifies it as a *viridarium* or *giardino* (Fiorini 1999B, 158-59), but there is no reference to the hydrological assets of the area. A later entry for the same fief dated to 14 February 1398 is more informative and describes the Dejr il-Bniet *giardino* as having water and watercourses.[62]

---

and have verbally recounted to the present writer how the gallery's ceiling is pierced by a series of shaft wells.

[51] For the duration of World War II, the British Government had a military presence stationed in the area and had sequestered the water resources originating from this gallery. Personal communication by V. Tabone.

[52] See chapter three, section 3.1.6.

[53] See chapter two, section 2.6.2.

[54] [...] *Jmperpetuum cum omnibus suis Juribus, racionibus, proprietatibus, reditibus, proventibus, edificijs, aquis, aque ductibus, aquarum decursibus, pascuis, pratis, molendinis* [...]. See Fiorini 1999B, 226.

[55] Malta Survey Sheet 73, School of Military Engineering, Chatham, 1921. Shafts probably pierce the ceiling of this perched aquifer gallery, thus allowing water to be retrieved directly from within the Mdina settlement (MD4). See chapter five, section 5.3.

[56] [...] *Jn perpetuum cum omnibus suis Juribus, racionibus, proprietatibus, reditibus, proventibus, hedificijs, aquis, aque ductibus, aquarum decursibus* [...]. See Fiorini 1999B, 234. Mid-seventeenth-century documentation reveals that a portion of the Ta' Brija estate obtained a perennial water supply from a *mina*, or perched aquifer gallery (RA25), at Wied l-Isqof in the Rabat (Malta) territory. See chapter five, section 5.3.

[57] This entry dates to 16 February 1398: [...] *aquis, aque ductibus, aquarum decursibus* [...]. See Fiorini 1999B, 243.

[58] Refer to Malta Survey Sheet 73, School of Military Engineering, Chatham, 1921; and Malta Survey Sheet 101, School of Military Engineering, Chatham, 1919.

[59] [...]*La Saccaya suis certis finibus limitatum in quo oritur quidam <phe.> fons aque cum predicto fonte* [...]. See Fiorini 1999B, 18-19.

[60] The entrance to the Saqqajja perched aquifer gallery has been completely screened off by St Paul's niche and fountain since at least the eighteenth century. Abdilla 2003, 20-21. The perched aquifer gallery at Saqqajja is accessed by means of a small door located to the right of the fountain.

[61] Documentation dating to 1530 mentions cleaning works at the Saqqajja *fontana*, possibly referring to periodic maintenance works being carried out in the water gallery feeding the Saqqajja *fontana* with water. See Fiorini 1992, 38-40. A *bando* issued in 1626 also relates how the waters yielded by the Saqqajja *fontana* had been significantly diminished due to the practice of retrieving water from *condotti* or shafts which tapped into the fountain system. See chapter five, section 5.3.

[62] [...]*<Jn perpetuum> quoddam viridarium <s.> cum aquis, aquarum cursibus, spacijs, terris cultis et Jncultis, Juribus, tenimenis et pertinencijs suis vocatum de Jrbinet*. See Fiorini 1999B, 230-33. Another mention of water and the presence of watercourses at Dejr il-Bniet is

7.3.5 Hydrology-related evidence from Gozo

Gozo is only occasionally represented in *Cancelleria Regia* registers, but the available data portray a hydrological situation similar to that of Malta. The documentary source relating to the granting of the Tal-Majmuni estate in the *Rangisija* district mentions the presence of water and water outlets.[63] Another water-related reference dating to 13 November 1372 deals with the perpetual concession of half the rights to an unnamed fountain on Gozo.[64] The concession of an unlocated parcel of land called Ta' Ħamitu on Gozo also contains details of the hydrological assets of the area that consisted of waters, aqueducts and water canals (Fiorini 1999B, 316). The 1399 reference to the spring of Għajn il-Qasab probably corresponds with a spring of water still feeding the Għajn Qasab (GO12) estate at Nadur (Gozo) with a perennial water supply (Fiorini 1999B, 324-25; Figs. 2.17, 6.104, 6.110).[65] Field-research has shown this water source to originate from within a perched aquifer gallery at map reference 36081 90314 (Figs. 6.107, 6.108, 6.109, 6.111).

7.4 Documents of the Cancelleria Regia *at the State Archives of Palermo: The fifteenth century*

Hydrology-related *Cancelleria Regia* documentation for the early fifteenth century is less informative than the preceding fourteenth-century entries. A number of late fourteenth and early fifteenth-century entries for Malta and Gozo are the result of a stocktaking exercise, commissioned by Martino, King of Sicily, involving the general reorganisation of fiefs and other royal concessions of the *demanium*. It was during such an undertaking that old privileges were confirmed and confiscated properties awarded to trusted individuals (Luttrell 1978, 168). Important Late Medieval fiefs on Malta, such as Għajn Tuffieħa (MG15), Fiddien (RA11) and Il-Qlejgħa (RA57) feature prominently in the *Cancelleria Regia* documents for the fifteenth century, but the majority of the documentary sources do not shed direct evidence on the water sources of these landholdings during the period in question.[66] It is interesting to note, however, that a 1404 document on the fief of Dejr il-Bniet (DI1) describes it as a *viridarium*, thus indicating that a *giardino* framework was in place (Fiorini 2004, 65). A 1478 entry on the fief of Il-Qlejgħa also gives an indication that fruits were being produced, underlining the impression of a well-established *giardino* framework as existing within (Fiorini 2007, 362-65). A substantial portion of notarial entries for fief landholdings show that it was normal for these properties to be leased out either directly to farmers or to intermediaries, who would in turn lease the land to farmers.[67]

7.4.1 The Għeriexem spring[68]

A June 1416 document records the transfer of rights of the small fief of Għeriexem (RA26) in Malta in exchange for half of the Sicilian fief of *Inbaccari*, in the limits of Piazza in central Sicily. Incorporated in the rights and privileges over the Għeriexem estate is access to water, aqueducts or water canals and watercourses.[69] It should be noted that a perched aquifer gallery at Għeriexem still supplies the area with a perennial water supply (Figs. 5.108, 5.109, 5.110, 5.111, 5.113). This water system was partially investigated in November 2012 and found to consist of a linear gallery the width of which varies from 60 to 80 cm. The height of the gallery ceiling varies considerably but in sections incorporates a series of rectangular shafts which at intervals pierce the roof of this man-made subterranean water conduit. The shafts are located at a distance of ca. 30 m from each other and most contain handholds and footholds, used to facilitate access to, and exit from, the gallery (Figs. 5.111, 5.112).[70]

Tool markings preserved on the rock-face of both the gallery and shafts appear to be identical, thus suggesting that both formed an integral part of a project aimed at supplying the Għeriexem *giardino* with a perennial water supply. Taking this evidence into consideration it is safe to propose that this finely excavated perched aquifer gallery is, therefore, typologically identical to *qanat*-type water extraction systems. In *qanat* systems, shafts were necessary for the removal of the excavated material from within the gallery and for establishing an appropriate floor gradient in order to encourage efficient water extraction. On completion of the gallery excavation project, shafts were often also utilised for water extraction.[71] That this was the case with the Għeriexem perched aquifer gallery

found in the concluding section of this same document: [...] *dictumque viridarium cum aquis, aquarum cursibus, spacijs, terries cultis et Jncultis* [...]. See also chapter five, section 5.2.1.

[63] This *Cancelleria Regia* entry dates to 16 February 1398: [...] *tenimentis, aquis, aquarum decursibus, spacijs, terries cultis et Jncultis* [...]. See Fiorini 1999B, 240.

[64] [...] *Facte sunt patentes lictere <de> Licterio de Barbara de Gaudisio de concessione medietatis cujusdam fontane curie cujus altera medietas est dictj Licterij site et posite in dicta Jnsola Gaudisij* [...]. See Fiorini 1999B, 55-56.

[65] See also chapter six, section 6.1.11.2.

[66] On the granting for lease of land in Marsa known as Tabursa see Fiorini 2004, 9-11. For the exchange of the fiefs of Il-Qlejgha and Għajn Tuffieħa for a pre-specified annual income see Fiorini 2004, 25-27. For other entries concerning Għajn Tuffieħa see Fiorini 2004, 112-13; 126-27; 164; 194-95; 207; 225; 263-64; 302; 352-53. See also Fiorini 2007, 129, 145, 174; 197, 369-72. In the entry dated 9 September 1467, Għajn Tuffieħa is once again referred to as a *viridarium*. For entries concerning the Il-Qlejgħa *feudo* see Fiorini 2004, 130; 230-31; 265; 275; 299; 355-56; 398-400; 480-81; 704. See also Fiorini 2007, 253-55; 306-8; 375-76.

[67] On the subleasing of rented property during the fifteenth and sixteenth centuries see Wettinger 1982, 8-9.

[68] For further information on the Għeriexem estate and spring see also chapter five, section 5.3.

[69] [...] *Juribus, prerogativis, aquis, aqueductibus, aquarum decursibus, proprietatibus, redditibus* [...]. See Fiorini 2004, 221. A subsequent entry for the fief of Għeriexem contains no information on the hydrological sources of this property. See Fiorini 2004, 222-23.

[70] Documentation dating to the 1690s shows that towards the end of the seventeenth century, this water system was investigated by the French military engineer and architect Mederico Blondel des Croisettes, who specifically mentions the *mina*, or gallery, and the *spiragli*, or shafts. See Univ. 187, ff. 1-14; and chapter five, section 5.3.

[71] See chapter three, section 3.1.6.

was evident by the fact that the floor area immediately underlying each shaft was lowered in order to allow water accumulation, thus making it possible to lower a water bucket and allow it to fill up with water. Currently abutting the gallery entrance at Għeriexem is a monumental fountain and water collection basin which appears to date to the seventeenth century.

7.4.2 *Għajn Qajjed*

Another entry dating to October 1457 refers to a small spring of water or watercourse located within property at Fiddien in Rabat (Malta) (Fiorini 2004, 676). Interestingly, the expression used to describe this source is *fonticulum*, a term substantially different from the other water-related terminology used in other *Cancelleria Regia* documents. The nature of the mentioned water source is difficult to ascertain. As this entry does make mention of a cliff or ravine located to its south side, as well as the mention of a territory in its vicinity having the place-name of Il-Ħafa, it is possible to deduce that the land in question corresponds to a parcel of land which in seventeenth-century documentation was identified as *Jitassun*, a member of the *giardino di due fontane* (RA14) located in the Għajn Klieb (RA18) area (NLM. Treas. B. 289, Plan 9; Fig. 6.8 "D"). Should this be the case, it is likely that in this instance reference is being made to spring water originating from the Għajn Qajjed (RA19) water gallery by means of partly-subterranean and partly-open water canals.

7.4.3 *The Saqqajja and Qattara areas of Rabat (Malta)*

The *Cancelleria Regia* entry documenting the granting of land called Is-Saqqajja (RA40) in Rabat (Malta) dating to August 1443 also makes mention of water within: [...] *cum dicte pecie terre Juribus, aquis, aquarumque decursibus* [...] (Fiorini 2004, 522). This possibly refers to water flowing out of a fountain at Saqqajja which originates from within a perched aquifer gallery. Importantly, this entry makes a double emphasis on water presence in this property. The hydrological resources of Saqqajja are also mentioned in a *Cancelleria Regia* entry dating to the latter half of the fourteenth century (Fiorini 1999B, 18-19; see section 7.3.4).[72]

Water availability is similarly reported in a document dating to February 1440, chronicling the granting of land at Il-Qattara (MO1) located on the north-eastern flank of the *Civitas* (Fiorini 2004, 454). As the Il-Qattara area consists of an exposed deposit of Middle Globigerina Limestone, it is likely that this entry alludes to a localised perched aquifer source which exists in this area (Fig. 5.101).[73] The Il-Qattara place-name is also synonymous with water availability (Wettinger 2000, 439).

7.4.4 *Gozo*

A *Cancelleria Regia* entry for 3 November 1428 reports the granting of land called Dar Ħamra(?) in the district of L-Egħjun (modern day Għejun meaning "of the fountains or springs") and Pachi in Gozo (Fiorini 2004, 257-58). The physiognomy of this water source is not known. Water availability to the bestowed property is in part hinted at through the mention of trees growing within this agricultural installation (Fiorini 2004, 258). No evidence is given, however, on the species of the trees in question, in this way making the hydrological needs of the cultivated crop difficult to assess. This entry is the first documented instance for the Tal-Għejun (GA20) place-name, which can be associated with agricultural property in the Xagħra region of Gozo (Wettinger 2000, 221, 550). Alternatively referred to as *Tewnit* in a 1465 notarial deed, this territory is in 1428 described as *viridarium* or *giardino.*[74] Another notarial deed dating to 1565 is more illuminating and reveals this property to be: *loyun, viridariolum arboribus plantatum cum aquis aqua decursibus maceriis circumdatum in contrata tal cache* (Wettinger 2000, 221), making ample reference to a secure water presence within. The *Tewnit* place-name, which refers to a district at Il-Għejun, also known as Wied ir-Riħanija, has an obscure meaning, but it was first recorded in 1399 (Wettinger 2000, 550).

Another entry concerning Crown Property from Gozo involves the emphyteutical lease of two *salmata* of land, including a small spring called *De Avitu* which Fiorini tentatively associates with the Taż-Żnuber spring at Id-Dwejra (Fiorini 2007, 263-65). As this entry lacks any further hydrology-related data, it is difficult to reach any conclusions on the nature of this water source. Two other *fontanelle*, or small springs, are listed in another *Cancelleria Regia* entry in connection with the emphyteutical lease in perpetuity of four *salmata* of land in the Qala district.[75] The springs in question are Għajn il-Ħarim and Għajn il-Ħaġra.[76] The location of Għajn il-Ħarim remains unknown, but Għajn il-Ħaġra can perhaps be associated with an area at Wied ta' Grejgel close to Daħlet Qorrot, the current place-name of which is Għajn Ħaġar (GO11).[77] The nature and physiognomy of these water sources is not known.[78]

[72] For Knights' period documentation on the Saqqajja perched aquifer gallery see chapter five, section 5.3.
[73] For the water retention potential of Middle Globigerina Limestone deposits refer to chapter two, section 2.6.2.
[74] *Jl heyun, viridarium, in Jnsule Gaudisii*. See Wettinger 2000, 221.
[75] [...] *in quibus sunt due fontanelle vocate An Alyarim et <Anag.> An Alyagira in contrata dila Cala* [...]. See Fiorini 2007, 273-74.
[76] Neither of these two water sources is listed in Wettinger's place-name study of 2000.
[77] The Għajn Ħaġar (GO11) area is located at map reference 37902 89381.
[78] See also chapter six, section 6.1.11.3.

7.5 *The Notarial Documents: Notaries Paulo Bonello and Giacomo Zabbara*

A more wide-ranging representation of the hydrological situation in Malta during the fifteenth century emerges through a study of the notarial deeds drawn up by Notaries Paulo Bonello and Giacomo Zabbara.[79] In the available fifteenth and sixteenth-century documentation, data on fiefs, several of which featured commonly in *Cancelleria Regia* documents, generally fade into the shadows. The general exceptions to this are the estates of Għajn Riħana (MO4), Għajn Tuffieħa (MG15) and Il-Qlejgħa (RA57). In contrast, the water situation in smaller, private landholdings randomly scattered throughout Malta and Gozo comes to the fore. There are only a few surviving pre-fifteenth-century notarial deeds, but the available documentation shows that by about 1300, a few notaries were already active in Malta and Gozo.[80] A study of the notarial deeds by Notaries Paulo Bonello and Giacomo Zabbara shows that the contract formulae used bear significant resemblance to other coeval contracts on the continent (Fiorini 1996, x). The deeds drawn up by both notaries adhere to the framework stipulated by the *Capitula Tebellionum* of 1440, the aim of which was to regulate the activities of the notarial profession.[81] A matter of consideration in the context of hydrology-related data is that the surviving contracts drawn up by Notary Paulo Bonello only span a three-month period, namely September – November 1467 (Fiorini 1996, x). Notary Giacomo Zabbara's contracts cover a much broader time-frame and date from March 1486 to May 1501.[82]

Paulo Bonello was issued with the warrant of notary public in March 1464. He was involved in various business ventures, one of which was the undertaking of procuratorships on behalf of individuals who owned property in Malta but resided abroad (Fiorini 2005, xxxviii). Various entries drawn up by Notary Bonello mention water cisterns and the availability of springs feeding arable land with a perennial water supply.

Existing documentation suggests that Giacomo Zabbara was trained in Sicily, from where he obtained the warrant necessary to exercise his profession.[83] The language used in the drawing up of most documents was generally Latin. When, on the other hand, Zabbara was short of the right phrase or of the precise Latin term, he used the Sicilian equivalent (Fiorini 1996, xiii-xv). The use of the Sicilian vernacular was frequently resorted to when drawing up agrarian-related contracts or when in need of technical farming terms for which the use of Latin was inadequate. Apart from showing strong Sicilian connections, it also reflects an analogous situation in Sicily during this period (Fiorini 1996, xv). In instances where even the use of Sicilian was deemed inadequate, such as when dealing with place-name-related evidence or in the description of technicalities, the use of Late Medieval Maltese was resorted to in order to allow greater precision of expression (Fiorini 1996, xv).

Entries from Notary Zabbara's deeds cite the right of access to water flowing through the valley course in an area close to Qattara, to the north-east of the *Civitas*,[84] and to another located close to a field called Ta' Ħajta near Żebbuġ (Malta) (Fiorini 1999A, 18). In another entry, a valley watercourse is noted as a specific designation for the division of property (Fiorini 1996, 149). The same appears to apply to half a field in the district of Tewnit in Gozo, to the east of which was a *vallonus aque pluvialis* (Fiorini 2005, 173).

An amicable settlement regarding the use of cistern water at Raħal Għaxaq gives an insight into water sharing agreements which were being resorted to in the fifteenth century. The agreement stipulated that one party had rights to two-thirds and the other to one-third of the cistern water. The drains leading to and from the cistern were also to remain as they had been in the past (Fiorini 1996, 215-16). Similarly, an entry dating to September 1496 makes reference to shares in cisterns located in an alleyway called Ta' Raħab at Ħabel Qoton (Fiorini 1999A, 332). Some entries show that the right to a share in the water of a cistern could be inherited (Fiorini, 1999A, 343). In another entry there is mention of land called *Bjr Hugile*, which perhaps can be transliterated to Bjar Ħaġel.[85] A separate entry dating to September 1496 records access to the water of what appears to be a pond identified as *Demus*, possibly located in the territory of Marsa.[86]

---

[79] An investigation of the notarial deeds of Zabbara and Bonello shows that in fifteenth-century Malta, properties could be leased out directly to a farmer or through the employment of an intermediary, who would then sublet the land in smaller proportions amongst a number of different farmers. It was also common for peasants to lease land belonging to the inhabitants of Mdina for periods of three, four or six years. The lease period was possibly determined by the type of crop under cultivation. See Wettinger 1982, 8.

[80] For a historical account on the evolution of notarial practices in Malta till 1500 see Fiorini 2005, xi-l.

[81] In brief, this imposed the use of a *bastardello*, or notebook. Apart from a notary and a judge, two witnesses were needed when the contract being drawn involved a value of up to a gold *libra*. For any value above this, three witnesses were necessary. To avoid the danger of having false contracts inserted illegally, notaries were urged not to leave any blank space between one contract and the next. The registration of contracts had to take place in the space of one month from the date of the drawing up of the act. Each notary had to compile an annual alphabetical index of all names of parties in his contracts and make this available for reference purposes. See Fiorini 2005, xxxi-xxxiii.

[82] It appears that Notary Zabbara passed away sometime between May 1501 and October 1503. See Fiorini 1996, x. It is worth pointing out that there is a lapse of six years between Part 1 and Part 2 of the Zabbara register. The last entry in the Part 1 section of the register dates to beginning of March 1488. The first entry in the Part 2 section of the register dates to September 1494. See Fiorini 1999A, xiii.

[83] Giacomo Zabbara's warrant to practise as notary public was registered in the Chancery of Palermo on 28 September 1462. See Fiorini 1999, xi.

[84] [...] *dicitur la Cactara conJunctum cum viis publicis et cum vallono aque pluvialis cum omnibus et singulis Jpsarum terrarium venditarum Juribus* [...]. See Fiorini 1996, 3.

[85] The Bjar Ħaġel place-name is not given in Wettinger 2000.

[86] Fiorini, 1999A, 338-39. Also refer to the Ta' Demus entry in Wettinger 2000, 111.

Several deeds also make reference to *giardini* and "Għajn" water sources, such as the mention of [...] *viridarium quod dicitur Ayn Culliye* [...], recorded in a May 1486 document (Fiorini 1996, 48).[87] A deed drawn up by Notary Bonello in September 1467 should be noted (Fiorini 2005, 26-27). This documents the transfer of property between Antonius Busalib and Antonius Callus, both residents of the Civitas. Busalib renounced in favour of Callus all rights over a field and a meadow at Wied ir-Rum (RA56) in the territory of Rabat (Malta). Callus, on the other hand, renounced his rights over half the water-yield of a *fonte aquarum* called Il-Fawwara ta' Wied ir-Rum. A brief topographical description of the property in question makes it possible to identify this landholding with the *giardino* Tal-Callus (RA39) at Wied ir-Rum, discussed in chapter six, section 6.1.7 above (Figs. 6.46, 6.63).[88] Field-investigation carried out by the present writer has determined the presence of two *fontane* or springs, both of which originate from within perched aquifer galleries. It is possible that the 1467 entry mentioning the spring of Il-Fawwara ta' Wied ir-Rum is a reference to one of these two water sources. The presence of a perched aquifer gallery feeding the Callus estate with water in the latter half of the fifteenth century would seem to be most probable.

The mention of another landholding at Wied ir-Rum in a deed dating back to October 1467 also merits discussion. It concerns the lease of a *viridarium* called *di Cursu* (RA8), made up of fields and trees and which had water availability. Highly significant is the mention of the cultivation of trees and that of a spring or fountain having the name of Għajn Qamar, which the lessee was promising to clean (Fiorini 2005, 99-100). The land in question could not be located.[89] Taking into account the *giardino* setup of the *di Cursu* territory, it is likely that the Għajn Qamar spring originated from within a perched aquifer gallery. If this were so, the cleaning works mentioned in the deed of lease probably refer to the maintenance which such water systems necessitate periodically. Maintenance works primarily involve the removal of sediment, mud and at times even clay infiltration from within the gallery system in order to ensure an unhindered water flow from the gallery interior towards its entrance.

A number of entries mentioning Għajn Riħana (MO4) in the territory of Bidnija make no reference to the water resources of this landholding. Nonetheless, Għajn Riħana is specifically mentioned as containing three *viridarij* (Fiorini 1996, 77-78),[90] namely Ġnien il-Kbir, Il-Ġnejna and Il-Għalqa tal-Ġnien (Fiorini 1999A, 161-62). A particular clause for the lease of a section of the Għajn Riħana territory bound the lessee to plant twenty-five trees of an unspecified species (Fiorini 1996, 243-44). Both these entries are indicative of a water presence within the Għajn Riħana estate.[91] A separate entry on Għajn Riħana dating to September 1495 regulates usage of land called Il-Wilġa or L-Għalqa tal-Ġnien, "below the reeds", where the lessee could draw all the water needed for irrigation provided that this was not detrimental to the trees planted within.[92] The L-Għalqa tal-Ġnien *giardino* appears to belong to the Għajn Riħana territory,[93] and the specific mention to the area "below the reeds" indicates the presence of a perennial water source and that cultivation was taking place on terraced land. This hydrological arrangement again implies the presence of a perched aquifer gallery feeding the L-Għalqa tal-Ġnien *giardino* with a perennial water source. The *giardino* setup of the Għajn Riħana territory is also specified in a separate entry dating to September 1495, and it involves the pledge to deliver the fruits produced by the land leased in this *giardino* and the regular planting of trees (Fiorini 1999A, 157-58).

A notarial deed documenting the sale of the fief of Għajn Tuffieħa (MG15) in May 1487 reveals the hydrological assets of the area: [...] *Jn perpetuum predictum pheudum de Ayn Tuffaha cum suo viridario et omnibus et singulis suis Juribus, menbrjs, spacijs, aquis, aquarum decursu, griptis* [...] (Fiorini 1996, 195). Particularly interesting is the mention of the presence of subterranean features at the Għajn Tuffieħa estate. It is not known if this refers to the Razzett tal-Qasam (MG18) section of the Għajn Tuffieħa territory, where caves dating back to the Palaeochristian period have been documented (Cardona 2002, 161; Figs. 5.68, 6.94) or if this might refer to a subterranean water capture system.[94] Another entry documenting the lease of the Għajn Tuffieħa *giardino* for six years once again touches upon the water resources present within, namely the "waters" and the "water outlets" of the *giardino*.[95] A right to water access is inserted in another lease of the Għajn Tuffieħa *giardino* dating to August 1487 (Fiorini 1996, 269).

[87] Għajn Kullija (RA3) was identified by Wettinger as horticultural land located to the west of Mdina, the first recorded instance for which is a will dating to July 1452. See Wettinger 2000, 185-86.

[88] Stanley Fiorini reached similar conclusions. See Fiorini 2003, 24.

[89] Various farmers owning agricultural land at Wied ir-Rum and Wied Ħażrun who in the past have provided me with information concerning their properties could not recall ever hearing the mention of Għajn Qamar. It is probable that this place-name, which is not even listed in Wettinger 2000, has since the fifteenth century faded from popular memory.

[90] The three *giardini* forming part of the Għajn Riħana landholding are once again documented in an entry dating to January 1495. See Fiorini 1999A, 60.

[91] Other entries mentioning Il-Bidni, one of the *giardini* of Għajn Riħana estate, date to 17 August 1487 and 5 March 1488. The cultivation of twenty trees in this *giardino* is mentioned in the latter instance. Another entry specifically mentions the existence of a *giardino* engaged in fruit cultivation at Għajn Riħana. See Fiorini 1996, 248-49, 261, 333-34.

[92] [...] *Jnsuper Jdem Nicolaus ad cuctuneriam aquaticiam concessit dicto Georgio terras vocatas Il Guilge sive La Galca de il Ginen subta li canni tantu quantu abastira la acqua Jta quod non preJudicetur arboribus Jn eorum Jrrigacione necessaria et hoc Jn anno vicisitudinis cuctori* [...]. Fiorini 1999A, 156-57. In 1499 income from the Għajn Riħana estate amounted to ca. 70 *uncie* – a rather hefty amount when compared to the income generated by some of the other royal estates. See Wettinger 1982, 5-6.

[93] This is seemingly confirmed by a subsequent entry dated to 21 September 1495. See Fiorini 1999A, 161-62.

[94] For a discussion of the hydrological assets of the Għajn Tuffieħa *feudo* see chapter six, section 6.1.9.

[95] [...] *cum suo viridario, aquis aquarum decursibus, spacijs, ridumjs, griptis* [...]. See Fiorini 1996, 219.

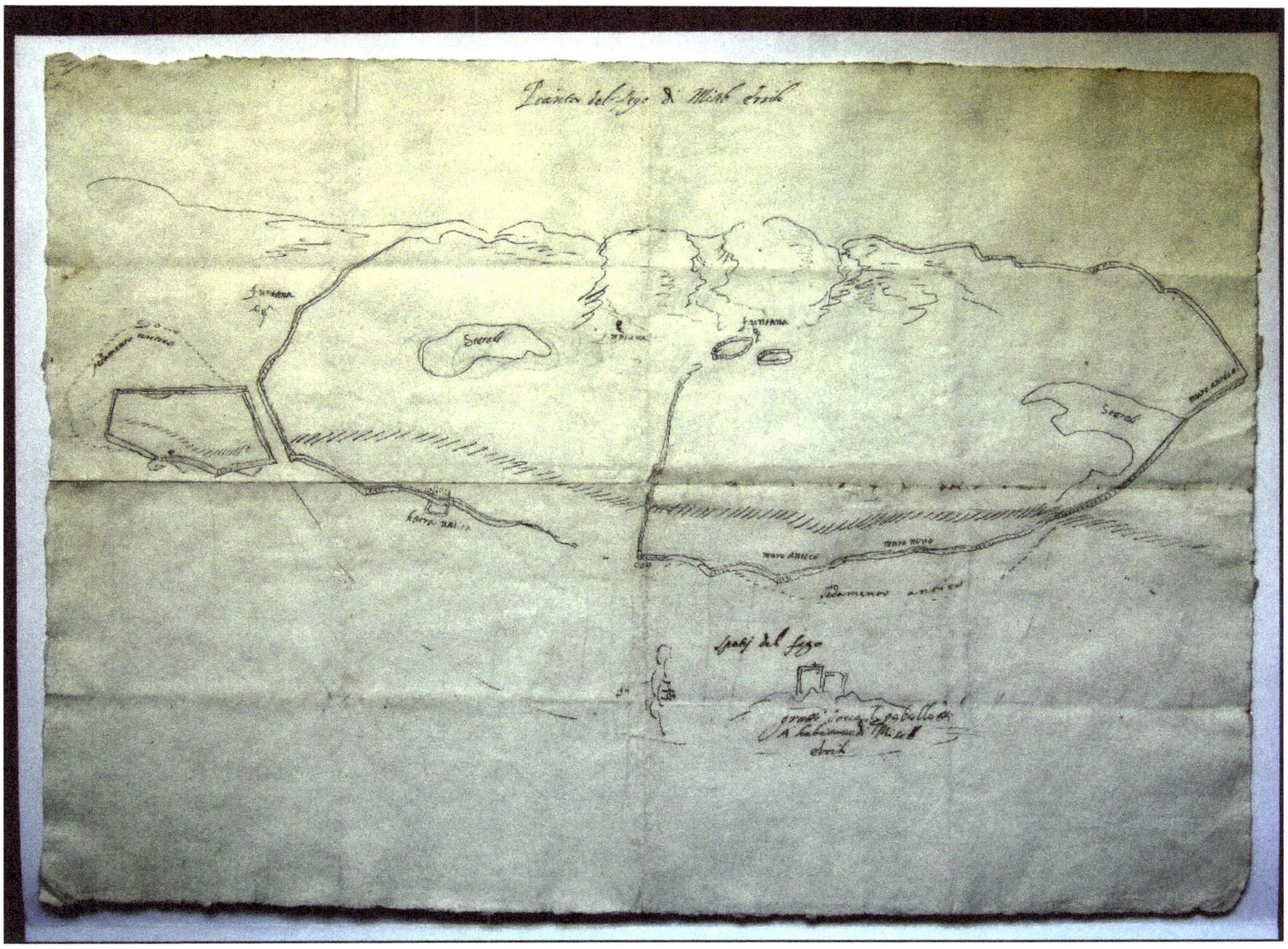

Fig. 7.3 Plan of the fief of Miżieb ir-Riħ (MG17), possibly located in the Pwales territory dating to ca. 1564. Marked on the plan are three *fontane* and remains probably of archaeological interest. Source: Extracted from the Cathedral Archives, Beni Catedrale, vol. 4, f. 6. Filed under: Drawing Inventory no. 67.

Dating to 13 August 1487 is a document recording the lease of a parcel of land known as Ġnien il-Far (Fiorini 1996, 244-45). Whilst making no reference to the availability of a perennial water source, the conditions stipulated in this contract make it likely that the life source of the *giardino* was a spring of water possibly originating from within a perched aquifer gallery. In this case the lessee had to plant twenty-five fruit trees in each year of the lease, had to ensure that the trees within the *giardino* were watered properly and that water was not to be used for the irrigation of more than two *modia*[96] of the land.[97] A similar picture emerges from the document recording the lease of a *giardino* known as Ta' Binġemma (MG19) on 22 August 1487. The contract bound the lessee to plant a total of twenty-five trees, five of which had to be orange trees, and to supply all the water required for the farm on that estate (Fiorini 1996, 251-52).[98]

A comparable situation is portrayed in other accounts documenting the lease of a *giardino* having the name of in-Nixxiegħa at Pwales (Fiorini 1996, 258-59).[99] There is no direct reference to water in this entry, but place-name evidence together with the fact that the lessee was bound to plant twenty-five trees each year strongly suggests a water presence. The same applies to the lease of the estate of Għajn Astas (MG14), also at Pwales (Figs. 5.68, 5.88).[100] Apart from the toponomastic evidence, a water presence is detected through the mention of reeds growing within.[101] Similarly, a contract on the leasing of a *giardino* at Ġebel Ċiantar refers to the presence of water sources and the planting of twelve new trees annually (Fiorini 2005, 50).

[96] The singular of *modia* is *modd*. *Modd* is a unit of land measurement equivalent to a *salma*. See Fiorini 1996, xviii.

[97] Another entry dating to September of 1487 also mentions fruit being grown in the Ġnien il-Far *giardino*. See Fiorini 1996, 289.

[98] Field-research in the Binġemma area has revealed the existence of a couple of dried-up perched aquifer galleries. As the Late Medieval extent of the fief of Binġemma is not known, it is difficult to establish if these were located within.

[99] In a separate entry dating to October 1494, the lessee is contractually bound to plant twenty-five trees each year. See Fiorini 1999A, 23-24.

[100] Fiorini, 1996, 260. An entry dating to August 1494 again bound the lessee to plant twenty-five trees each year. See Fiorini 1999A, 150.

[101] Għajn Astas originates from within a perched aquifer gallery. See chapter five, section 5.2.1.4.

A notarial deed dating to September 1495 documenting the division of the *giardino* of Ta' Santa Marija (MG22), also at Pwales (Fiorini 1999A, 166-67), is particularly revealing, and it indicates that the water source feeding the *giardino* must have originated from a perched aquifer gallery. This entry decidedly states that a church, fountain and its waters and a reservoir are to remain common property so that the owners of the co-owned estate could make use of them. The deed also emphasises the fact that the reservoir was to remain co-owned by the two tenants and when repairs were necessary, the brunt of the expenses had to be shared. The water yielded by the fountain was to remain common property and was equally divided between both parties through the use of a time-sharing arrangement, which is still currently employed in many of the *giardini* investigated in north and north-west Malta and Gozo. Each of the land tenants was to have alternate access to the water of the fountain for the duration of a whole night and day.[102] This 1495 document also makes reference to *anticaglj e petrj grandj* located in this *giardino*, a matter which probably relates to remains of archaeological interest, thus suggesting an even earlier occupation of this site.[103]

"Bir" toponyms provide an indirect reference to a water presence. The main problem concerning the interpretation of these is that it is difficult to discern at face value, if "Bir" place-names refer to wells or cisterns.[104] The ambiguity concerning the interpretation of "Bir" terminology in the Maltese language is typically exemplified by a notarial deed dating to February 1487 in which is pledged the use of a *cisterna*, or cistern, called Bir it-Taħtieni (Fiorini 1996, 129-30). In this particular instance it is possible to determine that the water-related feature in question is a cistern only because this is explicitly specified.

The Xagħret il-Għadira place-name (Fiorini 1996, 193-94), an area close to Raħal Bisqallin and Bulebel, can be linked with the presence of a Middle or Lower Globigerina Limestone outcrop, which due to its poor water percolation properties retained a seasonal pool of water. The Tas-Swieqi place-name, mentioned in an entry dating to November 1494, possibly alludes to the presence of water channels (Fiorini 1999A, 42-43; Wettinger 2000, 542). The name of an arable field called Tas-Safsafa, in the district of Bufula (Fiorini 1999, 124-25), is also indicative of a water presence.[105]

7.6 The Place-Name Evidence

The availability of water resources both for agricultural and domestic consumption in Malta during the Late Medieval period can be also detected through an analysis of the place-name evidence. For the purpose of this section of the study, only toponyms mentioned in historical documentation till 1560 are taken into consideration.[106] In matters relating to rural Malta, the first thirty years of the Order's presence locally can indeed be considered as an extension of the Late Medieval period (Luttrell 1975B, 15-16, 55). That the Hospitaller Knights were not showing much involvement and interest in the countryside resources of Malta between 1530 and 1560 is apparent from the minutes of the Council of the Order of 1643. One particular entry reported that rent on a number of fiefs which had become the Order's property with the annexation of Malta in 1530 had never been collected (Blouet 1963, 71). Furthermore, it is only in the early seventeenth century that a stocktaking exercise started taking the rural assets of the Order into proper account.[107]

7.6.1 "Għajn" and "Għewejna" place-names

Place-names which most obviously refer to a water presence and availability are those containing the "għajn" designation.[108] A total of eighty-seven "għajn" toponyms feature in Wettinger's study (Wettinger 2000, 181-93), thirty-six of which predate 1560. One place-name of great interest is that of Għajn Tuffieħa (MG15), the first recorded instance of which dates to 1299 (Wettinger 2000, 191). Field-research led to the identification of three water galleries, all located in the Għajn Tuffieħa *feudo* (Figs. 5.68, 6.72, 6.94). One of these probably dates to the late nineteenth or early twentieth century, but the other two possibly form part of hydrology-related High or Late Medieval efforts in increasing the agricultural output of the Għajn Tuffieħa *giardino*.[109]

[102] The most significant section of this entry reads: [...] *Jnprimjs la dicta ecclesia, funtana e soy acquj et gebia remanino Jn comunj Jta quod omnj uno li po usarj et che la gebia predicta si divj conczarj et repararj quando serra bisogno a spisi comunj; una dominical j dictj acquj su di luna partj e lautra dominica di lautra, li acquj di la dicta funtana licet restano Jn comunj quanto alo bivirj et usarilj tamen si dividino avichenda videlicet una noctj e lo Lornu sequentj su di luna parj e laltra noctj e Lorno sequentj sus t divino servirj alaltra partj,* [...]. See Fiorini 1999A, 166-67.

[103] The sixteenth-century description and plan of the otherwise unlocated fief of Miżieb ir-Riħ (MG17) bears some resemblance to the data imparted in this notarial entry. On the Miżieb ir-Riħ plan are marked three *fontane* – one of which appears to be fronted by a circular reservoir. What appear to be wall features are marked as *pedamento antico* and *muro antico* (Fig. 7.3). Caves inhabited by peasants are also documented within this site. Refer to ACM Beni Cattedrale (INV. No. 891) Vol. 4, f. 6. The plan is now archived in: ACM Drawing Archives, File no. 67.

[104] See the entry for "Bir" in Serracino-Inglott 2003, 56. See also section 7.6.2.

[105] The Safsafa, or White Willow tree, (also referred to as Żafżafa in Maltese), typically grows directly within a watercourse and is a strong indicator of water. See Borg 2002, 102.

[106] This section is primarily based on the place-names for the Maltese islands published by G. Wettinger. See Wettinger 2000.

[107] The urgency for this undertaking was probably conditioned by a rapid population increase on the domestic front and an increasingly volatile international food-importation situation. See Blouet 1963, 69.

[108] The English equivalent for "Għajn" (pl. "għejun") is fountain – a terminology that is indicative of an artificial water source. Aquilina, 1990, 947.

[109] For a discussion of the Għajn Tuffieħa water resources see chapter six, section 6.1.9.

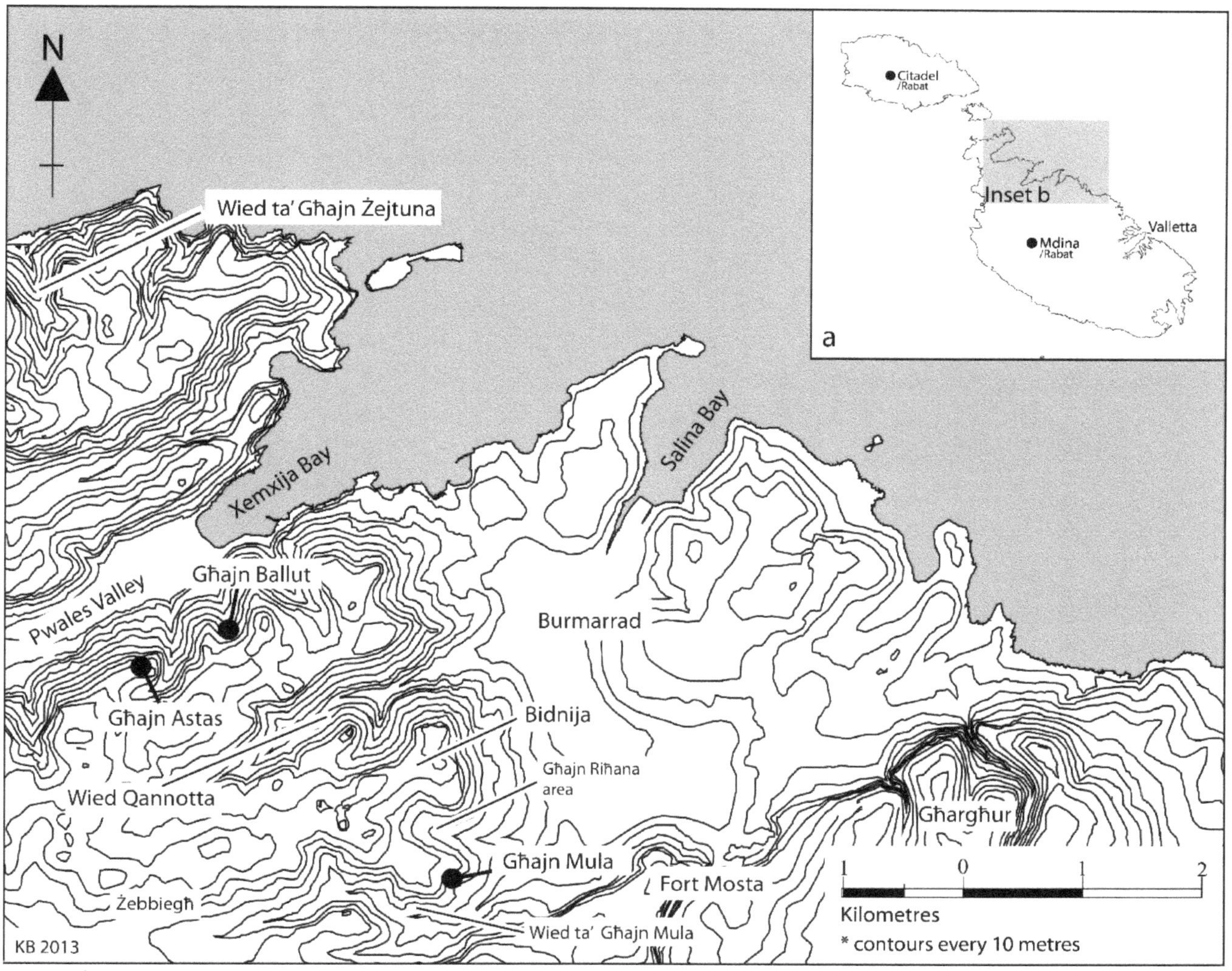

Fig. 7.4 Map of the Maltese islands (a), with inset (b) showing the spatial distribution of the springs and place-names mentioned in the text in the Bidnija area. Base map data: MEPA.

7.6.1.1 Fourteenth-century "għajn" place-names

A larger number of "għajn" toponyms first appear in fourteenth-century documentation. Interestingly, a significant number of these refer to springs in Gozo. The Gozitan springs of Għajn Għabdun (GO8), Għajn Ħosna, Għajn il-Kbira, Għajn Mhalhal and Għajn Naħrin (GO11) all date to 1373 (Wettinger 2000, 183-88). Għajn Qasab (GO12) at Nadur, Gozo, was first mentioned in 1374 as *fons yl casaba* (Wettinger 2000, 188). The Għajn Għabdun and the Għajn il-Qasab place-names, both of which formed part of Crown Property estates which in 1530 were transferred to the Hospitaller Order are of relevance to this study.[110] Field-research has shown that both springs currently originate from within perched aquifer galleries. The Għajn Għabdun estate formed part of the *Fego di San Georgio*. For a discussion on the Għajn Għabdun and the Għajn Qasab springs see chapter six, sections 6.1.11.1 and 6.1.11.2 respectively.

Other fourteenth-century "għajn" place-names include Għajn Tewżien (RA21), Għajn Astas (MG14), Għajn Tarġa (MG16) and Għajn Riħana (MO4). The first mention of Għajn Tewżien is in 1361 *Real Cancelleria* documentation in which it is described as *ayn tosen, fons in viridario vocato de lu Re* (Wettinger 2000, 191). Għajn Tewżien is best documented in the seventeenth-century *cabrei* entries for the Ġnien is-Sultan (RA13) and its neighbouring areas, in which there is the mention of a *mina*, or perched aquifer gallery, which was ca. 1.4 km long, and which in the first half of the seventeenth century fed the S. Michaele *giardino* with a perennial water source (NLM. Treas. B. 289, f. 63).[111] It is difficult to establish if a perched aquifer gallery framework was already in place in the fourteenth century. The 1361 description of the area as a *viridario* or *giardino* is nonetheless significant, and gives an indication of a perennial water presence in this estate as early as the latter half of the fourteenth century.

[110] In the *Real Cancelleria* manuscript dating to 1373 this water source is mentioned as *fons abiduni*. Wettinger 2000, 183.

[111] See also chapter six, section 6.1.3.

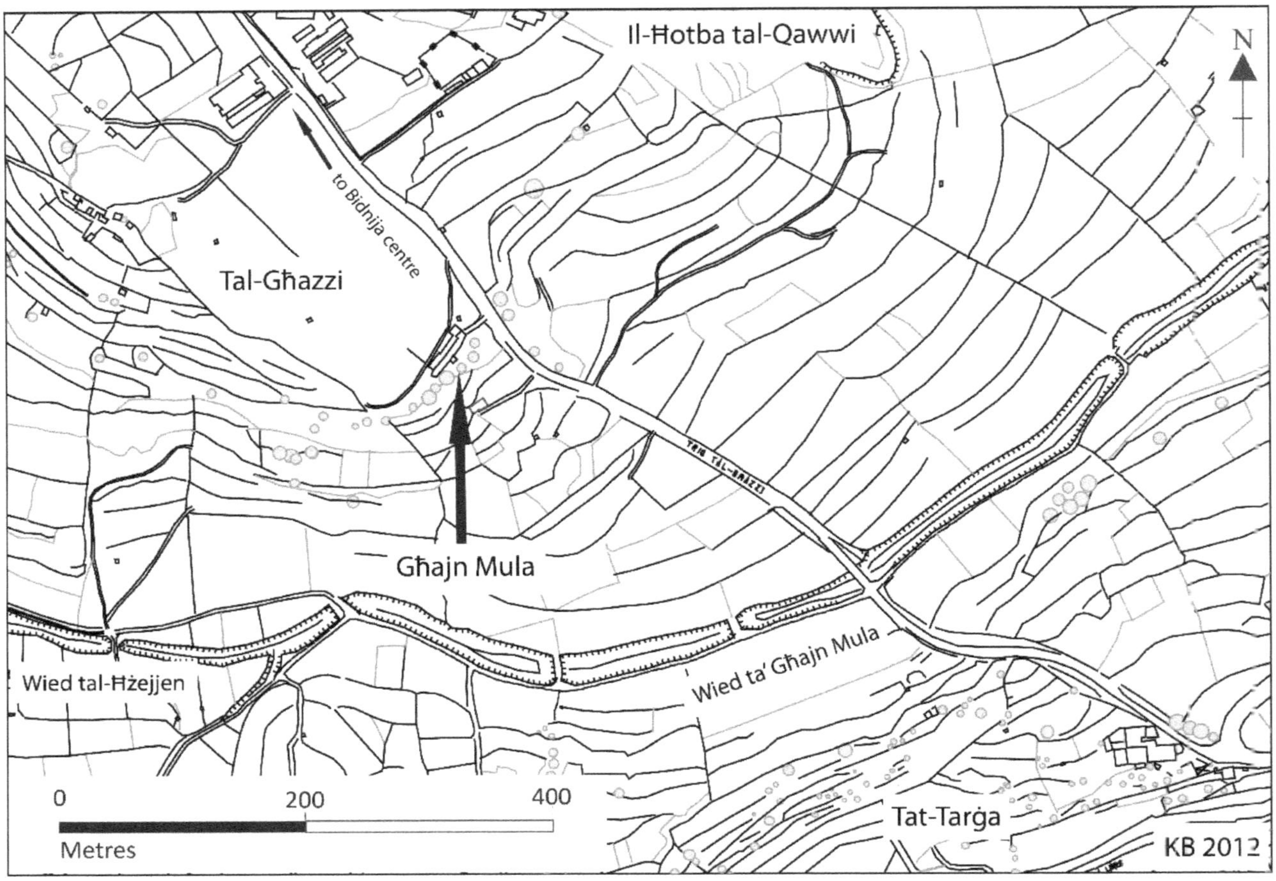

Fig. 7.5 Plan of the Tal-Għazzi promontory at Bidnija, into which tunnels the Għajn Mula perched aquifer gallery (MO3). Topographical map image adapted from MEPA Map Server data.

Għajn Astas (MG14) was first mentioned in 1372 and described as: [...] *haynastasi, tenimentum in territorio Meliveti in contrata pualis* (Wettinger 2000, 190). Personal field-research has successfully located Għajn Astas or Stas in the Pwales territory at map reference 44022 77365 (Fig. 5.68, 5.88). This spring originates from within a finely-excavated perched aquifer gallery. Calcite deposits cover substantial portions of the vertical gallery walls, making it likely that this subterranean water retrieval system is centuries old (Figs. 5.89, 5.90, 5.91, 5.92). The water retrieved from Għajn Astas is gravity-fed through the use of masonry canals into an open water collection reservoir.[112] It is still unclear if Għajn Tarġa (MG16), which in a *Real Cancelleria* entry dating to 1399 is described as *ayn targe, contrata* (Wettinger 2000, 191), originates from a perched aquifer gallery or more simply, a water-yielding crevice (Figs. 5.68, 5.99).[113]

The same applies to Għajn Riħana (MO4), the exact location of which remains unknown. Since the early seventeenth century, the Għajn Riħana estate was owned by the Mdina Cathedral Chapter,[114] and it covered a considerably large expanse of land which from the Victoria Lines escarpment extended northwards to Bidnija, eastwards till Fort Mosta and north-eastwards till Burmarrad (Fig. 7.4).[115] The Għajn Riħana place-name is first encountered in the *Cancelleria Regia* documentation. The earliest known entry is that of 24 February 1398 and refers to the Għajn Riħana fountain as well as the water sources feeding this fief, listed as *aquis*, *aqueductibus* and *aquarum* (Fiorini 1999B, 256). Another entry dates to 1399 and describes the Għajn Riħana fief as: [...] *aynrichana, feudum confinatum ab uno latere cum feudo Benarrato* (Wettinger 2000, 189). A late fifteenth-century entry for Għajn Riħana states that this estate contains three *viridarijs* or *giardini* (Fiorini 1996, 77-78),[116] namely

[112] See also chapter five, section 5.2.1.4.

[113] See chapter five, section 5.2.1.4.

[114] The Għajn Riħana estate was acquired by the Cathedral Chapter on 4 May 1615. See Cabreo Cattedrale 1838, PR 2536, f. 13. Nineteenth-century documentation on the Għajn Riħana estate is rather extensive and includes evidence of the availability of water sources. Blue dots on plans in ff. 28-30, 31 and 36 possibly indicate the location of either cisterns or wells. A dotted line in the *parte tramontana* on f. 36 possibly indicates a spring or water course.

[115] Personal communication by N.C. Vella. For a discussion of the Late Medieval notarial sources of evidence on the Għajn Riħana estate see section 7.5.

[116] The three *giardini* forming part of the Għajn Riħana landholding are once again documented in an entry dating to January 1495. See Fiorini 1999A, 60.

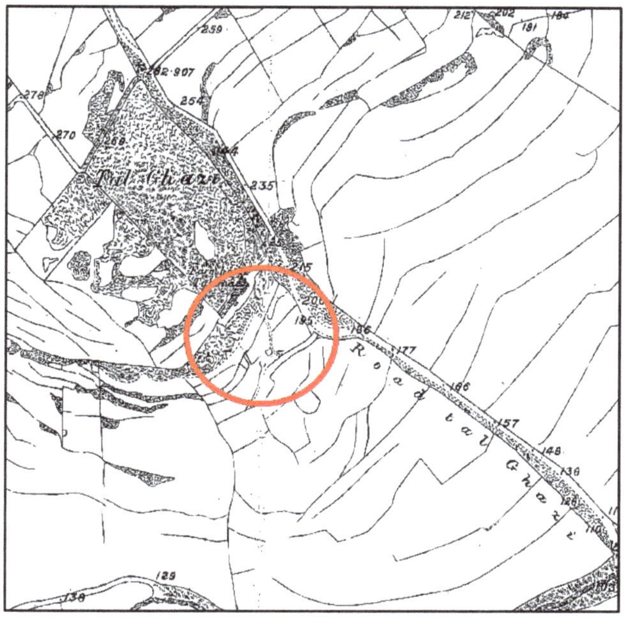

Fig. 7.6 Map detail of Malta Survey Sheet 38 dating to 1898. Even though not captioned, this clearly indicates the Għajn Mula fountain and its fronting reservoir (MO3).

Ġnien il-Kbir, Il-Ġnejna and Il-Għalqa tal-Ġnien (Fiorini 1999A, 161-62).[117]

Taking into consideration the geology and the topography of the area covered by the Għajn Riħana estate, two distinct types of water sources were detected within. Land faulting makes the geology of the area occupied by the Għajn Riħana estate complex and diverse. Mtarfa Member, Blue Clay, Upper, Middle and Lower Globigerina deposits are all present. Lower Coralline deposits of the Xlendi Member and Attard Member flank the southernmost boundary of the Għajn Riħana estate (Debono and Xerri 1993, Geological Map of Malta, Sheet 1). A scrutiny of the Geological Map for the Bidnija-Burmarrad area makes it more than evident that any perennial water feeding the Għajn Riħana estate had to originate from the Żebbiegħ-Bidnija areas, where substantial perched aquifer deposits are stored in the exposed Mtarfa Member strata that characterise the topography of this area.

The first water source detected at Bidnija consisted of a seasonal water flow running through the Wied ta' l-Imselliet, Wied tal-Ħżejjen and the Wied ta' Għajn Mula valleys. The second water source is perennial in nature and originates from a number of perched aquifer galleries. It is the present writer's belief that the water galleries were strategically located in key areas of the Għajn Riħana estate. All are excavated in exposed tracts of Mtarfa Member deposits. Being level with the uppermost terraced tracts of cultivable land meant that the extracted water could be gravity-fed to the various portions of the estate by means of open masonry canals. The Għajn Riħana *fontana* was a primary landmark in the area as shown by the fact that this water source had already given the Għajn Riħana fief its name by the latter half of the fourteenth century.

Field-research in the Bidnija-Burmarrad territory has resulted in the identification of four water galleries.[118] The first is located in the Tal-Għazzi area at map reference 46473 75445 and can be probably identified with Għajn Mula (MO3; Figs. 7.4, 7.5).[119] This water source together with its fronting reservoir is clearly represented on an 1898 survey sheet of the Bidnija area even though there is no caption indicating that the feature in question is a spring (Figs. 7.6, 7.7, 7.8, 7.9, 7.10).[120] Terraced land underlying Għajn Mula was found to contain surface ceramic scatters and other archaeologically significant material datable from between the Phoenician/Punic to the Early Medieval periods (Docter *et al* 2012, 115-19). Another gallery known amongst the Bidnija locals as Għajn Bidnija (MO2),[121] tunnels into the south-facing side of Ġebel Għawżara at map reference 46475 76155. Għajn Bidnija was already indicated in an 1838 cadastral map of the Ġebel Għawżara area (Docter *et al* 2012, 119, 121). Another perched aquifer gallery is known to exist at map reference 46088 76091.

Fig. 7.7 Narrow passageway partly cut into sediment accumulation gives access to the Għajn Mula (MO3) perched aquifer gallery at Tal-Għazzi, Bidnija. Source: the author.

[117] Other place-names for land in the Għajn Rihana locality are those of Il-Ħmieri ta' Għajn Riħana, first recorded in 1600 and Tarġa ta' Għajn Riħana, dating to 1534, but these are of no direct hydrological relevance. See Wettinger 2000, 321, 548.

[118] N.C. Vella brought to my attention the existence of a fifth gallery, which in nineteenth-century documentation is marked as a water source, but which has since dried up.

[119] The name of this perched aquifer gallery was given to me by C. Buhagiar, an accountant and part-time farmer hailing from and owning farming land at Bidnija. The Tal-Għazzi area is mentioned as: [...] *xara tal Ghasi sopra Hayn Rihana* in a 1651 entry. See Wettinger 2000, 604.

[120] Malta Survey Sheet 38, School of Military Engineering, Chatham, 1898.

[121] Personal communication C. Buhagiar.

Fig. 7.8 The partly collapsed initial section of the Għajn Mula (MO3) perched aquifer gallery at Tal-Għazzi, Bidnija. Source: the author.

Fig. 7.9 Għajn Mula (MO3) bifurcates into two separate perched aquifer galleries. The above is a detail of the eastern gallery interior. Source: the author.

The Għajn Riħana spring likewise originated from within a perched aquifer gallery, the entrance of which must have been at an approximate linear distance of some 290 m from that of Għajn Mula at the estimated map reference of 46589 75751 (Fig. 7.11).[122] Considering that the Għajn Riħana gallery tunnelled into the same headland at Tal-Għazzi within which Għajn Mula is located, it follows that it was likewise tapping the same perched aquifer source. Għajn Riħana appears to have been composed both of a rock-cut gallery as well as a fronting subterranean masonry-built gallery extension which channelled the tapped perennial water further downhill into an area known as Wied ta' l-Arkata. Water exited the Għajn Riħana gallery at map reference 46680 75877. A masonry-built shaft with steps was also present in the vicinity and gave access to the masonry-built section of the Għajn

Fig. 7.10 General view of the reservoir into which the water retrieved by the Tal-Għazzi perched aquifer gallery is collected. The access passageway to the Għajn Mula gallery (MO3) is located right in the middle of the dense reed growth visible in the background. Source: the author.

Riħana gallery. This shaft presumably still survives but as the area is overgrown with reeds its entrance could not be located. The gallery setup at Għajn Riħana has close parallels with a similar hydrological situation encountered at Għajn Qasab in the territory of Nadur, Gozo.[123] Similarly, a rock-cut perched aquifer gallery is fronted by a masonry-built underground section, which tunnels below terraced fields. Access to this subterranean built section of the gallery is by means of a rectangular-shaped shaft, which in this instance lacks a staircase facilitating access to the shaft and gallery.

C. Buhagiar pointed out the still extant remains of an aqueduct which channelled water towards agricultural land located further downstream (Fig. 7.11). The present writer believes that the main purpose behind the setup of Għajn Riħana described above was that of furnishing a perennial water supply to agricultural territory located further downstream on the Burmarrad plain, physically located on Upper and Middle Globigerina deposits. It is probable that at this stage the spring water would have been stored in an open water reservoir for later distribution. In the absence of such a hydrological setup the agricultural terrain in this section of the Għajn Riħana fief would be directly reliant on seasonal rainfall, thus having only limited water backup for irrigation purposes.

7.6.1.2 Fifteenth-century "għajn" place-names

Fifteenth-century "għajn" toponyms for the island of Malta include Għajn Ħamiem (RA16), Għajn Kulija (RA3), Għajn il-Klieb and Għajn Żejtuna (ME3). Għajn

[122] The Għajn Riħana spring was located thanks to the assistance of Mr C. Buhagiar, who accompanied me on site and provided me with a detailed description of the Għajn Riħana gallery, which is composed of both rock-cut and masonry-built sections.

[123] For a discussion of the documentary sources and the hydrological setup at Għajn Qasab see chapter six, section 6.1.11.2.

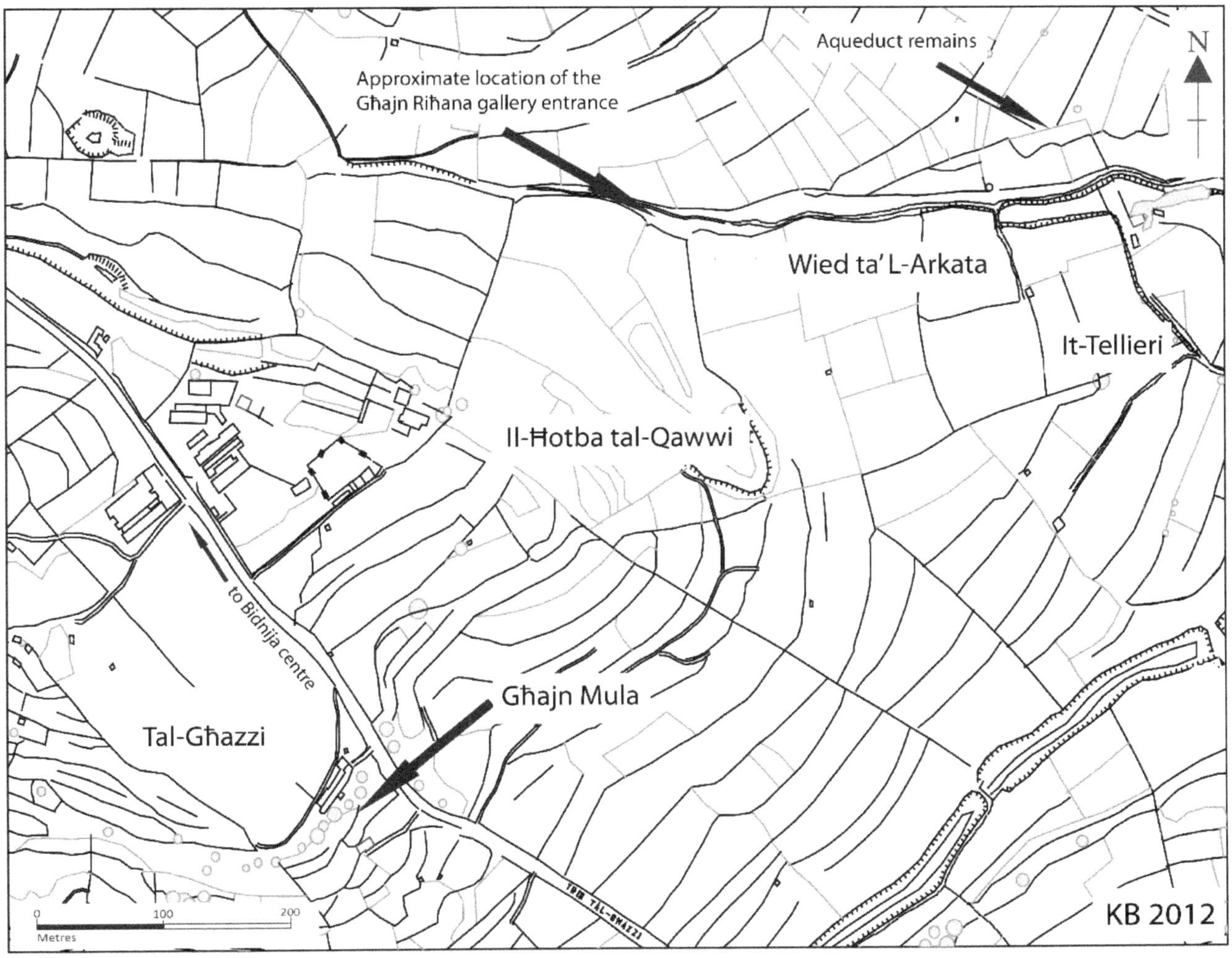

Fig. 7.11 Plan showing the approximate location of the Għajn Riħana perched aquifer gallery at Bidnija (MO4). The gallery water was channelled further downstream by means of a partly subterranean, partly masonry-built aqueduct in order to irrigate agricultural land in the Burmarrad area. Topographical map image adapted from MEPA Map Server data.

Ħamiem, first mentioned in 1436, is located below the east side of the Mdina fortifications. Described as *Aijn Hammem, benificium* (Wettinger 2000, 184), the water of the fountain apparently originated from one or more perched aquifer galleries. It is unlikely that the name of this water source stemmed from a personal name (Wettinger 2000, 184), and might well be a corruption of the Arabic *ḥammām*,[124] meaning bath(s). Toponomastic evidence does indeed suggest an earlier utilisation of this site. Archaeologically-related knowledge on the land-use of this area is limited and based on Themistocles Zammit's partial investigation of the area behind the Għajn Ħamiem eighteenth-century arcaded loggia undertaken in August and September 1922. The excavation of a deep trench revealed a ca. 18 m long rock-cut vaulted tunnel, 1.50 m high and 0.90 m wide. Its roof was made of coping-stones resting on upright pilasters placed at regular intervals of ca. 1.22 m. Of relevance is Zammit's description of the tunnel walls being lined with "a kind of pozzolana" mortar in order to make this water-related feature waterproof. Based on this description the tunnel in question can be associated with a subterranean aqueduct, probably channelling perched aquifer water derived from one or more subterranean perched aquifer galleries which have, so far, not been identified.[125]

Knowledge about the location Għajn Kullija (RA3) is limited. An entry dating to 1452 describes this water source as: *ayn cullie, terrae*, probably located in agricultural land to the west of Mdina (Wettinger 2000, 185-86).

The first mention of Għajn il-Klieb (RA18) dates to 1488 (Wettinger 2000, 185), but its exact location remains shrouded in mystery.[126] In 1501 Għajn Klieb is described as: *hayn kileb, territorium sive vineale* (Wettinger 2000,

[124] Medieval *ḥammām* typically consisted of a dressing room, an unheated room, a warm room and a steam room. See Powers 1979, 654.
[125] According to T. Zammit, this water tunnel did not date to the Roman phase of Maltese history. See Zammit T. 1922, 132-34.

[126] The hydrological setup of Għajn Klieb is best documented in NAM. Descriptive Plans of the Crown Property, 1867, f. 471. See chapter six, section 6.1.3.

185), and in 1537 as: *hayn culeb, vinea* (Wettinger 2000, 185). It can therefore be assumed that a *giardino* framework was already in place during the opening decades of the sixteenth century.[127] It was claimed by Themistocles Zammit that Għajn Klieb was formerly known as Għajn Ċirani (Zammit T. 1922, 9). A 1536 entry for the latter describes it as: *Hayn chirani, clausura in contrata ginen isoltan* (Wettinger 2000, 182).[128] This makes Zammit's assertions possible, but they still await further verification.

Għajn Żejtuna (ME3) was first documented in 1479 (Wettinger 2000, 193). It is not known if this spring located in Għajn Żejtuna valley, in the territory of Mellieħa, originates from within a perched aquifer gallery.[129] A *fonte aquarum* called Il-Fawwara ta' Wied ir-Rum is referred to in 1467 (Fiorini 2005, 26-27). This can be probably identified with one of the two perched aquifer galleries originating within the Tal-Callus estate (RA39) at Wied ir-Rum (RA56).[130]

Fifteenth-century "għajn" toponyms for the island of Gozo include Tal-Għejun (GA20), Għajn Għajxa (GO9), Għajn Fekruna (GO6) and Għajn Futni (GO7). Tal-Għejun consists of agricultural land in the Tal-Kaċċa (Xagħra) territory, first mentioned in 1428 documentation (Wettinger 2000, 221).[131] Għajn Għajxa, first documented in 1489, is described as: *hayn ayxe, gabella*, but its location remains unknown (Wettinger 2000, 184). Għajn Futni and Għajn Fekruna which date to 1492 and 1497 respectively are similarly unidentified (Wettinger 2000, 182-83).

7.6.1.3 Sixteenth-century "għajn" place-names

Amongst the "għajn" toponyms from the first half of the sixteenth century is Għajn Btejtes (RA20), the earliest reference occurring in 1535. It consisted of a vineyard at Wied ir-Rum described as: [...] *hayn biteytes, viridarium in contrata gued irrum* (Wettinger 2000, 181). In a separate entry dating to 1539 this same water source is mentioned as *Hayn bitatis* (Wettinger 2000, 181). Both can be safely associated with the Għajn Tajtes or Tejtes area of Wied ir-Rum located at map reference 42697 70109 (Fig. 4.10). Personal field-research in Wied ir-Rum has demonstrated that a perennial water supply feeding an open-air reservoir, which the farming community of the area identify as Għajn Tejtes (RA20), originates from what appears to be a series of three interconnected perched aquifer galleries located at map reference 42744 70245. These are sited further uphill from the reservoir and tunnel into Mtarfa Member deposits level with the uppermost terraced fields below the Ta' San Ġakbu (RA49) promontory at Wied ir-Rum.[132]

Għajn Naħrin is a place-name connected with agricultural territory in the whereabouts of Daħlet Qorrot in Gozo, first mentioned in 1374 documentation. Its hydrological setup is described in more detail in chapter six, sections 6.1.11.3 (Wettinger 2000, 187).

Even though it is likely that most of the "għajn" toponyms listed so far in this section of the study probably reflect the presence of perched aquifer galleries located within the context of Late Medieval Maltese fiefs, there are exceptions to this. Għajn Sfurija (Wettinger 2000, 190), located in Mġarr (Malta) territory, encountered in 1536 documentation, does not originate from within a perched aquifer gallery (Figs. 5.67, 5.68).[133]

Tal-Għajn and Għajn Filep tal-Qortin in the territory of Marsa, first documented in 1533 and 1543 respectively, also refer to a water source which was located in Middle Globigerina Limestone formations and which tapped a mean-sea-level aquifer source.[134] It is likely that Għajn Filep (MA8) was a natural spring, forming as a result of the pressure exerted by the mean-sea-level aquifer when it came into contact with the largely impervious Middle and Lower Globigerina Limestone deposits which perimeter the shoreline of the Grand Harbour region in the Marsa area (Fig. 5.102). The sheer mass of the mean-sea-level aquifer pressing against these geological deposits forced water upwards, thus providing a probable explanation for the existence of surface springs in selected areas of the Marsa shoreline.[135]

The diminutive of "għajn" is "għewejna" (sing.) or "għewejjen" (pl.), meaning a little spring or fountain (Aquilina 1990, 948). While such place-names are difficult to come by in documentary sources, there are nonetheless several such entries dating to the first half of the sixteenth century. The Tal-Għawejna toponym first dates to 1526 and probably refers to a district of Birgu (Wettinger 2000, 220). Other stretches of land having such a place-name were located in the territory of Għajn Xejba or Ġnien Miriħ in Gozo and Żebbuġ (Gozo) (Wettinger 2000, 220). The Tal-Għawejjen toponym dates to 1559 and refers to agricultural territory at Lippija near Mġarr (Malta): [...] *ta lahueyn, Clausura in contrata Jllippie* (Wettinger 2000, 220).

As a concluding note to this section, two toponyms both dating to 1629 deserve special mention. These post-date

[127] For the association of vineyards with *giardini* in the Late Medieval period see Wettinger 1982, 23-27.
[128] Għajn Ċirani is also mentioned in an entry dating to 1535: [...] *hayn girani, contrata in territorio insulae Melitae*. See Wettinger 2000, 182.
[129] The field-investigation of land to the immediate east of Għajn Żejtuna valley revealed at least two perched aquifer galleries.
[130] See also chapter six, section 6.1.7.
[131] See also section 7.4.4.
[132] See also chapter five, section 5.2.1.1.
[133] See chapter five, section 5.2.1.4.
[134] Also refer to the Ħabel il-Għajn place-name for a field in the Marsa area dating to 1550: [...] *habel il hayn, Clausura in contrata di lamarsa*. Wettinger 2000, 239. The dynamics of localised perched aquifer deposits collecting within Globigerina Limestone formations are discussed in chapter two, section 2.6.2.
[135] Personal communication by M. Sapiano.

the self-imposed 1560 cut-off date mentioned in the opening paragraph of this section, but they succeed in giving a more distinct indication as to the nature and physiognomy of the water source they represent. Għajn Xorok (ME5) is a water spring located in the Għajn Żejtuna territory of Mellieħa and is described as follows: *ta hayn xioroch, fontana nel territorio di hayn zeituna in contrata della Melleha* (Wettinger 2000, 193). This toponym intimates the existence of a subterranean water gallery, the outer section of which was possibly masonry-constructed and roofed-over by means of either flat stone slabs, known in Maltese as "xorok" or by means of gabled slabs. The early seventeenth description of Għajn Ħommed (ME6), a spring at Miżieb ir-Riħ (MG17), is similarly interesting. This spring is described as: *oltre i colli e roccami che guardano verso tramontane in cui vi è una concavità dalla quale esce la fontana sopradetta ta Hommet vicino a cui vi è il colle detto Gebel erraff* (Wettinger 2000, 184). The mention of a "cavity" from which a spring of water flowed can here be taken to mean either a natural hollow or a man-made gallery.

7.6.2 *"Bir" place-names*

A substantial number of place-names related to either water retrieval or storage dating from the fifteenth century onwards are "bir"-related (Wettinger 2000). Keeping in mind that in the Maltese language the word "bir" (pl. "bjar") is used to denote both wells and cisterns, the interpretation of the majority of these toponyms is a difficult task, unless specified in the available documentation. The range of different water-related features that can be denoted by "bir" are: (a) a subterranean or above-ground water reservoir; (b) a sizeable canal in which water is stored; and (c) a well which in the case of the Maltese geological scenario is either tapping a perched or a mean-sea-level aquifer source.[136]

Birkirkara and Bir Bitut are the earliest "bir" entries listed in Wettinger dating to 1402 and 1417 respectively (Wettinger 2000, 37-38; 45). In both instances no further evidence is given to help distinguish whether these refer to a cistern or well. Similarly, the Bir il-Miġlis and Ta' Dar il-Bir entries date to 1503 and 1523 respectively and indicate the use of wells or cisterns either within or close to a domestic context (Wettinger 2000, 47, 103). A 1499 entry making reference to Bajjada [ta'] Bir Jaħlef – a field in the Qormi district – is similarly uninformative (Wettinger 2000, 15). It is only through further evidence given by G.F. Abela in 1647 describing this water source as a *pozzo d'acqua viva* (Abela 1647, 92),[137] that it can be perhaps deduced that at the end of the fifteenth century, this was already tapping the mean-sea-level aquifer.

More "bir" place-names which possibly tapped an aquiferous source include Bir il-Wiesa (LU1): [...] *bir ilhuasa, puteum in via que ducit de casali Luca ad feudum de la marsa* (Wettinger 2000, 52); and Bir il-Mielaħ (WI1) in the locality of Marsaskala, which seems to indicate the existence of a well tapping a mean-sea-level brackish water source (Wettinger 2000, 47). The Ġebel ta' Bir Għawejna place-name, a small field in the Ħandaq territory, again alludes to the presence of a well. "Għawejna" can be taken to represent a small spring or fountain (Aquilina 1990, 948; Wettinger 2000, 136-37).

Other "bir" entries can allow some preliminary conclusions. The Bir Said place-name dating back to 1492 is one such example. The accompanying text explicitly describes a cistern: [...] *bir sayd, cisterna in contrata meġin.*[138] Bir il-Wilġa in the Mrieħel territory, dating to 1501, likewise refers to a cistern: [...] *bir ilhuilge, cisterna in contrata mirehel* (Wettinger 2000, 52).[139] The Bir il-Ħerza place-name was first recorded in 1517: [...] *bir ilblat sive ilchirse, cisterna in contrata Jl mineca* (Wettinger 2000, 44),[140] possibly a locality in the Naxxar-Għargħur area. Particularly noteworthy in this 1517 entry is the mention of *ħerża*, which in a Maltese context commonly consists of a hollowed out stone head placed over the access point of either a well or cistern.[141] Other sixteenth-century "Bir" place-names which can be safely associated with cisterns include: Bir ir-Rum, Bir Ċikku, Bir il-Għajn, Bir Miksur, Bir il-Qasab, Bir ta' Ras Torbiet, Bir ir-Riħ, Bir is-Sieqja, Bjar il-Wilġa and Bjar Rqajja'.[142] Bir l-Imwieġel in the locality of either Qrendi or Żurrieq also appears to point towards water storage in either cisterns or reservoirs (Wettinger 2000, 44).[143] "Meġin" or "meġil" place-names[144] are similarly worth some consideration and again may suggest the presence of either

[136] See entry for "Bir" in Serracino-Inglott, 2003, 56. It appears that in Gozo, or at least parts of it, contemporaneous "bir" terminology is specifically associated with a water cistern. Wells are specifically referred to as "spiera" (pl. "spejjer"). Personal communication by J. Attard Tabone. A recent publication classifies "bir" as a bell-shaped rock-cut water tank. See Gatt G. 2005, 1-2.

[137] See also chapter five, section 5.2.2.

[138] Meġin was a locality probably located close to Żurrieq. Wettinger 2000, 50.

[139] Wettinger 2000, 52.

[140] The "mineq" place-name (dimin. "mnejqa" or "mnejqat") likewise refers to probable presence of water enclosures and reservoirs. See Wettinger 2000, 384, 396.

[141] See entry for "herża" in Serracino-Inglott 2003, 224.

[142] Bir ir-Rum (1517), located in the Naxxar territory: [...] *bir Jrum, cisterna in contrata casalis Miselmet.* Bir Ċikku (1524) located in the Żurrieq territory: [...] *talhuarrata seu bir chiccu, cisterna in casali zurric.* Bir il-Għajn (1534) was located at Qrendi: [...] *bir il chain, cisterna in casali kierndi.* Bir Miksur (1520): [...] *puczu ruttu sive bir miksur at casali tarxen.* Bir il-Qasab (1556): [...] *bir el casaba, cisterna in contrata casalis gregorii.* Bir ta' Ras Torbiet (1541): [...] *Jl bir ta ras torbet, cisterna aquae pluvialis in contrata ta ras torbet.* Bir ir-Riħ (1541): [...] *Jlric, cisterna aquae pluvialis in casali Naxar videlicet in contrata casalis birchircara.* Bir is-Sieqja (1559): [...] *birisekie, cisterna fossa in contrata sancte sophie.* Bjar il-Wilġa (1585): [...] *biar il guilge, contrata* [...] *cisterna aque pluvialis.* Bjar Rqaja' (1544): [...] *ta biar irrocaja, Clausura cum cisterna in contrata casalis Antun.* See Wettinger 2000, 39, 41, 47-50, 60.

[143] Also refer to the "Mwieġen" entry in Aquilina 1990, 875; and Wettinger 2000, 406.

[144] The first recorded "meġin" place-name dates back to 1496. Wettinger 2000, 368-69.

a cistern or water storage reservoir. The Tal-Ġiebja place-name (1505) likewise indicates the storage of rainwater in a reservoir (Wettinger 2000, 147).

The Bjar Qatran (Wettinger 2000, 59) and Bir Qatran (Wettinger 2000, 48-49) place-names from 1539 and 1545 respectively, are also worthy of inclusion. "Qatran" is the Maltese equivalent of tar-pitch (Aquilina 1990, 1144), probably used as a waterproofing agent in cisterns. Tal-Qatran is the place-name of a field at Bir Jaħlef, probably located in the Qormi territory, and four fields at Bir Għisa (Wettinger 2000, 438). This place-name might indicate the presence of cisterns made watertight through the application of a tar lining on their interior.

"Ħawt" place-names, for which the earliest recorded instance is 1436 (Wettinger 2000, 302), are usually associated with either watering or feeding troughs (Aquilina 1990, 521). Nevertheless, there is the possibility that "Ħawt" place-names can also be indicative of a water tank, cistern, or reservoir (Wettinger 2000, 302-3), though in the absence of more concrete evidence, this distinction is difficult to make. Another associated feature is the "ġifna" for which there is some early seventeenth-century documentation (Wettinger 2006, 50). This is comparable to a water basin or "ħawt" having the form of a large concave vessel carved out of stone, placed close to wells or cisterns and filled with water retrieved from within (Aquilina 1990, 403; Fig. 6.127).

7.6.3 "Ġnien" place-names

An analysis of *giardino* or ġnien (pl. ġonna; dimin. ġnejna, ġnejjen or ġnejniet) place-names given in G.F. Abela's *Della Descrittione di Malta* of 1647 has already brought to prominence the fact that such toponyms do not necessarily represent the availability of a water source.[145] On the other hand, the "ġnien" place-names discussed by Abela in the uninhabited part of Malta section frequently refer to spring water sources.[146] It was within such *a giardino*-typology that fruit trees and other crops, the successful cultivation of which necessitated a perennial water source, were cultivated. Various other *giardini* mentioned by Abela, located in the inhabited part of Malta, generally contained plantations in which the cultivation of carob, almond, fig and olive trees prevailed. These are more drought-resistant trees and their successful cultivation and harvest yield does not require the availability of a perennial water source.[147]

The above framework is again confirmed through a perusal of the Late Medieval place-names published by Wettinger (Wettinger 2000, 149-64). It is particularly significant that late fourteenth-century "ġnien" place-names can be closely linked to Crown Property in Malta and Gozo and in their majority refer to the availability of arable land. This is certainly the case with Ġnien Romen in Gozo, the first reference to which dates to 1373, and which consisted of a small field at Il-Għajn (GO14): [...] *ginen rumen, fons.*[148] From the same year the Ġnien Mrik place-name (GO4) refers to royal property in Gozo at Għajn Xejba in the territory of Xagħra (Wettinger 2000, 159). Reference to the Ġnien is-Sultan (RA13) horticultural territory located to the west of Rabat (Malta) first appears in 1376 documentation: [...] *lu Jardinu di lu Re, viridarium magnum* (Wettinger 2000, 162). Ġnien il-Firien, the first reference to which dates to 1391, is land adjoining the fief of Ħemsija (RA58) (Wettinger 2000, 154). A section of the fief of Għajn Tuffieħa (MG15) was in 1446 also identified as Ġnien il-Firien: [...] *ginen firren, territorii et phegii di ayntifa cum la Clausura di la captara ginen firren* (Wettinger 2000, 154).

Fifteenth-century "ġnien" place-names include Ġnejjen tad-Dwieli, a locality at Xewkija ta' Bonsignuri (Wettinger 2000, 150), and Il-Ġnejna, made up of horticultural land at the fief of Għajn Riħana (MO4): [...] *il gineyne, viridarium* (Wettinger 2000, 150). Ġnien il-Kbir, first mentioned in 1495, also refers to territory forming part of the Għajn Riħana estate: [...] *ginen ilkybir, viridarium* (Wettinger 2000, 157). A 1493 entry refers to Ġnejna ta' l-Imġarr in the Mġarr (Malta) territory: [...] *lu iardino piculo, viridarium in contrata muxari* (Wettinger 2000, 150). This can be probably identified with the Ġnien tal-Ġnejna (MG24) area at map reference 42100 75190 in which at least three perched aquifer sources originating from within galleries are present (Figs. 5.68, 5.69, 5.70, 5.71, 5.72, 5.73, 5.74, 5.75). Also significant is the "Imġarr" place-name which can be tentatively taken to mean "water courses"[149] or at least to indicate a water presence. Ġnien il-Far was first documented in 1487 and consists of horticultural land at Tabrija and Dejr Saf (Wettinger 2000, 153). The presence of a water source at Ġnien il-Għarib is perhaps revealed by the place-name of the area (Għadir Bordi), in which it was located. "Għadir" can be cautiously associated with a pool or pond.[150] Similarly, one can safely assume a water presence since Ġnien Ħaġġiem consisted of land at Dejr is-Safsaf (Wettinger 2000, 155), and the "safsaf" place-name is a strong sign of White Willow existence (Borg 2002, 102).[151]

7.6.4 Other water-related place-names

Place-name evidence only records a handful of "nixxiegħa"-type toponyms. The oldest of these is In-Nixxiegħa, a parcel of horticultural land in the Pwales territory, the first mention of which dates back to 1487

[145] See chapter five, section 5.2.
[146] See chapter five, section 5.2.1.
[147] See chapter five, section 5.2.
[148] This can be tentatively identified with Ħondoq ir-Rummien in Gozo. See Wettinger 2000, 161.
[149] See entries for *mġajr* and *mġarra* in Aquilina 1990, 818-19.
[150] See entry for *għadira* in Aquilina 1990, 938.
[151] The first mention of the Is-Safsafa place-name refers to a territory at Wied Bufula and dates to 1495. See Wettinger 2000, 485; and section 7.5.

(Wettinger 2000, 417). Other "nixxiegħa" entries refer to land at Wied Qannotta and Ġebel Ċiantar in Malta and a district in Rabat (Gozo). All these localities contain perched aquifer water sources and it is possible that in most instances, "nixxiegħa" place-names refer to a natural water source (Wettinger 2000, 417).[152] "Nixxiegħa" toponyms appear to contrast sharply with "għajn"-type place-names, a significant number of which were already determined above to originate from within perched aquifer galleries.

A significant number of Late Medieval place-names suggest the presence of pools or ponds of water which at times were stagnant in nature, but which nonetheless might have still been used in agriculture. Such a situation is indicated by the "għodor", "għadira" and "għadir" toponyms. The spatial location of a number of such place-names appears to coincide with the presence of either Upper Globigerina and Middle Globigerina deposits, the marly component of which limits water percolation and infiltration. These place-names might also point to the location of localised perched aquifer deposits within the various Globigerina formations. The "Il-Għadir" toponym is found in various localities in south-eastern and south-western Malta, the earliest recorded instance of which dates back to 1417 and refers to an area known as *lu gadir* located close to Bengħisa (Wettinger 2000, 177). Other Il-Għadir place-names refer to tracts of land at Ħal Tmin and Ħal Safi (Wettinger 2000, 177) – both Late Medieval Maltese settlements. Il-Għadir toponyms dating to the sixteenth century make explicit mention of the availability of water, such as is the case with an entry dating to 1542: [...] *tel gadir, lu abunaturi seu stagnoni di lu lino* (Wettinger 2000, 178); and a 1596 entry: [...] *ta gadir, fontana* (Wettinger 2000, 178).

The Tal-Għadira place-name first recorded in 1506, also meaning a pond or pool of stagnant water, once again refers to various areas falling within the territory of Late Medieval settlements such as Ħal Fuqani, Ħal Kbir and Ħal Safi (Wettinger 2000, 178). Another "għadira" place-name refers to fields at Mrieħel in Malta (Wettinger 2000, 154, 179), an area in which significant deposits of Middle Globigerina Limestone formations are known to be present (Debono and Xerri, 1993, Geological Map of Malta, Sheet 1). Tal-Għodor is the toponym of various localities, the earliest entry of which dates to 1513 (Wettinger 2000, 224), and might likewise indicate the formation of pools of water over Globigerina formations.

Maqgħad Ilma and Tal-Maqgħad place-names can similarly be associated with stagnant water (Wettinger 2000, 361). These represent territory in Għaxaq, Birgu (BI1) and Tal-Ġawhar amongst other areas, and might again be indicative of water deposits gathering above Globigerina formations due to poor infiltration rates. Similarly significant are a string of "menqa" toponyms, which indicate the presence of a water pond, pool or swampy ground and were used for field-strips at Ħal Millieri, Għargħur and Ġebel Ċiantar (Wettinger 2000, 373). Tan-Noqra place-names probably allude to water collecting in Globigerina Limestone areas having a poor permeability and can be associated with small, hollow cavities in which water gathers and stagnates (Wettinger, 2000, 418). Fields in the Tan-Noqra area of Bengħisa were mentioned in 1505 documentation. It is also worthwhile noting that a sizeable outcrop of exposed Middle Globigerina Limestone deposit is present in this area. Noqra place-names were similarly recorded for agricultural territory near Ħal Manin, Santa Dominika ta' Rindieħ near Żabbar and at Ħal Farruġ (Wettinger 2000, 418).

Such place-names can possibly be linked with areas of flax cultivation. Pools of water were necessary for the retting of flax, the processing of which involved the rotting away of the inner stalk, leaving the outer fibres intact in order to manufacture the fibre into lace, damask or linen amongst other products (Wettinger 1982, 29-30). In other instances, the collected water might have been used in agriculture.

A surface water presence, either in valleys or in areas containing exposed Upper and Middle Globigerina Limestone deposits, can perhaps also be detected through the Tal-Ħsejrija and the Tal-Ħsieri place-names (1516 and 1544 respectively), used for land at Ħal Kbir and Ħal Dwin and it may be indicative of reeds growing within.[153] "Marġ" place-names, the earliest of which dates to 1488, indicate the presence of a flood-plain or marshy land,[154] denoting flood-prone areas following a heavy downpour. There are instances, however, where "marġ" toponyms can be linked directly to areas synonymous with perched aquifer sources, such as at Il-Marġa ta' Djar Ħandul and Il-Marġa ta' Għajn Qajjied. The first mention of Il-Marġa ta' Djar Ħandul toponym appears in 1621 (Wettinger 2000, 362). Dejr Ħandun or Ħandul is well-documented and dates back to the late fourteenth century, and throughout its history it was always synonymous with a perched aquifer water presence.[155] Similarly, it is likely that the Il-Marġa ta' Għajn Qajjied (Wettinger 2000, 363) toponym owes its origin to a perennial water presence in the Għajn Qajjed *giardino* (RA19) which field-research has shown to originate from within a perched aquifer gallery.[156]

Għar Ilma, meaning water cave, is another place-name worth mentioning. This name is present in districts at Qallelija,[157] Wardija, Tal-Ħandaq and Msida in Malta and Għajn Għabdun (GO8) in Gozo (Wettinger 2000, 206).

[152] Aquilina defines "nixxiegħa" as a spring of water. See Aquilina 1990, 918. See also the San Ġakbu tan-Nixxiegħa toponym in Wettinger 2000, 493.

[153] Wettinger 2000, 331. Aquilina, on the other hand, translates "ħsieri" to mean "distress". Aquilina 1990, 553.

[154] Wettinger, 2000, 362.

[155] See chapter five, section 5.2.4.

[156] See chapter six, section 6.1.3.

[157] See the Għar Ilma (RA22) entry in Appendix 1.

The Għar Ilma place-name first appears in 1470 documentation in which it is listed as: [...] *callelia alias gar ilme, contrata* (Wettinger 2000, 206). Qallelija is an area in the territory of Rabat (Malta) which in the 1419 militia list is described as: [...] *callelia, casali* (Wettinger 2000, 432). As perched aquifer deposits are present in this area, the Għar Ilma place-name is a likely reference to a perched aquifer gallery in which the retrieved aquifer water was stored. The Għar Ilma entry for Wardija dates to 1523 (Wettinger 2000, 206) and likewise hints at a similar situation.[158] The first mention of Għar Ilma in Gozo dates to 1585: [...] *gar Jlme, viridariolum vitibus et arboribus plantatum in contrata hain habdun in territorio huius insulae Gaudisii* (Wettinger 2000, 206). This can be probably associated with the perched aquifer gallery located in the Santa Luċija (Gozo) territory at the approximate map reference of 29216 89461 and in which the water retrieved by this gallery system is still stored (GO13; Figs. 2.17, 7.12, 7.13).[159]

The Qbiela ta' Għar Ilma (DI2) place-name (1550) is applied to horticultural land at the fief of Dejr il-Bniet (DI1) located in the territory of Rabat (Malta) (Wettinger 2000, 440). It may be linked, at least indirectly, to a perched aquifer gallery which still feeds this *giardino* with a perennial water source. Għar Ilma (RA23) is also the place-name of an area in the Mtaħleb territory of Rabat (Malta).[160] It appears that this place-name was never formally documented in the historic sources, but it too indicates the existence of a subterranean perched aquifer gallery in which the retrieved water collects.[161] That Għar Ilma at Mtaħleb was a major toponym is clear from the place-name of the underlying valley area – Wied ta' Għar Ilma.[162]

Maltese place-name evidence also yields information on the historic employment of water lifting machinery. The first recorded instance of a *sienja* dates to 1673 (Wettinger 2006, 115),[163] where it appears that such water-lifting devices were a more or less common feature in the Qormi area (MA10). As in the case of Sicily, where *sienja* terminology was already in use in 1286, it is possible that within a local perspective this term was embraced through a direct contact with Arabic (Wettinger 2006, 115).[164] Interestingly, Maltese *sienja* toponyms, such as Ħabel is-Sienja l-Kbira and Ħabel is-Sienja ż-Żgħira, both date to the mid-eighteenth century (Wettinger 2000, 256). On the other hand, the sixteenth-century place-names of Tal-Miġbed and Miġbed il-Għajn indicate the presence of a water-wheel retrieving water from within a well and cistern (Wettinger 2000, 381).[165] The employment of

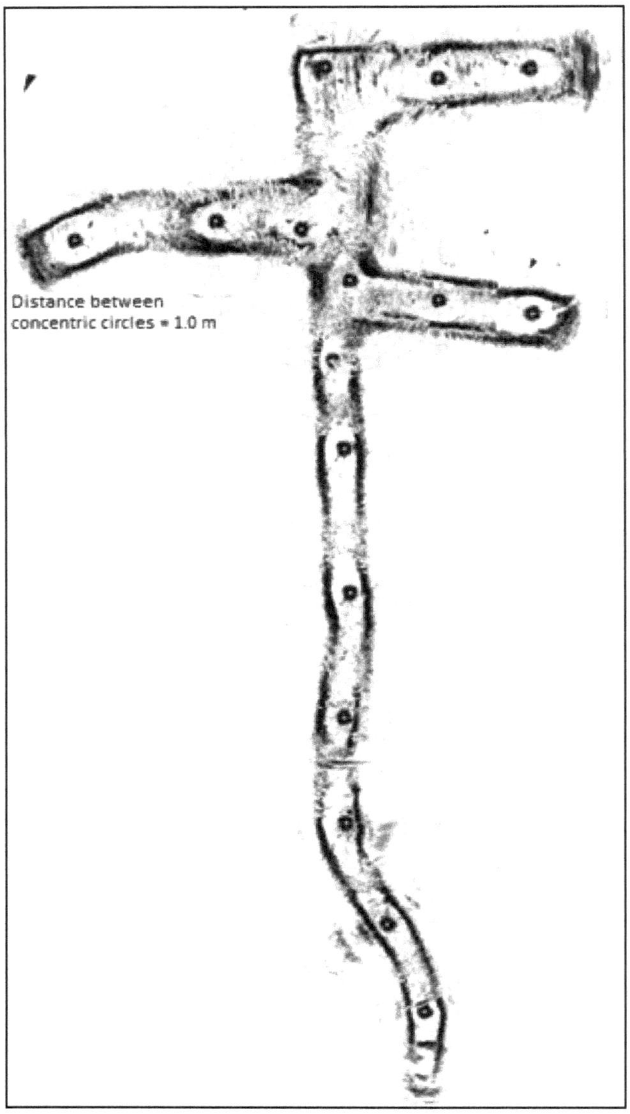

Fig. 7.12 Sonar data map image of the Għar Ilma perched aquifer gallery (GO13) at Santa Luċija, Gozo. Source: Cistern Exploration Project (Malta/Gozo) http://users.csc.calpoly.edu/~zwood/MaltaMapping/site46.html

[158] Field-research conducted by the present writer led to the identification of a perched aquifer gallery in Wardija at map reference 44710 77530, but it is not known if this can be tentatively associated with the above-mentioned Għar Ilma place-name. Even though not labelled as a spring, this water source is clearly indicated in Survey Sheet 28, School of Military Engineering, Chatham, 1903.
[159] See chapter four, section 4.2.
[160] The Għar Ilma toponym indicates an area at Mtaħleb having the map reference of 42301 70772. Also refer to Malta Survey Sheet 84, School of Military Engineering, Chatham, 1903.
[161] Personal communication by A. Camilleri. The presence of a perched aquifer gallery at Għar Ilma is further hinted at by the place-name of the adjoining promontory – Ix-Xagħra tal-Bir located at map reference 42314 70862. This indicates the possible existence of one or more shafts piercing the ceiling of this subterranean gallery.
[162] The Wied ta' Għar Ilma is located at the approximate map reference of 42189 70530.
[163] The English equivalent of *Sienja* is a water-wheel. See Aquilina 1990, 1309.
[164] It has been previously proposed that the *sienja* and the *noria* technology was introduced locally during the Muslim period. See Blouet 2004, 44.
[165] "Miġbed" is the wheel by which a water-wheel or windmill is set in motion. See Aquilina 1990, 827.

Fig. 7.13 The Għar Ilma perched aquifer gallery (GO13) in the territory of Santa Luċija, Gozo. The first mention of this water feature dates back to 1585. Source: the author.

water-lifting machinery is similarly hinted at by the Ta' Qadus place-name dating to 1467. This can be associated with a water-wheel, bucket or scoop used as an aid to irrigation (Wettinger 2000, 429).

"Mġarr" is the plural of "Miġra" and indicates the presence of watercourses, possibly located in valley systems.[166] "Mġajra" is its diminutive form and thus represents the presence of small watercourses (Wettinger 2000, 376). "Miġra" place-names are similarly indicative of a probable natural water channel or watercourse (Wettinger 2000, 382). "Qana" and is the plural form "qonja" are two other man-made water-related features first recorded in 1513 and 1539 respectively (Wettinger 2006, 60). Ta' Qana is the place-name of a field at Ħal Saflieni, the first mention of which dates to 1523 (Wettinger 2000, 433). A typical "qana" setup consisted of a masonry canal used for conveying water or a small canal in the soil meant as an aid to irrigation (Aquilina 1990, 1119). It is probable that water drainage holes within field rubble walls can be likewise identified as *qonja* (Wettinger 2006, 60). *Sieqja* toponyms are similarly indicative of the presence of water canals (Wettinger 2000, 530). The Tal-Maħsel place-name, meaning "of the washing place", can also be linked to water availability.[167]

Qattara place-names (Wettinger 2000, 439), suggest the availability of a trickle or dripping stream of water for use in agriculture. In the case of the Il-Qattara *giardino* (RA32) at Wied ir-Rum in the territory of Rabat (Malta), a perennial water source originates from within a perched aquifer gallery (Figs. 4.3, 4.10, 5.40).[168] The Ċurċara (MG8) place-name similarly alludes to the availability of dripping water and can be identified with an area in the Mġarr (Malta) territory close to the Ġnien tal-Ġnejna (MG24), which corresponds to map reference 41840 75420 (Wettinger 2000, 100-1; Fig. 5.71). Is-Saqqajja (RA40), first recorded in 1372, refers to a spring and square at Rabat (Malta) and actually denotes a watering place (Wettinger 2000, 522).[169] *Saqwi* toponyms may

[166] The word "Mġarr" is first used in the 1419 militia roster.

[167] The first entry for this place-name dates back to 1495 and concerns unidentified land in Malta and two field strips in Gozo. See Wettinger 2000, 357.

[168] See chapter five, section 5.2.1.1.

[169] It has already been demonstrated how seventeenth-century documentation considers the Saqqajja system as going back to the pre-Knights period. See chapter five, section 5.3.

signify either perennial water sources or the irrigation of a field or agricultural area with rainwater harvested in cisterns or reservoirs. "Fawwara" place-names likewise indicate a water presence (Wettinger 2000, 522). Amongst other areas, Tal-Fawwara toponyms refer to areas at Ġebel Ċiantar in the territory of Siġġiewi (SI6) and an area in the fief of Għajn Tuffieħa (MG11; MG15).[170]

7.7 Concluding Remarks

The methodology followed in the carrying out of this research, together with its limitations is discussed in chapter one, sections 1.6 and 1.7. Section 7.1 deals with the Knight Commissioners' report of 1524, but this is largely uninformative and not much can be gathered on the prevailing hydrological situation in Malta during this period. This was probably due to an extremely short sojourn lasting approximately a week, during which, the Knight Commissioners did not have enough time to gain familiarity with the hydrological potential of agricultural territories and *giardini* in north and north-west Malta included in this study. The only indication for the presence of *giardini* emanates from a passing reference to irrigated orchards located west of Mdina, but no further information is disclosed.

Section 7.2 covers the Quintinus Haeduus report on Malta of 1536. The availability of springs and gardens or *giardini* features more prominently in this account and is probably based on direct personal knowledge of what the archipelago had to offer. It is likely that many of the *giardini* were located in valleys and that the overlying garigue plains still had not been reclaimed by agriculture. The nature of the springs feeding these fruit-orchards was not expounded upon, however, and there is no mention of Upper Coralline Limestone perched aquifer galleries in this account. Similarly, none of the *giardini* were mentioned by name. In the accompanying cartographic representation of the Maltese archipelago, there is an indication of the Marsa *hortus* and what seemingly represents a water canal draining water into the inner reaches of the Grand Harbour, but nothing else is water-related.

The relevance of the *Cancelleria Regia* documentation to the hydraulic state of affairs in Late Medieval Malta is explored in sections 7.3 and 7.4. This provides limited, yet important data in making a case for the existence of a series of established *giardini* containing well-managed perched aquifer sources by 1399. Crown property estates located to the west and north-west Malta, as is the case with the fiefs of Gomerino (RA15), Għajn Tuffieħa (MG15), the Għajn Qajjed (RA19) and Ġnien is-Sultan (RA13) territory, Fiddien (RA11), Għajn Riħana (MO4) and Għajn Qasab (GO12), amongst other royal land-holdings, reveal such a situation.[171] The majority of these entries stop short of giving a detailed description of the *fontane* feeding these *giardini* with a perennial water supply, but a number of texts hint, albeit indirectly, at the presence of perched aquifer galleries. Extensive field-research has provided the location and identification of many perched aquifer galleries that still supply these land estates with a perennial water supply. The context of such perched aquifer gallery systems appears to be directly related to the Late Medieval estates they were feeding with water. That a number of *giardini* were during the Late Medieval period already engaged in the cultivation of various crops, including fruit production, is revealed by a *Cancelleria Regia* entry dating to 7 June 1393, wherein the rights of the land-holder over the acquired *giardini* including the fruit produced by the land are indicated (Fiorini 2004, 713, 715).

A philological appreciation of the employed Latin water-related terminology is attempted in Appendix 5 below, but any discussion on related pitfalls must take into consideration that Latin words may well have come totally adrift from their Classical Latin meaning. Furthermore, no one Latin dictionary or word-list will solve difficulties likely to be encountered when dealing with Medieval Latin texts. Another major drawback with *Cancelleria Regia* documents is the lack of direct reference to the nature of the water sources mentioned in this section. An additional factor which might have jeopardised the integrity of the data presented in section 7.3 is the apparent slackness of the Sicilian chancery in registering drafts of royal mandates (Luttrell 1977, 33).

Water-related notarial data for Late Medieval Malta are presented in section 7.5. Particularly revealing is a notarial deed dating to 1467 documenting the transfer of a section of the Tal-Callus estate (RA39) at Wied ir-Rum. This property and its hydrological assets are well documented in seventeenth and eighteenth-century *cabrei*,[172] and the 1467 mention of a *fonte aquarum* called Il-Fawwara ta' Wied ir-Rum can be probably associated with one of the two Upper Coralline Limestone perched aquifer galleries known to exist in this estate (Fiorini 2005, 26-27). Similarly, the 1467 mention of periodic maintenance works at the Għajn Qamar water source at Wied ir-Rum (Fiorini 2005, 99-100) again possibly refers to the occasional cleaning works necessary in order to ensure an efficient water-yield by the perched aquifer gallery systems. The 1495 notarial deed mentioning the *giardino* ta' Santa Marija (MG22) at Pwales provides sufficient evidence to show that the described water source, reservoir and the agreed-upon water-sharing framework collectively

[170] For data on the Fawwara area at Għajn Tuffieħa see chapter six, section 6.1.9.

[171] For a discussion on Malta's royal landholdings in the fifteenth century see Wettinger 1982, 4. In central and eastern Malta were located the fiefs of Buleben, Marnisi, Dejr Limara, Petralonga and Marsa. Excluding a few exceptions, the fiefs in question were isolated from the ca. fifty settlements which existed in fifteenth century Malta. It appears that the *giardini* of the Marsa and Qormi areas (MA9; MA10) were developed further during the course of the seventeenth century, mainly with the aim of being agricultural production centres for the Valletta settlement. See Blouet 1963, 76.

[172] See chapter six, section 6.1.7.

point to the presence of a perched aquifer gallery (Fiorini 1999A, 166-67).

Notarial data also make mention of "wells" and "cisterns", but unless specified, the ambiguity concerning the interpretation of "Bir" terminology in the Maltese language already discussed in chapter five section 5.4, does not permit any further conclusions to be reached.[173] Other place-name evidence encountered in notarial documentation allowed a tentative correlation to be made with either a localised perched aquifer placed within Globigerina Limestone outcrops or poor water percolation within these same geological deposits. Further research is nonetheless necessary in this respect. A major shortcoming with any scholarly work attempting an analysis of the data contained in Maltese notarial records is the fact that these are very fragmentary in nature, especially when dealing with fifteenth-century documentation. The oldest register does not predate 1467 and for the 1467-1500 period there are only five surviving registers (Wettinger 1983, 1-2).

Section 7.6 attempts an analysis of Maltese place-name evidence till 1560 and is principally based on G. Wettinger's examination of toponomastic evidence published in 2000. It has to be pointed out from the onset that a major pitfall of this exercise, especially when attempting an examination of "għajn" and "ġnien" place-names for the fourteenth, fifteenth and sixteenth centuries, lies in the fact that the dates accompanying each place-name only represent their first recorded instance. These might actually date to an earlier period. Nonetheless, it is interesting to note that the springs of Għajn Tewżien (RA21), Għajn Għabdun (GO8) and Għajn il-Qasab (GO12), first mentioned in the fourteenth century, were all located within Crown Property estates and originate from within Upper Coralline Limestone perched aquifer galleries.[174] Similarly, the occurrence of fourteenth-century "ġnien" toponomastic evidence can be linked to Crown Property estates.[175]

The interplay between *għajn*-type place-names and *fontana* has already been pointed out in chapter six, section 6.2.[176] More interesting, however, is the fact that in a Sicilian document dating to 1187, of which there is both an Arabic and Latin version, the High Medieval Sicilian equivalent of "għajn" is in Latin given as *fons* or *fonte* (Wettinger 2006, 15, 18-19). The association of the Medieval Sicilian "għajn" place-names with perched aquifer galleries still has to be satisfactorily ascertained, but it is indeed significant to note that the original central Sicilian place-name of the present-day town of Aidone was *Ayndun* (Lombardo 1999, 5; Trovato, 1999, 27). Aidone is still fed with numerous springs which capture a perennial water supply by means of perched aquifer galleries which tunnel into the ravine below which this settlement is located. Furthermore, it is also significant that the 1187 Sicilian equivalent of the Maltese "saqqajja" is in Latin given as either *aqueductus* or *rivulum* (Wettinger 2006, 15, 35). The association of "saqqajja" with *aqueductus* therefore, either refers to water channels or, in the instance to the Is-Saqqajja (RA40) place-name at Rabat (Malta), might be indicative of the availability there of an Upper Coralline perched aquifer gallery as early as 1372. Place-name evidence can also be indicative of water accumulation in areas containing exposed Globigerina Limestone deposits, but more research in this respect is still necessary.[177] Similarly significant is that through the employment of Medieval Sicilian Arabic, Maltese "miġra" or "mġarr" place-names can be safely associated with water-related features (Wettinger 2006, 15).[178]

[173] For possible definitions of "Bir" see Appendix 4. See also section 7.6.2.
[174] See chapter five, section 5.2.4 and chapter six, sections 6.1.3, 6.1.11.1 and 6.1.11.2 respectively.
[175] See section 7.6.3.
[176] There is also historical documentation to show that in a number of instances a *fontana* was also described as a *mina* or perched aquifer gallery. See also Appendix 4.
[177] See section 7.6.4 above.
[178] See also Appendix 4.

Chapter Eight

Conclusions

8.0 Introduction

When I first set out exploring the possibility of researching Maltese medieval water management strategies as a research topic for this publication, the availability, or otherwise, of relevant data was a recurrent concern. Little was I aware of the wealth of water-related documentation preserved in the local archives and libraries which still awaited a proper study and interpretation. The adoption of a retrogressive analytical approach permitted me to sieve through the collected data in a rational and unbiased manner in order to reconstruct the various modes of water management technology resorted to during the different phases of Maltese archaeology discussed below.

An interpretation of the data presented in chapters two to seven is attempted in section 8.1 of this chapter. Section 8.2 reflects upon the main objectives of this study and evaluates if these were successfully addressed. An evaluation of the strengths and weaknesses of this work and its contribution to the understanding of medieval water management beyond the Maltese archipelago is undertaken in sections 8.3 and 8.4. The need of granting legal protection status to sites, whose Late Medieval potential for future research was outlined by this study, is deliberated in section 8.5. This section also identifies questions which could not be answered by this study.

8.1 Interpreting the Evidence

A principal aim of this book was to place Maltese water capture systems into their appropriate cultural context. Without entering into questions of water allocation and distribution, this study also attempts a review of water management in Malta during the Modern, Early Modern and Late Medieval period.[1] Wells and Upper Coralline Limestone perched aquifer galleries were the main mode of water capture discussed in this work. Water storage was also touched upon, and involved the utilisation of above-ground reservoirs and subterranean cisterns for the harvesting of the captured water and rain runoff.

Section 3.2 (chapter three), presents ample evidence to show that wells and cisterns were used locally since antiquity. However, this section (8.1) will concentrate exclusively on the distribution and employment of water capture and storage systems during the Late Medieval period. As specified in section 5.4, *bir* place-names and listings, can in the absence of further evidence, represent both wells and cisterns. This makes it frequently difficult to distinguish between one and the other when attempting an interpretation of British, Early Modern and Late Medieval documentation. Nonetheless, incontestable evidence was presented to show how a number of wells, discussed in chapters three to seven, captured water from perched aquifers located in Upper Coralline and Globigerina Limestone formations, as well as the mean-sea-level aquifer.

The availability of wells and cisterns in the Late Medieval period is much better documented for the rural sections of Malta. Fifteenth-century documentation shows that between one-third and one-half of Maltese villagers owned agricultural property, which in some cases was quite distant from their dormitory settlement. In 1522, peasant landowners from Naxxar owned land that was spread over the whole northern region of Malta (Wettinger 1982, 4). Even though a number of such properties were in notarial deeds documented as being arable in nature, in the majority of the cases, water availability was not due to the presence of a perched aquifer or mean-sea-level water sources, but through the collection of rain runoff in cisterns and reservoirs. This pivotal variable would mean that such areas were still limited in the type of crop that could be cultivated. Significantly, the majority of the "well" and "cistern" entries mentioned above are found in areas of central, southern, south-eastern and south-western Malta which do not have access to perched aquifer resources, where rain water is harvested above Blue Clay deposits.[2] A number of such areas could alternatively have access to a series of localised perched aquifers obtaining water deposits through a series of Upper, Middle and Lower Globigerina deposits within which water retention is made possible as a consequence of the high marly content available in these rock deposits.[3]

Prior to this research, the relationship between archaeological site/settlement location and the availability of Globigerina Limestone localised aquifers was largely misunderstood and underestimated. Section 3.2 of chapter three demonstrates how the potential for water retention and yielding of Globigerina formations was recognised since antiquity. Data presented in chapters four to seven disclose how Globigerina strata containing marly deposits

[1] Water distribution was momentarily touched upon in chapters four and five only to demonstrate that aqueduct water was in between the seventeenth and nineteenth centuries conveying water captured by perched aquifer galleries which predated the aqueduct system.

[2] See chapter seven, section 7.6.2.

[3] See chapter two, section 2.6.2.

were utilised for water capture and storage in the British, Early Modern and Late Medieval periods. The investigation of the hydrological properties of the Maltese Globigerina Limestone formations and the manner in which these conditioned medieval settlement location in central and southern Malta is an ongoing process. This is an area of study which entails further research and evaluation and is a matter that will be again touched upon and expounded in section 8.2.

A principal mode of water capture discussed in detail throughout this study involves the utilisation of subterranean galleries tunnelling into Upper Coralline Limestone deposits in order to provide the availability of a perennial water supply. The data presented in chapters five to seven together with Appendices 3 to 5 illustrate how the interplay between *għajn*-type place-names and *fontana* terminology can be frequently associated with water originating from these subterranean galleries. Furthermore, an examination and discussion of the archival, historical and field data, an analysis of which is presented in chapters five to seven, has already permitted the association of a number of these water extraction systems with the Late Medieval period. Agricultural terrain with access to a perennial water source is in Maltese classified as *saqwi* and has the potential to produce three crops per year, if adequately managed.

An evaluation of the different Upper Coralline Limestone gallery typologies and a consideration of the possible time-frame during which this technology was imported locally, has until this point, not been attempted. Personal field-research has determined the existence in Malta of two distinct types of perched aquifer galleries: (a) those containing shaft-wells, which at intervals pierce the ceiling of the principal *qanat*-type gallery; and (b) more crudely excavated perched aquifer gallery systems in which shaft wells are conspicuously absent. Field-research has shown how the majority of the perched aquifer galleries containing shaft-wells are either located below a principal Late Medieval settlement – as is the case with the *Civitas* and its suburb of Rabat – or are closely associated with Crown Property estates. Hence, perched aquifer galleries containing shafts were detected at Gomerino (RA15), Rabat (Malta), Fiddien (RA11), Għajn Qajjed (RA19), Għajn Klieb (RA18), Ġnien is-Sultan (RA13) and its adjoining agricultural territories and Tas-Santi (MG23).

It is my hypothesis that the perched aquifer galleries located in Crown Property landholdings are the work of a specialised workforce, specifically commissioned in order to improve the hydrological assets of these designated areas during the High Medieval period. In such a scenario, it is unlikely that the technical knowhow necessary for the successful detection of a reliable water source and the excavation of the *qanat*-type gallery system would have been available locally. In the absence of a competent local workforce, it is possible that technical expertise would have had to be brought over from abroad, probably Sicily. Should Maltese *qanat*-type water capture systems be the undertaking of a centralised Sicilian authority, this would give partial credence to Wittfogel s hydraulic theory, the focus of which is centralised management for the construction of water systems, maintenance, water allocation, crop harvesting and storage.[4] It is likely that the principal motivation for the introduction in Malta of *qanat*-type technology was agricultural intensification in an otherwise water-deficient framework. Following an initial period of capital investment, the management and maintenance of Maltese Crown Property estates, was, on the other hand, transferred to fief holders and the agrarian workforce.

Shaft-less perched aquifer galleries are in general shorter in length and are probably the work of a non-specialised workforce. Such subterranean galleries may have well been sited in areas that were already yielding small quantities of perched aquifer water, and were intended as a measure through which the yielded water could be increased. Cases in point appear to be Wied ir-Rum (RA56) and its adjoining Wied Ħażrun (RA55). It is probable that the workforce behind these water systems was made up of farmers, who with their limited understanding of the local geology, could still in many instances boost the water-yield of specifically designated

Fig. 8.1 Interior view of a perched aquifer gallery at Ferla in the territory of Syracuse, Sicily. Source: the author.

[4] See chapter one, section 1.2.

areas.[5] This scenario recalls, amongst others, Timothy Earl's hypothesis. This asserted that small-scale irrigation systems would not require any form of centralised control and could be easily operated and maintained by the farming community.[6] It is also possible that the Maltese shaft-less galleries were inspired by the *qanat*-type systems, which in the context of rural Late Medieval Mata, must have certainly been prominent features within the agricultural landscape.

Malta's close ties with Sicily during the High Medieval and the Late Medieval periods[7] make it tempting to propose that perched aquifer gallery technology filtered into Malta through the archipelago's intimate ties with Sicily since the twelfth century AD (Buhagiar K. 2007, 118-22). As from the Norman period, Sicily and presumably, its smaller satellite island territories were administered directly from the royal court, thus providing a context for organised agriculture-related investment (Houben 2002, 147-48; Lofrano 2013, 1665-67). Personal field trips to Sicily and South Italy have led to the identification of various perched aquifer galleries, such as at Ferla in the territory of Syracuse, at Morgantina and Enna, both forming part of the territory of Enna, S. Lucia di Mendola in the territory of Palazzolo Acreide, Giardino della Kolymbetra in Agrigento and at the Late Medieval troglodytic site of Zungri in the Calabria region (Figs. 3.12, 3.14, 4.10, 8.1, 8.2, 8.3, 8.4, 8.5, 8.6, 8.7).[8]

Fig. 8.2 Detail of the innermost section of the Ferla perched aquifer gallery in the territory of Syracuse, Sicily. Aquifer water seeping into the gallery at this point is channelled to the exterior by means of a rock-excavated canal. Source: the author.

A matter worthy of consideration is that a substantial portion of political appointees to Malta and landholders had a direct link with Sicily, a state of affairs explained by the *Cancelleria Regia*, the notarial entries and other sources of evidence. In the 1230s, Frederick II s proctor at Malta, Paolino de Malta, had close ties between both Malta and Sicily. Paolino de Malta was either Sicilian or Italian and in 1235 he was granted as fief the Sicilian *casale* of Stafenda.[9] The same applies to Giliberto Abbate, Frederick II s governor in Malta. Giliberto was a member of a wealthy Sicilian and influential Abbate family from Trapani.[10] Further contacts with Sicily are exposed in contracts such as the one dated 5 June 1459 recording the endowment of Johannes Landolina with the fiefs of *Fringintini* (better known as Għafrid il-Ġinwi) in Malta, and *Stafenda* in Sicily (Fiorini 2004, 701). The same applies to another contract documenting the lease of the Għajn Riħana estate (MO4) (Fiorini 1999A, 60), the owner of which hailed from Syracuse. Another deed reveals that a Jew from Syracuse, Master Jacobus Il Mucatil, owned land in Malta called Tal-Għawdxi at Bieb il-Gżira (Fiorini 2005, 89-90). Similarly, in June 1416, the fief of Għeriexem (RA26) in the territory of Rabat (Malta) was exchanged with half the fief of *Inbaccari* in the territory of Piazza Armerina in Sicily (Fiorini 2004, 221-23). Further ties, familiarity with and knowledge of Sicily s hydro-related management techniques might have stemmed from the close associations various of Malta's *secreti* had with the semi-urban, semi-feudal nobility in Palermo, Catania, Syracuse and Noto.[11] In the fourteenth century, Raħal

[5] Such an undertaking was brought to my attention by J. Micallef, a farmer owning property in the Wied Ħażrun area. In the late 1980s he tried, albeit unsuccessfully, to boost the potential of a natural vein yielding a small water trickle by excavating a short gallery. The gallery was excavated into Mtarfa Member deposits, was ca. fifteen metres long and took a little more than ten working days to complete. As far as he can remember, his was the only attempt during the past half a century to excavate a perched aquifer gallery in the Wied Ħażrun/Wied ir-Rum area. Since the 1950s it was more feasible for anyone wanting a perennial water source to drill a borehole in order to tap the mean-sea-level water source.

[6] See chapter one, section 1.2.

[7] For Malta's connection with Sicily throughout the medieval period between ca. 500 to 1500 see Dalli 2008, 243-58.

[8] Water galleries yielding a perennial water supply appear to abound in the Calabria region of South Italy. Personal communication by R. Farrugia, a paediatrician by profession, who until recently joined various trekking expeditions to Calabria.

[9] Stafenda is today better described as a *contrada* and is located in south-east Sicily between the settlements of Noto and Ispica. Luttrell 1976, 405-8.

[10] Giliberto was also probably the brother of Enrico Abbate, a leading functionary of Frederick II who in 1240 was assigned consul to Tunis. Furthermore, Enrico was appointed *secreto* in Sicily from ca. 1253 to 1256. Luttrell 1997, 3; Sciascia 1989, 1174-81.

[11] The *secreto* of Malta and the *vicesecreto* of Gozo exercised local authority which was only comparable to that of the Castellan and Captain of the islands. The *secreti* of Malta usually hailed from amongst the nobility of the *Civitas* and possessed interests in both Malta and Sicily. Bresc 1975, 144.

Fig. 8.3 A perched aquifer gallery excavated into a limestone deposit at Zungri in the territory of Vibo Valentia, Calabria, Italy. This deposit is visually similar to the Maltese Mtarfa Member stratum. Source: the author.

Niklusi was similarly in the ownership of a noble family from Noto and at a later stage came to form part of the Maltese properties held by the Benedictine monastery of San Niccolò d Arena at Catania (Wettinger 1975, 194).

The time-frame accounting for the local appearance of perched aquifer galleries is a complex matter which necessitates the consideration of a number of variables. In Sicily the dating of galleries was similarly problematic. The belief that this water capture infrastructure dates back to at least the Norman occupation of Sicily was confirmed by the identification of twelfth-century AD ceramic finds in a *qanat* system in Palermo (Biancone and Tusa 1997). H. Goblot advocates a ninth-century date for the introduction of these water capture systems in Sicily (Goblot 1979, 134; Todaro 1989, 115-17). Likewise, it has been recently claimed that the *Uscibene qanat* at Palermo dates to the Muslim period (Lofrano 2013, 1666). A number of water galleries in the province of Granada, Spain have been tentatively dated to the Late Medieval period (Cressier *et al* 1989, 545-50). On the other hand, the lack of concrete historic and archaeological evidence for Malta's High Medieval period certainly does not facilitate matters in this respect. A brief discussion of the known sources of evidence is necessary at this point.

A reconstruction of the Maltese landscape in Late Antiquity and the medieval period at large is still a work in progress. The available archaeological and historical documentation hints at a clear-cut break between the Byzantine period, which ended in 870 AD, and the Norman occupation of Malta, which commenced in 1091 AD. The Muslim conquest of 870 was marked by bloodshed and destruction, probable retaliations against the Christian inhabitants and an orchestrated demographic shifting programme, which included the death or exile of the local bishop, with the island, or its main settlements, being reduced to a *hirba* or an uninhabited ruin (Brown 1975, 81-84; Luttrell 1975A, 21-28, 92, 100; Wettinger 1986, 90-91; Wettinger 2012, 368-69).[12] It was a common early Ifrīqiyan military policy to demolish principal architectural monuments in newly-conquered territory and reutilise this fabric elsewhere (Metcalfe 2009, 26), and the surviving documentation for Malta leaves no reason to conclude that it was spared such treatment. There is a possibility that the principal aim behind the Muslim conquest of Malta was not colonisation, but the neutralisation of formerly Byzantine-held territory, including its military-related infrastructure.[13] Malta, in the latter part of the ninth century, appears to have been located outside the main Sicilian–Ifrīqiyan shipping route. Additionally, political instability in Palermo caused by the death of six governors within a three-year period might have detracted attention from the newly-acquired Maltese archipelago soon after its annexation within the Muslim sphere of influence.[14]

The post-870 phase is sparsely documented and most of the tenth-century Muslim sources are silent on the period between 870 and 1048. One of the handful of known sources dating to this period, Ibn Hawqal, implies that Malta was an uninhabited island containing flocks of sheep, wild donkeys and an abundance of honey (Gabrieli

[12] The expedition for the conquest of Malta was launched from Sicily and has to be viewed in the context of Muslim territorial expansion taking place in south-east Sicily following the fall of the settlement of Castrogiovanni (Enna), in 859 AD. It was recently proposed that the Muslim Maltese expedition of 870 AD is to be interpreted in the context of failed military campaigning by the Muslim camp. Until the conquest of Malta, an episode of military success which probably greatly boosted Muslim morale, military success in south-east Sicily only consisted in the capitulation of four lesser settlements, the military strategic value of which was insignificant. Metcalfe 2009, 25-26.

[13] The capture of Syracuse in 877-878 by the Muslims and the ensuing course of events might epitomise the situation in Malta soon after its fall to the Muslim forces in 870 AD. Soon after the breach of the walls of the Syracusan settlement in May 878, the Christians and Jews who escaped massacre were rounded up and led to Palermo. A number of captives were released, but only when ransoms were negotiated and paid seven years later. Much of the infrastructure of the city of Syracuse, including its walls, churches and houses were damaged and the settlement was stripped bare of most of its transferable wealth in two months of post-conquest looting. See Metcalfe 2009, 26-28.

[14] Coeval to the Muslim conquest of Malta was a failed Muslim siege on Salerno and the recapture of the Muslim emirate of Bari by the Byzantine navy. Muslim sources for 872-878 are silent on the unfolding course of events in Sicily. Metcalfe 2009, 26-27.

1961, 245-53; Dalli 2006, 27; Wettinger 2012, 369-71). According to a later source, Al Himyari, Malta was, moreover, visited by ship-builders because its wood was of the strongest kind, and by fishermen, for the tastiness of its fish. He also mentions trees of pine, juniper and olive (Brincat 1995, 11-12). It is unlikely that Malta was totally depopulated during this period (Luttrell 2002, 100; Molinari and Cutajar 1999, 9-16; Cutajar 2004, 58; Carroll *et al* 2012, 37-38), but a drastic decline in population numbers is not improbable.

By way of hypothesis, it is likely that Muslim retaliation against the local population was primarily directed towards the urban centres of the archipelago, namely Mdina and Birgu in Malta, and the Citadel in Gozo. Even if an ethnic-cleansing policy was employed, it is doubtful that this would efficiently target more remote countryside locations that could still harbour small communities accustomed to living at subsistence or near-subsistence level. Cliff-face settlements excavated in remote areas of the Maltese countryside provide an excellent case in point. Anyone unfamiliar with the topography and terrain is almost certain to overlook the presence of inconspicuous troglodytic settlements which blend extremely well with their natural surroundings. Nonetheless, even if several remote countryside troglodytic installations remained in operation, the remaining inhabitants appear to have been too few in number to influence the subsequent course of events – even to leave any trace of their existence in the spoken language (Wettinger 1986, 95). A linguistic analysis of the Maltese language excludes any signs of language stratification, with the linguistic base appearing to be of solely Arabic origin (Brincat 1995, 1-7). There are also close parallels between Sicilian and Maltese toponyms, which suggest intimate Sicilian–Maltese linguistic connections (Brincat 1995, 27; Brincat 2011, 33-62).

Historic documentary sources claim that Malta was once more repopulated by the Muslims in 1048-1049 in time to ward off a Byzantine invasion – a fact which has been cautiously interpreted not as a sure sign of the island s depopulation in earlier centuries, but from the perspective of a large-scale Muslim colonisation which possibly took place as late as the early eleventh century (Dalli 2006, 58-62). The fact that women and daughters formed part of this wave of migration makes it improbable that the sole purpose behind the colonisation of Malta during this period was the establishment of a garrison. A number of variables might have conditioned a demographic intensification at this point: (a) demographic expansion connected with a period of economic prosperity experienced by Sicily during the first half of the eleventh century; (b) the Sicilian civil war between the different Muslim caliphates which commenced in 1038; and (c) fear of the Normans, whose territorial ambitions in the Southern Italian peninsula certainly made them a force to be reckoned with. The uncertainty caused by the impending Norman conquest of Sicily in the eleventh century forced a section of the wealthier Muslim families to emigrate from Sicily (Brincat 1995, 20-21; Von Falkenhausen 2002, 262; Metcalfe 2003, 28), and there is the remote possibility that several opted to settle in Malta.

Malta was far from being a deserted place around a century later, when the Muslim geographer Al Idrisi speaks in terms of the archipelago as being, "away from the island of Pantelleria at a distance of 100 miles towards the east one finds the island of *Gozzo* (Gozo) with a secure port. From *Gozzo* one goes to a small isle named *Kamuna* (Comino). From there going eastwards one finds the island of Malta. It is large and has a sheltered place on the east side. Malta has a town and abounds in pastures, sheep, fruit and honey." (Wettinger 1986, 97)[15] Investment in agricultural intensification, which seemingly took place in the post-1127 period,[16] appears to have been generally fruitful, and succeeded in placing to the forefront cotton cultivation, an item of luxury trade, the mention of which is made in a property inventory drawn up in Genoa in June 1164 (Dalli 2006, 158; Buhagiar M. 2007, 18, 40). Furthermore, it is significant that twelfth-century pollen samples retrieved from the Marsa plain adjoining the east coast of the island indicate an increase in wood, cereal and flax vegetation (Fenech 2007, 112), and appear to confirm the agricultural intensification process the island was experiencing during this period.

Any interpretation of Maltese landscape development must take into consideration contemporaneous developments taking place in South Italy and Sicily. Landscape evolution in South Italy during the Norman

Fig. 8.4 Zungri, Calabria, Italy. A small rock-excavated water collection tank located close to the perched aquifer gallery illustrated in Fig. 8.3. Source: the author.

[15] For a slightly different translation of the Al Idrisi passage see Amari 1880, vol. i, 53, 75.

[16] It was during this period that a Latin garrison was presumably stationed on the island together with clergy responsible for attending to the spiritual needs of the soldiers. This probably marked the initial stages of the long-drawn Christianisation process of the island – a strategy which was to meet a final success during the course of the thirteenth century, and which was possibly unduly delayed due to a strengthening in maritime ties with North Africa in the first half of the twelfth century. See Luttrell 1975B, 31; Buhagiar M. 2007, 82.

Fig. 8.5 Castello di Lombardia at Enna, Sicily. A perched aquifer gallery tunnels into the rock below the western-facing bastion walls of the castle. Source: the author.

period appears to have first centred on the development of *giardini*-type cultivations, then dry-farming, and only at a later stage, the utilisation of waste-land for rough grazing and wood gathering (Martin 2002, 19, 37).[17] Sicily, having been exposed to a Muslim influence since the mid-eighth century, offered different agricultural possibilities, where Muslim agrarian technology made it possible to grow citrus fruit, date palm and sugar cane (Martin 2002, 20; Lofrano 2013, 1664-67).[18] *Castra* (plural) or *castrum* (singular), small fortified settlements which were often perched on ridges and hilltops, and within the context of South Italy appeared in the landscape before the turn of the eleventh century, were probably a direct response to an increasing number of Muslim incursions. In the regions of Benevento and Capua, these not only provided protection, but also served as a centre of lordship and as a point of agricultural consolidation (Ramseyer 2006, 31; Martin 2002, 27). In many ways, this appears to be the case of the emerging Maltese model, where in the majority of aspects related to Maltese rural economy and the administration of Crown estates, these were under the direct control of the urban inhabitants of the *Civitas*, who also received much of the profits (Wettinger 1982, 4).

It is possible that fertile valleys in the north-west sector of Malta, such as Wied Ħażrun, Wied ir-Rum, Ġnien is-Sultan (RA13) and Wied Liemu, together with other valleys which contain extensive field terraces, water galleries and rock-cut settlements[19] all formed part of this post-1127 agricultural intensification process. As is the case with the perched aquifer galleries, many of Malta's cliff-face troglodytic settlements were excavated into the

[17] From a precipitation point of view, the western section of South Italy enjoys abundant rainfall and a blend of hilly and mountainous terrain. This is completely unlike the prevailing situation in the Puglia region which is afflicted by a semi-arid type climate, river-carved valleys, garigue areas of karst formation and terra rossa soil deposits which in general closely parallel the Maltese scenario. The climatic reality of the Puglia region meant that during the Late Medieval period only a limited range of agricultural produce could be cultivated. This consisted primarily of cereals, vines, olives and a few varieties of fruit – mainly figs, almonds and pears. Cereal cultivation was limited to wheat and a winter crop of barley.

[18] During the course of the ninth and tenth centuries, Muslims from neighbouring Sicily conducted raids and established centres of political power in southern Italy. Even though a Muslim emirate was founded in Bari between 848-868, and Muslim colonies were established in Garigliano and Agripoli, it appears that in the latter two instances, these mainly functioned as bases from which raids into neighbouring regions were launched, thus exerting only a minimum influence when it came to the diffusion of Muslim agrarian technology in the Puglia region. See Ramseyer 2006, 22.

[19] For a discussion on Maltese Late Medieval troglodytic settlements see Buhagiar K. 2012A, 153-65; and Buhagiar K. 2012B, 95-118. There are close parallels between the Maltese troglodytic model and the Sicilian one, where the location of most cave-settlements was likewise conditioned by the available geological profile. Sicilian cave sites were often located in naturally defendable, difficult to reach places and troglodytic settlements at Scicli, Modica, Ragusa, Cava d Ispica, San Corrado (Noto) and Pantalica were observed to be excavated in a friable sedimentary rock deposit which is visually identical to the Maltese Mtarfa Member rock stratum. See also Messina 1989, 109-11.

brittle Mtarfa Member deposit, a practice that suggests that their occupants had a sound knowledge of the local geology (Buhagiar K. 2012A, 160). Mtarfa Member deposits are commonly located only a few metres above the perched aquifer, making it possible for water to be successfully tapped by means of an underlying perched aquifer gallery, thus ensuring the agricultural estate embracing this setup a perennial water source. The majority of the Maltese cliff-face troglodytic settlements were probably subject to an organic type of development,

Fig. 8.6 Castello di Lombardia at Enna, Sicily. Detail of the perched aquifer gallery entrance. A rock-cut canal channels aquifer water to the gallery's exterior. Source: the author.

which appears to be closely associated with the development of *giardini* during the period in question.[20]

Field-terracing construction coupled with the hydrological intensification of an area entails a labour intensive input, and it may take decades to transform completely a previously uncultivated landscape into an agriculturally productive one that caters for an arable type of agriculture.[21] As mentioned previously, the *Gumerin* (Gomerino – RA15) and *Deir Ilbniet* (Dejr il-Bniet - DI1) estates, both in the territory of Rabat (Malta), were already listed as *giardini* in 1317 and 1350 respectively (Bresc 1975, 152). Due to the availability of perennial water springs, *giardini* are capable of producing a summer crop in an otherwise arid season, thus increasing the economic value of such land.

The Maltese *giardino* framework was possibly already in place by the thirteenth century. In 1241, the *secrezia* received a total of 11.457 *tarì* from the leasing of gardens in Malta, but as the *masserie* of the *curia* were administered directly, the exact revenue from them is not known (Bresc 1975, 129-30).[22] Angevin-period[23] sources dating to the latter half of the thirteenth century document the presence of a *massarius*, whose role was to look after the local interests of the *Curia* (Fiorini 1993, 129), the Maltese revenues of which were derived from *masserie*.[24] These consisted of large tracts of land estates mainly centred in the countryside areas surrounding Rabat and Mdina, which were formerly worked by *villani curie* or serfs, and appear to have been mainly engaged in cereal production (Fiorini 1993, 129). Other documentary evidence shows how in 1270, the castellan and the captain of Malta were instructed to look into the reinstatement of lands, including a *tenimentum* and "garden" at Pwales in the place called *liburgnatum*, which had been confiscated during the reign of King Frederick II (Dalli 2006, 125-26). This evidence hints at the presence of a *giardino* setup at Pwales dating to the first half of the thirteenth century.[25] Similarly, a reference to *masserie* might likewise point towards the fact that in specifically designated areas, the *giardino* framework was already in place by the latter half of the thirteenth century. Serf emancipation brought about the disappearance of the *masserie* estates, which were

[20] The nearly complete absence of *rahal*-type place-names in the north and north-west sectors of Malta can probably be explained by the fact that the exposed Mtarfa Member deposits and the easily accessible perched aquifer sources tapped by means of galleries, encouraged the inhabitants to resort to rock-cut, rather than aboveground masonry-built, settlements. See Buhagiar K. 2012A, 161. On *rahal*-type settlements see Wettinger 1975, 181-216.

[21] A time-frame model involving the intensification of a local landscape is so far unavailable, but the creation over ten square kilometres of terrain of six kilometres of new cultivable surface can take a part-time work force of twenty to fifty people anything between eight and twenty years to complete. See Horden and Purcell 2000, 234-35. In twelfth-century Sicily, it was the Muslim sector of the population that was primarily involved in the agricultural industry. See Metcalfe 2002, 292. This situation was quite probably also reflected in Malta during the same period.

[22] In the thirteenth century, Maltese *masserie* were a direct result of royal fiefs or *latifundia* which were devoted to the growing of cereals. It appears that these estates were primarily dependent on a labour force of *villani*. 1241 documentation reveals that eighty-four *servi* from Gerba were employed within the *masserie* framework in Malta. See Bresc 1975, 131.

[23] The Angevin occupation of the archipelago only lasted from 1268 to 1283. See Dalli 2006, 121.

[24] During the course of the fifteenth century, the *masserie* were replaced by fiefs. Following this change, the *massarius curie* became the collector of the *terratici*. In 1438 this office was both known as *massarius Regie Curie* and *Recollectionis massarium Regie Curie*. See Bresc 1975, 131.

[25] The Pwales area is known to contain various *giardini*, the water source of a number of which is currently derived from a series of perched aquifer galleries. Ghajn Astas (MG14) is one particular case in point. See chapter five, section 5.2.1.

Fig. 8.7 A rock-excavated water gallery and adjoining water collection tank at the Giardino della Kolymbetra in Agrigento, Sicily. Source: the author.

replaced by fiefs leased out for military service or against the payment of a *census* (Bresc 1975, 134).

In the fourteenth century, the *curia* did make an agricultural distinction when dealing with its Maltese agricultural territories. It is during this period that the cereal-growing *latifundia*, primarily located to the north-west of Malta, were upgraded to cater for the more specialised economy of the *galce* and thus became characterised by closed and irrigated fields that were directly dependent on the urban market (Bresc 1975, 131).[26] Taking the above into account, it is probably no coincidence that a whole string of *giardini* and *għajn* territories date to the fourteenth century. Even though a number of *giardini* were probably already present locally, it is possible that the hydrological resources of such areas were intentionally boosted during this period in order to cater for a changing agricultural landscape. Late fourteenth-century documentation for Gozo strongly suggests that the *giardino*/perched aquifer gallery framework was similarly already in place during this period. An account for the sums due from Gozo in 1374 shows that a rather hefty 53 *uncie* 19½ *tarì* were due from income generated by the "garden" and "fountains" (Bresc 1975, 129). Similarly, on 9 May 1375 the *secreto* Gaddus Cuskieri was ordered to re-annex "gardens" which had been leased or granted out as fiefs, to the administration of the *secrezia* (Bresc 1975, 141).

Further indirect evidence of Malta's agriculture-related hydrological resources can be obtained through cotton production – an agricultural product on which the Maltese economy was dependent during the Late Medieval period.[27] Whilst wheat, barley, *maħlut* and cumin were also being cultivated locally, it appears that it was only through the cultivation of cotton and revenues obtained

[26] It was through the creation of such an agricultural framework that many of the fief holdings in Malta were transferred into the hands of aristocratic families who became directly responsible for the administration of these estates.

[27] For a discussion on the cultivation of Maltese cotton during the Late Medieval period see Wettinger 1982, 15-21. Cotton plant cultivation thrives in tropical-type climates with high temperature and sufficient levels of moisture and, generally speaking, is common in areas having light, loose soils which permit the deep penetration of the roots. Through the centuries, however, cotton cultivation was successfully introduced in areas which only partially fulfill these requirements. A determining factor permitting this was the efficient use of irrigation and selective breeding which led to the development of new strains with shorter growing periods. In southern Europe, the cultivation of cotton on inhospitable soil was only made possible through the employment of advanced agricultural practices. See Mazzaoui 2008, 8, 25; Watson 1974, 26.

from its exportation and that of cumin[28] that the expenses incurred in the importation of wheat could be offset (Wettinger 1982, 13-5; 18).[29] In the second half of the twelfth century, Maltese cotton had already found a niche market in Genoa, where it was apparently held in high esteem.[30] This coincides with the dramatic expansion in cotton trading that Europe experienced as from the twelfth century.[31] Fourteenth-century documentation also shows cotton being exported to Syracuse, through which it was redirected to Tripoli, Bari, Venice, Ancona and even Barcelona.[32] In the fourteenth century, Maltese cotton was considered superior in quality to cotton produced in Calabria and Sicily, but inferior to the Apulian-grown cotton (Mazzaoui 2008, 31).

If the necessary agricultural infrastructure was available, cotton cultivation did have its advantages. In the Mediterranean region, cotton was essentially a summer crop. As long as simple crop rotation techniques were implemented, an annual cultivation of the cotton crop could take place. Arable fields in which cotton was planted between spring and summer could also support the cultivation of wheat or barley during the winter season (Mazzaoui 2008, 8, 20). The fact that local cotton was sown in March-April and harvested in September-October meant that its growing season largely coincided with the Maltese summer period, which on average is characterised by a lack of rainfall (Wettinger 1982, 18, 20, 45). Such a situation possibly limited cotton cultivation to areas which either had access to water reservoirs or cisterns[33] or perennial water sources which permitted its cultivation in the generally rainless and warm summer season. While this is a matter which needs further research, it is possible that apart from Gozo, the hydrological resources of which are more abundant than those of Malta, much of the cotton was being produced in the fief landholdings of north and north-west Malta,[34] the life source of which appears to have been water retrieved from perched aquifer galleries.

Personal research experience has indicated that many fief landholdings in this part of Malta containing access to perched aquifer deposits would have been composed of *giardini* areas geared towards fruit production, as well as fields.[35] It was within the latter areas that cotton cultivation also probably took place.[36] Cotton cultivation in fief landholdings is perhaps reflected in a notarial entry dating to 1493 when fields forming part of the inheritance of nobleman Petrus de Vaccaro were producing *bombici* or cotton.[37] Another entry dating to April 1497 shows a similar situation. This involves a contract mentioning the nobleman Antonius Gatt Desguanes in which there is the mention of the sale of more than thirteen *cantara* of ginned cotton.[38] In a separate entry, the nobleman Antonius Lancie Desguanes also leased land to a peasant for the purpose of cotton cultivation.[39]

Another Late Medieval agricultural crop that might likewise yield evidence on water availability is viticulture. Immediate Maltese post-Muslim period documentation mentions vineyards in a rather vague and general manner (Wettinger 1982, 23-27). Apart from possibly reflecting the general Muslim attitude of disapproval towards viticulture, Late Medieval documentation gives the impression of a marginal intensification in the cultivation of this crop in later centuries. The authorities of the *Civitas* registered a complaint in 1434 claiming that the importation of foreign wine was harming the Maltese viticulture industry, which provided a livelihood for a thousand Maltese inhabitants (Wettinger 1982, 23). Vineyards were primarily located in areas which either experienced a reliable supply of spring water or fields with access to cistern and well water. It is therefore possible that amongst the areas engaged in viticulture were several with *giardini* located in north and north-west Malta.[40]

The extent to which Maltese Late Medieval agriculture was influenced by Muslim agricultural practices still needs to be satisfactorily assessed. Muslim expansion into the western Mediterranean region brought about the importation of crops from Asia and the eastern shores of the Mediterranean basin. Crop importations consisted of several species of cereals – namely rice, sorghum and hard wheat, sugar cane and cotton, the latter slowly replacing wool and linen as the principle textile fibre. A range of

---

28 On the cultivation of cumin during the Late Medieval period see Wettinger 1982, 22.

29 Cumin was another cash crop, the importance of which was second only to that of cotton. See Wettinger 1982, 21-23.

30 During the twelfth century, Malta was apparently also in use as a transit port for cotton originating from the Puglian, Calabrian, Sicilian, Syrian and Byzantine markets. Blouet 1963, 57. That cotton production was held in high esteem during the Norman period is evidenced from the fact that a cotton cloth factory based in Palermo known as the *ṭirāz* was taken over by the Normans. Furthermore, the earliest use of the word "cotton" occurs in a diploma of King Roger II dating to 1144-1145. Mazzaoui 2008, 26.

31 Mazzaoui 2008, 29.

32 Ships loaded with Maltese cotton were already putting in at Syracuse before 1375. The processing of cotton was in the hands of Maltese and Syracusan Jews. See Bresc 1975, 131-32; Mazzaoui 2008, 31; Wettinger 1982, 15-16; Buhagiar M. 2005. Malta, along Sicily, remained the main cotton exporter to Barcelona till the late eighteenth century. Improvements in ship construction and new navigation techniques must have aided the growth of the Mediterranean cotton industry from the thirteenth century onwards. The use of larger ships permitted trade to take place all year round with the exception of the period between November 15 and January 20. Mazzaoui 2008, 48-49.

33 For an example of late fifteenth-century cotton cultivation through the use of cistern water see Fiorini 1999A, 329.

34 The association between cotton cultivation and *giardini* is also hinted at in Bresc 1975, 131.

35 It was common for a fief to be divided into a number of different *giardini*. A case in point is Għajn Riħana (MO4) for which there is fourteenth-century documentation which shows that this was composed of three different *giardini*, each of which was probably provisioned by means of a perched aquifer gallery.

36 The cultivation of cotton in *giardini* has also been proposed by Wettinger, who pointed out that sixty-three localities having a *ġnien* place-name are either recorded as having been used as a *viridarium* or as vineyards. See Wettinger 1982, 15-22.

37 Will of Nobilis Petrus de Vaccaro, 8 June 1493: deeds of Notary Bartholomeus Sillato, NAV, Ms. 1069, f. 38.

38 Contract, 8 April 1497 by Notary G. Zabbara. See NAV, R 494/2.

39 Contract, 23 November 1499 by Notary G. Zabbara. See NAV, R 494/2.

40 The location of vineyards in *giardini* is also suggested by documentation showing that these often also contained fruit trees. See Wettinger 1982, 25. Similarly, vineyards in early fifteenth-century Spain were frequently grown in areas with access to irrigated water. See Glick 1970, 28-29.

fruits and vegetables, including oranges, lemons, lime, banana, watermelons, spinach, artichokes, colocasia and eggplant were also introduced in Muslim occupied territories including Sicily (Watson 1974, 9-10; Watson 1995, 62; Zaimeche 2004, 4; Lofrano 2013, 1666-67).[41] The cultivation of fruit trees was presumably introduced in Malta during the Muslim or the post-Muslim period.[42] As hefty water resources are needed for the successful cultivation of fruit tree plantations, it is possible that within a local context, the *giardino* framework in the north-west sector of Malta was specifically devised to cater for such circumstances. Citrus cultivation left no traceable mark in the pollen samples retrieved from Marsa and elsewhere (Fenech 2007, 112).[43] This is possibly because citrus pollen has a very low chance of survival in the hostile matrix of Marsa from medieval times and later.[44]

That during the course of the fourteenth-century agricultural-related investment in the Maltese countryside bore the desired fruits can be perceived through the exportation of local agricultural produce to famine-stricken areas abroad. In 1356, the local administration was directed by the royal authorities to export grain to Messina in order to ease the famine experienced there. Other such instances were registered in 1371, 1377 and 1399 respectively.[45] A reversal of this situation was experienced locally as from 1439, when drought afflicted the archipelago's cereal, cotton and cumin crops. From this point onwards, it became a common occurrence for local settlements to complain of a shortage of grain in almost each alternate year. The deteriorating food situation led Maltese officials in 1447 and 1448 to sequester ships carrying grain to Africa in order to force the merchants on board to sell their cargo to the local market (Bresc 1975, 133). Nevertheless, until 1480 there was hardly ever the need to import more than 1,000 or 2,000 *salme* of crops in order to wave off famine.[46] It was only from 1480 onwards that a serious deterioration in matters related to agricultural production took place. By 1530, the archipelago had to import some 9,000 *salme* of wheat annually from Sicily.

The available archaeological and historical evidence makes it difficult to determine the dynamics behind the deteriorating Maltese agricultural scenario. The island which thirteenth and fourteenth-century documentation depicts as being rather prosperous, fell prey during the fifteenth century, to drought, poverty and in some cases, also a demographic decrease (Blouet 1963, 43). This is a scenario which might be explained, at least in part, by a lack of agriculture-related investment during this period. A point worthy of some consideration is that following the collapse of Frederick II's empire in the mid-thirteenth century, the disintegration of authority in the central Mediterranean region had substantial social and economic implications for Malta.

New agricultural investment related to arable land in north and north-west Malta appears to date to the fourteenth century, but was possibly counterbalanced by attempts made by the Aragonese in the early fifteenth century at territorial expansion. This may have resulted in a further neglect of the Maltese outpost and possibly motivated the Moorish reprisal of 1429 and a series of smaller raids. The fact that the 1434-1439 period was reported to be totally sterile from an agricultural perspective (Blouet 1963, 44, 56) possibly implies large-scale damage to Malta's agrarian infrastructure, including the peasant workforce.[47] The 1429 reprisal similarly corresponds with an accelerated dilapidation of Royal patrimony, which the archipelago experienced as from the 1430s.[48] Furthermore, it is significant that in 1506 *giardini*, *census* and fiefs generated a total income of only 511 *uncie* to the *secrezia*. Of this amount, only 63 *uncie*, 18 *tarì* and 18 *grani* were generated by the *giardini*. Such figures either hint at land mismanagement or else the alienation of substantial portions of land formerly forming part of the royal fiefs. Towards the beginning of the sixteenth century it was common for fief-holders not to pay the *secrezia* the *census* due (Bresc 1975, 131, 142).[49]

It is a well-documented fact that throughout most of the Late Medieval period Malta did not produce enough wheat to meet its local market demands. The importation of duty-free Sicilian wheat and other cereals on a practically yearly basis became the norm, the revenue for which was

[41] The Muslim agricultural revolution was not only restricted to the introduction of new crops. Areas with a geologic and climatic unity similar to what the Muslim occupants were accustomed to, also experienced a hydrological transformation which facilitated the cultivation of new crops. See Watson 1995, 63.

[42] The Muslim agricultural revolution peaked in ca. 1200 mainly influencing North Africa, Spain and Sicily amongst other Mediterranean islands. Christian Europe appears to have been largely unreceptive to the availability of this new reality and only sorghum cultivation was mastered in some districts of Italy. It was only during the thirteenth century that lemon, sour oranges and limes became familiar fruit trees in Spain and Italy. Similarly, the cultivation of sugar cane was reintroduced in Sicily also during the thirteenth century and in Spain, the fourteenth century. See Watson 1995, 63-64.

[43] The medieval period is also not represented in core samples retrieved from the Salina Bay area. See Carroll *et al* 2012, 33-34.

[44] Personal communication by A. Bonanno.

[45] In the former case, permission was granted for the export of redundant grain to an unspecified location, in the latter, the exportation of a quantity of wheat and barley crop of two years of age from Gozo to North Africa. See Wettinger 1982, 13. A bad harvest in 1374 necessitated the importation of 560 *salme* of Sicilian wheat. Bresc 1975, 132.

[46] Gozo produced a surplus of wheat in 1454, but again suffered from severe drought in 1468 during which crops failed. See Wettinger 1982, 14, 22; Blouet 1963, 43. Such a situation was also blamed, at least in part, to the ravages of the Hafsid fleets. See Bresc 1975, 133.

[47] The Hafsid army invading Malta in 1429 was composed of 18,000 troops and it is claimed that between 3,500 to 4,500 Maltese were carried away into captivity. The subsequent ransom of an unknown number of Maltese citizens probably caused a further drain on local resources. See Dalli 2006, 235-36.

[48] The fragmentation of Royal patrimony in Malta was during this period conditioned by careless spending, corruption and the difficulties associated with controlling foreign-owned territory at a distance. See Bresc 1975, 148.

[49] A *census* consists of agricultural property given in emphyteusis.

generated by the exportation of Maltese cotton and cumin (Wettinger 1982, 2). A deficiency of locally grown wheat cannot be taken to signify that Malta lacked the appropriate infrastructural-related agricultural framework. It is probable that Late Medieval agronomics simply dictated that better revenue could be generated through the cultivation of specialised crops targeting specific niche markets abroad. This certainly applies in the case of cotton and cumin cultivation, and it has already been argued in this section that the agricultural framework since the fourteenth century AD took this reality into account. Proof of this is the fact that apart from providing a livelihood for the peasant community, Maltese agriculture formed the basis of the local economy, which included several hundreds of craftsmen, masons, carpenters, blacksmiths, the nobility, at least one physician, a surgeon, notaries and priests (Wettinger 1982, 2-3).

The evidence presented here makes it reasonably safe to propose that the identified perched aquifer galleries formed an integral component of the *giardini* framework of north and north-west Malta. *Giardini* were already present locally by 1399 and appear to have formed part of a well-concerted centralised effort aimed at improving the water-yield and agricultural output of an otherwise semi-arid environment which was only dependent on seasonal rainfall. The available evidence suggests that by 1530 the *giardino* system had been overlooked and half-forgotten, except by the rural communities which were directly dependent on it for their livelihood.

As a concluding comment to this section, it is to be realised that in any discussion on the water management framework of medieval Malta, fresh water reliability and resilience against episodes of drought are factors which have to be given their due consideration. In essence, the water capture and storage systems included in this study would have effectively provided some relief in this respect and permitted a more reliable crop production framework.

8.2 Revisiting the Objectives of this Study

The purpose of this section is to refer back to main objectives listed in chapter one, section 1.1 and attempt to determine if these were satisfactorily addressed when related to the results attained by this study.

The first objective involved the employment of a retrogressive analytical approach to identify Late Medieval water capture systems. As specified in chapter one, section 1.6, this mode of research is largely conditioned by the available/known historical data, and only on completion of a landscape study will it be known how far back one is able to go.[50] Retrogressive analysis formed the backbone of this study and, as demonstrated in chapters four to seven, was indeed successful in singling out a number of Late Medieval sites and their associated water-related features. An analysis of the *Acque (Seminario)* documentation[51] in chapter five, section 5.3 had already revealed how the Għeriexem and the Saqqajja perched aquifer galleries predated the Knights period. From a hydrological perspective, however, a retrogressive analytical approach yielded significantly important results for the Tal-Callus estate (RA39) at Wied ir-Rum (RA56). Adequate historical coverage for the Tal-Callus estate showed that a *fonte aquarum* called Il-Fawwara mentioned in a 1467 notarial deed can be associated with one of the perched aquifer galleries which still feeds this estate with a perennial water supply.[52] Furthermore, an analysis of *cabreo* documentation and other sources made it possible to correlate a number of *fontana* sources with *għajn*-type springs originating from perched aquifer galleries within Upper Coralline Limestone. The retrogressive investigation of available sources also involved a philological appreciation of key water-related terminology in Italian, Maltese and Latin in an attempt to better understand the Early Modern and Late Medieval meaning of these terms.[53] A principal limitation which hindered the application of this research method involved the fact that historical coverage for most sites included in this study, did not predate the opening decades of the seventeenth century. Furthermore, Gozo was very poorly covered by the available historical record. It is hoped that future archival research will help address these short-comings, and lead to the identification of further documentation which can be integrated into the retrogressive analytical framework already presented in this book.

Relating Late Medieval water capture technology to Maltese geological stratigraphy was another objective of this study. The characteristics of Maltese geological stratification are discussed in chapter two, sections 2.4 and 2.6. The sites included in this study were, whenever possible, correlated to the geological formations in which they are placed. The case studies presented in chapters four to seven amply showcase how perched aquifer galleries were the primary mode of water capture and extraction in areas where Upper Coralline Limestone deposits prevailed. Water capture in Globigerina Limestone areas was primarily achieved through the excavation of wells from which water had to be lifted to the surface. Whilst carrying out research for this study it was also realised that springs in Marsa and other localities in south-east Malta, even though yielding mean-sea-level aquifer water, appear to owe their origin to impermeable marly formations in Globigerina rock strata.[54] In other instances, the water retention properties of marly formations in Globigerina deposits were also utilised for water storage.

Another objective was the contextualisation of the mean-sea-level and perched aquifers in terms of their

[50] See chapter one, section 1.6.
[51] CEM, 114/7.
[52] See chapter six, section 6.1.7 and chapter seven, section 7.5.
[53] See Appendices 3 to 5.
[54] See chapter five, section 5.2.2 and chapter seven, section 7.6.1.3.

archaeological and historical significance. This objective is closely interrelated to the former one just discussed. Data presented in chapters four to seven indicate that perched aquiferous sources, especially those forming in Upper Coralline Limestone deposits were much easier to intercept than the mean-sea-level aquifer. It was only in the latter half of the nineteenth century that the dynamics behind the workings of the mean-sea-level aquifer were properly understood. Furthermore, this study presents data to show how low-lying areas located close to mean-sea-level such as Qormi and Marsa (MA9; MA10), together with Birgu (BI1) were certainly tapping this water source in the early Knights period through the excavation of wells. Indeed, this probably reflects an earlier period mode of water capture, the roots of which lie in the medieval period.

On the other hand, this study does not attempt to correlate the spatial distribution of raħal-type settlements in central and southern Malta with that of localised aquifers forming within Globigerina Limestone deposits.[55] Even though this might be considered to be a flaw by some, it should be made clear that it was originally the author's intention of doing so. It was realised quite early on, however, that for this undertaking to succeed, better knowledge was necessary on: (a) the exact location of Malta's Late Medieval abandoned settlements; and (b) more accurate geological data pinpointing the location of highly localised Globigerina Limestone aquifers. With regard to the latter point, Adrian Mifsud s prospective Ph.D. thesis will, it is hoped, help clarify matters.

Another objective consisted in the categorisation of water sources mentioned in the historical documentary sources. This was undertaken in an attempt to correlate these to existing ones, and was primarily seen to in chapter five, but chapters six and seven also provide data in this respect. A principal difficulty encountered when trying to locate the springs and water sources mentioned by G.F. Abela revolved around the fact that the exact location of these sources was not specified. This might have resulted in the wrong association of several springs due to changing place-name evidence or inaccuracies in popular knowledge.

The final objective involved the investigation of the Maltese *giardino* agricultural framework. The data presented in chapters five to eight contribute towards a better understanding of the *giardino* framework in the Early Modern and the Late Medieval periods. They strongly indicate that *giardini* located to the north and north-west sector of Malta contained access to perennial water sources originating from perched aquifer galleries. *Giardini* located in central and southern Malta were generally reliant on seasonal rainfall and catered for the cultivation of drought resistant trees and crops. Exceptions to this are a number of *giardini* in the Qormi and Marsa (MA9; MA10) localities which must have had direct access to a mean-sea-level aquiferous source, and a *giardino*, the location of which, has in this study, been correlated with the Għajn Dwieli (PA2) area of Paola. Agricultural activity at Għajn Dwieli was probably aided by the tapping of water resources originating from localised aquifers present in the Globigerina Limestone formations of this locality.

8.3 Strengths and Weaknesses: An Evaluation

As is the case with any scholarly research work, this study encompasses both strengths and weaknesses. The amalgamation of field-research and the detailed analysis of historical and archival data, as well as the adoption of a retrogressive analytical approach are a major strength of this work. The methodology embraced in carrying out this research is specified in chapter one, section 1.6, and allowed the correlation of various modes of water capture and storage to the Late Medieval period. It is only through the adoption of a retrogressive approach that British, Knight and Late Medieval period intervention in the landscape could be identified and assessed.

Prior to the completion of this work, the antiquity of perched aquifer galleries was, quite justifiably, questioned in some academic quarters. On the other hand, there is now sufficient evidence to demonstrate that this mode of water capture was already present in a pre-Knights period Maltese landscape. Field, historical and archival research not only permitted the identification of two distinct gallery typologies, but further uncovered the association of a whole string of sites, with the hydraulic and agrarian context of the Late Medieval period. The most important of these are Tal-Callus (RA39) and Il-Għars (RA30). Other sites are listed in the table hereunder.

Name of Site	Identification Number (see Appendix 1)
Dejr il-Bniet	DI1
Fiddien	RA11
Gomerino estate	RA15
Ġnien is-Sultan	RA13
Ġnien Mrik	GO4
Ġnien tal-Ġnejna	MG24
Għajn Astas	MG14
Għajn Ballut	MG2
Għajn Għabdun	GO8
Għajn Klieb	RA18
Għajn Mula	MO3
Għajn Qajjed	RA19
Għajn Qasab	GO12
Għajn Riħana	MO4
Għajn Tewżien	RA21
Għajn Tuffieħa	MG15
Għajn Tuta	ME2

[55] Exceptions to this are Ħal Millieri and *Hal Caprat*. In both instances, associations with close-by localised perched aquiferous deposits available in Middle Globigerina formations, have been proposed. See chapter two, section 2.6.2 and chapter five, section 5.22.

Għar Ilma	GO13
Għar ta' l-Isqof	RA25
Gћeriexem	RA26
Il-Għars	RA30
Mġarr *giardino*	GO16
Mtaħleb *giardino*	RA35
Saqqajja	RA40
Sqaq Ħandu	RA42
Ta' Brija estate	RA44
Tal-Callus	RA39
Tas-Santi	MG23
Wied Ħażrun	RA55
Wied ir-Rum	RA56
Wied tal-Lunzjata	GO21
Wied tal-Qlejgħa	RA57

The principal weaknesses of this study have already been reviewed in sections 5.4, 6.2 and 7.7 of chapters five to seven. To these, one must include the fact that the dating for the introduction of *qanat*-type water capture systems in the Maltese archipelago presented in section 8.1 lacks scientific backing. The reasons why no Carbon-14 dating of calcite deposits was carried out are given in chapter one, section 1.6. The methodological limitations discussed in detail in chapter one, section 1.7 also have to be given their due consideration. A number of key water galleries should have ideally been surveyed and interior plans and elevation drawings of key features produced. On the other hand, this was an almost impossible task to carry out by a lone researcher.[56]

The following is a synthesis of other possible shortcomings, which may be due to:

a. limited cartographic evidence and inaccurate topographic details present in the sixteenth and seventeenth-century maps;
b. the difficulty in making a distinction between natural springs and artificial water sources, as well as the absence of the exact location of the mentioned sources in G.F. Abela's account;
c. errors resulting from the wrong interpretation and analysis of the points mentioned in "b" above;
d. the difficulty encountered when trying to distinguish between "well" and "cistern" hydrological features mentioned in the *cabreo* documentation, as well as in interpreting *cabreo* plan data because of the lack of an accompanying textual description of properties or because of the absence of standardised norms in representing water-related features and in the adopted colour-coding procedures;
e. difficulty in identifying properties included in *cabreo* documentation for which there was a textual description, but no accompanying plan;
f. the possibility that an interpretation of the Latin terminology attempted in chapters five to seven and in Appendix 5 is flawed due to the fact that Latin words may well have come totally adrift from their Classical Latin meaning;
g. a lack of direct reference to the nature of the water sources mentioned in the *Cancelleria Regia* documents as well as the fragmentary nature of Maltese notarial documentation; and
h. a lack of access to key sites in which field-research should have been ideally undertaken.

8.4 Contribution to Knowledge

This work's contribution to knowledge is twofold. At a local level, this study led to the identification of Late Medieval water capture and storage systems, the context of which was previously uncertain.[57] Whilst it has to be realised that Malta's context is different from that of neighbouring areas, few other territories outside the archipelago have been examined in the same way and with the same intensity.

This research can also contribute towards a better understanding of medieval Mediterranean water management beyond the Maltese archipelago in areas experiencing similar geological stratifications as well as semi-arid climatic norms. It is in this context that south Sicily and the Puglia region of Italy come to the fore. Personal field-research in Sicily has already succeeded in the identification of a number of perched aquifer galleries.[58] On the other hand, in both regions, a comprehensive investigation of medieval hydrological systems contained within the landscape still has to be attempted. This will probably lead to the exposure of more adit and *qanat*-type water capture systems, allowing for better analogies with the Maltese situation to be drawn. In Puglia, geological sedimentation visually identical to the Maltese Upper Coralline Limestone and Globigerina Limestone deposits prevails.[59] The equivalent of Maltese Globigerina is in the area of Lecce commonly referred to as *pietra Leccese*, and it would be interesting to verify if any localised perched aquiferous sources form within, and if so, whether these were intercepted for use by agriculture and settlement since antiquity.

Furthermore, the methodology adopted in carrying out of research work for this publication[60] is in many ways a reflection of non-invasive hydro-archaeological field surveys previously carried out in the al-Andalus region of Spain.[61] Both these methods of research can inspire and benefit any such future undertaking in south Sicily and Puglia.

[56] See section 8.5.
[57] See sections 8.1, 8.2 and 8.3.
[58] See chapter three, section 3.1.6 and chapter eight, section 8.1.
[59] See also footnote 17 in section 8.1.
[60] See chapter one, section 1.6.
[61] See chapter one, section 1.2.

8.5 *The Future Beckons: Site protection and prospects for future research*

Through this study various water-related perched aquifer systems were safely associated with the Late Medieval period of Maltese archaeology and history. On reflecting over the significance of the "new" source material currently available, two immediate diametrically opposed, but related, courses of action emerge as necessary. The first involves the granting of legal protection to a selected number of water galleries included in this study through their inclusion in the national list of scheduled monuments. Ideal candidates for this purpose are the two perched aquifer galleries at the Tal-Callus estate at Wied ir-Rum (RA39; Figs. 6.46, 6.63), and water gallery "C" (Fig. 4.10) at Wied Ħażrun (RA55). A 1467 deed mentioning a *fonte aquarum* called Il-Fawwara ta' Wied ir-Rum in property pertaining to Antonius Callus is a probable reference to one of these two galleries at the Tal-Callus *giardino* (Fiorini 2005, 26-27).[62] That water gallery "C" at Wied Ħażrun quite probably dates to a pre-Knights' period is indicated by the presence of significant quantities of calcite deposits on its interior floor and sides.[63]

Even though the Government is already the legal holder of these properties, the granting of legal protection to these sites should not lead to the current land tenants relinquishing the lease granted to them. On the contrary, efforts should be undertaken to make the land holders of these agricultural estates aware of the archaeological and cultural importance of these water extraction systems they still make use of on a daily basis. Development which could jeopardise the archaeological integrity of the site and the landscape within which these water systems are located should also be supervised and controlled. The Saqqajja (RA40) and the Għeriexem (RA26) perched aquifer systems at Rabat (Malta) should be similarly granted legal protection status. Late seventeenth-century documentation on both water systems clearly indicates that these predate the Knights' period.[64]

The second course of action entails prospects for future research, based on questions that could not be fully answered by the present work. A more detailed examination and surveying of the water capture and storage systems, other than that afforded in this study, is necessary. The 3D laser photogrammetry scanning of the Saqqajja, Għeriexem and Gomerino estate (RA15) *qanat*-type galleries is essential. This will make available data on gallery and shaft dimensions and make it possible to carry out more detailed comparative analysis between the various gallery systems. Two separate shaft typologies were detected in this study: the circular and the rectangular-shaped ones. Any correlation or differences between the two still has to be satisfactorily addressed and needs to be compared with research data yielded through the study of the Palermo *qanat* water capture systems (Lofrano 2013, 1667).[65]

The occurrence of calcite deposits was identified inside a number of Maltese Upper Coralline Limestone perched aquifer galleries. Technically speaking, it is possible to date calcite samples through the application of Carbon-14 dating. On the other hand, no attempt at the scientific dating of calcite samples was embarked upon in this project. This is because of the difficulties in the interpretation of the Carbon-14 results, primarily due to the lack of availability of calcite dating calibration models for Malta.[66] As further work on the chronological dating of these subterranean water capture systems is essential, the prospect of scientific dating using Carbon-14 and other methods will be given priority in future research work. Ideal "dating" candidate sites would, in this respect be, calcite samples originating from water gallery "C" at Wied Ħażrun (RA55) and another gallery at the Ġnien tal-Ġnejna (MG24).

Future research work also needs to continue assessing the relationship of a wide-ranging number of archaeological sites to localised perched aquifer deposits within their vicinity. The physiognomy and water-bearing character of the Maltese Globigerina Limestone deposits similarly needs further specialised study and evaluation. Efforts should in addition be undertaken to increase public awareness of these water systems and Malta's hydrology-related heritage in general.

The primary scope of this book was to contextualise water capture and storage systems in the Maltese landscape and their tentative dating. Nonetheless, the data presented in this publication can be easily integrated within the framework of future studies involving the investigation of irrigated areas and agricultural estates through the adoption of the "historic landscape characterisation" methodology discussed in chapter one, section 1.5. This would enable a more holistic investigation of Maltese landscape archaeology and history to be carried out, especially when widening the scope of future research work to also consider water allocation and distribution.[67]

In conclusion, it is hoped that future sessions of the Malta/Gozo Cistern Exploration Project will be organised. Past research sessions have already exposed the importance of this mode of research within a local context. New ground-breaking data had already been obtained even when sensor, image and video data were as yet of an unsatisfactory quality. The continuous upgrading of the small remotely controlled submersible with higher

[62] See also chapter six, section 6.1.7; and chapter seven, section 7.5.
[63] See chapter six, section 6.1.6.
[64] See chapter five, section 5.3 and chapter seven, section 7.4.1.
[65] The recent investigation of several Palermitan *qanat* systems revealed three different shaft typologies: (a) the square-shaped; (b) the rectangular-shaped; ad (c) the circular-shaped shafts. See Lofrano 2013, 1667.
[66] See also chapter one, section 1.6.
[67] This data can also be integrated within the framework of a web-based information system. See Mamassis and Koutsoyiannis 2010, 103-14.

resolution image and data sensors will undoubtedly reward any forthcoming research sessions with more comprehensive data coverage of these subterranean spaces. This will definitely enable better documentation and interpretation of the archipelago's subterranean water storage tanks and perched aquifer galleries on which this new technology will be tested.

References

A. Archival Sources

National Archives (Malta)

NAM. GOV 01.3/6
NAM. GOV 01.3/7
NAM. GOV 01.3/11
NAM. GOV 01.3/12
NAM. GOV 01.3/14

NAM. PW 270

NAM. *Descriptive Plans of the Crown Property in Gozo*, 1861, Surveyed under the administration of Lieutenant General Sir John Gaspard Le Marchant G.C.M.G. Governor and Commander-in-Chief, The Forces – The Hon. Giovanni Vella, Collector of Land Revenue, Works Division, Floriana.

NAM. *Descriptive Plans of the Crown Property in Malta*, 1862, Surveyed under the administration of Lieutenant General Sir John Gaspard Le Marchant G.C.M.G. Governor and Commander-in-Chief, Works Division, Floriana.

NAM. *Descriptive Plans of the Crown Property in Malta*, 1866, The Hon. Giovanni Vella, Collector of Land Revenue, Works Division, Floriana.

NAM. *Descriptive Plans of the Crown Property in Malta*, 1867, The Hon. Giovanni Vella, Collector of Land Revenue, Works Division, Floriana.

NAM. Government Property Survey Sheet Links, Sheet 26.
NAM. Government Property Survey Sheet Links, Sheet 71.
NAM. Government Property Survey Sheet Links, Sheet 72.
NAM. Government Property Survey Sheet Links, Sheet 85.
NAM. Government Property Survey Sheet Links, Sheet 87.
NAM. Government Property Survey Sheet Links, Sheet 101.
NAM. Government Property Survey Sheet Links, Sheet 125.

Notarial Archives (Valletta)

NAV. R494/2

National Library (Malta)

NLM. Univ. 187
NLM. Treas. A. 74
NLM. Treas. B. 289
NLM. Treas. B. 290
NLM. Treas. B. 291
NLM. Treas. B. 292
NLM. Lib. Ms. 155.

Cathedral Museum (MCM)

AIM Acta Civilia vol. 171
ACM Beni Catedrale (INV. No. 891), vol. 4
Cabreo del Cattedrale 1838, PR 2536

Curia Episcopalis Melite (CEM)

CEM 114/7

B. Published Sources

ABDILLA 2003
Abdilla, O., *St Paul's Monument at Saqqajja: Its Architectural and Socio-Cultural Significance and a Proposal for its Conservation* (unpublished undergraduate dissertation, University of Malta, 2003).

ABELA 1647
Abela, G.F., *Della Descrittione di Malta Isola nel Mare Siciliano con le sue Antichità ed altre Notitie* (Malta, 1647).

ABULAFIA 1975
Abulafia, D., 'Henry Count of Malta and his Mediterranean activities: 1203-1230', in *Medieval Malta: Studies on Malta Before the Knights*, ed. by A.T. Luttrell (London, 1975), 104-25.

ADAMS, A. L. 1870
Adams, A.L., *Notes of a naturalist in the Nile Valley and Malta* (Edinburgh, 1870).

ADAMS, R. 1966
Adams, R. McC., *The Evolution of Urban Society: Early Mesopotamia and Prehistoric Mexico* (Chicago, 1966).

AGIUS 2013
Agius, J.K., *Aspects of Demographic Trends in Nineteenth-Century Malta 1841-1901* (unpublished undergraduate dissertation, University of Malta, 2013).

AGIUS-VADALA and GANADO 1986
Agius-Vadalà, M. and Ganado, A., *The Pre-Siege Maps of Malta 1536-1563* (Malta, 1986).

AGIUS-VADALA and GANADO 1994
Agius-Vadalà, M. and Ganado, A., *A Study in depth of 143 Maps representing the Great Siege of Malta of 1565*, vol. 1 (Malta, 1994).

ALBERTI 2014
Alberti, G., 'Modeling group size and scalar stress by logistic regression from an archaeological perspective', *Plos One*, vol. 9, no. 3 (2014), 1-15.

ALLEN 2001
Allen, H. D., *Mediterranean Ecogeography* (Harlow, 2001).

ALSHAWABKEH and BAL'AWI 2010
Alshawabkeh, Y. and Bal'awi, F., '3D digital documentation, assessment, and damage quantification of the Al-Deir monument in the ancient city of Petra, Jordan', *Conservation and Management of Archaeological Sites*, vol. 12, no. 2 (2010), 124-45.

AMARI 1889
Amari, M., *Biblioteca Arabo-Sicula*, 3 vols. (reprinted in 1982 with appendix of 1889).

ASHBY 1915
Ashby, T., 'Roman Malta', *Journal of Roman Studies* (1915), 23-83.

ASME and AWWA 2010
ASME Innovative Technologies Institute and AWWA, *Risk Analysis and Management for Critical Asset Protection (RAMCAP) Standard for Risk and Resilience Management of Water and Wastewater Systems* (Denver, 2010).

ASTON 1997
Aston, M., *Interpreting the Landscape – Landscape Archaeology and Local History* (London, 1997).

ATTEMA 2002
Attema, P., 'Two challenges for landscape archaeology', in *New Developments in Italian Archaeology*, ed. by P. Attema, G., Burgers, E. Joolen, M. Leusen and B., Mater, BAR International Series 1091 (Oxford, 2002), 18-27.

ATTEMA *et al* 2002
New Developments in Italian Archaeology, ed. by P. Attema, G., Burgers, E. Joolen, M. Leusen and B., Mater, BAR International Series 1091 (Oxford, 2002).

AQUILINA 1990
Aquilina, J., *Maltese-English Dictionary*, vols. 1 & 2 (Malta, 1990).

AQUILINA and FIORINI 2001
Aquilina, G. and Fiorini, S., *Documentary Sources of Maltese History, Part IV – Documents at the Vatican, Malta: Visita Apostolica no. 51, Mgr Petrus Dusina, 1575* (Malta, 2001).

AZZOPARDI 1995
Azzopardi, A., *A New Geography of the Maltese Islands* (Malta, 1995).

BALBI DA CORREGGIO 1965
Balbi da Correggio, F., *Diario dell'assedio di Malta: 18 maggio-8 settembre 1565* (Rome, 1965).

BARCELÓ 1986
Barceló, M., 'La qüestió de l'hidraulisme andalusí', in *Les aigües cercades: Els qanât(s) de l'illa de Mallorca*, ed. by M. Barceló *et al* (Palma de Mallorca, 1986), 9-36.

BARCELÓ 1995
Barceló, M., 'De la congruencia y la homogeneidad de los espacios hidráulicos en Al-Andalus', in *El agua en la agricultura de al-Andalus* (Barcelona, 1995), 25-39.

BARCELÓ 1996
Barceló, M., 'Acerca de nada: consideraciones sobre dos artículos de S. Gutiérrez', *Arqueologia y Territorio Medieval*, vol. 3, 21-36.

BARCELÓ 1998
Barceló, M, 'Sol puesto. Estado, terror y agua. La hipótesis de la sociedad hidràulica de K.A. Wittfogel', in *Una arqueologia gigantesca. El estudio de las antiguas sociedades hidráulicas en las repúblicas centroasiáticas de la extinta URSS*, ed. by L. Vea (Bellaterra, 1998), 5-13.

BARKER and BINTLIFF 1999
Barker, G. and Bintliff, J., 'Geoarchaeology in Mediterranean landscape archaeology: concluding comments', in *Environmental Reconstruction in Mediterranean Landscape Archaeology*, ed. by P. Leveau, F. Trément, K. Walsh and G. Barker (Oxford, 1999), 207-10.

BARKER and GILBERTSON 2000
Barker, G. and Gilbertson, D., 'Living at the margin: themes in the archaeology of drylands', in *The Archaeology of Drylands – Living at the margin*, ed. by G. Barker and D. Gilbertson (London, 2000), 3-18.

BAXTER and JOHNSON 1947
Baxter, J.H. and Johnson, C., *Medieval Latin World-List from British and Irish Sources* (London, 1947).

BAZZA 2006
Bazza, M., 'Overview of the history of water resources and irrigation management in the Near East Region', in *Water Science and Technology: Water Supply*, vol. 7, no. 1 (2007), 201-9.

BEEKMAN *et al* 1999
Beekman, C.S., Weigand, P.C. and Pint, J.J., 'Old World irrigation technology in a New World context', *Antiquity*, vol. 73 (1999), 440-46.

BELL 1994
Bell, J.A., *Reconstructing Prehistory: Scientific Method in Archaeology* (Philadelphia, 1994).

BELOW *et al* 2007
Below, R., Grover-Kopec, E. and Dilley, M., 'Documenting drought-related disasters: a global reassessment', *The Journal of Environment Development*, vol. 16 (2007), 328-44.

BERTRAND and SÁNCHEZ VICIANA 2009
Bertrand, M. and SánchezViciana, J.R., 'Canalizos y tajeas, dos sistemas de captación de agua mediante galerías subterráneas en las altiplanicies granadinas. Andalucía Oriental', *Arqueologia y Territorio Medieval*, vol. 16 (2009), 151-78.

BINACONE and TUSA 1997
Biancone, V. and Tusa, S., 'I Qanat dell'area centrosettentrionale della piana di Palermo', *Archeologia e territorio* (1997), 375-389.

BLAGG *et al* 1990
Blagg, T.F.C., Bonanno, A. and Luttrell, A.T., *Excavations at Ħal Millieri, Malta* (Malta, 1990).

BLOUET 1963
Blouet, B., *The Changing Landscape of Malta during the rule of the Order of St John of Jerusalem 1530-1798* (unpublished Ph.D. thesis, University of Hull, 1963).

BLOUET 2004
Blouet, B., *The Story of Malta* (Malta, 2004).

BOLLE 2003
Bolle, H.J., 'Climate, climate variability, and impacts in the Mediterranean area: an overview', in *Mediterranean Climate: Variability and Trends*, ed. by H.J. Bolle (Heidelberg, 2003), 5-86.

BONANNO 1984
Bonanno, A., 'Giovanni Francesco Abela's legacy to the Jesuit College', *Proceedings of History Week*, The Historical Society (Malta, 1984), 27-37.

BONANNO 2005
Bonanno, A., *Malta: Phoenician, Punic and Roman* (Malta, 2005).

BONANNO 2009
Bonanno, A., 'L'acqua nella cultura dei temple megalitici di Malta', *Mare Internum: Archeologia e Cultura del Mediterraneo*, vol. 1 (2009), 25-31.

BONDI 1988
Bondì, S.F., 'City planning and architecture', *The Phoenicians*, ed. by S. Moscati (Milan, 1988), 248-283.

BONELLO 1988
Bonello, S., *Engineering properties of rocks and soils of the Maltese Islands* (unpublished undergraduate dissertation, University of Malta, 1988).

BORG D. 2011
Borg, D., *The Mgarr Area: Studies of an Early Modern Landscape* (unpublished M.A. thesis, University of Malta, 2011).

BORG J. 2002
Borg, J., *Siġar u Arbuxxelli f'Malta* (Malta, 2002).

BOSIO 1604
Bosio, G., *Historia della Sacra Religione Militare di S. Giovanni Gersolomitano detta di Malta* (Malta, 1604).

BOWEN-JONES *et al* 1961
Bowen-Jones, H., Dewdney, J.C. and Fisher, W.B., *Malta: Background for Development* (Durham, 1961).

BRESC 1975
Bresc, H., 'The secrezia and the royal patrimony in Malta: 1240-1450', in *Medieval Malta: Studies on Malta Before the Knights*, ed. by A.T. Luttrell (London, 1975), 126-62.

BRINCAT 1995
Brincat, J., *Malta 870-1054: Al-Himyarī's Account and its Linguistic Implications* (Malta, 1995).

BRINCAT 2011
Brincat, J., *Maltese and other Languages: A Linguistic History of Malta* (Malta, 2011).

BROWN 1975
Brown, T.S., 'Byzantine Malta: a discussion of the sources', in *Medieval Malta: Studies on Malta Before the Knights*, ed. by A.T. Luttrell (London, 1975), 71-87.

BRUINS 1986
Bruins, H.J., *Desert Environment in the Central Negev and Kadesh-Barnea during Historical Times* (Netherlands, 1986).

BRUINS 1993
Bruins, H.J., 'Drought risk and water management in Israel: planning for the future', in *Drought Assessment, Management, and Planning: Theory and Case Studies*, ed. by D. A. Wilhite (New York, 1993), 133-56.

BRUNO and CUTAJAR 2002
Bruno, B. and Cutajar, N., 'Archeologia Bizantina a Malta: primi risultati e prospettive di indagine', in *Da Pyrgi a Mozia: Studi sull'Archeologia del Mediterraneo in memoria di Antonia Ciasca*, ed. by M.G. Amadasi, M. Liverani and P. Matthiae (Rome, 2002), 109-38.

BRUNO and CUTAJAR 2013
Bruno, B. and Cutajar, N., 'Imported amphorae as indicators of economic activity in Early Medieval Malta', in *The Insular System of the Early Byzantine Mediterranean: Archaeology and History*, ed. by D. Michaelides, P. Pergola and E. Zanini (Oxford, 2013), 15-29.

BUHAGIAR, K. 1997
Buhagiar, K., *The Ghar il-Kbir Settlement and the Cave Dwelling Phenomenon in Malta* (unpublished undergraduate dissertation, University of Malta, 1997).

BUHAGIAR, K. 2000
Buhagiar, K., 'The Roman baths at the Ta' Baldu tenement', *Treasures of Malta*, vol. 7, no. 2 (2000), 47-51.

BUHAGIAR, K. 2002
Buhagiar, K., *Medieval and Early Modern Cave-Settlements and Water Galleries in North-West Malta South of the Great Fault – A Field Survey and Gazetteer* (unpublished M.A. thesis, University of Malta, 2002).

BUHAGIAR, K. 2007
Buhagiar, K., 'Water management strategies and the cave-dwelling phenomenon in Late-Medieval Malta', *Medieval Archaeology*, vol. 51 (2007), 103-31.

BUHAGIAR, K. 2008A
Buhagiar, K., 'Comino: historical and archaeological observations', *Malta Archaeological Review 2004-2005*, Issue 7 (2008), 8-14.

BUHAGIAR, K. 2008B
Buhagiar, K., 'Baħrija: its archaeological significance', *Melita Historica*, vol. 14 no. 4 (2007), 357-374.

BUHAGIAR, K. 2012A
Buhagiar, K., 'Caves in context: the Late Medieval Maltese scenario', in *Caves in Context: The Cultural Significance of Caves and Rockshelters in Europe*, ed. by K.A. Bergsvik and R. Skeates (Oxford, 2012), 153-65.

BUHAGIAR, K. 2012B
Buhagiar, K., 'Malta, an Island satellite in the lee of Sicily: investigating the troglodytic context for the Late Medieval and Early Modern periods', in *L'Insediamento Rupestre di Monte S. Antonio a Regalbuto: Alle Origini del Raḥal di 'Abbûd*, ed. by I. Contino and F. Buscemi (Enna, 2012), 95-118.

BUHAGIAR, K. 2013
Buhagiar, K., 'Revisiting Wied ir-Rum: some recent archaeological discoveries', *Melita Historica*, vol. 16, no. 1 (2012), 77-108.

BUHAGIAR, M. 1975
Buhagiar, M., 'Medieval Churches in Malta', in *Medieval Malta: Studies on Malta Before the Knights*, ed. by A.T. Luttrell (London, 1975), 163-80.

BUHAGIAR, M. 1988
Buhagiar, M., 'Two archaeological sites: Ras ir-Raħeb, Malta, and Ras il-Wardija, Gozo, *Melita Historica*, vol. 10, no. 1 (1988), 69-87.

BUHAGIAR, M. 2000
Buhagiar, M., 'Four new Late Roman – Early Byzantine burial sites in the Island of Malta', *Melita Historica*, vol. 13, no. 1 (2000), 23-37.

BUHAGIAR, M. 2005
Buhagiar, M., *The Late Medieval Art and Architecture of the Maltese Islands* (Malta, 2005).

BUHAGIAR, M. 2007
Buhagiar, M., *The Christianisation of Malta: Catacombs, Cult Centres and Churches in Malta to 1530*, BAR International Series 1647 (Oxford).

BUHAGIAR and FIORINI 1996
Buhagiar, M. and Fiorini, S., *Mdina the Cathedral City of Malta. A Reassessment of its History and a Critical Appreciation of its Architecture and Work of Art*, vol. 2 (Malta, 1996).

BURNS 1974
Burns, A., 'Ancient Greek water supply and city planning: a study of Syracuse and Acragas,' *Technology and Culture*, vol. 15, no. 3 (1974), 389-412.

BUTZER *et al* 1985
Butzer, K.W., Mateu, J.F., Butzer, E.K. and Kraus, P., 'Irrigation Agrosystems in Eastern Spain: Roman or Islamic origins?', *Annals of the Association of American Geographers*, vol. 75, no. 4 (1985), 479-509.

CACHIA 1988
Cachia, J., *The mechanical and physical properties of the Globigerina Limestone as used in local masonry construction* (unpublished A. & C.E. thesis, University of Malta, 1988).

CALATRAVA and GARRIDO 2005
Calatrava, J. and Garrido, A., 'Modelling water markets under uncertain water supply', *European Review of Agricultural Economics*, vol. 32, no. 2 (2005), 119-42.

CAMILLERI 2003
Camilleri, A., *Fejn jinżamm in-naħal l-imġiebaħ jitrawmu: a case study in industrial archaeology* (unpublished undergraduate dissertation, University of Malta, 2002).

CAMUFFO *et al* 2010
Camuffo, D., Bertolini, C., Barriendos, M., Dominiguez-Castro, F., Cocheo, C., Enzi, S., Sghedoni, M., della Valle, A., Garnier, E., Alcoforado, M.J., Xoplaki, E., Luterbacher, J., Diodato, N., Maugeri, M., Nunes, M.F. and Rodriguez, R., '500-year temperature reconstruction in the Mediterranean Basin by means of ducumentary data and instrumental observations', *Climatic Change*, vol. 101, no. 1-2, 169-199.

CANCELLIERE *et al* 2009
Cancelliere, A., Nicolosi, V. and Rossi, G., 'Assessment of drought risk in water supply systems', in *Coping with Drought Risk in Agriculture and Water Supply Systems: Drought Management and Policy Development in the Mediterranean*, ed. by A. Iglesias, L. Garrote, A. Cancelliere, F. Cubillo and D. Wilhite, vol. 26 (2009), 93-110.

CAPRINO 1966
Caprino, C., 'Lo scavo del complesso rupestre sulla V terrazza e della IV terrazza', in *Missione Archaeologica Italiana a Malta: Rapporto preliminare della Campagna 1965*, ed. by M. Cagiano de Azevedo (Rome, 1966), 125-42.

CAPRINO 1967
Caprino, C., 'Lo scavo del complesso rupestre sulla V terrazza', in *Missione Archaeologica Italiana a Malta: Rapporto preliminare della Campagna 1966*, ed. by M. Cagiano de Azevedo (Rome, 1967), 81-86.

CARDONA 2002
Cardona, M., 'Għajn Tuffieħa and Manikata', in *Mellieha Through the Tides of Time*, ed. by J. Catania (Malta, 2002), 155-68.

CARMAN 1998
Carman, J., 'Interpretation, writing and presenting the past', in *Interpreting Archaeology: Finding meaning in the past*, ed. by I. Hodder, M. Shanks, A. Alexandri, V. Buchli, J. Carman, J. Last and G. Lucas (London, 1998), 95-99.

CARROLL *et al* 2012
Carroll, F.A., Hunt, C.O., Schembri, P.J. and Bonanno, A., 'Holocene climate change, vegetation history and human impact in the Central Mediterranean: evidence from the Maltese Islands', *Quaternary Science Reviews*, vol. 52 (2012), 24-40.

CARUANA 1881
Caruana, A.A., *Recent Discoveries at Notabile* (Malta, 1881).

CARUANA 1882
Caruana, A.A., *Report on the Phoenician and Roman Antiquities in the Group of the Islands of Malta* (Malta, 1882).

CARUANA SCICLUNA 1888
Caruana Scicluna, G., *Le Diverse Qualità dell' Acqua di Malta* (Malta, 1888).

CASADO SÁENZ and MONTOYA 2009
Casado Sáenz, M. and Montoya, F.F., 'The role of groundwater during drought', in *Coping with Drought Risk in Agriculture and Water Supply Systems: Drought Management and Policy Development in the Mediterranean*, ed. by A. Iglesias, L. Garrote, A. Cancelliere, F. Cubillo and D. Wilhite, vol. 26 (2009), 221-42.

CASSAR 2004
Cassar, J., 'Composition and property data of Malta's building stone for the construction of a database', in *Architectural and sculptural stone in cultural landscape*, ed. by R. Přikryl and P. Siegl (Prague, 2004), 11-28.

CASSAR 2010
Cassar, J., 'The use of limestone in a historic context – the experience of Malta', in *Limestone in the Built Environment: Present-Day Challenges for the Preservation of the Past*, ed. by B.J. Smith, M. Gomez-Heras, A.H. Viles and J. Cassar (London, 2010), 13-25.

CELAYA *et al* 2009
Celaya, I., Rodríguez, P. and Carbonell, X., 'Methods for social participation and conflict resolution', in *Coping with Drought Risk in Agriculture and Water Supply Systems: Drought Management and Policy Development in the Mediterranean*, ed. by A. Iglesias, L. Garrote, A. Cancelliere, F. Cubillo and D. Wilhite, vol. 26 (2009), 161-73.

CENTRAL OFFICE OF STATISTICS 1958
Central Office of Statistics (Malta), *Census 1957 – The Maltese Islands: Report on Population and Housing* (Malta, 1958).

CHADWICK 1884
Chadwick, O., *Report on the Water-Supply of Malta* (Malta, 1884).

CHADWICK 1894A
Chadwick, O., *Report on the Water Supply of Malta with Recommendations as to the Extension Thereof* (Malta, 1894).

CHADWICK 1894B
Chadwick, O., *Report on the Filtering Water Supplies Derived from Springs in Malta* (Malta, 1894).

CHADWICK 1897
Chadwick, O., *Report on the Water Supply of Malta and Gozo* (Malta, 1897).

CHAPMAN DAVIES 2008
Chapman Davies, H., *The Archaeology of Water* (Gloucester, 2008).

CIANTAR 1772-1780
Ciantar, G., *Malta Illustrata overo Descritione di Malta, Isola del Mare Siciliano e Adriatico con le sue Antichità ed altre Notizie* (Malta, 1772-1780).

CLARK *et al* 2008
Clark, C.M., Osltad, C., Buhagiar, K. and Gambin, T., 'Archaeology via underwater robots: mapping and localization within Maltese cistern systems', *Proceedings of the 10th International Conference on Control, Automation, Robotics and Vision (ICARCV 2008)* (2008), 662-67.

COOK 2013
Cook, E.R., 'Megadroughts, ENSO, and the invasion of Late-Roman Europe by the Huns and Avars', in *The Ancient Mediterranean Environment between Science and History*, ed. by W.V. Harris (Leiden, 2013), 89-102.

COOKE 1896
Cooke, J.H., 'Notes on the Globigerina Limestone of the Maltese Islands', *Geological Magazine*, vol. 33 (1896), 502-11.

COSGROVE and DANIELS 1988
Cosgrove, D. and Daniels, S., 'Introduction: iconography and landscape', in *The Iconography of Landscape: Essays on the symbolic representation, design and use of past environments*, ed. by D. Cosgrove and S. Daniels (Cambridge, 1988), 1-10.

COUSINS 2010
Cousins, S.M., 'The Prudhoe landscape history project: a retrogressive study of the landscape history of part of Sothern Northumberland', in *Northern Landscapes: Representations and Realities of North-East England*, ed. by T. Faulkner, H. Berry and J. Gregory (Woodbridge, 2010), 25-39.

CRESSEY 1958
Cressey, G.B., 'Qanats, Karez, and Foggaras', *Geographical Review*, vol. 48, no. 1 (1958), 27-44.

CRESSIER 1983
Cressier, P., 'L'Alpujarra médiévale: une approche archéologique', *Mélanges de la Casa de Velázquez*, vol. 19 (1983), 89-124.

CRESSIER 1988
Cressier, P., 'Fonction et évolution du réseau castral en Andalousie orientale: le cas de l'Alpujarra', *Castrum 3. Guerre, fortification et habitat dans le monde méditerranien au Moyen-Age* (1988), 123-34.

CRESSIER 1993
Cressier, P., 'Le Catastro de la Ensenada (v. 1752): La structure des petites 'villes' islamiques d'Andalousie, *Environmental Design: Journal of the Islamic Environmental Design Research Centre 1-2*, no. 13-14, 'Urban Morphogenesis: Maps and Cadastral Plans' (1993), 38-43.

CRESSIER *et al* 1989
Cressier, P, Bertrand, M., Carbonero, M.A., Diaz, A., Malpica, A. and Quesada, T., 'Agricultura e hidraulica medievales en el antiguo Reino de Granada. El caso de la alpuljarra costera', in *El Agua en zonas áridas: arqueología e historia (II)*, ed. by L. Cara (Almería, 1989), 545-53.

CROUCH 1984
Crouch, D.P., 'The Hellenistic water system of Morgantina, Sicily', *American Journal of Archaeology*, vol. 88 (1984), 353-64.

CROUCH 1993
Crouch, D.P., *Water Management in Ancient Greek Cities* (Oxford, 1993).

CROUCH 2001
Crouch, D.P., 'Geology and ancient water supply in two Greco-Roman cities: Syracuse and Agrigento', in *Water Use and Hydraulics in the Roman City*, ed. by A.O. Koloski-Ostrow (Iowa, 2001), 17-26.

CROUCH 2004
Crouch, D.P., *Geology and Settlement: Greco-Roman Patterns* (Oxford, 2004).

CUTAJAR 2001
Cutajar, N., 'Recent discoveries and the archaeology of Mdina', *Treasures of Malta*, vol. 8, no. 1 (2001), 79-85.

CUTAJAR 2004
Cutajar, N., 'The Archaeology of Malta's Middle Ages: documenting cultural change between the Tenth and the Thirteenth century', in *Malta: Roots of a Nation: the Development of Malta from an Island People to an Island Nation*, ed. by K. Gambin (Malta, 2004), 55-64.

DALLI 2006
Dalli, C., *Malta the Medieval Millennium* (Malta, 2006).

DALLI 2008
Dalli, C., 'Satellite, sentinel, stepping stone. Medieval Malta in Sicily's orbit', in *Malta in the Hybleans, the Hybleans in Malta*, ed. by A. Bonanno and P. Militello (Palermo, 2008), 245-58.

DALMIGLIO 2003
Dalmiglio, P., 'L'abitato del Fosso Formicola', in *Insediamenti rupestri medievali della Tuscia – I: le abitazioni*, ed. by E. De Minicis (Rome, 2003), 35-62.

DART *et al* 1993
Dart, C.J., Bosence, D.W.J. and McClay, K.R., 'Stratigraphy and structure of the Maltese graben system', *Journal of the Geological Society, London*, vol. 150 (1993), 1153-66.

De Feo *et al* 2010
De Feo, G., De Gisi, S., Malvano, C., Capolongo, D., Del Prete, S., Manco, M., Maurano, F. and Tropeano, E., 'Historical, biological and morphological aspects of the Roccarainola qanat in the district of Naples, Italy', *Water Science & Technology: Water Supply – WSTWS*, vol. 10, no. 4 (2010), 647-55.

DE MEULEMEESTER 2005
De Meulemeester, J., 'Granaries and irrigation: archaeological and ethnological investigations in the Iberian Peninsula and Morocco', *Antiquity*, vol. 79 (2005), 609-15.

DOCTER *et al* 2012
Docter, R.F., Vella, N.C., Cutajar, N., Bonanno, A. and Pace, A., 'Rural Malta: first results of the joint Belgo-Maltese survey project', in *BABESCH,* 87 (2012), 107-49.

DOLGOFF 1996
Dolgoff, A., *Physical Geology* (New York, 1996).

EARLE 1978
Earle, T., *Economic and Social Organisation of a Complex Chiefdom: The Halelea District, Kaua'i, Hawaii* (Michigan, 1978).

EARLE 1980
Earle, T., 'Prehistoric irrigation in the Hawaiian Islands: an evaluation of evolutionary significance', *Archaeology & Physical Anthropology in Oceania*, vol. 15, no. 1 (1980), 1-28.

ENGLISH 1968
English, P.W., 'The origin and spread of qanats in the Old World', *Proceedings of the American Philosophical Society*, vol. 112, no. 3 (1968), 170-81.

ENGLISH 1997
English, P.W., 'Qanats and lifeworlds in Iranian plateau villages', in *Proceedings of the Conference on Transformation of Middle Eastern Natural Environment: Legacies and Lessons* (Austin, 1997), 187-205.

ERCOLI 2000
Ercoli, L., 'Ancient systems of hydraulic supply in Sicily and surrounding small islands', in *Cura Aquarum in Sicilia: Proceedings of the tenth International Congress on the History of Water Management and Hydraulic Engineering in the Mediterranean Region (Syracuse, May 16-22, 1998)*, ed. by G.C.M. Jansen (Leiden, 2000), 57-67.

ERCOLI *et al* 2008
Ercoli, L., Rizzo, G., Megna, B. and Parlapiano, M., 'Characterization of mortars from ancient and traditional water supply systems in Sicily', *Journal of Thermal Analysis and Calorimetry*, vol. 92, no. 1 (2008), 323-30.

FANTAR 1988
Fantar, M.H., 'North Africa', in *The Phoenicians*, ed. by S. Moscati (Milan, 1988), 166-85.
FARRUGIA RANDON R. and FARRUGIA RANDON S. 1993
Farrugia Randon, R. and Farrugia Randon, S., *Comino, Filfla and St Paul's Island* (Malta, 1993).

FAULKNER *et al* 2010
Faulkner, T., Berry, H. and Gregory, J., *Northern Landscapes: Representations and Realities of North-East England* (Woodbridge, 2010).

FAVIA 2012
Favia, P., 'Disciplinary territories and disciplines of the territory: paths and intersections between the archaeological research and other scientific approaches to the landscape', *Plurimondi*, vol. 10 (2012), 97-108.

FENECH 2007
Fenech, K., *Human-Induced Changes in the Environment and Landscape of the Maltese Islands from the Neolithic to the 15th Century AD*, BAR International Series 1682 (Oxford, 2007).

FIORINI 1983
Fiorini, S., 'Status Animarum I: a unique source for 17th and 18th century Maltese demography', *Melita Historica*, vol. 8, no. 4 (1983), 325-43.

FIORINI 1986
Fiorini, S., 'Status Animarum II: a census of 1687', *Proceedings of History Week 1984* (1986), 41-100.

FIORINI 1992
Fiorini, S., *The 'Mandati' Documents at the Archives of the Mdina Cathedral, Malta 1437-1539* (Malta, 1992).

FIORINI 1993
Fiorini, S., 'Malta in 1530', in *Hospitaller Malta 1530-1798*, ed. by V. Mallia-Milanes, 111-98.

FIORINI 1996
Fiorini, S., *Documentary Sources of Maltese History, Part 1 Notarial Documents, No. 1 Notary Giacomo Zabbara, R494/1(I): 1486-1488* (Malta, 1996).

FIORINI 1999A
Fiorini, S., *Documentary Sources of Maltese History: Part 1 Notarial Documents – No. 2, Notary Giacomo Zabbara, R494/1(II-IV) 1494-1497* (Malta, 1999).

FIORINI 1999B
Fiorini, S., *Documentary Sources of Maltese History: Part II Documents in the State Archives, Palermo, No. 1, Cancelleria Regia: 1259-1400* (Malta, 1999).

FIORINI 2003
Fiorini, S., 'X'nafu dwar Callus?', in *Min kien Callus?*, ed. by C. Cassar and Y. Vella (Malta, 2003), 15-53.

FIORINI 2004
Fiorini, S., *Documentary Sources of Maltese History: Part II Documents in the State Archives, Palermo, No. 2, Cancelleria Regia: 1400-1459* (Malta, 2004).

FIORINI 2005
Fiorini, S., *Documentary Sources of Maltese History: Part I Notarial Documents – No. 3, Notary Paulo Bonello MS.588:1467-1517; Notary Giacomo Zabbara MS.1132: 1471-1500* (Malta, 2005).

FIORINI 2007
Fiorini, S., *Documentary Sources of Maltese History: Part II Documents in the State Archives, Palermo, No. 3, Cancelleria Regia: 1460-1485* (Malta, 2007).

FLORIO 1598
Florio, I., *A Worlde of Wordes, or Most copious, and exact Dictionarie in Italian and English* (London, 1598).

FLORIO 1611
Florio, I., *Queen Anna's New World of Words, or Dictionarie of the Italian and English tongues* (London, 1611).

FORBES 1955
Forbes, R.J., *Studies in Ancient Technology*, vol. 1 (Leiden, 1955).

FOWLER 1998
Fowler, P.J., 'Writing on the countryside', in *Interpreting Archaeology: Finding meaning in the past*, ed. by I. Hodder, M. Shanks, A. Alexandri, V. Buchli, J. Carman, J. Last and G. Lucas (London, 1998), 100-9.

FRIESEN 1999
Friesen, T.M., 'Resource structure, scalar stress, and the development of Inuit social organisation', *World Archaeology*, vol. 31, no. 1 (1999), 21-37.

GABRIELI 1961
Gabrieli, F., 'Ibn Hawqal e gli Arabi di Sicilia', *Rivista Studi Orientali*, vol. 35 (1961), 245-53.

GAMBIN 2008
Gambin, T., 'The maritime cartography of the Malta-Sicily Channel', in *Malta and Sicily: Miscellaneous Research Projects*, ed. by A. Bonanno (Palermo, 2008), 126-99.

GARRIDO and GÓMEZ-RAMOS 2009
Garrido, A. and Gómez-Ramos, A., 'Risk management instruments supporting drought planning and policy', in *Coping with Drought Risk in Agriculture and Water Supply Systems: Drought Management and Policy Development in the Mediterranean*, ed. by A. Iglesias, L. Garrote, A. Cancelliere, F. Cubillo and D. Wilhite, vol. 26 (2009), 133-52.

GATT, G. 2005
Gatt, G., *Qiegħda fil-Ponta ta' Lsieni* (Malta, 2005).

GATT, P.A. 2006
Gatt, P.A., 'Model of limestone weathering and damage in masonry: sedimentological and geotechnical controls in the Globigerina Limestone formation (Miocene) of Malta', *Xjenza*, vol. 11 (2006), 30-39.

GERRARD 2011
Gerrard, C., 'Contest and co-operation: strategies for medieval and later irrigation along the upper Huercha valley, Aragón, north-east Spain', *Water History*, vol. 3 (2011), 3-28.

GERRARD and PETLEY 2013
Gerrard, C.M. and Pettley, D.N., 'A risk society? Environmental hazards, risk and resilience in later Medieval Europe', *Natural Hazards*, vol. 69, no. 1 (2013), 1051-79.

GILBERTSON and CHISHOLM 1996
Gilbertson, D.D. and Chisholm, N.W.T., 'ULVS XXVIII: Manipulating the desert environment: ancient walls, floodwater farming and territoriality in the Tripolitanian pre-desert of Libya', *Libyan Studies 27* (1996), 17-52.

GILBERTSON and HUNT 1996
Gilbertson, D.D. and Hunt, C.O., 'Roman-Libyan agriculture: walls and flood water farming', in *Farming the Desert – The UNESCO Libyan Valleys Archaeological Survey*, 1, ed. by G. Barker (London, 1996), 191-225.

GILBERTSON *et al* 2000
Gilbertson, D., Hunt, C. and Gillmore, G., 'Success, longevity and failure of arid-land agriculture: Romano-Libyan floodwater farming in the Tripolitanian pre-desert', in *The Archaeology of Drylands – Living at the Desert Margin*, ed. by G. Barker and D. Gilbertson (London, 2000), 137-59.

GINORI LISCI 1978
Ginori Lisci, L., *Cabrei in Toscana: raccolte di mappe, prospetti e vedute sec. XVI-sec. XIX* (Lucca, 1978).

GLICK 1970
Glick, T.F., *Irrigation and Society in Medieval Valencia* (Cambridge, 1970).

GLICK 1979
Glick, T.F., *Islamic and Christian Spain in the Early Middle Ages* (Princeton, 1979).

GLICK 2007
Glick, T.F., *Paisajes de conquista: Cambio cultural y geográfico en la España medieval* (València, 2007).

GLICK and KIRCHNER 2000
Glick, T.F. and Kirchner, H., 'Hydraulic systems and technologies of Islamic Spain: history and archaeology', in *Working with Water in Medieval Europe: Technology and Resource-Use*, ed. by P. Squatriti (Leiden, 2000), 267-329.

GOBLOT 1979
Goblot, H., *Les Qanats: Une Technique d'Acquisition de l'Eau* (Paris, 1979).

GOSDEN and HEAD 1994
Gosden, C. and Head, L., 'Landscape: a usefully ambiguous concept', *Archaeology in Oceania*, vol. 29, no. 3 (1994), 113-16.

GRECH 1999
Grech, C.F., 'The Maltese Islands as a classical example of man-induced desertification and land degradation', in *National Awareness Seminar on Desertification and Land Degradation: Proceedings of Seminar Foundation for International Studies, Malta*, ed. by A. E. Baldacchino and C.M. Tanti (Malta, 1999), 89-104.

GRECH 2001
Grech, C., *A Forest History of the Maltese Islands to AD 1800* (unpublished Ph.D. thesis, University of Aberdeen, 2001).

GREENE and MOORE 2010
Greene, K. and Moore, T., *Archaeology: An Introduction* (London, 2010).

GRIMA 2004
Grima, R., 'The landscape context of megalithic architecture', in *Malta before History: the world's oldest free-standing stone* architecture, ed. by D. Cilia (Malta, 2004), 327-45.

GRIMA 2005
Grima, R., *Monuments in Search of a Landscape: the landscape context of monumentality in Late Neolithic Malta* (unpublished Ph.D. thesis, University College London, 2005).

GROVE and RACKHAM 2001
Grove, A.T. and Rackham, O., *The Nature of Mediterranean Europe: an ecological history* (London, 2001).

HAHS 2010
Hahs, D.G., *Medieval Malta: Abandoned Villages, Chapels and Farmhouses* (unpublished M.A. thesis, Florida State University, 2010).

HARDEN 1980
Harden, D., *The Phoenicians* (New York, 1980).

HARROWER 2009
Harrower, M.J., 'Is the hydraulic hypothesis dead yet? Irrigation and social change in ancient Yemen', *World Archaeology*, vol. 41, no. 1 (2009), 58-72.

HODDER *et al* 1998
Interpreting Archaeology: Finding meaning in the past, ed. by I. Hodder, M. Shanks, A. Alexandri, V. Buchli, J. Carman, J. Last and G. Lucas (London, 1998).

HODGE 2002
Hodge, A.T., *Roman Aqueducts and Water Supply* (London, 2002).

HOLT 2000
Holt, R., 'Medieval England's water-related technologies', in *Working with Water in Medieval Europe: Technology and Resource-Use*, ed. by P. Squatriti (Leiden, 2000), 51-100.

HORDEN and PURCELL 2000
Horden, P. and Purcell, N., *The Corrupting Sea: A Study of Mediterranean History* (Oxford, 2000).

HUNT 1978
Hunt, R.C., 'Irrigation, conflict and politics: a Mexican case', in *Origins of the State: A Symposium* (revised version), ed. by R. Cohen and E. Service (Philadelphia, 1978), 69-123.

HUNT 1988
Hunt, R.C., 'Size and the structure of authority in canal irrigation systems', *Journal of Anthropological Research*, vol. 44, no. 4 (1988), 335-55.

HUNT and VELLA 2005
Hunt, C.O. and Vella, N.C., 'A view from the countryside: pollen from a field fill at Mistra valley', *Malta Archaeological Review*, vol. 7 (2005), 61-69.

HYDE 1955
Hyde, H.P.T., *The Geology of the Maltese Islands – With Special Reference to Water Supply and Possibilities of Oil* (Malta, 1955).

IGLESIAS, A. 2003
Iglesias, A., 'Climate changes in the Mediterranean region: physical aspects and effects on agriculture', in *Mediterranean Climate: Variability and Trends*, ed. by H. J. Bolle (London, 2003), 107-11.

IGLESIAS, A. *et al* 2009
Iglesias, A., Moneo, M. and Quiroga, S., 'Methods for evaluating social vulnerability to drought', in *Coping with Drought Risk in Agriculture and Water Supply Systems: Drought Management and Policy Development in the Mediterranean*, ed. by A. Iglesias, L. Garrote, A. Cancelliere, F. Cubillo and D. Wilhite, vol. 26 (2009), 153-60.

IGLESIAS, E. *et al* 2003
Iglesias, E., Garrido, A. and Gómez-Ramos, A., 'Evaluation of drought management in irrigated area', *Agricultural Economics*, vol. 29 (2003), 211-29.

JOHNSON 2007
Johnson, M., *Ideas of Landscape* (London, 2007).

JONES, B. 1985
Jones, B., 'Introduction', in *Archaeology Field Survey in Britain and Abroad*, ed. by S. Macready and F. H. Thompson (London, 1985), 1-7.

JONES, W.H. 1948
Jones, W.H., *Hippocrates*, The Loeb Classical Library, 1 (Massachusetts, 1948).

JUDSON and KAHANE 1963
Judson, S. and Kahane, A., 'Underground drainage ways in southern Etruria and northern Latium', *Journal of the British School at Rome*, vol. 31, no. 74, 1963.

KIRCHNER 2008
Kirchner, H., 'Archeologia degli spazi irrigati medievali e le loro forme di gestione sociale', in *L'Acqua nei secoli Altomedievali* (Spoleto, 2008), 471-502.

KIRCHNER 2009
Kirchner, H., 'Original design, tribal management and modifications in medieval hydraulic systems in the Balearic Islands (Spain)', *World Archaeology*, vol. 41, no. 1, 151-68.

KIRCHNER and NAVARRO 1993
Kirchner, H. and Navarro, C., 'Objetivos, métodos y práctica dela arqueología hidráulica', *Archeologia Medievale*, vol. 20 (1993), 121-50.

KOLOSKI-OSTROW 2001A
Water Use and Hydraulics in the Roman City, ed. by A. O. Koloski-Ostrow (Iowa, 2001).

KOLOSKI-OSTROW 2001B
Koloski-Ostrow, A.O., 'Water as a symbol of wealth? An overview of the Roman evidence', in *Water Use and Hydraulics in the Roman City*, ed. by A.O. Koloski-Ostrow (Iowa, 2001), 1-15.

LATHAM 1965
Latham, R.E., *Revised Medieval Latin World-List from British and Irish Sources* (London, 1965).

LAYTON and UCKO 1999
Layton, R. and Ucko, P.J., 'Gazing on the landscape and encountering the environment', in *The Archaeology and Anthropology of the Landscape: Shaping your Landscape*, ed. by R. Layton and P.J. Ucko (London, 1999), 1-20.

LEACH 1979
Leach, H., Rev. of *Economic and Social Organisation of a Complex Chiefdom: The Halelea District, Kaua'i, Hawaii*, by T. Earle, *Journal of the Polynesian Society*, vol. 88, no. 4 (1979), 473-75.

LEVEAU *et al* 1999
Environmental Reconstruction in Mediterranean Landscape Archaeology, ed. by P. Leveau, F. Trément, K. Walsh and G. Barker (Oxford, 1999).

LIGHTFOOT 2000
Lightfoot, D.R., 'The origin and diffusion of qanats in Arabia: new evidence from the northern and southern peninsula', *The Geographical Journal*, vol. 166, no. 3 (2000), 215-26.

LIGHTFOOT and MILLER 1996
Lightfoot, D.R. and Miller, J.A., 'The rise and fall of a walled oasis in medieval Morocco', *Annals of the Association of American Geographers*, vol. 86, no. 1 (1996), 78-101.

LIONELLO *et al* 2006
Lionello, P., Malanotte-Rizzoli, P., Boscolo, R., Alperti, P., Artale, V., Li, L., Luterbacher, J., May, W., Trigo, R., Tsimplis, M., Ulbrich, U. and Xoplaki, E., 'The Mediterranean climate: an overview of the main characteristics and issues', in *Mediterranean Climate Variability*, ed. by P. Lionello, P. Malanotte-Rizzoli and R. Boscolo (Heidelberg, 2006), 1-26.

LIONELLO *et al* 2012
Lionello, P., Abrantes, F., Congedi, L., Dulac, F., Gacic, M., Gomis, D., Goodess, C., Hoff, H., Kutiel, H., Luterbacher, J., Planton, S., Reale, M., Schröder, Struglia, M.V., Toreti, A., Tsimplis, M., Ulbrich, U. and Xoplaki, E., 'The climate of the Mediterranean region in historical times', in *The Climate of the Mediterranean Region: From the Past to the Future*, ed. by, P. Lionello (London, 2012), xlv-xlvi.

LLORIS 2006
Lloris, F.B., 'An irrigation decree from Roman Spain: The Lexi Rivi Hiberiensis', *The Journal of Roman Studies*, vol. 96 (2006), 147-97.

LOCKWOOD 1952
Lockwood, J.F., *A Smaller Latin-English Dictionary* (London, 1952).

LOMBARDO 1999
Lombardo, R., '"Ayduni" with its four rivers', in *Aidone and Morgantina*, ed. by V. Alabiso (Palermo, 1999), 4-13.

LOFRANO *et al* 2013
Lofrano, G., Carotenuto, M, Maffettone, R., Todaro, P., Sammataro, S., Kalavrouziotis, I.K., 'Water collection and distribution systems in the Palermo plain during the Middle Ages', *Water*, vol. 5 (2013), 1662-76.

LÓPEZ-BARRERO and IGLESIAS 2009
López-Barrero, E. and Iglesias, A., 'Soft law principles for improving drought management in Mediterranean countries', in *Coping with Drought Risk in Agriculture and Water Supply Systems: Drought Management and Policy Development in the Mediterranean*, ed. by A. Iglesias, L. Garrote, A. Cancelliere, F. Cubillo and D. Wilhite, vol. 26 (2009), 21-36.

LOUD and METCALFE 2002
Loud, G.A. and Metcalfe, A., *The Society of Norman Italy* (Boston, 2002).

LOWENTHAL 1978
Lowenthal, D., 'Finding valued landscapes', *Progress in Human Geography*, vol. 2 (1978), 373-418.

LOWENTHAL 1979
Lowenthal, D., 'Age and artefact', in *The Interpretation of Ordinary Landscapes*, ed. by D.W. Meinig (New York, 1979), 103-28.

LOWENTHAL 2003
Lowenthal, D., *The Past is a Foreign Country* (Cambridge, 2003).

LUTERBACHER and XOPLAKI 2003
Luterbacher, J. and Xoplaki, E., '500-year winter temperature and precipitation variability over the Mediterranean area and its connection to the large-scale atmospheric circulation', in *Mediterranean Climate: Variability and Trends*, ed. by H.J. Bolle (London, 2003), 133-53.

LUTERBACHER *et al* 2006
Luterbacher, J., Xoplaki, E., Casty, C., Wanner, H., Pauling, A., Küttel, M., Rutishauser, T., Brönnimann, S., Fischer, E., Fleitmann, D., González-Rouco, F.J., Garcia-Herrera, R., Barriendos, M., Rodrigo, F., Gonzalez-Hidalgo, J.C., Saz, M.A., Gimeno, L., Ribera, P., Brunet, M., Paeth, H., Rimbu, N., Felis, T., Jacobeit, J., Dünkeloh, A., Zorita, E., Guiot, J., Türkes, M., Alcoforado, M.J., Trigo, R., Wheeler, D., Tett, S., Mann, M.E., Touchan, R., Shindell, D.T., Silenzi, S., Montagna, P., Camuffo, D., Mariotti, A., Nanni, T., Brunetti, M., Maugeri, M., Zerefos, C., De Zolt, S., Lionello, P., Nunes, M.F., Rath, V., Beltrami, H., Garnier, E. and Ladurie, E.L.R., 'Mediterranean climate variability over the last centuries: a review', in *Mediterranean Climate Variability*, ed. by P. Lionello, P. Malanotte-Rizzoli and R. Boscolo (Heidelberg, 2006), 27-126.

LUTTRELL 1975A
Medieval Malta – Studies on Malta before the Knights, ed. by A.T. Luttrell (London, 1975).

LUTTRELL 1975B
Luttrell, A.T., 'Approaches to medieval Malta', in *Medieval Malta – Studies on Malta before the Knights*, ed. by A.T. Luttrell (London, 1975), 1-70.

LUTTRELL 1976
Luttrell, A.T., 'Frederick II and Paulino de Malta: 1235', *Quellen und Forschungen aus Italienischen Archiven und Bibliotheken 55/56* (Tübingen, 1976), 405-9.

LUTTRELL 1977
Luttrell, A.T, 'The earliest documents transcribed in the Cathedral Archives, Mdina: 1316-1372', in *Archives of the Cathedral of Malta Misc. 32A: 1313-1529*, ed. by J. Azzopardi (Malta, 1977), 29-51.

LUTTRELL 1978
Luttrell, A.T., 'The sale of Gumerin on Malta: 1318', *Estudios Históricos y Documentos de los Archivos de Protocolos 6* (Catalunya, 1978), 167-177.

LUTTRELL 1979
Luttrell, A.T., 'Malta troglodytica: Għar il-Kbir', *Heritage*, vol. 2 (1979), 461-65.

LUTTRELL 1991
Luttrell, A.T., 'Medieval Malta: the non-written and the written evidence', in *Malta: A Case Study in International Cross-Currents*, ed. by S. Fiorini and V. Mallia-Milanes (Malta, 1991), 33-45.

LUTTRELL 1997
Luttrell, A.T., 'Giliberto Abbate's report on Malta: circa 1241', in *Proceedings of History Week*, ed. by K. Sciberras (Malta, 1997), 1-29.

LUTTRELL 2002A
Luttrell, A.T., *The Making of Christian Malta: from the Early Middle Ages to 1530* (Hampshire, 2002).

LUTTRELL 2002B
Luttrell, A.T., 'Medieval Malta: approaches and reproaches', in *The Making of Christian Malta: from the Early Middle Ages to 1530* (Hampshire, 2002), 1-17.

MABRY 2000
Mabry, J.B., 'Wittfogel was half right: the ethnology of consensual and nonconsensual hierarchies in irrigation management', in *Hierarchies in Action: Cui Bono?*, ed. by M.W. Diehl (Carbondale, 2000), 284-94.

MACREADY and THOMPSON 1985
Archaeology Field Survey in Britain and Abroad, ed. by S. Macready and F.H. Thompson (London, 1985).

MAGNUSSON and SQUATRITI 2000
Magnusson, R. and Squatriti, P., 'The technologies of water in medieval Italy', in *Working with Water in Medieval Europe: Technology and Resource-Use*, ed. by P. Squatriti (Leiden, 2000), 217-66.

MAGRO CONTI 1996
Magro Conti, J., *Roman Bathing Establishments in Malta and Gozo* (unpublished undergraduate dissertation, University of Malta, 1996).

MAGRO CONTI and SALIBA 2007
The Significance of Cart-Ruts in Ancient Landscapes, ed. by J. Magro Conti and P. C. Saliba (Malta, 2007).

MALIVA and MISSIMER 2012
Maliva, R. and Missimer, T., *Arid Lands Water Evaluation and Management* (Heidelberg, 2012).

MAMASSIS and KOUTSOYIANNIS 2010
Mamassis, N. and Koutsoyiannis, D., 'A web based information system for the inspection of the hydraulic works in Ancient Greece', in *Ancient Water Technologies*, ed. by L.W. Mays (Heidelberg, 2010), 103-14.

MAR 1910
Annual Report on the Working of the Museum for the Financial Year 1909-1910, Government Printing Office (Malta, 1910).

MAR 1922
Annual Report on the Working of the Museum for the Financial Year 1921-1922, Government Printing Office (Malta, 1922).

MAR 1924
Annual Report on the Working of the Museum for the Financial Year 1923-1924, Government Printing Office (Malta, 1924).

MAR 1925
Annual Report on the Working of the Museum for the Financial Year 1924-1925, Government Printing Office (Malta, 1925).

MAR 1965
Report on the Working of the Museum Department for the Year 1965, Department of Information (Malta, 1965).

MARTIN 2002
Martin, J.M., 'Settlement and the agrarian economy', in *The Society of Norman Italy*, ed. by G.A. Loud and A. Metcalfe (Boston, 2002), 17-45.

MARTÍN-BUENO and Reklaityte 2008
Martín-Bueno, M. and Reklaityte, I., 'L'acqua nella Spagna del Medioevo: elemento primordiale per tre ambienti culturali diversi', in *L'Acqua nei secoli Altomedievali* (Spoleto, 2008), 201-42.

MATTINGLY 1995
Mattingly, D.J., *Tripolitania* (London, 1995).

MAYS 2010
Mays, L.W., 'A brief history of Roman water technology', in *Ancient Water Technologies*, ed. by L.W. Mays (Heidelberg, 2010), 115-37.

MAZZAOUI 2008
Mazzaoui, M.F., *The Italian Cotton Industry in the Later Middle Ages 1100-1600* (Cambridge, 2008).

MCCORMICK *et al* 2012
McCormick, M., Büntgen, U., Cane, M.A., Cook, E.R., Harper, K., Huybers, P., Litt, T., Manning, S.W., Mayewski, P.A., More, A.F.M., Nicolussi, K. and Tegel, W., 'Climate change during and after the Roman Empire: reconstructing the past from scientific and Historical evidence', *Journal of Interdisciplinary History*, vol. 43, no. 2 (2012), 169-220.

MERCER 1985
Mercer, R., 'Great Britain: a view of British Archaeological Field Survey', in *Archaeological Field Survey in Britain and Abroad*, ed. by S. Macready and F.H. Thompson (London, 1985), 8-24.

MERCIECA 1931
Mercieca, V., *Brevi Cenni sulle Acque Sorgive e Sugli Acquedotti dell'Isola di Malta* (Malta, 1931).

MESSINA 1989
Messina, A., 'Troglodytismo medievale a Malta', *Melita Historica*, vol. 10, no. 2 (1989).

METCALFE 2002
Metcalfe, A., 'The Muslims of Sicily under Christian rule', in *The Society of Norman Italy*, ed. by G. A. Loud and A. Metcalfe (Boston, 2002), 289-317.

METCALFE 2003
Metcalfe, A., *Muslims and Christians in Norman Sicily: Arabic Speakers and the End of Islam* (London, 2003).

MEZZUCCATO *et al* 1999
Mazzuccato, C., Mezzolani, A. and Morgi, A., 'Infrastrutture idriche a Tharros: note sul sistema fognario', in *Tharros Nomen*, ed. by E. Acquaro, M.T. Francisi, T.K. Kirova and A.M. Vaccaro (La Spezia, 1999), 117-34.

MICALLEF *et al* 2004
Micallef, P., Attard, G. and Mangion, J., 'Water resources management in Malta: cultural heritage and legal and administrative set-up', B/48, *Participatory water saving management and water cultural heritage. Proceedings of the first WASHMED (Water Saving in the Mediterranean)* (2004), 199-207.

MITCHELL 1962
Mitchell, P.K., 'Pioneer rainfall observations in the Maltese Islands', *Melita Historica*, vol. 3, no. 3, 59-64.

MOLINARI and CUTAJAR 1999
Molinari, A. and Cutajar, N., 'Of Greeks and Arabs and of feudal Knights', *Malta Archaeological Review*, vol. 3 (1999), 9-16.

MORGAN 1960
Morgan, M.H., *Vitruvius: The Ten Books on Architecture* (New York, 1960).

MORRIS 1952
Morris, T.O., *The Water Supply Resources of Malta* (Malta, 1952).

MUIR 1998
Muir, R., 'Landscape: a wasted legacy', *Area*, vol. 30, no. 3 (1998), 263-71.

MURRAY 1890
Murray, J., 'The Maltese Islands with special reference to their geological structure', *The Scottish Geographical Magazine*, vi. (1890), 449-89.

NATIONAL STATISTICS OFFICE 2011
National Statistics Office (Malta), *Demographic Review 2010* (Malta, 2011).

NESBIT and O'HARA 2000
Nesbit, M. and O'Hara, S., 'Irrigation agriculture in Central Asia: a long-term perspective from Turkmenistan', in *The Archaeology of Drylands – Living at the margin*, ed. by G. Barker and D. Gilbertson (London, 2000), 103-22.

NIERMEYER 1976
Niermeyer, J.F., *Mediae Latinitatis Lexicon Minus* (Leiden, 1976).

OOSTHUIZEN 2006
Oosthuizen, S., *Landscapes decoded. The origins and development of Cambridgeshire's medieval fields* (Hertfordshire, 2006).

ORTEGA ORTEGA 2010
Ortega Ortega, J.M., 'La agricultura de los vencedores y la agricultura de los vencidos: La investigación de las transformaciones feudales de los paisajes agrarios en el Valle del Ebro (siglos XII-XIII)', in *Por una arqueología agraria: Perspectivas de investigación sobre espacios de cultivo en las sociedades medievales hispánicas*, ed. by H. Kirchner, BAR International Series 2062 (Oxford, 2010), 123-46.

PARK 1977
Park, C., 'Island with few options', *Geographical Magazine*, vol. 49, no. 8 (1977), 508-9.

PATTON 1954
Patton, W.R., *The Histories – Polybius*, The Loeb Classical Library, vol. 4 (U.K., 1954).

PEDLEY *et al* 2002
Pedley, M., Clarke, M.H. and Galea, P., *Limestone Isles in a Crystal Sea* (Malta, 2002).

POWELL 2009
Powell, E.A., 'The Past in High-Def', *Archaeology*, vol. 62, no. 3 (2009), 20-25.

POWERS 1979
Powers, J.F., 'Frontier municipal baths and social interaction in Thirteenth-Century Spain', *The American Historical Review*, vol. 84, no. 3 (June, 1979), 649-67.

PRICE 1994
Price, D.H., 'Wittfogel's neglected hydraulic/hydroagricultural distinction', *Journal of Anthropological Research*, vol. 50, no. 2 (1994), 187-204.

PRITCHARD 1978
Pritchard, J.B., *Recovering Sarepta, A Phoenician City: Excavations at Sarafand, Lebanon, 1969-1974, by the University Museum of the University of Pennsylvania* (New Jersey, 1978).

QUINTINUS 1536
Quintinus Haeduus, J., *Insulae Melitae Descriptio* (Lyons, 1536).

RAMSEYER 2006
Ramseyer, V., *The Transformation of a Religious Landscape: Medieval South Italy 850-1130* (New York, 2006).

RETAMERO 2008
Retamero, F., 'Irrigated agriculture, risk and population. The Andalusi hydraulic systems of the Balearic Islands as a case study (10th – 13th c)', in *Marqueurs des paysages et systèmes socio-économiques: Proceedings of Le Mans COST Conference*, ed. by R. Compatangelo-Soussignan, J.R. Bertrand, J. Chapman and P.Y. Laffont (Rennes, 2008), 135-47.

RIPPON 2006
Rippon, S., *Landscape, community and colonisation: the North Somerset Levels during the 1st to 2nd millenia AD* (York, 2006).

RIPPON 2012
Rippon, S., *Historic Landscape Analysis: Deciphering the countryside* (Oxford, 2004).

RIPPON 2013
Rippon, S., 'Historic landscape character and sense of place', *Landscape Research*, vol. 38, no. 2, 179-202.

RIZZO 1932
Rizzo, C., *Report on the Geology of the Maltese Islands – including chapters on possible ground-water-tables and prospecting for mineral-oil and natural-gas* (Malta, 1932).

ROBERTS 1987
Roberts, B.K., 'Landscape archaeology', in *Landscape and Culture: Geographical and Archaeological Perspectives*, ed. by J.M. Wagstaff (London, 1987), 77-95.

ROSEN 2000
Rosen, S.A., 'The decline of desert agriculture: a view from the classical period Negev', in *The Archaeology of Drylands – Living at the margin*, ed. by G. Barker and D. Gilbertson (London, 2000), 45-62.

ROSSI *et al* 2009
Rossi, G., Bonaccorso, B. and Nicolosi, V., 'Characterizing drought risk in a Sicilian river basin', in *Coping with Drought Risk in Agriculture and Water Supply Systems: Drought Management and Policy Development in the Mediterranean*, ed. by A. Iglesias, L. Garrote, A. Cancelliere, F. Cubillo and D. Wilhite, vol. 26 (2009), 187-220.

ROWLAND and NOBLE HOWE 1999
Rowland, I. and Noble Howe, T, *Vitruvius: Ten Books on Architecture* (Cambridge, 1999).

RYNNE 2000
Rynne, C., 'Waterpower in medieval Ireland', in *Working with Water in Medieval Europe: Technology and Resource-Use*, ed. by P. Squatriti (Leiden, 2000), 1-50.

SAELI and SAELI 2015
Saeli, M. and Saeli E., 'Analytical studies of the Sirocco room of Villa Naselli-Ambleri: A XVI century passive cooling structure in Palermo (Sicily)', *Journal of Cultural Heritage*, vol. 16 (2015), 344-51.

SALIBA 2008
Saliba, P., *An Evaluation of Rubble Walls and Terracing in the Maltese Landscape* (unpublished M.A. thesis, University of Malta, 2008).

SALIBA *et al* 2002
Saliba, P.C., Magro Conti, J. and Borg, C., *A Study of Landscape and Irrigation Systems at Is-Simblija limits of Dingli, Malta & Conservation Project* (Malta, 2002).

SAMMUT CARBONE 2002
Sammut Carbone, E., *Environmental Education Across the Curriculum* (Malta, 2002).

SAPIANO 2008
Sapiano, M., 'Measures for facing water scarcity and drought in Malta', *European Water (EWRA)*, vol. 23/24 (2008), 77-84.

SAPIANO *et al* 2006
Sapiano, M., Mangion, J. and Batchelor, C., *Malta Water Resources Review*, FAO (Rome, 2006).

SAPIANO *et al* 2008
Sapiano, M., Micallef, P., Attard, G. and Zammit, M.L., 'The evolution of water culture in Malta: an analysis of the changing perceptions towards water throughout the ages', in *Water Culture and Water Conflict in the Mediterranean Area* - CIHEAM (Centre International de Hautes Etudes Agronomiques Méditerranéennes), ed. by M. El Moujabber, M. Shatanawi, G. Trisorio Liuzzi, M. Ouessar, P. Laureano and R., Rodriguez (Bari, 2008), 97-109.

SCARBOROUGH 1991
Scarborough, V.L., 'Water management adaptations in non-industrial complex societies: an archaeological perspective', in *Archaeological Method and Theory 3*, ed. by M.B. Schiffer (Arizona, 1991), 101-54.

SCERRI 2005
Scerri, S., *Ix-Xagħra l-Ħamra Golf Course, Mellieħa, Malta. Technical Appendix 6: Geo-environmental Baseline Studies* (Malta, 2005).

SCHEMBRI 1993
Schembri, P.J., 'Physical geography and ecology of the Maltese Islands: a brief overview', in *Options Méditerranéennes: Malta: Food, Agriculture, Fisheries and the Environment,* ed. by S. Busuttil, F. Lerin and L. Mizzi, Centre International d' Hautes Etudes Agronomiques Méditerranéennes (CIHEAM), ser. B:Etudes et Recherches 7 (Paris, 1993), 27-39.

SCHEMBRI and BALDACCHINO 1998
Schembri, P.J. and Baldacchino, A.E., *Ilma, Blat u Ħajja: Is-sisien ta' l-ambjent naturali Malti* (Malta, 1998).

SCIASCIA 1989
Sciascia, L., 'I cammelli e le rose: gli Abbate di Trapani da Federico II a Martino il Vecchio', *Mediterraneo Medievale: Scritti in onore di Francesco Giunta*, vol. 3 (Soveria Mannelli, 1989), 1174-81.

SERRACINO-INGLOTT 2003
Serracino-Inglott, M., *Id-Dizzjunarju Malti u Teżawru ta' Malti mħaddem* (Malta, 2003).

SHATANAWI *et al* 2008A
Shatanawi, M., Shammout, M. and Naber, S., 'Water conflicts among sectors and environmental uses in Jordan', in *Water Culture and Water Conflict in the Mediterranean Area* - CIHEAM (Centre International de Hautes Etudes Agronomiques Méditerranéennes), ed. by M. El Moujabber, M. Shatanawi, G. Trisorio Liuzzi, M. Ouessar, P. Laureano and R., Rodriguez (Bari, 2008), 159-72.

SHATANAWI *et al* 2008B
Shatanawi, M., Shammout, M., Naber, S., Aboul-Fotouh, N., Abdin, A.E., Sedky, S., Candela, L., Domingo, F., Berbel, J., Alarcón, J.J., Laureano, P., Tamoh, K., Von Igel, W., Soler, M., Trevisiol, E.R., Cambruzzi, T., Baldo, G., Pra Levis, F. and Rodriguez, R., in *Water Culture and Water Conflict in the Mediterranean Area* - CIHEAM (Centre International de Hautes Etudes Agronomiques Méditerranéennes), ed. by M. El Moujabber, M. Shatanawi, G. Trisorio Liuzzi, M. Ouessar, P. Laureano and R., Rodriguez (Bari, 2008), 205-12.

SHAW 1991
Shaw, B.D., 'The noblest monuments and the smallest things: wells, walls and aqueducts in the making of Roman Africa', in *Future Currents in Aqueduct Studies*, ed. by A.T. Hodge (Leeds, 1991).

SIDWELL 1995
Sidwell, K., *Reading Medieval Latin* (Cambridge, 1995).

SKINNER *et al* 1997
Skinner, M., Redfern, D. and Farmer, G., *The complete A-Z Geography Handbook* (London, 1997).

SPELLMAN 2000
Spellman, G., 'The dynamic climatology of drylands', in *The Archaeology of Drylands – Living at the margin*, ed. by G. Barker and D. Gilbertson (London, 2000), 19-44.

SPRATT 1854
Spratt, T.A.B., *On the Geology of Malta and Gozo* (Malta, 1854).

SULTANA and FALZON 2002
Sultana, J. and Falzon V., *Wildlife of the Maltese Islands* (Malta, 2002).

SULTANA and GAUCI 1999
Sultana, J. and Gauci, M., 'The changing landscape of Gozo', in *National Awareness Seminar on Desertification and Land Degradation: Proceedings of Seminar Foundation for International Studies, Malta*, ed. by A.E. Baldacchino and C.M. Tanti (Malta, 1999), 109-112.

TAMOH *et al* 2008
Tamoh, K., Von Igel, W., Soler, M. and Candela, L., 'Analysis of water use patterns in the Sa Pobla Plain and Alcudia Bay', in *Water Culture and Water Conflict in the Mediterranean Area* - CIHEAM (Centre International de Hautes Etudes Agronomiques Méditerranéennes), ed. by M. El Moujabber, M. Shatanawi, G. Trisorio Liuzzi, M. Ouessar, P. Laureano and R., Rodriguez (Bari, 2008), 131-42.

TEMPELHOFF *et al* 2009
Tempelhoff, J., Hoag, H., Ertsen, M., Arnold, E., Bender, M., Berry, K., Fort, C., Pietz, D., Musemwa, M., Nakawo, M., Ur, J., van Dam, P., Melosi, M., Winiwarter, V. and Wilkinson, T., 'Where has water come from?', *water history*, vol. 1, no. 1 (2009), 1-8.

TEUMA 2005
Teuma, E.P., 'Qanat, Saqqajja and Roman aqueduct system at Rabat, Malta', *Proceedings of History Week 2003* (2005), 75-100.

TODARO 1989
Todaro, P., 'Utilizzazioni del sottosuolo di Palermo in età medievale', in *Palermo Medievale*, Testi dell'VIII colloquio Medievale, Palermo 26-27 aprile 1989, ed. by C. Roccaro (Palermo, 1989), 109-128.

TODARO 2002
Todaro, P., *Guida di Palermo Sotterranea (*Palermo, 2002).

TROVATO 1999
Trovato, A., 'The balcony of Sicily', in *Aidone and Morgantina*, ed. by V. Alabiso (Palermo, 19999), 26-33.

TSAKIRIS 2009
Tsakiris, G., 'A paradigm for applying risk and hazard concepts in proactive planning', in *Coping with Drought Risk in Agriculture and Water Supply Systems: Drought Management and Policy Development in the Mediterranean*, ed. by A. Iglesias, L. Garrote, A. Cancelliere, F. Cubillo and D. Wilhite, vol. 26 (2009), 81-92.

TSAKIRIS and PANGALOU 2009
Tsakiris, A. and Pangalou, D., 'Drought characterisation in the Mediterranean', in *Coping with Drought Risk in Agriculture and Water Supply Systems: Drought Management and Policy Development in the Mediterranean*, ed. by A. Iglesias, L. Garrote, A. Cancelliere, F. Cubillo and D. Wilhite, vol. 26 (2009), 69-80.

TURNER and CROW 2010
Turner, S. and Crow, J., 'Unlocking historic landscapes in the Eastern Mediterranean: two pilot studies using Historic Landscape Characterisation', *Antiquity*, vol. 84 (2010), 216-29.

ULBRICH *et al* 2012
Ulbrich, U., Lionello, P., Belušić, D., Jacobeir, J., Knippertz, P., Kuglitsch, F.G., Leckebusch, G.C., Luterbacher, J., Maugeri, M., Maheras, P., Nissen, K.M., Pavan, V., Pinto, J.G., Seubert, S., Toreti, A., Xoplaki, E. and Ziv, B., 'Climate of the Mediterranean: synoptic patterns, temperature, precipitation, winds, and their extremes', in *The Climate of the Mediterranean Region: From the Past to the Future*, ed. by, P. Lionello (London, 2012), 301-34.

VELLA, E. 2010
Vella, E., *A Landscape Archaeology Approach to the Study of the Maltese Girna: A Case Study based on the areas of Ix-Xagħra L-Ħamra, Id-Dar Il-Bajda and Ta' L-Għansar in Northern Malta* (unpublished M.A. thesis, University of Malta, 2008).

VELLA, H.C.R. 1980
Vella, H.C.R., *The Earliest Description of Malta (Lyons 1536) by Jean Quentin d'Autun: Translation and Notes* (Malta, 1980).

VELLA and SPITERI 2008
Vella, N.C. and Spiteri, M., 'Documentary sources for the study of the Maltese landscape', in *Storja 30th Anniversary Edition 1978/2008*, ed. by H. Frendo (Malta, 2008), 16-29.

VERMEULEN 2004
Vermeulen, F., 'Roads for soldiers and civilians in the Civitas Menapiorum', in *Archaeology in Confrontation: Aspects of Roman Military presence in the Northwest*, ed. by F. Vermeulen, K. Sas and W. Dhaeze (Ghent, 2004), 125-42.

VETTER *et al* 2009
Vetter, T., Reiger, A.K. and Nicolay, A., 'Ancient rainwater harvesting systems in the north-eastern Marmarica (north-western Egypt)', *Libyan Studies*, vol. 40 (2009), 9-23.

VITA-FINZI 1969
Vita-Finzi, C., *The Mediterranean Valleys – Geological Changes in Historical Times* (Cambridge, 1969).

VITA-FINZI 1978
Vita-Finzi, C., *Archaeological Sites in their Setting* (London, 1978).

VON FALKENHAUSEN 2002
Von Falkenhausen, V., 'The Greek presence in Norman Sicily: the contribution of the archival material in Greek', in *The Society of Norman Italy*, ed. by G. A. Loud and A. Metcalfe (Boston, 2002), 235-87.

WARD-PERKINS 1962
Ward-Perkins, J.B., 'Etruscan engineering: road-building, water supply and drainage', in *Hommages à Albert Grenier*, ed. by M. Renard (Brussels, 1962), 1636-43.

WATSON 1974
Watson, A.M., 'The Arab agricultural revolution and its diffusion, 700-1100', *The Journal of Economic History*, vol. 34, no. 1 (1974), 8-35.

WATSON 1983
Watson, A.M., *Agricultural Innovation in the Early Islamic World: The diffusion of crops and farming techniques, 700-1100* (Cambridge, 1983).

WATSON 1995
Watson, A.M., 'Arab and European Agriculture in the Middle Ages: a case of restricted diffusion', in *Agriculture in the Middle Ages: Technology, Practice and Representation*, ed. by D. Sweeney (Philadelphia, 1995), 62-75.

WAUGH 1995
Waugh, D., *Geography: An Integrated Approach* (London, 1995).

WAUGH 2003
Waugh, D., *The New Wider World* (Cheltenham, 2003).

WERTIME 1968
Wertime, T.A., 'A metallurgical exploration through the Persian desert', *Science*, vol. 159 (1968), 927-35.

WETTINGER 1969
Wettinger, G., 'The militia list of 1419-20: a new starting point for the study of Malta's population', *Melita Historica*, vol. 2 (1969), 80-106.

WETTINGER 1975
Wettinger, G., 'The lost villages and hamlets of Malta', in *Medieval Malta – Studies on Malta before the Knights*, ed. by A.T. Luttrell (London, 1975), 181-216.

WETTINGER 1982
Wettinger, G., 'Agriculture in Malta in the Late Middle Ages', *Proceedings of History Week 1981*, ed. by M. Buhagiar, The Historical Society (Malta, 1982), 1-48.

WETTINGER 1986
Wettinger, G., 'The Arabs in Malta', in *Malta: Studies of its Heritage and History*, Mid-Med Bank Ltd. (Malta, 1986), 87-104.

WETTINGER 2000
Wettinger, G., *Place-Names of the Maltese Islands ca. 1300-1800* (Malta, 2000).

WETTINGER 2006
Wettinger, G., *Kliem Malti Qadim* (Malta, 2006).

WETTINGER 2012
Wettinger, G., 'Malta in the High Middle Ages', *Melita Historica*, vol. 15, no. 4 (2011), 367-90.

WHITE *et al* 2010
White, C., Hiranandani, D., Olstad, C., Buhagiar, K., Gambin, T. and Clark, C., 'The Malta cistern mapping project: underwater robot mapping and localization within ancient tunnel systems', *Journal of Field Robotics* (2010).

WILHITE 1993
Wilhite, D.A., 'The enigma of drought', in *Drought Assessment, Management, and Planning: Theory and Case Studies*, ed. by D.A. Wilhite (New York, 1993), 3-16.

WILHITE 2009
Wilhite, D.A., 'Drought monitoring as a component of drought preparedness', in *Coping with Drought Risk in Agriculture and Water Supply Systems: Drought Management and Policy Development in the Mediterranean*, ed. by A. Iglesias, L. Garrote, A. Cancelliere, F. Cubillo and D. Wilhite, vol. 26 (2009), 3-20.

WILKINSON 2010
Wilkinson, T.J., 'Empire and environment in the northern fertile crescent', in *Landscapes and Societies: Selected Case Studies*, ed. by I.P. Martini and W. Chesworth (Heidelberg, 2010), 135-52.

WILSON 2004
Wilson, A.I., 'Classical water technology in the early Islamic world', in *Technology, ideology, water: from Frontinus to the Renaissance and beyond*, ed. by C. Bruun and A. Saastamoinen (Rome, 2004), 115-41.

WILSON 2006
Wilson, A., 'The spread of Foggara-based irrigation in the ancient Sahara', in *The Libyan Desert: Natural Resources and Cultural Heritage*, ed. by D. Mattingly, S. McLaren, E. Savage, Y. al-Fasatwi and K. Gadgood (London, 2006), 205-16.

WILSON and MATTINGLY 2003
Wilson, A. and Mattingly, D., 'Irrigation technologies: foggaras, wells and field systems', in *The Archaeology of the Fazzān: Volume 1, Synthesis*, ed. by D. Mattingly (Hertford, 2003), 235-78.

WITTFOGEL 1955
Wittfogel, K.A., 'Developmental aspects of hydraulic societies', in *Irrigation Civilizations: A Comparative Study*, ed. by S. J. Haynes (Washington, 1955), 43-57.

WITTFOGEL 1967A
Wittfogel, K.A., *Oriental Despotism: A Comparative Study of Total Power* (Massachusetts, 1967).

WITTFOGEL 1967B
Wittfogel, K.A., Rev. of *The Evolution of Urban Society: Early Mesopotamia and Prehistoric Mexico* by R. McC. Adams, *American Anthropologist*, vol. 69, no. 1 (1967), 90-92.

WUTTMANN 2001
Wuttmann, M., 'Les qanats de 'Ayn-Manâwîr (oasis de Kharga, Égypte)', in *Irrigation et drainage dans l'Antiquité, qanāts et canalisations souterraines en Iran, en Égypte et en Grèce, séminaire tenu au Collège de France sous la direction de Pierre Briant*, ed. by P. Briant (Paris, 2001), 109-36.

ZAIMECHE 2004
Zaimeche, S., 'Sicily', in *Foundation for Science Technology and Civilisation*, ed. by L. Ball (November, 2004), 1-20.

ZAMMIT, A. 2009
Zammit, A., *Our Architects: A Private Archive Unveiled* (Malta, 2009).

ZAMMIT, T. 1922
Zammit, T., 'Excavations in Malta', *Antiquaries Journal*, vol. 2, no. 2 (April, 1922), 131-14.

ZAMMIT, T. 1924
Zammit, T., *The Water Supply Resources of the Maltese Islands* (Malta, 1924).

ZAMMIT LUPI 2002
Zammit Lupi, T., 'The Cabreo of Fra Mario Bichi', *Treasures of Malta*, vol. 8, no. 2 (23) (2002), 59-63.

ZAMMIT-MAEMPEL 1977
Zammit-Maempel, G., *An Outline of Maltese Geology* (Malta, 1977).

C. Unpublished Sources

WALTHALL 2008
Walthall, A., 'Qanats at Morgantina', unpublished course paper in-part fulfilment of a Ph.D. thesis, supervised by Dr Sandra Lucore, American School of Classical Studies at Athens.

D. Cartographic Sources

PEDLEY 1992
Pedley, M.H., *Geological Map of the Maltese Islands*, Sheets 1 and 2, revised at 1:25,000 scale by Debono, G., and Xerri, S., Oil Exploration Directorate, Office of the Prime Minister (Malta, 1993).

Malta Survey Sheet 26, School of Military Engineering, Chatham, UK, 1903.
Malta Survey Sheet 28, School of Military Engineering, Chatham, UK, 1903.
Malta Survey Sheet 35, School of Military Engineering, Chatham, UK, 1903.
Malta Survey Sheet 36, School of Military Engineering, Chatham, UK, 1899.
Malta Survey Sheet 38, School of Military Engineering, Chatham, UK, 1898.
Malta Survey Sheet 45, School of Military Engineering, Chatham, UK, 1903.
Malta Survey Sheet 56, School of Military Engineering, Chatham, UK, 1898.
Malta Survey Sheet 71, School of Military Engineering, Chatham, UK, 1922.
Malta Survey Sheet 72, School of Military Engineering, Chatham, UK, 1922.
Malta Survey Sheet 73, School of Military Engineering, Chatham, UK, 1921.
Malta Survey Sheet 84, School of Military Engineering, Chatham, UK, 1903.
Malta Survey Sheet 85, School of Military Engineering, Chatham, UK, 1922.
Malta Survey Sheet 87, School of Military Engineering, Chatham, UK, 1920.
Malta Survey Sheet 99, School of Military Engineering, Chatham, UK, 1903.
Malta Survey Sheet 101, School of Military Engineering, Chatham, UK, 1919.
Malta Survey Sheet 125, School of Military Engineering, Chatham, UK, 1907.

E. Internet Sources

BUHAGIAR, K. 2008
Buhagiar, K., 'St. Leonard Cave Church, Lunzjata l/o Rabat', in *Malta and Sicily: Miscellaneous research projects*, ed. by A. Bonanno (Palermo, 2008).
Available from: http://kasa.officinastudimedievali.it/cd/risorse/Libro/Malta_and_Sicily.pdf
Accessed: 30 June 2016

CIANTAR 2012
Ciantar, N., 'Rendering justice to Ġużeppi Callus', *Sunday Times of Malta*, June 3, 2012.
Available from: http://www.timesofmalta.com/articles/view/20120603/life-features/Rendering-justice-to-u-eppi-Callus.422598
Accessed: 30 June 2016

Cistern Exploration Project (Malta/Gozo).
Available from: http://users.csc.calpoly.edu/~zwood/MaltaMapping/index.html
Accessed: 30 June 2016

Cistern Exploration Project (Malta/Gozo), *Site 46: Għar Ilma, Gozo.*
Available from: http://users.csc.calpoly.edu/~zwood/MaltaMapping/site46.html
Accessed: 30 June 2016

European Landscape Convention, Florence, 2000.
Available from: http://conventions.coe.int/Treaty/en/Treaties/Html/176.htm
Accessed: 30 June 2016

Koperattiva Rurali Manikata, *The Historic Farmstead (Ir-Razzett tal-Qasam).*
Available from: http://www.manikatafarmers.com/historic_farmstead.html
Accessed: 30 June 2016

Oxford Dictionaries, *Refugium.*
Available from: http://oxforddictionaries.com/definition/english/refugium
Accessed: 30 June 2016

Legal Notice 241 of 2010, 27 April 2010.
Available from: http://mra.org.mt/wp-content/uploads/2012/08/40.Groundwater-Abstraction-Metering-Regulations.pdf
Accessed: 30 June 2016

Malta Environment and Planning Authority, *Il-Ballut tal-Wardija (l/o San Pawl il-Baħar).*
Available from: http://www.mepa.org.mt/wardija
Accessed: 30 June 2016

MEPA Map Server
Available from: http://www.mepa.org.mt/mepa-mapserver
Accessed: 30 June 2016

Online Unit Conversion, *Convert from meter to cana (canna, canne).*
Available from: http://www.onlineunitconversion.com/meter_to_cana.canna.canne.html
Accessed: 30 June 2016

THAMES 1993
Thames, J. L., *Role of Forestry in Combating Desertification*, FAO Conservation Guide, 1993.
Available from: http://www.fao.org/docrep/t0115e/T0115E0l.htm
Accessed: 30 July 2016

Appendix 1: Gazetteer of Sites Mentioned in the Text

This gazetteer database contains basic information on the sites mentioned in the text. These are listed and classified according to the district in which they are located and in order to facilitate usability, sites were assigned an identification number to allow cross-referencing with the text. Taking the site of *Ghadira di Torbet Gharfagna* in the Attard district as an example, its identification number is "AT1" and is referred to in the text as *Ghadira di Torbet Gharfagna* (AT1).

1. Attard district

Name of site: *Ghadira di Torbet Gharfagna* **Id. no:** AT1
Map reference: approx. 48864 73022
Description of site: Place-name hints at the collection of water within a specific area of Middle or Lower Globigerina Limestone deposit.
Tentative date: Early Modern.
Published source: Abela, 1647, 86.

Name of site: *Ghadira tal Bordi* **Id. no:** AT2
Map reference: unknown
Description of site: Place-name hints at the collection of water within a specific area of Middle or Lower Globigerina Limestone deposit.
Tentative date: Early Modern.
Published sources: Abela, 1647, 86; Wettinger, 2000, 178.

Name of site: Ta' Qali **Id. no:** AT3
Map reference: approx. 48529 72315
Description of site: Boreholes and galleries tapping the mean-sea-level aquifer.
Tentative date: "ancient system of underground canals" of an unknown dating; Modern.
Published source: Morris, 1952, 34, 37.

2. Birgu district

Name of site: Birgu **Id. no:** BI1
Map reference: unknown
Description of site: Cisterns and brackish wells contained within the urban fabric of Birgu.
Tentative date: Late Medieval; Early Modern.
Published source: Bosio, 1594-1602, 28-31.

Name of site: *Giardino* **Id. no:** BI2
Map reference: unknown
Description of site: Birgu locality; agricultural land enclosed by means of a high wall.
Tentative date: Early Modern.
Iconographic sources: Agius-Vadalà and Ganado, 1986, 466-69; Agius-Vadalà and Ganado, 1994, 11.

3. Birkirkara district

Name of site: *Hal Caprat* **Id. no:** BIR1
Map reference: approx. 52474 71998
Description of site: The documented presence of cisterns associated with the abandoned settlement of *Hal Caprat*.
Tentative date: Late Medieval; Early Modern.
Published sources: Abela, 1647, 91; Wettinger, 2000, 280-81.

4. Dingli district

Name of site: Dejr il-Bniet **Id. no:** DI1
Map reference: 44571 69238
Description of site: *Giardino/viridarium*; Upper Coralline Limestone perched aquifer gallery.
Tentative date: Late Medieval; Early Modern.
Published sources: Abela, 1647, 65; Fiorini, 1999B, 158-59; Wettinger, 2000, 108; Fiorini, 2004, 65.

Name of site: Qbiela ta' Għar Ilma **Id. no:** DI2
Map reference: approx. 44571 69238
Description of site: Horticultural land at the fief of Dejr il-Bniet; may be linked, at least indirectly to a perched aquifer gallery which feeds the Dejr il-Bniet *giardino* with a perennial water source.
Tentative date: Early Modern.
Published source: Wettinger, 2000, 440.

5. Gudja district

Name of site: Bir Miftuħ **Id. no:** GU1
Map Reference: 54665 67804
Description of site: The name of a district near Gudja. The area lends its name to the Late Medieval *cappella* dedicated to Santa Marija present there; associated water collection cistern.
Tentative date: Late Medieval; Early Modern
Published sources: Abela, 1647, 105; Wettinger, 2000, 47.

6. Għargħur district

Name of site: Wied id-Dis **Id. no:** GH1
Map reference: 51355 76030
Description of site: *Giardino*-type cultivation which in the past was documented to contain olive tree, vineyards and almond tree plantations.
Tentative date: Late Medieval; Early Modern.
Published sources: Abela, 1647, 85; Wettinger, 2000, 571-72.

7. Iklin district

Name of site: Ix-Xwieki **Id. no:** IK1
Map reference: 50698 74946
Description of site: *Giardini*-type cultivations containing olive tree, vineyards and almond tree plantations.
Tentative date: Early Modern.
Published sources: Abela, 1647, 85; Wettinger, 2000, 625.

8. Lija district

Name of site: Ħal Mann **Id. no:** LI1
Map reference: 49426 73506
Description of site: *Giardino*-type cultivation which in the past was documented to contain containing almond, carob and fig tree plantations.
Tentative date: Early Modern.
Published sources: Abela, 1647, 85; Wettinger, 2000, 283.

9. Luqa district

Name of site: Bir il-Wiesa **Id. no:** LU1
Map reference: unknown
Description of site: A well or cistern placed between the localities of Luqa and Marsa.
Tentative date: Late Medieval; Early Modern.
Published source: Wettinger, 2000, 52.

10. Marsa and Qormi districts

Name of site: *Bir Buhagiar* **Id. no:** MA1
Map reference: unknown
Description of site: Shaft well probably tapping the mean-sea-level aquifer.
Tentative date: Early Modern.
Published source: Abela, 1647, 93.
Archival source: NLM. Treas. B. 289, f. 20.

Name of site: *Bir Eebeyer* **Id. no:** MA2
Map reference: unknown
Description of site: Shaft well probably tapping the mean-sea-level aquifer.
Tentative date: Early Modern.
Published source: Abela, 1647, 93.

Name of site: *Bir el Meru* **Id. no:** MA3
Map reference: unknown
Description of site: Shaft well probably tapping the mean-sea-level aquifer.
Tentative date: Early Modern.
Published source: Abela, 1647, 93.

Name of site: *Bir il vuassa* **Id. no:** MA4
Map reference: unknown
Description of site: A well tapping a mean-sea-level aquiferous source, possibly located within the *Giardino della Marsa.*
Tentative date: Early Modern.
Archival sources: NLM. Treas. B. 289, f. 18; NLM. Treas. B. 289, f. 26.

Name of site: *Bir yahlef* or *Bir hialef* **Id. no:** MA5
Map reference: unknown
Description of site: Shaft well probably tapping the mean-sea-level aquifer.
Tentative date: Early Modern.
Published sources: Abela, 1647, 92; Wettinger, 2000, 15.
Archival source: NLM. Treas. B. 289, f. 30.

Name of site: *Clausura tal gadira* **Id. no:** MA6
Map reference: unknown
Description of site: Possibly derived from the Maltese word *għadira*; might indicate the presence of a shallow water pool.
Tentative date: Early Modern.
Archival source: NLM. Treas. B. 289, f. 22.

Name of site: *Clausura tal Hammieri* **Id. no:** MA7
Map reference: unknown
Description of site: Two wells or cisterns located in a road-side clearing adjoining the *Clausura tal Hammieri.*
Tentative date: Early Modern.
Archival source: NLM. Treas. B. 290, ff. 48, 50, 52-54.

Name of site: Għajn Filep **Id. no:** MA8
Map reference: unknown
Description of site: Mean-sea-level aquifer spring located close to the coastal zone of Marsa.

Tentative date: Late Medieval; Early Modern; Modern.
Published sources: Abela, 1647, 93; Balbi da Correggio, 1965, 27, 44, 52; Wettinger, 2000, 182.
Iconographic sources: Agius-Vadalà and Ganado, 1994, 7; Agius-Vadalà and Ganado, 1986, 16, 14, 45.

Name of site: Marsa *giardino* **Id. no:** MA9
Map reference: unknown
Description of site: Agricultural land enclosed by means of a high wall.
Tentative date: Early Modern.
Iconographic sources: Agius-Vadalà and Ganado, 1986, 2, 8, 14, 16; Agius-Vadalà and Ganado, 1994, Map 118.

Name of site: Qormi (locality) **Id. no:** MA10
Map reference: 52720 70440
Description of site: Seasonal spring flowing through valley bed; early seventeenth century shaft wells extracting water from the mean-sea-level aquifer and related water lifting machinery; shaft wells and mean-sea-level aquifer galleries dating to the nineteenth and the twentieth century.
Tentative date: Early Modern; Modern.
Published source: Abela, 1647, 92, 128.
Archival sources: NAM. GOV 01.3/11, Despatch To, no. 81, 1867, ff. 394-99; NAM. GOV 01.3/11, Despatch To, no. 83, 1867, ff. 400-1; NAM. GOV 01.3/11, Despatch To, no. 87, 1867, ff. 406-7; NAM. GOV 01.3/11, Despatch To, no. 101, 1867, ff. 423-30; NAM. GOV 01.3/11, Despatch To, no. 108, 1867, ff. 440-41.

Name of site: *Stagno detto ta' Sirina* **Id. no:** MA11
Map reference: unknown
Description of site: A pond or pool of water in the Qormi district; probably formed by rainwater surface runoff together with low rock porosity – a characteristic of some Globigerina Limestone formations.
Tentative date: Early Modern.
Published source: Abela, 1647, 87.

Name of site: Wied il-Kbir **Id. no:** MA12
Map reference: 52240 69190
Description of site: Seasonal water course; mean-sea-level aquifer galleries in the Wied is-Sewda area, several of which are connected to the surface by means of shaft wells.
Tentative date: Modern.
Published sources: Chadwick, 1894A, 6; Zammit, T., 1924, 43; Micallef *et al.*, 2004, 201-2.

Name of site: Wied is-Sewda **Id. no:** MA13
Map reference: 50870 71492
Description of site: Seasonal water course; shaft wells and galleries tapping the mean-sea-level aquifer; connected to the Wied il-Kbir mean-sea-level aquifer galleries.
Tentative date: Modern.
Published sources: Chadwick, 1897, 12-13; Zammit, T., 1924, 32; Mercieca, 1931, 22.

11. Marsaxlokk and Birżebbuġa districts

Name of site: *S. Giorgio di Marsa Scirocco* **Id. no:** MAR1
Map reference: Tentatively 59005 66616 or 57767 65676
Description of site: A brackish water source which can be tentatively identified with either: (a) the public fountain located on the coastal edge fronting Triq ir-Żejtun at Marsaxlokk; or (b) the whereabouts of St George's Bay in Birżebbuġa.
Tentative date: Early Modern.
Archival source: NLM. Treas. B. f. 13.

12. Mdina

Name of site: Bishop's Palace **Id. no:** MD1
Map reference: 46289 71682
Description of site: Subterranean shaft tapping the Upper Coralline Limestone perched aquifer which in 1725 was reported as being completely dry.
Tentative date: Early Modern.
Archival source: Univ. 187, f. 95, 100.

Name of site: De Redin Bastions **Id. no:** MD2
Map reference: 46307 71681
Description of site: Upper Coralline Limestone perched aquifer gallery tunnelling below the De Redin Bastion area at Mdina having a length of ca. 54 m; gallery ceiling pierced by shafts; possibly supplied the Il-Ħemsija estate with a perennial water source.
Tentative date: Late Medieval; Early Modern.
Archival source: Univ. 187, f. 95, 100.

Name of site: Magisterial Palace **Id. no:** MD3
Map reference: 46256 71522
Description of site: Water-yielding canal.
Tentative date: Early Modern.
Archival source: Univ. 187, f. 95, 100.

Name of site: Mdina settlement **Id. no:** MD4
Map reference: 46192 71687
Description of site: Subterranean shafts, a number of which yielded spring water possibly originating from Upper Coralline Limestone perched aquifer galleries; rainwater harvested in cisterns; wells; early sixteenth century mention of two unnamed springs located outside the settlement walls.
Tentative date: Late Medieval; Early Modern.
Published source: Bosio, 1594-1602, 30.
Archival sources: Univ. 187, f. 95, 100.

13. Mellieħa district

Name of site: Ġnien Ingraw **Id. no:** ME1
Map Reference: 42127 79938
Description of site: Water-carved valley; Upper Coralline Limestone perched aquifer gallery and associated water reservoirs; gallery interior possibly modified in 1849 and 1854 respectively; *giardino*; masonry and troglodytic dwellings.
Tentative date: Late Medieval; Early Modern.
Published sources: Abela, 1647, 73; Wettinger, 2000, 157.

Name of site: Għajn Tuta **Id. no:** ME2
Map reference: 40054 81621
Description of site: Upper Coralline Limestone perched aquifer gallery in the Mellieħa district containing at least one shaft and fronting water reservoir; outer section of the water gallery is masonry-built.
Tentative date: Late Medieval; Early Modern.
Published sources: Abela, 1647, 75; Wettinger, 2000, 192.
Iconographic sources: Agius-Vadalà and Ganado, 1994, 7, Map 118; Agius-Vadalà and Ganado, 1986, 45, 466-69.

Name of site: Għajn Żejtuna **Id. no:** ME3
Map reference: Approx. 43651 80755
Description of site: The source of the Għajn Żejtuna spring is not known but Upper Coralline Limestone perched aquifer galleries are present at the above indicated map reference.
Tentative date: Late Medieval; Early Modern.
Published sources: Abela, 1647, 73; Wettinger, 2000, 193.

Name of site: Għajn Żnuber **Id. no:** ME4
Map reference: 41580 78649
Description of site: Spring originating from within an Upper Coralline Limestone perched aquifer gallery the entrance section of which has collapsed; associated water reservoir.
Tentative date: Early Modern.
Published sources: Abela, 1647, 72; Morris, 1952, 101; Wettinger, 2000, 193.

Name of site: Għajn Xorok **Id. no:** ME5
Map reference: Approx. 43659 80758
Description of site: An Upper Coralline Limestone perched aquifer gallery and fronting water reservoir; masonry-built outer section of gallery; Wied ta' Għajn Żejtuna area.
Tentative date: Early Modern.
Published source: Wettinger, 2000, 193.

Name of site: Għajn Ħommed **Id. no:** ME6
Map reference: unknown.
Description of site: A spring in the Miżieb ir-Riħ locality which in the early seventeenth century was documented as emerging from within a cavity in the rock.
Tentative date: Early Modern.
Published source: Wettinger, 2000, 184.

Name of site: Wied San Niklaw **Id. no:** ME7
Map reference: 41686 79706
Description of site: Water-carved valley; Upper Coralline Limestone perched aquifer gallery and fronting water reservoir; outer section of the water gallery is masonry-built; *giardino*; masonry and troglodytic dwellings.
Tentative date: Late Medieval; Early Modern.
Published source: Abela, 1647, 73.

14. Mosta and Bidnija districts

Name of site: Il-Qattara **Id. no:** MO1
Map reference: 46751 73466
Description of site: Localised perched aquifer located within a specific area of either Upper and/or Middle Globigerina Limestone deposits.
Tentative date: Early Modern.
Published sources: Abela, 1674, 70; Wettinger, 2000, 439; Fiorini, 2004, 454.

Name of site: Għajn Bidnija **Id. no:** MO2
Map reference: 46475 76155
Description of site: Upper Coralline Limestone perched aquifer gallery tunnelling into the south-facing side of Ġebel Għawżara at Bidnija.
Tentative date: Early Modern; Modern.
Published source: Docter *et al.*, 2012, 119, 121.
Cartographic source: Malta Survey Sheet 38, School of Military Engineering, Chatham, 1898.

Name of site: Għajn Mula **Id. no:** MO3
Map reference: 46473 75445
Description of site: Upper Coralline Limestone perched aquifer gallery.
Tentative date: Late Medieval; Early Modern.
Cartographic source: Malta Survey Sheet 38, School of Military Engineering, Chatham, 1898.

Name of site: Għajn Riħana **Id. no:** MO4
Map reference: Approx. 46680 75877
Description of site: Upper Coralline Limestone perched aquifer gallery which is no longer accessible; *giardini* which in the late fifteenth century were known as Ġnien il-Kbir, Il-Ġnejna and Il-Għalqa tal-Ġnien.
Tentative date: Late Medieval; Early Modern, Modern.
Published sources: Abela, 1647, 70; Fiorini, 1996, 77-78, 243-44, 248-49, 261, 333-34; Fiorini, 1999A, 161-62; Fiorini, 1999B, 256; Wettinger, 2000, 150, 156, 189.

Archival source: Cabreo Cattedrale 1838, PR 2536, f. 13.

Name of site: *La Fonte* **Id. no:** MO5
Map reference: unknown
Description of site: Possibly located to the east of the Mdina settlement.
Tentative date: Early Modern.
Iconographic sources: Agius-Vadalà and Ganado, 1986, 10.

Name of site: Wied il-Għasel **Id. no:** MO6
Map reference: Approx. 48712 75644
Description of site: Mean-sea-level aquifer gallery.
Tentative date: Modern.
Published sources: Morris, 1952, 62-63; Mangion *et al.*, 2004, 202.

15. Mġarr (Malta) and Wardija districts

Name of site: *Aayn* **Id. no:** MG1
Map reference: tentatively 43334 75267
Description of site: Toponym can perhaps be associated with the locality known as Tal-Għajn, in which there is a water-yielding spot marked as 'fountain' on the 1899-1900 1:2,500 Malta Survey Sheet 36 and which is known as L-Għajn ta' Pietru.
Tentative date: Early Modern.
Published source: Abela, 1647, 69.

Name of site: *Aayn Ballut* **Id. no:** MG2
Map reference: Approx. 44710 77530
Description of site: Water spring originating from within a perched aquifer gallery; associated open-air water reservoir.
Tentative date: Late Medieval; Early Modern.
Published sources: Abela, 1647, 72; Wettinger, 2000, 181.

Name of site: *Aayn gefrà* **Id. no:** MG3
Map reference: unknown
Description of site: Possibly located in the Ras il-Pellegrin area.
Tentative date: Early Modern.
Published sources: Abela, 1647, 68.

Name of site: *Aayn Kasab* **Id. no:** MG4
Map reference: unknown but tentatively associated with the Tal-Għajn locality at 43339 75272
Description of site: not available
Tentative date: Early Modern.
Published sources: Abela, 1647, 70; Wettinger, 2000, 188.
Cartographic source: Malta Survey Sheet 36, School of Military Engineering, Chatham, 1899.

Name of site: *Aayn tal Carcara* **Id. no:** MG5
Map reference: 42160 72950
Description of site: A spring possibly originating from one of the two Upper Coralline Limestone perched aquifer galleries located in the upper reaches of Tas-Santi valley.
Tentative date: Late Medieval; Early Modern.
Published source: Abela, 1647, 68.

Name of site: *Aayn tal Calcara* **Id. no:** MG6
Map reference: unknown
Description of site: not available
Tentative date: Early Modern.
Published source: Abela, 1647, 73.

Name of site: Burnoħħala **Id. no:** MG7
Map reference: 43032 73551
Description of site: Upper Coralline Limestone perched aquifer gallery fronted by water reservoir.
Tentative date: Early Modern; Modern.
Published sources: Zammit, T., 1924, 39; Morris, 1952, 17, 19.

Name of site: Iċ-Ċarċar **Id. no:** MG8
Map reference: 41810 75350
Description of site: Probably a reference to a seasonal spring which drained rain water into the Ġnejna valley system.
Tentative date: Early Modern.
Published source: Abela, 1647, 68.
Cartographic source: Malta Survey Sheet 35, School of Military Engineering, Chatham, 1903.

Name of site: Il-Ballut tal-Ħoxni **Id. no:** MG9
Map reference: 42490 76850
Description of site: Three water springs, two of which originate from within Upper Coralline Limestone perched aquifer galleries known as Tal-Ġuz and Tal-Gandott ta' Għajn Ħamiem; water collection reservoirs; *giardino* framework.
Tentative date: Early Modern; Modern.
Archival source: NAM. Descriptive Plans of the Crown Property, 1866, f. 216.

Name of site: Birhauex **Id. no:** MG10
Map reference: unknown
Description of site: Water cistern which early seventeenth century documentation shows to be located in the middle of the road between Żebbiegħ and the Għajn Tuffieħa *giardino.*
Tentative date: Early Modern.
Published source: Abela, 1647, 73.

Name of site: Il-Fawwara **Id. no:** MG11
Map reference: 41449 76641
Description of site: Water collection reservoir; watering trough almost completely destroyed by the Għajn Tuffieħa road widening works
Tentative date: Early Modern.
Published sources: Abela, 1647, 73; Wettinger, 2000, 123.

Name of site: Il-Mina l-Kbira **Id. no:** MG12
Map reference: unknown
Description of site: Spring possibly originating from within a perched aquifer gallery.
Tentative date: Early Modern.
Published source: Abela, 1647, 72.

Name of site: Il-Mina ta' Maġemba **Id. no:** MG13
Map reference: unknown
Description of site: Spring possibly originating from within a perched aquifer gallery.
Tentative date: Early Modern.
Published source: Abela, 1647, 72.

Name of site: Għajn Astas/Stas **Id. no:** MG14
Map reference: 44022 77365
Description of site: Spring originating from within an Upper Coralline Limestone perched aquifer gallery; water storage reservoirs; British period pillbox.
Tentative date: Late Medieval; Early Modern; Modern.
Published sources: Abela, 1647, 72; Fiorini, 1996, 260; Fiorini, 1999A, 23-24; Wettinger, 2000, 190.

Name of site: Għajn Tuffieħa **Id. no:** MG15
Map reference: 41620 76280
Description of site: Roman baths and adjoining miniature catacomb; *giardino*; three Upper Coralline Limestone perched aquifer galleries; water storage reservoirs one of which is in early seventeenth century documentation

identified as *gebia grande*; British period water tunnel tapping Upper Coralline Limestone perched aquifer; masonry and troglodytic dwellings.
Tentative date: Roman; Late Medieval; Early Modern; Modern.
Published sources: Abela, 1647, 72; MAR, 1930, x; Morris, 1952, 92-93; Fiorini, 1996, 195, 219, 269; Fiorini, 1999B, 10-11, 14, 260, 309; Wettinger, 2000, 154, 191; Dalli, 2006, 152; Vella and Spiteri, 2008, 24.
Archival sources: NLM. Treas. B. 289, ff. 90-95; NAM. Descriptive Plans of the Crown Property in Malta, 1866, ff. 209-13.
Iconographic source: Survey Sheets 26, 45 School of Military Engineering, Chatham, 1903.

Name of site: Għajn Tarġa **Id. no:** MG16
Map reference: 43140 76810
Description of site: Perennial spring originating from either an Upper Coralline Limestone perched aquifer gallery or a water-yielding crevice; *giardino*.
Tentative date: Late Medieval; Early Modern.
Published sources: Abela, 1647, 72; Wettinger, 2000, 191.

Name of site: Miżieb ir-Riħ **Id. no:** MG17
Map reference: unknown but located in the Pwales district.
Description of site: Three *fontane*, one of which seems to be fronted by a circular reservoir; remains of a probable archaeological interest; troglodytic dwellings.
Tentative date: Late Medieval; Early Modern.
Archival source: ACM Beni Cattedrale (INV. No. 891) Vol. 4, f. 6.

Name of site: Razzett tal-Qasam **Id. no:** MG18
Map reference: 41199 77597
Description of site: Palaeochristian period tombs; *giardino*; masonry and troglodytic dwellings; dried up Upper Coralline Limestone perched aquifer gallery.
Tentative date: Late Antiquity; Late Medieval; Early Modern; Modern.
Published sources: Fiorini, 1996, 195; Cardona, 2002, 161.
Archival sources: NLM. Treas. B. 289, ff. 93-94; Abela, 1647, 72; NAM. Descriptive Plans of the Crown Property, 1866, f. 199.

Name of site: Ta' Binġemma **Id. no:** MG19
Map reference: Approx. 43840 73951
Description of site: *Giardino* containing fruit trees, water supply.
Tentative date: Late Medieval.
Published sources: Fiorini, 1996, 251-52; Wettinger, 2000, 33-34.

Name of site: Ta' Għasfura **Id. no:** MG20
Map reference: unknown but tentatively 41070 73960
Description of site: Localised perched aquifer gathering above/within Middle Globigerina Limestone deposit.
Tentative date: Early Modern; Modern.
Published sources: Abela, 1647, 68; Wettinger, 2000, 190.

Name of site: Ta' Safsafa **Id. no:** MG21
Map reference: Tentatively 45120 78020
Description of site: Exact location unknown but the area contains various perennial water sources originating from within Upper Coralline Limestone perched aquifer galleries.
Tentative date: Early Modern.
Published source: Abela, 1647, 72.

Name of site: Ta' Santa Marija **Id. no:** MG22
Map reference: unknown
Description of site: *giardino* in the Pwales district; water source possibly originating from within an Upper Coralline Limestone perched aquifer gallery; reservoir; church; archaeological remains.
Tentative date: Late Medieval.
Published sources: Fiorini, 1999A, 166-67; Wettinger, 2000, 518.

Name of site: Tas-Santi **Id. no:** MG23
Map reference: 41923 73634

Description of site: Large parcel of land rich in geological and ecological diversity; seasonal spring flowing through valley bed; *giardino*; two Upper Coralline Limestone perched aquifer galleries one of which can be possibly identified with *Aayn tal Carcara*; open air water reservoir; two water troughs; water distribution canals; masonry-built and troglodytic dwellings.
Tentative date: Late Medieval; Early Modern.
Published source: Wettinger, 2000, 520.
Archival sources: NLM. Treas. B. 289, ff. 84-88; Abela, 1647, 68; NLM. Treas. B. 290, ff. 98-99; NLM. Treas. B. 291, ff. 19, 66; NAR. Descriptive Plans of the Crown Property, 1867, f. 599.

Name of site: Wied/Ġnien tal-Ġnejna **Id. no:** MG24
Map reference: 42120 75160
Description of site: Seasonal spring flowing through valley bed; *giardino*; three Upper Coralline Limestone perched aquifer galleries; a subterranean water reservoir and an open air water reservoir; calcite deposition inside one of the perched aquifer galleries.
Tentative date: Late Medieval; Early Modern.
Published sources: Abela, 1647, 69; Wettinger, 2000, 150.
Iconographic source: Agius-Vadalà and Ganado, 1986, 8.

16. Mqabba districts

Name of site: Ta' Kandja **Id. no:** MQ1
Map reference: 50728 67186
Description of site: Mean-sea-level aquifer galleries.
Tentative date: Modern.
Published source: Mangion *et al.*, 2004, 202.

17. Naxxar districts

Name of site: Magħtab **Id. no:** NA1
Map reference: Approx. 50260 76810
Description of site: *Giardini*-type cultivations containing olive tree plantations.
Tentative date: Early Modern.
Published sources: Abela, 1647, 85; Wettinger, 2000, 356.

18. Paola district

Name of site: *Giardino del Gran Maestro/Giardino de la fontaina* **Id. no:** PA1
Map reference: unknown
Description of site: Għajn Dwieli/Wied Blandun area; agricultural land enclosed by means of a high wall.
Tentative date: Early Modern.
Iconographic sources: Agius-Vadalà and Ganado, 1986, 466-69; Agius-Vadalà and Ganado, 1994, 11, 14; Agius-Vadalà and Ganado, 1994, Maps 118, 119.

Name of site: *Torre di Aayn duyeli* **Id. no:** PA2
Map reference: Approx. 56033 70756
Description of site: Water spring probably originating from within a Globigerina Limestone perched aquifer.
Tentative date: Early Modern.
Published sources: Abela, 1647, 94; Wettinger, 2000, 182.

19. Rabat, Mtaħleb and Baħrija districts (Malta)

Name of site: *Aayn Bierda* **Id. no:** RA1
Map reference: tentatively associated with either 39760 73100 or 39800 73190

Description of site: Upper Coralline Limestone perched aquifer gallery; closely associated open-air reservoir.
Tentative date: Late Medieval; Early Modern.
Published source: Abela, 1647, 67.

Name of site: *Aayn Cirani* **Id. no:** RA2
Map reference: ca. 44531 71631
Description of site: A probable reference to one of the springs present in the Għajn Klieb territory of Rabat (Malta).
Tentative date: Early Modern.
Published sources: Abela, 1647, 111; Wettinger, 2000, 182.

Name of site: *Aayn Colia* **Id. no:** RA3
Map reference: unknown
Description of site: Probably located in the Rabat (Malta) district; Upper Coralline Limestone perched aquifer spring.
Tentative date: Early Modern.
Published sources: Abela, 1647, 81; Wettinger, 2000, 185-86.

Name of site: *Aayn Pietro* **Id. no:** RA4
Map reference: unknown
Description of site: Probably located in the Rabat (Malta) territory; Upper Coralline Limestone perched aquifer spring.
Tentative date: Early Modern.
Published sources: Abela, 1647, 81; Wettinger, 2000, 504.

Name of site: *Alferez* **Id. no:** RA5
Map reference: unknown
Description of site: Probably located in the Rabat (Malta) territory; Upper Coralline Limestone perched aquifer spring.
Tentative date: Early Modern.
Published source: Abela, 1647, 81.

Name of site: Buskett **Id. no:** RA6
Map reference: 45745 68552
Description of site: Archival documentation makes mention of a *gebia noua* and *la fontana grande*, possibly originating from within an Upper Coralline Limestone perched aquifer gallery in the Tas-Sala area.
Tentative date: Early Modern; Modern.
Archival sources: NLM. Treas. B. 289, ff. 101-3, 108, 112-13.
Iconographic source: Survey Sheet 101, School of Military Engineering, Chatham, 1919.

Name of site: Busugrilla **Id. no:** RA7
Map reference: Approx. 44792 71861
Description of site: Described by Abela as *pozzo d'acqua viua* and probably tapping an Upper Coralline Limestone aquiferous deposit; fed by the minor sources of *nescieaa tal borgi tal mitarfe* and *Megira tà Mesrah Mnusa*. The Busugrilla place-name was since the Early Modern period either an alternative name for Għajn Tewżien, or else a minor tributary supplementing the Għajn Tewżien source.
Tentative date: Late Medieval; Early Modern.
Published sources: Abela, 1647, 80, 111; Morris, 1952, 57; Wettinger, 2000, 53.
Archival source: NLM. Treas. B. 289, f. 63

Name of site: *di Cursu* **Id. no:** RA8
Map reference: unknown
Description of site: Land at Wied ir-Rum which 1467 documentation describes as *viridarium* containing tree cultivations and a spring or fountain known as Għajn Qamar.
Tentative date: Late Medieval.
Published sources: Wettinger, 2000, 347; Fiorini, 2005, 99-100.

Name of site: *el Megira tà Mesrah Mnusa* **Id. no:** RA9
Map reference: unknown
Description of site: A spring which in the early seventeenth century was connected to the Valletta aqueduct system; a minor source feeding the Busugrilla spring with water.

Tentative date: Early Modern.
Published sources: Abela, 1647, 110-11; Wettinger, 2000, 382.

Name of site: *Fego di beberrua* **Id. no:** RA10
Map reference: Approx. 43176 72121
Description of site: Archival documentation mentions the presence of two small springs the location of which is unknown. Field research has indicated the presence of an Upper Coralline Limestone perched aquifer gallery at map reference 43078 72088.
Tentative date: Early Modern.
Published source: Wettinger, 2000, 31.
Archival source: NLM. Treas. B. 289, f. 79.

Name of site: Fiddien **Id. no:** RA11
Map reference: Approx. 44010 71790
Description of site: An agricultural plain touching various *giardini* frameworks most of which contain Upper Coralline Limestone perched aquifer gallery systems; forms part of the watershed area for the adjoining Wied tal-Qlejgħa valley system.
Tentative date: Late Medieval; Early Modern; Modern.
Published sources: Fiorini, 1999B, 115-16; Wettinger, 2000, 125.
Iconographic source: Agius-Vadalà and Ganado, 1986, 2.

Name of site: *Ġnien il-Haggel* **Id. no:** RA12
Map reference: Approx. 463351 68336
Description of site: Exact location unknown but possibly touched agricultural land known as Ta' Rapa; there is the documented mention of a *fontanella,* a *gebia negra* and a *gebia fonda*; *giardino* possibly used for fruit tree cultivation.
Tentative date: Early Modern.
Archival source: NLM. Treas. B. 289, ff. 106-12.

Name of site: Ġnien is-Sultan **Id. no:** RA13
Map reference: 44918 71775
Description of site: Upper Coralline Limestone perched aquifer galleries; seasonal spring flowing through valley bed; *giardini*; early seventeenth century water distribution chamber and associated subterranean canals; Late Medieval church dedicated to San Mikiel is-Sinċir.
Tentative date: Late Antiquity; Late Medieval; Early Modern.
Published sources: Fiorini, 1999B, 12-15, 17-18, 115-16; Wettinger, 2000, 162.
Iconographic sources: Agius-Vadalà and Ganado, 1986, 466-69; Agius-Vadalà and Ganado, 1994, Map 118.

Name of site: *Giardino di due fontane/delle due fontane* **Id. no:** RA14
Map reference: 44615 72099
Description of site: *Giardino* neighbouring the Għajn Qajjed estate; water distribution canals. During the early seventeenth century this received water four times weekly from the Għajn Qajjed *giardino.*
Tentative date: Late Medieval; Early Modern.
Archival source: NLM. Treas. B. 289, ff. 56-57.

Name of site: Gomerino/Għemmieri **Id. no:** RA15
Map reference: 43200 72670
Description of site: An over 424 m long Upper Coralline Limestone perched aquifer gallery containing ca. thirteen shafts which appear to be spaced around 30 m apart; possible close-by water collection reservoir; *giardino.*
Tentative date: Late Medieval; Early Modern; Modern.
Published sources: Abela, 1647, 66; Luttrell, 1978, 172-77; Fiorini, 1999B, 212-19; Wettinger, 2000, 222-23; Fiorini, 2007, 74, 400-1.
Archival sources: NAM. GOV 01.3/6, Despatch To, no. 22, 1844, f. 303; Eighteenth century Gomerino estate *cabreo* in possession of the Tabone family.

Name of site: Għajn Ħamiem **Id. no:** RA16
Map reference: 45905 71688
Description of site: A washer woman's fountain placed within the shelter of an arcaded loggia. The perennial water source for the fountain setup was probably supplied from a close-by Upper Coralline Limestone perched aquifer gallery which so far has not been identified.

Tentative date: Late Medieval; Early Modern.
Published sources: Abela, 1647, 80; Zammit, T., 1922, 132-34; Wettinger, 2000, 184.

Name of site: Għajn il-Kbira **Id. no:** RA17
Map reference: 41840 71100
Description of site: Upper Coralline Limestone perched aquifer gallery accessed through a rock-cut cave at the Mtaħleb *giardino*; fronting water reservoir.
Tentative date: Late Medieval; Early Modern; Modern.
Published sources: Abela, 1647, 65-66; Wettinger, 2000, 185.
Archival source: AIM Acta Civilia vol. 171, Case 12.

Name of site: Għajn Klieb **Id. no:** RA18
Map reference: 44536 71593
Description of site: Possibly formerly known as *Aayn Cirani*; *giardino* supplied with water from three separate springs; the principal spring at Għajn Klieb appears to originate from within a ca. 286 m long subterranean Upper Coralline Limestone perched aquifer gallery; large rectangular-shaped rock-cut feature possibly an old reservoir.
Tentative date: Late Medieval; Early Modern.
Published sources: Abela, 1647, 111; Zammit, T., 1924, 9, 37; Wettinger, 2000, 187.
Archival source: NAM. Descriptive Plans of the Crown Property, 1867, f. 471.

Name of site: Għajn Qajjed **Id. no:** RA19
Map reference: 44739 71685
Description of site: Upper Coralline Limestone perched aquifer gallery the water supply of which was in the early sixteenth century connected to the Valletta aqueduct system; water collection reservoirs one of which is in early seventeenth century documentation described as the *gran gebia*; water transportation canals; early twentieth century pumping station; rural farmsteads.
Tentative date: Late Medieval; Early Modern; Modern.
Published sources: Abela, 1647, 111; Mercieca, 1931, 12; Morris, 1952, 57; Fiorini, 1999B, 12-15; Wettinger, 2000, 188; Fiorini, 2004, 676.
Archival sources: NLM. Treas. B. 289, f. 54; NLM. Treas. B. 290, f. 85.
Cartographic source: Survey Sheet 72, School of Military Engineering, Chatham, 1922.

Name of site: Għajn Tejtes **Id. no:** RA20
Map reference: 42746 70222
Description of site: Upper Coralline Limestone perched aquifer gallery and fronting subterranean water storage reservoir; associated masonry-built reservoir located further downhill.
Tentative date: Early Modern; Modern.
Published sources: Abela, 1647, 65; Wettinger, 2000, 191.
Archival source: AIM Acta Civilia vol. 171, Case 12.

Name of site: Għajn Tewżien **Id. no:** RA21
Map reference: Approx. 44792 71861
Description of site: Upper Coralline Limestone perched aquifer gallery (1.4 km long) the water supply of which was in the early sixteenth century connected to the Valletta aqueduct system; located in the Ta' San Mikiel *giardino*; possibly alternatively referred to as Busugrilla. In the early seventeenth century Għajn Tewżien fed the Għajn Qajjed *giardino* with a water supply.
Tentative date: Late Medieval; Early Modern; Modern.
Published sources: Fiorini, 1999B, 12-15; Wettinger, 2000, 191.
Archival source: NLM. Treas. B. 289, f. 54, 64; Abela, 1647, 65, 111.

Name of site: Għar Ilma **Id. no:** RA22
Map reference: unknown
Description of site: Water possibly originating from a within a water gallery at Qallelija in the district of Rabat (Malta).
Tentative date: Late Medieval.
Published source: Wettinger, 2000, 206, 432.

Name of site: Għar Ilma **Id. no:** RA23
Map reference: Approx. 42301 70772

Description of site: Water originating from a within an Upper Coralline Limestone perched aquifer gallery at Mtaħleb in the district of Rabat (Malta); ceiling possibly pierced by shafts.
Tentative date: Early Modern.
Cartographic source: Malta Survey Sheet 84, School of Military Engineering, Chatham, 1903.

Name of site: Għar ta' l-għarwenin **Id. no:** RA24
Map reference: 41760 71160
Description of site: Upper Coralline Limestone perched aquifer gallery located within a cave probably formerly known as *Aayn San Giovan*; *giardini*.
Tentative date: Early Modern; Modern.
Published source: Abela, 1647, 66.
Archival source: AIM Acta Civilia vol. 171, Case 12.

Name of site: Għar ta' l-Isqof/Wied l-Isqof **Id. no:** RA25
Map reference: 46210 69451
Description of site: Upper Coralline Limestone perched aquifer gallery having a length of ca. 160 m located within a cave; cobbled path; water distribution canals; *giardini*.
Tentative date: Late Medieval; Early Modern; Modern.
Published source: Zammit, T., 1924, 22.
Archival source: CEM, 114/7, ff. 10-11.

Name of site: Għeriexem **Id. no:** RA26
Map reference: 45531 71534
Description of site: Upper Coralline Limestone perched aquifer gallery containing shaft-wells; *giardino*; aqueducts or water canals and watercourses.
Tentative date: Late Medieval; Early Modern.
Published sources: Morris, 1952, 57; Wettinger, 2000, 215; Fiorini, 2004, 222-23.
Archival source: Univ. 187, ff. 1-14.

Name of site: Ħofret ir-Rizz **Id. no:** RA27
Map reference: 43470 70200
Description of site: Solution Subsidence Structure; Knights period and British period tapping of the Upper Coralline Limestone perched aquifer; cave dwellings; canals for the collection of rain surface runoff.
Tentative date: Late Medieval; Early Modern; Modern.
Published sources: Abela, 1647, 110-11; Zammit, T., 1924, 36; Wettinger, 2000, 326.
Cartographic sources: Malta Survey Sheet Links no. 85; Malta Survey Sheet 85, School of Military Engineering, Chatham, 1922.

Name of site: *Hayn Clep tal Callus* **Id. no:** RA28
Map reference: 44542 71773
Description of site: Knights period reservoir.
Tentative date: Early Modern.
Archival sources: NLM. Treas. B. 289, f. 66; NLM. Treas. B. 289, Plan 9; NLM. Treas. B. 290, f. 85.

Name of site: Il-Bruka **Id. no:** RA29
Map reference: 41810 70330
Description of site: Exposed Middle Globigerina Limestone deposit within which collects a localised perched aquifer; rock-excavated open-air reservoir.
Tentative date: Early Modern; Modern.
Published sources: Abela, 1647, 66; Wettinger, 2000, 69.

Name of site: Il-Għars (formerly known as Ta' Falzun) **Id. no:** RA30
Map reference: 43273 69859
Description of site: Arable land adjoining the Ta' Baldu *giardino* below which tunnel two Upper Coralline Limestone perched aquifer galleries probably containing shaft wells and the *scirocco* chamber.
Tentative date: Late Medieval, Early Modern; Modern.
Published sources: Buhagiar, K., 2000, 181-86; Buhagiar, K., 2013, 167-71.
Archival Source: NAM. Descriptive Plans of the Crown Property, 1867, f. 497.
Cartographic source: Malta Survey Sheet 85, School of Military Engineering, Chatham, 1922.

Name of site: Il-Mosfar/Musfar **Id. no:** RA31
Map reference: 44746 71763
Description of site: *Giardino* neighbouring the Għajn Qajjed estate; reference to what are probably vertical shafts piercing the ceiling of a subterranean canal connecting the Għajn Qajjed spring to the Valletta aqueduct system.
Tentative date: Late Medieval; Early Modern.
Archival source: NLM. Treas. B. 289, f. 62.

Name of site: Il-Qattara **Id. no:** RA32
Map reference: 42620 69910
Description of site: Upper Coralline Limestone perched aquifer gallery at Wied ir-Rum; partly subterranean fronting reservoir; *giardino*; rock-cut *centimolo* and other caves.
Tentative date: Late Medieval; Early Modern; Modern.
Published sources: Abela, 1647, 65; Wettinger, 2000, 439; Buhagiar, K., 2013, 96-97.

Name of site: Li Sball **Id. no:** RA33
Map reference: 43208 70279
Description of site: Arable land located close to the Sqaq Ħandu in the district of Rabat (Malta).
Tentative date: Early Modern.
Published source: Abela, 1647, 66.

Name of site: *Marget Neuia/Chibula ta turba* **Id. no:** RA34
Map reference: 43180 71952
Description of site: Qanat-type Upper Coralline Limestone perched aquifer gallery, the ceiling of which is pierced by at least nine shafts; water distribution canals; reservoir. The gallery's outermost section is lined by dry-stone walls and is roofed over by means of a gabled roof-slab ceiling; known as the Tal-Fiskal fountain.
Tentative date: Early Modern.
Published sources: Abela, 1647, 112; Wettinger, 2000, 363.
Archival source: NLM. Treas. B. 289, f. 70.

Name of site: Mtaħleb *giardino* **Id. no:** RA35
Map reference: 41714 71077
Description of site: Upper Coralline Limestone perched aquifer galleries two of which were identified as *Fontana di Giardino Grande* and *Fontana di S. Giovanni*; seasonal spring flowing through valley bed; *giardini*; masonry and troglodytic dwellings; the unidentified spring of Għajn Tajba is probably located close-by.
Tentative date: Late Medieval; Early Modern; Modern.
Published sources: Abela, 1647, 65-66; Wettinger, 2000, 402-3.
Archival source: AIM Acta Civilia vol. 171, Case 12.
Iconographic source: Agius-Vadalà and Ganado, 1986, 466-69.

Name of site: *Nescieaa tal borgi* **Id. no:** RA36
Map reference: unknown
Description of site: A spring at Mtarfa which in the early seventeenth century was connected to the Valletta aqueduct system; a minor source feeding the Busugrilla spring with a water supply.
Tentative date: Early Modern.
Published sources: Abela, 1647, 110-11; Wettinger, 2000, 417.

Name of site: Qasam il-Kbir **Id. no:** RA37
Map reference: 43005 71402
Description of site: Various shafts giving access to an Upper Coralline Limestone perched aquifer gallery accessed at map reference 42944 71343 and known by the name of Għajn Żdejda. The gallery is thought to be ca. 300 m long.
Tentative date: Late Medieval; Early Modern.
Archival sources: NLM. Treas. B. 290, f. 91 ; NLM. Treas. B. 291, f. 18.
Cartographic source: See Survey Sheet 71, School of Military Engineering, Chatham, 1922.

Name of site: Ras il-Għajn **Id. no:** RA38
Map reference: unknown
Description of site: Għar Barka (Rabat, Malta) whereabouts; documented reference to a vertical shaft probably giving access to the innermost section of an Upper Coralline Limestone perched aquifer gallery.
Tentative date: Late Medieval; Early Modern; Modern.
Published source: Abela, 1647, 80.

Name of site: Razzett San Pawl/Tal-Callus **Id. no:** RA39
Map reference: 42408 69997
Description of site: *Giardino*; water distribution canals; Late Medieval chapel; two Upper Coralline Limestone perched aquifer galleries one of which is probably the *fonte aquarum* il-Fawwara mentioned in a 1467 notarial deed. One gallery contains at least a shaft well, the other a subterranean water storage reservoir.
Tentative date: Late Medieval; Early Modern; Modern.
Published sources: Buhagiar and Fiorini, 1996, 502; Aquilina and Fiorini, 2001, 166-73; Fiorini, 2003, 23-25; Fiorini, 2005, 26-27; Buhagiar, K., 2013, 96-97.
Archival sources: NLM. Treas. B. 289, f. 84; Abela, 1647, 65; NLM. Treas. B. 290, f. 77; NLM. Treas. B. 291, ff. 15, 86-87; NAM. Descriptive Plans of the Crown Property, 1867, f. 606.

Name of site: Saqqajja **Id. no:** RA40
Map reference: 46192 71308
Description of site: Upper Coralline Limestone perched aquifer gallery containing shafts fronted by a Knights period water fountain.
Tentative date: Late Medieval; Early Modern; Modern.
Published sources: Morris, 1952, 57; Fiorini, 1992, 38-40; Fiorini, 1999B, 13-14, 18-19; Wettinger, 2000, 522; Abdilla, 2003, 20-21; Fiorini, 2004, 522.
Archival source: Univ. 187, ff. 1-14.

Name of site: Simblija **Id. no:** RA41
Map reference: 42924 69862
Description of site: *Giardino*; Upper Coralline Limestone perched aquifer gallery and adjoining water collection reservoir excavated into a Blue Clay deposit; masonry and troglodytic dwellings; cave utilised as *centimolo*; cistern.
Tentative date: Late Medieval; Early Modern; Modern.
Published sources: Saliba *et al.*, 2002; Fiorini, 2003, 23-26; Buhagiar, M., 2005, 51-53; Saliba, 2008, 17-19; 66.
Archival sources: NLM. Treas. A. 74, f. 136; NLM. Treas. B. 292, ff. 27, 106, 108; NAM. Descriptive Plans of the Crown Property, 1867, f. 496.
Web source: http://www.timesofmalta.com/articles/view/20120603/life-features/Rendering-justice-to-u-eppi-Callus.422598

Name of site: Sqaq Ħandu **Id. no:** RA42
Map reference: 43518 70472
Description of site: Upper Coralline Limestone perched aquifer located close to surface level; old water-related features of unknown dating and other remains of archaeological interest; early seventeenth century canalisations forming part of the Valletta aqueduct system; subterranean gallery dating to the Knights period; nineteenth/early twentieth century wells and related water storage reservoirs.
Tentative date: Late Medieval; Early Modern; Modern.
Published sources: Abela, 1647, 66, 110-11; Fiorini, 1999B, 10-11, 14, 300; Wettinger, 2000, 109.
Cartographic source: Survey Sheet 85, School of Military Engineering, Chatham, 1922.

Name of site: Ta' Baldu **Id. no:** RA43
Map reference: 43274 69921
Description of site: *Giardini* for fruit tree cultivation; three Upper Coralline Limestone perched aquifer galleries, one containing a subterranean reservoir (Għar ta' Baldu); *camera di scirocco*; masonry and troglodytic dwellings.
Tentative date: Late Medieval; Early Modern; Modern.
Published sources: Abela, 1647, 65; Ciantar, 1772-1780, 149; Caruana, 1882, 93; Buhagiar, K., 2000, 181-86; Buhagiar, K., 2013, 167-71.
Cartographic source: Malta Survey Sheet Links no. 85.

Name of site: Ta' Brija **Id. no:** RA44
Map reference: Approx. 46209 69494
Description of site: *Giardino* framework receiving water from a close-by Upper Coralline Limestone perched aquifer gallery at map reference 46210 69451; described in Late Medieval documentation as having water, water outlets and water channels.
Tentative date: Late Medieval; Early Modern.
Published source: Fiorini, 1999B, 233-35.
Archival source: CEM, 114/7, ff. 10-11.

Name of site: Ta' Dekozzu **Id. no:** RA45
Map reference: Rabat district; exact location unknown.
Description of site: Spring, probably originating from an Upper Coralline Limestone perched aquifer gallery.
Tentative date: Early Modern; Modern.
Published source: Zammit, T., 1924, 41.

Name of site: Ta' Koronja **Id. no:** RA46
Map reference: 44196 71982; 44369 72112
Description of site: Surface and partly subterranean canals pertaining to the Valletta aqueduct system.
Tentative date: Early Modern.

Name of site: Ta' Liemu **Id. no:** RA47
Map reference: unknown.
Description of site: Probably a reference to one of the Upper Coralline Limestone perched aquifer galleries flanking the eastern side of Wied Liemu in the Rabat district.
Tentative date: Early Modern; Modern.
Published source: Zammit, T., 1924, 41.

Name of site: Ta' Rapa **Id. no:** RA48
Map reference: 46149 68474
Description of site: Upper Coralline Limestone perched aquifer gallery recorded as being ca. 81 m deep and originating from within a hill known as *La Roca* due south of Buskett valley.
Tentative date: Early Modern
Published sources: Zammit, T., 1924, 35, 38-39.
Archival source: NAM. Descriptive Plans of the Crown Property, 1867, ff. 505-6.
Cartographic Source: Malta Survey Sheet 101, School of Military Engineering, Chatham, UK, 1919.

Name of site: Ta' San Ġakbu **Id. no:** RA49
Map reference: 42827 70175
Description of site: *Giardini* for fruit tree cultivation at Wied ir-Rum; four Upper Coralline Limestone perched aquifer galleries, one of which is Għajn Tejtes; all galleries contain fronting subterranean reservoirs; masonry and troglodytic dwellings.
Tentative date: Late Medieval; Early Modern; Modern.
Published sources: Abela, 1647, 65; Wettinger, 2000, 492; Buhagiar, K., 2013, 98-107.

Name of site: Ta' San Mikiel **Id. no:** RA50
Map reference: 44792 71861
Description of site: *Giardino* giving access to the Għajn Tewżien spring; Late Medieval Church dedicated to San Mikiel is-Sinċir.
Tentative date: Late Medieval; Early Modern; Modern.
Archival source: NLM. Treas. B. 289, f. 63.

Name of site: *Tal Mitarfa* **Id. no:** RA51
Map reference: unknown
Description of site: A spring which in the early seventeenth century was connected to the Valletta aqueduct system.
Tentative date: Early Modern.
Published source: Abela, 1647, 110-11.

Name of site: Tas-Sala/Ta' Sala **Id. no:** RA52
Map reference: 45836 68589
Description of site: Upper Coralline Limestone perched aquifer gallery being 44 m deep originating from within a hill known as *La Roca* due south of Buskett valley; possibly referred to as *La fontana grande* in early seventeenth century documentation.
Tentative date: Early Modern.
Published source: Zammit, T., 1924, 35, 38-39.
Archival sources: NLM. Treas. B. 289, ff. 101-3, 108, 112-13; NAM. Descriptive Plans of the Crown Property, 1867, ff. 505-6.
Cartographic source: Malta Survey Sheet 101, British School of Military Engineering, Chatham, 1919.

Name of site: Tas-Senia (Malta) **Id. no:** RA53
Map reference: unknown
Description of site: Upper Coralline Limestone perched aquifer gallery located in the Buskett territory, the ceiling of which was pierced by a series of shafts.
Tentative date: Early Modern.
Published source: Zammit, T., 1924, 35, 38-39.

Name of site: Tax-Xieref **Id. no:** RA54
Map reference: 42217 70038
Description of site: Masonry and troglodytic dwellings; associated agricultural land at Wied ir-Rum.
Tentative date: Early Modern; Modern.
Published sources: Abela, 1647, 65; Wettinger, 2000, 618-19.

Name of site: Wied Ħażrun **Id. no:** RA55
Map reference: 42948 69625
Description of site: *Giardino* for fruit tree cultivation; two Upper Coralline Limestone perched aquifer galleries, fronting water storage reservoirs; masonry and troglodytic dwellings; *centimolo*. The spring at map reference 42970 69510 contains noticeable quantities of calcite deposition.
Tentative date: Late Medieval; Early Modern; Modern.
Published sources: Abela, 1647, 65; Wettinger, 2000, 577.
Archival sources: NLM. Treas. A. 74, f. 136; NLM. Treas. B. 292, ff. 27, 106, 108; NAM. Descriptive Plans of the Crown Property, 1867, f. 496.
Cartographic source: Malta Survey Sheet 99, School of Military Engineering, Chatham, 1903.

Name of site: Wied ir-Rum **Id. no:** RA56
Map reference: 42880 70110
Description of site: Water-carved valley; Upper Coralline Limestone perched aquifer galleries; seasonal spring flowing through valley bed; *giardini*; masonry and troglodytic dwellings.
Tentative date: Late Medieval; Early Modern; Modern.
Published sources: Abela, 1647, 65; Wettinger, 2000, 586; Fiorini, 1999B, 252.
Iconographic sources: Agius-Vadalà and Ganado, 1986, 2, 466-69; Agius-Vadalà and Ganado, 1994, 3, 5.

Name of site: Wied tal-Qlejgħa **Id. no:** RA57
Map reference: Approx. 44598 72185
Description of site: Water-carved valley; Seasonal spring flowing through valley bed; adjoins various *giardini* frameworks many of which contain Upper Coralline Limestone perched aquifer galleries; fruit cultivation.
Tentative date: Late Medieval; Early Modern; Modern.
Published sources: Fiorini, 1999B, 115-16, 252; Wettinger, 2000, 584; Fiorini, 2007, 362-65.
Iconographic source: Agius-Vadalà and Ganado, 1986, 2.

Name of site: Wilġa tal-Ħemsija **Id. no:** RA58
Map reference: Approx. 46634 71802
Description of site: Probably associated with the Late Medieval fief of Ħemsija; documented availability of water, water channels and water outlets; probably obtained a perennial water supply from an Upper Coralline Limestone perched aquifer gallery tunnelling below the De Redin Bastion curtain at Mdina. Adjoined agricultural property known as Ġnien il-Firien.
Tentative date: Late Medieval; Early Modern; Modern.
Published sources: Fiorini, 1999B, 225-28; Wettinger, 2000, 154.

20. Siġġiewi district

Name of site: *Aaym Ghliem Alla* **Id. no:** SI1
Map reference: unknown
Description of site: Probably located in the Girgenti/Gebel Ciantar locality; Upper Coralline Limestone perched aquifer spring.
Tentative date: Early Modern.
Published source: Abela, 1647, 96.

Name of site: *Aayn el Kibira* **Id. no:** SI2
Map reference: unknown
Description of site: Probably located in the Girgenti/Gebel Ciantar locality; Upper Coralline Limestone perched aquifer spring.
Tentative date: Early Modern.
Published sources: Abela, 1647, 95; Wettinger, 2000, 185.

Name of site: *Fontana de' Laurenti* **Id. no:** SI3
Map reference: unknown
Description of site: Upper Coralline Limestone perched aquifer spring; Probably located in close proximity to the Buskett woodland area and Wied l-Isqof.
Tentative date: Early Modern.
Published source: Abela, 1647, 95.

Name of site: *Noscieaa* **Id. no:** SI4
Map reference: unknown
Description of site: Probably located in the Girgenti/Gebel Ciantar locality; Upper Coralline Limestone perched aquifer spring.
Tentative date: Early Modern.
Published sources: Abela, 1647, 96; Wettinger, 2000, 417.

Name of site: Ta' San Ġorġ **Id. no:** SI5
Map reference: 46058 66605
Description of site: Upper Coralline Limestone perched aquifer gallery fronted by water collection reservoir.
Tentative date: Early Modern; Modern.
Published sources: Zammit, T., 1924, 35; Morris, 1952, 3.

Name of site: Tal-Fawwara **Id. no:** SI6
Map reference: unknown
Description of site: Probable Upper Coralline Limestone perched aquifer gallery.
Tentative date: Early Modern; Modern.
Published sources: Zammit, T., 1924, 35; Morris, 1952, 3.

Name of site: Tal-Lunzjata/Annunzjata **Id. no:** SI7
Map reference: Approx. 46706 66503
Description of site: Upper Coralline Limestone perched aquifer gallery fronted by water collection reservoir in the district of Siġġiewi.
Tentative date: Early Modern; Modern.
Published sources: Zammit, T., 1924, 35; Morris, 1952, 3, 19.

21. Wied il-Għajn/Marsaskala district

Name of site: Bir il-Mielaħ **Id. no:** WI1
Map reference: unknown
Description of site: A well in the locality of Marsaskala the name of which intimates the tapping of a mean-sea-level water source.
Tentative date: Late Medieval; Early Modern.
Published source: Wettinger, 2000, 47.

Name of site: Wied il-Għajn **Id. no:** WI2
Map reference: Approx. 60566 68806
Description of site: Salt marsh; water pond; spring.
Tentative date: Early Modern.
Published sources: Abela, 1647, 106; Wettinger, 2000, 573.

22. Żabbar district

Name of site: *Ghadira ta Rez* **Id. no:** ZA1
Map reference: unknown
Description of site: Place-name hints at the collection of water within an area of Middle or Lower Globigerina Limestone deposit.
Tentative date: Early Modern.
Published source: Abela, 1647, 106.

23. Żejtun district

Name of site: Bulebel *minore* (Malta) **Id. no:** ZA1
Map reference: Approx. 58060 69790
Description of site: The probable tapping of a localised Lower Globigerina Limestone aquiferous deposit; six cisterns two of which were in the first half of the seventeenth century back-filled with stones and earth.
Tentative date: Early Modern, Modern period.
Archival sources: NLM. Treas. B. 289, f. 5; NLM. Treas. B. 290, f. 31; NAM. GOV 01.3/6, Despatch To, no. 63, 1841, ff. 56-58.

24. Gozo and Comino

Name of site: *Aayn Tili* **Id. no:** GO1
Map reference: unknown
Description of site: Spring.
Tentative date: Early Modern.
Published source: Abela, 1647, 121.

Name of site: Comino **Id. no:** GO2
Map reference: 40400 85710
Description of site: Early eighteenth century open-air water storage reservoir positioned between Santa Marija tower and its adjoining quarters known as Il-Palazz; large subterranean cistern located below Sant Marija tower; bell-shaped cistern at Santa Marija Bay which due to its proximity to mean-sea-level functioned as a well tapping this aquiferous source.
Tentative date: Early Modern.
Archival sources: NLM. Treas. B. 289, ff. 156-59; NLM. Treas. B. 290, f. 104; NLM. Treas. B. 291, f. 21.

Naġe of site: *Fego ta Meimun* **Id. no:** GO3
Map reference: Approx. 32285 87913
Description of site: Possibly associated with land to the south-south-east of Rabat (Gozo) for which there is the early seventeenth century mention of a *Fontana Grande*.
Tentative date: Early Modern.
Archival source: NLM. Treas. B. 289, ff. 138-39.

Name of site: Ġnien Mrik **Id. no:** GO4
Map reference: Approx. 33300 90324
Description of site: Royal property in Gozo at Għajn Xejba in the territory of Xagħra.
Tentative date: Late Medieval; Early Modern.
Published source: Wettinger, 2000, 159.

Name of site: Għajn Barrani **Id. no:** GO5
Map reference: Approx. 34296 91928
Description of site: Spring probably originating from an Upper Coralline Limestone gallery at the Mġarr *giardino*.
Tentative date: Late Medieval; Early Modern.
Published sources: Abela, 1647, 122; Zammit, T., 1924, 31; Wettinger, 2000, 181.

Name of site: Għajn Fekruna **Id. no:** GO6
Map reference: unknown
Description of site: A district in Gozo first documented in 1497.
Tentative date: Late Medieval.
Published source: Wettinger, 2000, 182.

Name of site: Għajn Futni **Id. no:** GO7
Map reference: unknown
Description of site: A district in Gozo first documented in 1492.
Tentative date: Late Medieval.
Published source: Wettinger, 2000, 183.

Name of site: Għajn Għabdun **Id. no:** GO8
Map reference: 28709 89565
Description of site: Upper Coralline Limestone perched aquifer gallery; close-by water trough and related farmsteads; water transportation canals; troglodytic dwellings.
Tentative date: Late Medieval; Early Modern; Modern.
Published sources: Zammit, T., 1924, 30-31; Wettinger, 2000, 183.
Archival sources: NLM. Treas. B. 289, ff. 134, 148-50 and Plan 21; NLM. Treas. B. 291, f. 38; NAM. Descriptive Plans of the Crown Property, 1861, ff. 113-14.

Name of site: Għajn Għajxa **Id. no:** GO9
Map reference: unknown
Description of site: Land belonging to the royal court in Gozo first documented in 1428.
Tentative date: Late Medieval.
Published source: Wettinger, 2000, 184.

Name of site: Għajn Melel **Id. no:** GO10
Map reference: Approx. 31505 92936
Description of site: Spring possibly originating from within an Upper Coralline Limestone perched aquifer gallery.
Tentative date: Early Modern.
Published source: Wettinger, 2000, 186.
Archival source: NLM. Treas. B. 289, ff. 142-43 and Plan 21.

Name of site: Għajn Naħrin/Għajn Ħaġar **Id. no:** GO11
Map reference: 38268 89531/37902 89381
Description of site: Two springs, each of which is fronted by a water collection reservoir.
Tentative date: Early Modern; Modern.
Published sources: Wettinger, 2000, 187; Fiorini, 2007, 273-74.
Archival sources: NLM. Treas. B. 290, f. 106; NLM. Treas. B. 291, f. 22; NAM. Descriptive Plans of the Crown Property, 1861, f. 82.

Name of site: Għajn Qasab **Id. no:** GO12
Map reference: 36081 90314
Description of site: Upper Coralline Limestone perched aquifer gallery the outermost section of which is masonry-built and having at least one shaft piercing its ceiling; related water storage reservoirs; *giardino*.
Tentative date: Late Medieval; Early Modern.
Published sources: Fiorini, 1999B, 324-25; Wettinger, 2000, 188.
Archival sources: NLM. Treas. B. 289, f. 141 and Plan 15; NLM. Treas. B. 291, f. 22; NAM. Descriptive Plans of the Crown Property, 1861, f. 63.

Name of site: Għar Ilma **Id. no:** GO13
Map reference: 29259 89472
Description of site: Upper Coralline Limestone perched aquifer gallery; associated *giardino* framework.
Tentative date: Late Medieval; Early Modern; Modern.
Published sources: Zammit, T., 1924, 30-31; Wettinger, 2000, 206.
Web source: http://users.csc.calpoly.edu/~zwood/MaltaMapping/site46.html

Name of site: Il-Għajn **Id. no:** GO14
Map reference: Approx. 38809 87701
Description of site: *Giardino* described as *ginen rumen* containing a water spring; tentatively identified with Ħondoq ir-Rummien in Gozo.
Tentative date: Late Medieval.
Published source: Wettinger, 2000, 161.

Name of site: It-Taflija **Id. no:** GO15
Map reference: 32392 88620
Description of site: Probable location of the *Fego della taflia* mentioned in early seventeenth century documentation and which contained a spring known as *hain Catet.*
Tentative date: Late Medieval; Early Modern, Modern.
Published source: Wettinger, 2000, 188.
Archival source: NLM. Treas. B. 289, ff. 121-23.

Name of site: Mġarr *giardino/Feudo della Madia* **Id. no:** GO16
Map reference: 36870 87400
Description of site: Upper Coralline Limestone perched aquifer galleries one of which is mentioned as a *mina* in early eighteenth century documentation; seasonal spring flowing through valley bed; *giardino.* The water springs of the Mġarr *giardino* are: (a) Għajn Barrani; (b) Għajn ta' fejn il-Barri; (c) Għajn Ħluk; (d) Għajn Sielem; (e) Għajn ta' Franġisk Spiteri; (f) Għajn Rajes; and (g) Għajn Fawwara.
Tentative date: Late Medieval; Early Modern, Modern.
Published sources: Abela, 1647, 122; Zammit, T., 1924, 31; Wettinger, 2000, 181.
Archival sources: NLM. Treas. B. 289, ff. 151-53; NLM. Treas. B. 290, f. 105; NLM. Treas. B. 291, f. 21; NAM. GOV 01.3/6, Despatch To, 1841, f. 30; NAM. Descriptive Plans of the Crown Property, 1861, f. 54.

Name of site: Mġarr ix-Xini **Id. no:** GO17
Map reference: 33787 86799
Description of site: Seasonal spring flowing through valley bed; mean-sea-level aquifer gallery.
Tentative date: Modern.
Published source: Zammit, T., 1924, 32.

Name of site: *Territorio ta Kamet* **Id. no:** GO18
Map reference: Unknown
Description of site: A locality at Xewkija and a couple of other parcels of land in its neighbourhood containing cisterns.
Tentative date: Early Modern.
Archival sources: NLM. Treas. B. 289, ff. 126-27; NLM. Treas. B. 290, ff. 114-16; NLM. Treas. B. 291. f. 35.

Name of site: Ta' Ċimblu Rasu **Id. no:** GO19
Map reference: unknown
Description of site: A *Sorgente* which documentary sources identify as *Hain Kanac*, kitchen and a cistern described as *una cisterna con aqua sorgente.*
Tentative date: Early Modern.
Archival sources: NLM. Treas. B. 290, ff. 108-9; NLM. Treas. B. 291, f. 22.

Name of site: Tal-Għejun **Id. no:** GO20
Map reference: Approx. 34645 89580
Description of site: Agricultural property described as *viridarium* or *giardino* containing water courses in the Xagħra region of Gozo; alternatively referred to as *Tewnit.*
Tentative date: Late Medieval; Early Modern.
Published source: Wettinger, 2000, 221, 550.

Name of site: Wied tal-Lunzjata/*Giardino della gran Fontana* **Id. no:** GO21
Map reference: 30979 88814
Description of site: Perennial spring flowing through valley bed; Upper Coralline Limestone perched aquifer galleries one of which is known as Għajn Tuta; *giardini*; water collection reservoirs; drinking troughs; masonry and troglodytic dwellings.
Tentative date: Late Medieval; Early Modern.
Archival source: NLM. Treas. B. 289, ff. 143-45.
Iconographic source: Agius-Vadalà and Ganado, 1986, 2.

Name of site: Wied Żewwieqa **Id. no:** GO22
Map reference: Approx. 36821 87456
Description of site: Seasonal water course flowing through valley bed; excavation of subterranean galleries for the interception of ground water.
Tentative date: Modern.
Published sources: Chadwick, 1897, 12-13; Zammit, T., 1924, 32.

Appendix 2: Comparative Table of Place-Names

Divergent place-name transliteration in Late Medieval and Early Modern documentary sources of evidence, necessitated the drawing up of this comparative table of place-names, summarising the occurrence of the same toponym in different centuries. Place-names are listed in alphabetical order.

Place-Name (Contemporary)	High/Late Medieval Period	Knights Period	British Period
Bir Jaħlef (MA5)	*Biryahlef* (Wettinger, 2000, 15)	*Bir Yahlef/hialef* (Abela, 1647, 92; NLM. Treas. B. 289, f. 30)	
Bir Miftuħ (GU1)		*Birmiftuh* (Abela, 1647, 105)	
Bulebel (ZA1)		*Buleben* (NLM. Treas. B. 289, f. 5; NLM. Treas. B. 290, f. 31)	
Burmarrad	*Benwarrad* (Fiorini, 1999B, 256)		
Buqana (MO1)		*Bucana* (Abela, 1647, 82) *Bocana* (Univ. 187. f. 99)	
Busugrilla/Bir Żigrilla (RA7)	*Birzikirille* (Wettinger, 2000, 53)	*Bir Szegrella/ Birzegrella* (Abela, 1647, 80, 111)	
Dejr il-Bniet (DI1)		*Deyr el Binet* (Abela, 1647, 65)	
Ġnien Ingraw (ME1)		*Giardino d'Angarao* (Abela, 1647, 73)	
Ġnien il-Kbir (MO4) (Bidnija territory)	*Ginen ilkybir* (Wettinger, 2000, 157)		
Gomerino/Għemmieri (RA15)	*Iardinum de gumerin* (Luttrell, 1978, 173)	*S. Antonio tal Ghimeri/Gumerini* (Abela, 1647, 66)	
Għajn Għabdun (GO8)	*Fons Abiduni* (Wettinger, 2000, 183)	*Hain Habdun* (NLM. Treas. B. 289, ff. 134, 148-50)	
Għajn Astas/Stas (MG14)	*Haynastasi* (Wettinger, 2000, 190)	*Aayn Estas* (Abela, 1647, 72)	
Għajn Ballut (MG2)	*Hayn Ballut* (Wettinger, 2000, 181)	*Aayn Ballut* (Abela, 1647, 72)	
Għajn Barrani (GO5)		*Aayn Barrani* (Abela, 1647, 122) *Hayn Barrani* (Wettinger, 2000, 181)	
Għajn Bierda (RA1)		*Aayn Bierda* (Abela, 1647, 67)	
Għajn Dwieli (PA1; PA2)		*Aayn Duyeli* (Abela, 1647, 94)	
Għajn Filep (MA8)		*Aayn Filep* (Abela, 1647, 93)	

Place-Name (Contemporary)	High/Late Medieval Period	Knights Period	British Period
Għajn Ħaġar (GO11)	*An Alyagira* (Fiorini, 2007, 273-74)		*Għajn Ħaġar* (NAM. Descriptive Plans of the Crown Property, 1861, f. 82)
Għajn Ħamiem (RA16)	*Aijn Hammem* (Wettinger, 2000, 184)	*Aayn Hammem* (Abela, 1647, 80)	
Għajn Ħosna	*Fons chusune* (Wettinger, 2000, 184)	*Hain hosne* (Wettinger, 2000, 184)	
Għajn il-Kbira (Mtaħleb *giardino* – RA17)		*Aayn il Kbira* (Abela, 1647, 66) *Fontana di Giardino Grande* (AIM Acta Civilia vol. 171, Case 12)	
Għajn il-Kbira (Gozo)	*Fons magnus* (Wettinger, 2000, 185)		
Għajn Klieb (RA2; RA18)	*Hayn ilkileb; hayn kileb* (Wettinger, 2000, 185)	*Hayn culeb* (Wettinger, 2000, 185) *Aayn Clieb* (Abela, 1647, 80) Possibly formerly known as *Aayn Cirani/Hayn girani* (Abela, 1647, 111; Wettinger, 2000, 182)	
Għajn Kullija (RA3)	*Ayn cullie* (Wettinger, 2000, 185-86)	*Aayn Colia* (Abela, 1647, 81)	
Għajn Melel (GO10)	Possibly *Fons melchel* (Wettinger, 2000, 186)	*Aayn Mghelghel* (Abela, 1647, 122) *Hain Mihelhel* (NLM. Treas. B. 290, f. 132)	
Għajn Mielaħ		*Ghajn Melech* (NLM. Treas. B. 289, ff. 142-43, Plan 21; Wettinger, 2000, 187)	
Għajn Naħrin (GO11)	*Fons nacharen* (Wettinger, 2000, 187)	*Hayn nahrin* (Wettinger, 2000, 187) *Hain nacharin* (NLM. Treas. B. 289, Plan 15)	
Għajn Qajjed (RA19)	*Ayn l Cayd* (Fiorini, 1999B, 12-13)	*hayn Cayet* (NLM. Treas. B. 289, ff. 53-54) *Aayn Cayed/Aayn Kaied/Aayn Kayed* (Abela, 1647, 80, 111)	

Place-Name (Contemporary)	High/Late Medieval Period	Knights Period	British Period
Għajn Qasab (GO12)	*Ayn al Casat* (Fiorini, 1999B, 324-25) *fons yl casab* (Wettinger, 2000, 188)	*Hain Casab* (NLM. Treas. B. 289, f. 141) *Hajn Casap* (NLM. Treas. B. 291, f. 22)	
Għajn Rajes (GO16)		*Ta Rajes* (NLM. Treas. B. 291, f. 21)	*Għajn Rajes* (Zammit, T., 1924, 31)
Għajn Riħana (MO4)	*Aynrichana* (Wettinger, 2000, 189)	*hayn Richana* (NLM. Treas. B. 289, ff. 45-46) *Aayn Rihana* (Abela, 1647, 70)	
Għajn Tarġa (MG16)	*Ayn Targe* (Wettinger, 2000, 191)		
Għajn Tejtes/Btejtes (RA20)		*Hayn biteytes/Hayn bitatis* (Wettinger, 2000, 181) *Aayn Teytes* (Abela, 1647, 65)	
Għajn Tewżien (RA7; RA21)	*Ayn Tosen* (Fiorini, 1999B, 12-13)	*Hayn Tausen* (NLM. Treas. B. 289, f. 54) *Aayn Teuzien* (Abela, 1647, 111)	
Għajn Tuffieħa (MG15)	*Ayntufaha* (Wettinger, 2000, 191)	*hayn toffeha* (NLM. Treas. B. 289, ff. 89-91) *Aayn Toffieha* (Abela, 1647, 72)	
Għajn Tuta (Gozo – GO21)		*Hain tuta* (NLM. Treas. B. 289, f. 144)	
Għajn Żejtuna (ME3)	*Hayn Zeytune* (Wettinger, 2000, 193)	*Aayn Zeituna* (Abela, 1647, 73)	
Għajn Żnuber (ME4)		*Aayn Znuber* (Abela, 1647, 72)	
Għeriexem (RA26)	*Garexumu/Garhesim* (Wettinger, 2000, 215)	*Ghariexem* (Univ. 187, ff. 1-14)	
Ħal Mann (LI1)		*Vyed Hal Màn* (Abela, 1647, 85)	
Ħofret ir-Rizz (RA27)		*Hofret el Rez* (Abela, 1647, 110-11)	
Ħondoq ir-Rummien (GO14)	*Ginen rumen* (Wettinger, 2000, 161)		
Il-Ballut tal-Ħoxni (MG9)			*Il-Ballut ta' Għajn Tuffieħa* (NAM. Descriptive Plans of the Crown Property, 1866, f. 216)

Place-Name (Contemporary)	High/Late Medieval Period	Knights Period	British Period
Iċ-Ċarċar/Ċurċara (MG8)	*Churchara* (Wettinger, 2000, 100-1)	*Ciorciara* (Abela, 1647, 68)	*Ic-ciarciark* (Malta Survey Sheet 35, School of Military Engineering, Chatham, 1903)
Il-Fawwara (Siġġiewi – (SI6)		*Faüara* (Abela, 1647, 73)	
Il-Għars (RA30)			*Ta Falzun* (Malta Survey Sheet 85, School of Military Engineering, Chatham, 1922; NAM. Descriptive Plans of the Crown Property, 1867, f. 497)
Il-Mosfar/Musfar (RA31)		*Ginen Mosfar* (NLM. Treas. B. 289, f. 62)	
Il-Bruka (RA29)		*Bruca* (Abela, 1647, 66)	
Il-Qattara (RA32) (Wied ir-Rum)		*Giardino la Kattara* (Abela, 1647, 65)	
Il-Qattara (Buqana area – MO1)		*El Kattara* (Abela, 1647, 70)	
Is-Safsafa (MG21)		*Ta' Safsafa* (Abela, 1647, 72)	
Ix-Xwieki (IK1)		*El Sciüeki* (Abela, 1647, 85)	
Kemmuna/Comino (GO2)		*Cumino* (NLM. Treas. B. 289, ff. 156-59, Plan 22)	
Li Sball (RA33)		*Tà l'Estbal* (Abela, 1647, 66)	
L-Għar ta' l-għarwenin (RA24)		*Aayn San Giovan* (Abela, 1647, 66)	
Mġarr (Gozo – GO16)		*Feudo della Madia* (NLM. Treas. B. 291, f. 21)	
Magħtab (NA1)		*El Maatab* (Abela, 1647, 85)	
Tal-Fiskal (RA34)		*Marget Meüya/Margel üasa* (Abela, 1647, 112) Marget Neuia/Marget meuia (NLM. Treas. B. 289, f. 70)	
Tal-Għazzi		*xara tal Ghasi* (Wettinger, 2000, 604)	
Mtaħleb (RA17; RA23; RA35)		*Monte Alip* (Matteo Perez d'Aleccio frescoes in Agius-Vadalà and Ganado, 1994, Map 118).	

Place-Name (Contemporary)	High/Late Medieval Period	Knights Period	British Period
Qasam il-Kbir/ Il-Mejjieli iż-Żgħir (RA37)		*Kasam il Kbir dello Fiddeni* (NLM. Treas. B. 290, f. 91 ; NLM. Treas. B. 291, f. 18)	
Ras il-Pellegrin (MG3)		*Monte Pellegrino* (Matteo Perez d'Aleccio frescoes in Agius-Vadalà and Ganado, 1994, Map 118).	
Razzett San Pawl (RA39)		*Tal Callus* (NLM. Treas. B. 289, f. 84; Abela, 1647, 65) *Wied Ferin* (NLM. Treas. B. 290, f. 77)	
Sqaq Ħandu (RA42)	*Dejr Ħandun* (Wettinger, 2000, 109)	*Diar Handul*; *Diar Chandul/Chandut* (Abela, 1647, 66, 110-11)	
Saqqajja (RA40)	*La Saccaya* (Fiorini, 1999B, 13-14, 18-19)	*Saccaya* (Univ. 187, f. 99)	
Ta' San Ġakbu (RA49)		*Di S. Giacomo* (Abela, 1647, 65)	
Ta' San Mikiel (RA50)		*Terreno di S. Michaele* (NLM. Treas. B. 289, f. 63)	
Ta' Baldu (RA43)		*Di Baldu* (Abela, 1647, 65)	
Ta' Brija (RA44)	*Tabrje/Tabrie* (Fiorini, 1999B, 233-35)	*Tabria* (CEM, 114/7, ff. 10-11.	
Ta' Għasfura (MG20)		*Hain sifurie* (Wettinger, 2000, 190) *Aayn Sfuria* (Abela, 1647, 68)	*Ta Ghasfuria* (Malta Survey Sheet 56, School of Military Engineering, Chatham, 1898)
Tas-Santi (MG23)		*Fego del Santi* (NLM. Treas. B. 289, ff. 84-88)	
Tax-Xieref (RA54)		*Ta' Scieref* (Abela, 1647, 65)	
Wied il-Busbies		*Vyed el Besbies* (Abela, 1647, 112)	
Wied id-Dis (GH1)		*Vyed el Dis* (Abela, 1647, 85)	
Wied tal-Ġnejna (MG24)		*Ved tal Gineyna* (Abela, 1647, 69)	
Wied tal-Lunzjata (GO21)		*Giardino della gran Fontana* (NLM. Treas. B. 289, ff. 143-45)	
Wied il-Għajn (Marsaskala – WI1)		*Vyed el Aayn* (Abela, 1647, 106)	
Wied Ħażrun (RA55)		*Vyed Casrun* (Abela, 1647, 65)	

Place-Name (Contemporary)	**High/Late Medieval Period**	**Knights Period**	**British Period**
Wied ir-Rum (RA56)		*Guiderum* (Matteo Perez d'Aleccio frescoes in Agius-Vadalà and Ganado, 1994, Map 118). *Hued iron* (NLM. Treas. B. 289, ff. 84) *Vyed el Rum* (Abela, 1647, 65).	
Wied San Niklaw (ME7)		*Giardino di S. Nicola* (Abela, 1647, 73).	

Appendix 3: A Philological Appreciation of Water-Related Italian Terms Used in Chapters 5 to 7

In the compilation of the table hereunder, the 1598 and 1611 Italian to English dictionaries by I. Florio were extensively made use of. Being contemporaneous, or almost, with most documents discussed in chapters 5 and 6, the word-lists by Florio provide a good indication of the meaning Italian words had in the late sixteenth and early seventeenth centuries. Another Italian dictionary which was widely made use of is that by S. Battaglia. This takes into account Italian historical word development and was brought to my attention by Prof. J. Pace Asciak of the University of Malta.

Italian word/phrase	Late 16th and early 17th-century English equivalent	Historic and contemporary meaning in Italian
Acqua viva	Running spring water (Florio, 1598, 6; Florio, 1611, 9)	[Acqua] *che scaturisce da fonte perenne* (Battaglia, 1962-2002, vol. 1, 125)
Canale (pl. *canali*)	Channel; gutter; water pipe; [water] course (Florio, 1598, 56; Florio, 1611, 78)	*Corso d'acqua artificiale costruito per l'irrigazione* (Battaglia, 1962-2002, vol. 2, 613)
Condotto (pl. *condotti*)	Ditch (Florio, 1611, 116)	*Canale chiuso o costituito da tubi, attraverso il quale circolano fluidi o liquidi; canaletto scavato nel terreno per lo scolo delle acque piovane* (Battaglia, 1962-2002, vol. 3, 506-7)
Cisterna	A cesterne [sic.] to keepe [sic.] water in; a cesterne [sic.] for water (Florio, 1598, 74; Florio, 1611, 104)	*Serbatoio in muratura, per lo più di forma circolare, chiuso e coperto, di solito interrato, destinato a raccogliere e conservare l'acqua piovana* (Battaglia, 1962-2002, vol. 3, 195)
Cisterna d'acqua sorgiua/sorgente	No data available. Presumably a cistern/well of running/flowing spring water.	
Fiumicello	A little brooke [sic.]; little river (Florio, 1598, 133; Florio, 1611, 190)	Diminutive of *fiume*: *corso d'acqua a carattere permanente; ogni corso d'acq'a di qualsiasi grandezza* (Battaglia, 1962-2002, vol. 6, 52-53)
Fontana	Fountain; spring; well (Florio, 1598, 135; Florio, 1611, 192)	*Vena d'acqua a getto continuo; sorgente; fonte; polla* (Battaglia, 1962-2002, vol. 6, 142)
Fontana viva		*Sorgente attiva* (Battaglia, 1962-2002, vol. 6, 142)
Fontanella	Little fountain (Florio, 1598, 135; Florio, 1611, 192)	*Piccola fonte; piccola sorgente; zamillo d'acqua sorgiva* (Battaglia, 1962-2002, vol. 6, 143)
Fonte	Source; spring; fountain; fount; well (Florio, 1598, 135; Florio, 1611, 192)	*Vena d'acqua continua che sgorga spontaneamente da un'apertura del terreno; sorgente* (Battaglia, 1962-2002, vol. 6, 144)
Fonte viva		*Sorgente sempre attiva* (Battaglia, 1962-2002, vol. 6, 144)
Fonti d'acqua sorgente	No data available.	
Gebia	No data available, but presumably meaning water reservoir, probably of the uncovered type.	

Italian word/phrase	Late 16th and early 17th-century English equivalent	Historic and contemporary meaning in Italian
Giardino	Gardin or orchard [sic.]; garden; hort-yard [sic.] (Florio, 1598, 148; Florio, 1611, 209)	*Superfice di terreno, per lo più cintata (da un muro, da uno steccato o da una cancellata), coltivata con piante ornamentali e fiori, con prati erbosi, fontane, ecc.* (Battaglia, 1962-2002, vol. 6, 144)
Mina (pl. *mine*)	Digging underground; mineral mine; sapped mine (Florio, 1598, 226; Florio, 1611, 315)	*Galleria* (Battaglia, 1962-2002, vol. 10, 421)
Pozzo	Deep pit or well of water (Florio, 1598, 288; Florio, 1611, 395)	*Cavità del terreno di piccole dimensioni, poco profonda e piena d'acqua; fossa in cui viene defluire l'acqa di scolo o piovana per poterla bere o per bonificare un terreno* (Battaglia, 1962-2002, vol. 13, 1139)
Pozzo d'acqua viva	See meaning of *aqua viva* above. Tentative meaning: well of running/flowing spring water.	
Sorgente or *acqua sorgiva*	Rising up (Florio, 1611, 514)	*che sgorga; che scaturisce; che zampilla per lo più copiosa d'acqua; un fonte* (Battaglia, 1962-2002, vol. 14, 488)
Spere d'acqua corrente	No data available, but presumably meaning shaft-well of running/flowing spring water.	
Spiráglio (pl.) *Spirágli*	Vent; hole (Florio, 1598, 390; Florio, 1611, 524)	*Piccola apertura o fessura* (Battaglia, 1962-2002, vol. 14, 936)
Stagno	Pool; pond; standing water (Florio, 1598, 394)	*Raccolta di acqua stagnante dolce o slamastra, per lo più poco estesa e poco profonda, e talora di durata limitata, il cui fondo spesso si presenta melmoso e ricoperto di vegetazione palustre; mare chiuso* (Battaglia, 1962-2002, vol. 20, 46)
Torrente	Land flood; rain flood; swift-running stream (Florio, 1598, 425; Florio, 1611, 569)	*corso d'cqua, in partic. Montano, caretterizzato da portata irregolare con alternanza di periodi di secca e di piene violente in relazione alle precipitazioni atmosferiche; scorre a forte pendenza, in genere incassato strettamente fra gole e interrotto talora da salti e cascate* (Battaglia, 1962-2002, vol. 21, 66)

Appendix 4: A Philological Appreciation of Water-Related Maltese Words/Terms Used in Chapters 5 to 7

In the compilation of the table hereunder, place-name data by G. Wettinger published in 2000 were widely made use of. Furthermore, Sicilian documentation dating to 1187, of which there is both an Arabic and Latin version, and an analysis of which is published in G. Wettinger in 2006, was also taken into consideration. Maltese-English dictionaries by G. Aquilina and M. Serracino-Inglott were also made use of.

Maltese word/phrase	**Possible Medieval Sicilian equivalent in Latin (where applicable)**	**English Meaning**
Bir	*Puteum* (Wettinger, 2006, 20)	Subterranean or above-ground water reservoir; sizeable canal in which water is stored; well (Aquilina, 1990, 124; Serracino-Inglott, 2003, 56)
Fawwara	*Favaira* (Wettinger, 2003, 21)	Spring; source [of water] (Aquilina, 1990, 314)
Għadira/Għadir /Għodor		pool; pond (Aquilina, 1990, 938; Serracino-Inglott, 2003, 198)
Għajn (pl. Għejun)	*Fons*; *fonte* (Wettinger, 2006, 15, 18-19)	Fountain [one made artificially] (Aquilina, 1990, 947; Serracino-Inglott, 2003, 199)
Għewejna (pl. Għewejjen)		Little spring; fountain (Aquilina, 1990, 948)
Ħawt		Watering or feeding trough (Aquilina, 1990, 521; Serracino-Inglott, 2003, 221)
Miġra (pl.) Mġarr	*(cum) ductu aque*; *aqueductus*; *cursum aque* (Wettinger, 2006, 15)	[Mġarra] Watercourse; small water course (Aquilina, 1990, 818-19; Wettinger, 2000, 376)
Maqgħad [Ilma]		Stagnant water; a place where water collects (Aquilina, 1990, 783; Wettinger, 2000, 361)
Marġ		Meadow; grassland; flood-plain; marshy land (Aquilina, 1990, 786; Wettinger, 2000, 326)
Menqa		Water pond; pool; swampy ground (Aquilina, 1990, 807; Wettinger, 2000, 373)
Nixxiegħa		Spring of water (Aquilina, 1990, 918) Might also be taken to mean 'Għajn' (Serracino-Inglott, 2003, 411)
Noqra		Ditch of spring water; spring-hole (Aquilina, 1990, 923; Wettinger, 2000, 418)
Qattara		Dripping/dropping liquid [water] (Aquilina, 1990, 1142)
Qonja (pl.) Qana		Masonry canal used for conveying water; small canal in the soil meant as an aid to irrigation (Aquilina, 1990, 1119; Serracino-Inglott, 2003, 472)
Saqa/Sieqja Saqqajja	*Aqueductus*; *rivulum* (Wettinger, 2006, 15, 35)	To water; irrigate; water channel; aqueduct (Aquilina, 1990, 1270-71)

Appendix 5: A Philological Appreciation of Water-Related Latin Terms Used in Chapter 7

No one dictionary will solve difficulties likely to be encountered when dealing with Medieval Latin texts. Readers of this book should be alerted to the fact that Latin words may well have come totally adrift from their Classical Latin meaning, so it depends on medieval usage, which, of course, could well be different in Italian (including Sicilian), Spanish flavoured and other contexts. The Latin dictionaries and word-lists quoted in the table below were singled out by either Prof. S. Fiorini of the University of Malta or adhere to the suggested procedure best followed in such circumstances by K. Sidwell.[1] In drafting the hereunder, the advice of both Prof. S. Fiorini and Prof. A. Bonanno was solicited.

Latin word/phrase	Nominative/Genitive/Compound forms	English meaning
Aqueductibus	*Aqueductus* *Aqua + ducere*	[water] conduit (Latham, 1965, 27) Artificial channel, conduit (Personal communication by S. Fiorini)
Aquis	*Aqua(e)*	Water (Baxter and Johnson, 1947, 25; Lockwood, 1952, 55; Latham, 1965, 27)
Aquarum Decursibus or *Aquarum Cursibus*	*Aquarum decurs(us)* or *cursus*	Water-course (Baxter and Johnson, 1947, 116, 122; Latham, 1965, 127; Niermeyer, 1976, 292.
Arabilis		Arable land (Niermeyer, 1976, 54)
Decursibus	*Decurs(us) or cursus*	Water-course (Baxter and Johnson, 1947, 122)
Ductibus	*Ductum* or *ductus*	Conduit; channel; water-course (Baxter and Johnson, 1947, 144-45; Lockwood, 1952, 218; Latham, 1965, 158)
Fontana/fons /font		Source [of water]; spring; well; fount; fountain (Baxter and Johnson, 1947, 177; Lockwood, 1952, 279; Latham, 1965, 196; Niermeyer, 1976, 439)
Fonticulum		[Font]*iculum* is the diminutive of *Fontana* (Personal communication by S. Fiorini)
Pluvialis	*Pluviale*	Pluvial; rain [water] (Baxter and Johnson, 1947, 315; Lockwood, 1952, 549; Latham, 1965, 356-57)
Puteum	*Puteus*	Well; pit [for rain water] (Baxter and Johnson, 1947, 341; Lockwood, 1952, 600; Latham, 1965, 384)

[1] Sidwell, 1995.

www.ingramcontent.com/pod-product-compliance
Ingram Content Group UK Ltd.
Pitfield, Milton Keynes, MK11 3LW, UK
UKHW061845041026
15591UKWH00015B/921

9 781407 316291